Studies in Flood Geology

A Compilation of Research Studies Supporting Creation and the Flood

by

John Woodmorappe
M.A., Geology
B.A., Biology

Studies in Flood Geology
John Woodmorappe

© 1999 by the Institute for Creation Research
Second Edition

ISBN 0-932766-54-4

Institute for Creation Research
P.O. Box 2667
El Cajon, California 92021

No part of this publication may be reproduced, stored in a retrieval system, or transmitted in any form or by any means—electronic, mechanical, photocopy, recording, or otherwise—without the express prior permission of the Institute for Creation Research, with the exception of brief excerpts in magazine articles and/or reviews.

Printed in the United States of America

Table of Contents

Foreword 1

Introduction: "Studies in Creationism and Flood Geology," *Acts and Facts* Impact Article #238 (April 1993) 3

1. "Causes for the Biogeographic Distribution of Land Vertebrates after the Flood," *Proceedings of the 2nd International Conference on Creationism*, vol. 11 (1990), pp. 361-367 5

2. "The Antediluvian Biosphere and its Capability of Supplying the Entire Fossil Record," *Proceedings of the 1st International Conference on Creationism*, vol. 11 (1986), pp. 205-213 13

3. "A Diluviological Treatise on the Stratigraphic Separation of Fossils," *Creation Research Society Quarterly*, vol. 20, no. 3 (December 1983), pp. 133-185 21

4. "An Anthology of Matters Significant to Creationism and Diluviology: Report 2," *Creation Research Society Quarterly*, vol. 18, no. 4 (March 1982), pp. 201-223 77

5. "The Essential Nonexistence of the Evolutionary-Uniformitarian Geologic Column: a Quantitative Assessment," *Creation Research Society Quarterly*, vol. 18, no. 1 (June 1981), pp. 46-71 103

6. "An Anthology of Matters Significant to Creationism and Diluviology: Report 1," *Creation Research Society Quarterly*, vol. 16, no. 4 (March 1980), pp. 209-219, 227 131

7. "Radiometric Geochronology Reappraised," *Creation Research Society Quarterly*, vol. 16, no. 2 (September 1979), pp. 102-129, 147, inside back cover 145

8. "The Cephalopods in the Creation and the Universal Deluge," *Creation Research Society Quarterly*, vol. 15, no. 2 (September 1978), pp. 94-112 177

9. "A Diluvian Interpretation of Ancient Cyclic Sedimentation," *Creation Research Society Quarterly*, vol. 14, no. 4 (March 1978), pp. 189-208 199

Study Questions 221

Index 225

Foreword

I have known John Woodmorappe for over twenty years and know him to be a very gifted and motivated creationist geologist, as well as an assiduous reader and analyst of geological literature. His writings are all comprehensively documented and persuasively reasoned, comprising a highly valuable but little known resource in the study of the "Flood Model" of historical geology—or, simply, "Flood Geology."

During the period 1978-83, he published seven important papers in this field in the *Creation Research Society Quarterly*. These were followed by two more papers published in the *Proceedings of the International Conference on Creationism*, one in 1986, the second in 1990.

We believe that it will be a very useful service to serious creationist scientists, as well as others, to make all these papers available in one volume. No changes have been made in the papers as originally published. They have simply been photocopied and then assembled together in one useful volume. I believe anyone who will take the time to read them carefully will be firmly convinced, not only that the evolutionary approach to geology (whether uniformitarian or punctuational) is completely false, but also that flood geology is an eminently satisfying model within which to correlate all geological data.

In the very nature of things, it is not possible to *prove* either model to be true. When a scientist attempts to decipher the prehistoric past, he can never actually observe what happened—not at least until someone invents a time machine! Past events are *historical* events, not repeatable in a laboratory at the whim of an experimenter. One can always imagine an evolutionary scenario to explain anything, but imagination is not demonstration. The same is true as far as creation and the flood are concerned, neither of which can be observed in action.

The Flood Model, however, will always be found to explain the observable data more naturally and directly than the Evolution Model, and Woodmorappe's *Studies in Flood Geology* beautifully illustrates this fact.

The author has also written a brief summary of the nine chapters (originally the nine separate articles) which comprise this book, thus giving the reader the overview of the detailed studies he can read later. This overview, which is found in the Introduction, was originally published in ICR's *Acts & Facts* (April 1993). This publication, sent free of charge each month to all who request it, can be obtained by writing the Institute for Creation Research, P.O. Box 2667, El Cajon, California, 92021.

As mentioned earlier, seven of the articles were first published in the *Creation Research Society Quarterly*. Those interested in subscribing to this publication or in actual membership in the Society ($18 annually) may write to the membership secretary at P.O. Box 8263, St. Joseph, MO 64508.

The two most recent papers are part of the *Proceedings of the First and Second International Conferences on Creationism*, held in 1986 and 1990, respectively. Many other papers by other authors are included in these volumes, which can be ordered from the Creation Science Fellowship, Inc., 362 Ashland Avenue, Pittsburgh, PA 15228.

While neither I nor my colleagues at ICR necessarily endorse everything contained in this remarkable Woodmorappe collection, I can heartily endorse its basic thrust as well as its motivation and dedicated scholarship. The serious student of flood geology and scientific creationism will find the careful study of these papers stimulating and rewarding, answering many key questions and greatly strengthening the scientific case for true creation.

Henry M. Morris
President Emeritus, Institute for Creation Research

Introduction

Studies in Creationism and Flood Geology

JOHN WOODMORAPPE

Over the last 15 years, I have engaged in intensive scholarship in scientific creationism, with which I would like to acquaint the lay creationist reader. Many questions and issues in flood geology have been given at least a tentative answer as a result of my little-known research, which has been written for scientists, especially geologists. The purpose of this *Impact* article is to summarize my research in everyday language for the average reader.

In my "Causes for the Biogeographic Distribution of Land Vertebrates After the Flood" (*Proceedings of the 2nd International Conference on Creationism*, 1990, Vol. II, pp. 361-370), I explain why the animals on different continents are so different from each other if they originated from one point (Noah's Ark in the mountains of Ararat).

The interior regions of the continents were very cold for some time after the Flood, due to blockage of sunlight by volcanic aerosols released during the Flood, and animals did not freely spread in all directions upon their release from the Ark, but were shunted across narrow bands of land warm enough to support life. This ultimately caused very different animals to end up on different continents.

The postdiluvian peoples, after their post-Babel dispersion, probably introduced different animals to different continents (such as Australian marsupials, South American mammals, and Madagascaran primates). I point out that South America, Australia, and the island of Madagascar are all in direct line of maritime routes emanating from the Middle East, and hence are natural stopping points for the postdiluvian peoples.

Flightless birds on islands possibly resulted through microevolution (or, better, variation) from birds which had flown there. I present evidence that this can happen in a short time. Also because of this, we need not suppose that God created birds with useless wings.

In my "The Antediluvian Biosphere and its Capability of Supplying the entire Fossils Record" (*Proceedings of the 1st International Conference on Creationism*, 1986, Vol. II, pp. 205-218), I refute anti-creationists who have claimed that the material found in the fossil record could not possibly all have been alive on a recently created earth. I prove that the world's coal, oil, fossil crinoids, Karoo vertebrates, limestone components, etc., could all have come from the remains of creatures having lived in the short time between creation and the Flood and then buried by the Flood.

In "A Diluviological Treatise on the Stratigraphic Separation of Fossils" (*Creation Research Society Quarterly* 20(3): 133-185; December 1983), I examine, in great detail, how one flood accounts for the fact that different fossils are found in different layers of rock. I test, using over 9,500 global locations of fossils, the tendencies of over 30 different types of fossils to overlie each other in rock. Then I propose and test a new mechanism to explain, through one flood, the relatively few cases where rocks bearing many different kinds of fossils overlie each other.

This mechanism, which combines biogeographic zones of living things with a tendency for crustal rock to downwarp also is used to explain why, in the lower layers of fossils, there are fewer fossil types that have any representatives still alive today. I demonstrate that evolution with geologic ages is not the sole (or even the best) explanation for this trend.

I address the fact that there are few, if any, human remains in lower fossiliferous rock. According to evolution, it is because humans did not appear until very recently. I provide a diluvian explanation for this, showing through actual calculations that the antediluvian humans were so dispersed in the great volumes of sedimentary rock that it is extremely improbable that any of them ever would have been discovered. Alternatively, such discoveries are so infrequent that any such find could be easily ignored or discounted by evolutionists.

In "An Anthology of Matters Significant to Creationism and Diluviology: Report 2" (*Creation Research Society Quarterly* 18(4): 201-23, 239; March 1982), I discuss various topics, including further evidences against organic evolution, against the existence of ancient reefs in ancient rock, and against the usual claim of overthrusts (rock strata mechanically pushed over each other) to explain away instances of fossils overlying each other in wrong order, according to evolution. I also provide 200 examples of fossils occurring in "wrong" rock strata, according to evolution, and show that there usually is no evidence to support the usual evolutionary rationalization that these are situations where fossils from older rock were washed out and redeposited in younger strata.

In "The Essential Nonexistence of the Evolutionary-Uniformitarian Geologic Column: A Quantitative Assessment" (*Creation Research Society Quarterly* 18(1): 46-71; June 1981), I show, by overlying world maps of rocks attributed by evolutionary geologists to different ancient geologic periods, just how small a percentage of the earth's land surface has rocks of many alleged geologic periods all in one place. I also show, through calculations, that rocks of geologic periods supposed to have succeeded each other in time, rarely succeed each other as layers of rock.

In "An Anthology of Matters Significant to Creationism and Diluviology: Report 1" (*Creation Research Society Quarterly* 16(4): 209-19; March 1980), I cover many topics. For example, I document recent discoveries which show that many fossils once thought by evolutionists to have been restricted to certain layers of rock strata have now been found in many other layers of rock. I also provide evidence against the usual claim of evolutionary geologists that certain processes, whose effects are seen in rock, must have taken a long time to happen.

In my "Radiometric Geochronology Reappraised" (*Creation Research Society Quarterly* 16(2): 102-29, 147; September 1979), I engage in a thorough and systematic refutation of the dating methods used by evolutionary geologists to support their claim that the earth's fossil-bearing rock formed gradually over hundreds of millions of years, supposedly indicating that the earth must be billions of years old. Whereas other creationists have questioned the assumptions underlying isotopic dating, I provide numerous geologic demonstrations of the invalidity of radiometric dating. This includes over 400 published instances of serious discrepancies between isotopic age and the expected age of the rock based on its fossils, according to standard evolutionary thought. I also show that, contrary to intuitively held beliefs, internal consistence in dates obtained by these methods, and even agreement between results of different methods, are not proof for their validity.

I refute the claim that various dating methods agree that the earth is 4.5 billion years old. I demonstrate that there are gross contradictions in billion-year values from earth's rock, and that there are even some values obtained which are much greater than the 4.5-billion-year accepted age of the earth.

Most creationists research on the fallacies of evolution (for example, that of Dr. Duane Gish of the Institute for Creation Research) has focused on vertebrates. In "The Cephalopods in the Creation and the Universal Deluge" (*Creation Research Society Quarterly* 15(2): 94-111; September 1978), I focus on a group of invertebrate animals which include the modern squid and octopus. This group of animals is used by evolutionary geologists to a greater extent than any other fossil animal, to subdivide rock strata into alleged different spans of time. I show, in detail, the fallacies of these practices, as well as the fact that there is an even greater absence of expected evolutionary transitions among cephalopods than is the case among the vertebrates surveyed by Dr. Gish. Finally, I demonstrate how the ecological differences among cephalopods explain why all the living and fossil cephalopods were buried by one flood in the order in which they are found in rock strata. A popular-level version of this work on cephalopods, entitled "Cephalopod Conches," appeared in *Ministry*, January-February 1980.

In "A Diluvian Interpretation of Ancient Cyclic Sedimentation" (*Creation Research Society Quarterly* 14 (4): 189-208; March 1978), I show how one flood explains the fact that most of the world's coal deposits occur in sandwich-like layers interbedded with rock, and that standard evolutionary geology has a difficult time explaining this. I then develop a model to show how vast sheets of rising and falling flood waters buried floating vegetation (which later became coal) in between layers of mud (later shale) and sand (later sandstone).

Currently, I am working on several creationist projects which I anticipate publishing in the future. I would hope that creationists will make full use of this research, and that it will serve as a springboard for further research by other creationist scholars. It is only through careful and intense scholarship that creationism can grow in explanatory power, which is the goal of all scientific research.

Acts & Facts Impact Article #238 (April 1993)

Causes for the Biogeographic Distribution of Land Vertebrates after the Flood

Causes for the Biogeographic Distribution of Land Vertebrates After the Flood

John Woodmorappe

This study evaluates patterns in the global spread of land animals after their release from the Ark, and shows that: 1) most families have a heterogeneous biogeographic distribution; 2) causes for this include sweepstakes routes caused by the Ice Age and selective anthropogenic introductions. This distribution of problematic groups (e.g., Australian marsupials) appears to be explicable in a creationist context.

Introduction

The (imagined) inability of the creation model to explain the biogeographic distribution of living things was a major factor in its 19th century rejection in favor of organic evolution (Laferriere 1989). Although, as pointed out by the anti-creationist Jeffery (1983), it is untrue that modern creationists have ignored biogeography, the global distribution of animals has never been systematically studied from a modern creationist perspective. This work is a pilot study designed to investigate some of these factors. It is of direct relevance to the young-earth concept in showing that millions of years of organic evolution (i.e., in isolated populations) are not necessary to explain the peculiar biogeographic distribution of certain land vertebrates.

As in the case with most sciences, biogeography as a discipline was largely founded by scientific creationists (Browne 1983):

> The idea of an Ark in which pairs of animals were preserved during the Deluge had been a concept of far-reaching significance, as had the disembarkation on Mount Ararat and the subsequent dispersal of animals over the unoccupied globe. The biblical story, in fact, had done a great deal to stimulate investigations into the natural world and, among other things, provided the first systematic explanation for the phenomena of biogeography. Far from being the intellectual impediment ridiculed by Darwin and his circle, . . the idea of an Ark focused scholarly attention on the topographic arrangements of species, as well as encouraging naturalists to build up a repertoire of theoretical commitments and practical expertise in the analysis of organic distribution.

Methodology

This work is limited to animals released from the Ark. It does not consider the biogeography of living things before the Flood (a subject considered elsewhere (Woodmorappe 1983) as part of the explanation for the stratigraphic separation of fossils). Only land vertebrates are recognized as having been on the Ark for the reasons given in Jones (1973). Non-volant vertebrates are emphasized, since the birds and the bats have fossil records too fragmentary (see Carroll 1988) for a meaningful paleobiogeographic analysis of their extant families. At the same time, it should be remembered that most extant avian families are not endemic to particular continents (see Fig. 31 in Rich and Van Tets 1984), while some avian families have near-hemispheric distributions (see Table 1 in Keast 1984).

Throughout this work I assume only naturalistic causes for biogeographic patterns and reject the notion, advocated by some, that post-Flood vertebrates were guided back supernaturally to their former locations on the antediluvian earth. Only Late Tertiary rock contains faunas similar to extant life, but this is not evidence for such a return. Miocene/Pliocene rock is qualitatively different (in terms of thickness, areal distribution, and other features: see Ronov 1982) from earlier rock, so there is ample reason for concluding that Late Tertiary rock and its fauna are mostly post-Flood).

Most biogeographic studies to date have been at the specific level, yet it is almost universally recognized by creationists that the original created kind is broader than this. There are numerous instances of interbreeding between species, including those throughout large portions of families (for example, species within Anatidae: Scherer 1986), to say nothing of interbreeding between members of different genera (see Van Gelder 1977 for mammalian examples). Of course, many types of living things must have lost the capability of interbreeding at some time since the Creation. Jones (1972), using Biblical and scientific evidence, has concluded that the original created kind most closely corresponds to the family level of current taxonomy. This is accepted here. Since biogeographic distributions within kinds (i.e., usually within families) must have resulted from "microevolution" since the Flood (see Lester and Bohlin 1984 for examples of rapid speciation), they are not considered further.

This work approaches biogeography on an intercontinental, not subcontinental, scale. It should be noted, however, that biogeographic differentiation of families on a subcontinental scale is not great. Raup (1982), using computer-based randomly-chosen points on earth (as centers of circular areas of specified radius), has shown that a randomly-chosen hemisphere encompasses, on average, all living individuals of only 12% (maximum of 25%) of terrestrial families.

The paleontological record shows that many, if not most, living things have had a more widespread distribution than they do today (for example, consider tortoises: Auffenberg 1974). A comprehensive source for the biogeography of extant families as seen from both extant and fossil distributions (Carroll 1988) was therefore used as the primary source throughout this work. Since we cannot know which families have gone extinct only since their disembarkation from the Ark, no extinct families (except for extinct Australian marsupials) are considered here. It should be added that biogeographic differentiation at all levels (but especially lower taxa) has been overstated because of "chauvinotypy" (Rosen 1988): the tendency to generate synonyms by naming taxa from one's nation, biogeographic unit, etc., as unique.

This work assumes that continents have always been fixed. However, if continental drift took place during the Food, it is irrelevant to post-Flood biogeographic distributions. If it took place at the time of Peleg (Genesis 10:25), then all the factors discussed here remain valid. Only their sequence and timing would change.

Analysis

The biogeography of extant (Nowak & Paradiso 1982) and extinct (Carroll 1988) mammalian families, as well as that of reptiles (Carroll 1988), has been examined for biogeographic heterogeneity. Large areas of high endemicity (e.g., Australia, Madagascar) are considered separately below, while the initial focus is on the families native (or once native) to Eurasia/Africa versus North/South America.

The table gives the number of families particular to a given group of continents. Of the 40 families common to both blocs of continents, four are families presently restricted to one bloc but one living also on the other (as seen from the Miocene/Pliocene: hence post-Flood sediments). We see that 81 of the 112 families occur in at least one of the continents proximate to Ararat. Whereas the remaining 31 occur only in North and/or South America. This latter group demands an explanation.

Table 1

	Eurasia/ Africa	N. & S. America	All 5 Continents
Reptilian Orders			
Chelonia	0	0	1
Squamata	6	6	14
Mammalian Orders			
Rodentia	10	12	6
Carnivora	2	0	5
Insectivora	4	2	3
Primates	11	3	0
Edentata	0	5	1
Artiodactyla	4	1	4
Sum of Families	41	31	40

Factors in Post-Flood Distribution of Land Vertebrates

Since animals left the Ark after their kinds (Jones 1973), there was ample opportunity for vicariance (splitting) of faunas in the Middle East, even to some extent without sweepstakes routes. Yet the key to the dispersal of animals from Noah's Ark is the many sweepstakes situations in existence. The Ararat region is mountainous, generating nonrandom routes for migrating animals. The geography includes the Caspian and Black seas as barriers. The fauna, already separated by these local and regional sweepstakes routes, was in a position to be separated on the intercontinental scale.

Ice Age and Climate

The Ice Age after the Flood (Oard 1986) must have closed off large portions of the Northern Hemisphere to the animals originally spreading from the Ararat region. But an ice cover is not even necessary. If Oard's hypothesis is correct, volcanic dust caused a reduction in surface land temperatures. By analogy with nuclear winter models (Covey *et al.* 1984), interior portions of continents (especially Eurasia) would have been too cold to support life for some time after the Flood.

Consider the situation depicted in top, left. Except for coastal regions, where oceanic warming is a factor, Eurasia and North America are inhospitably cold (i.e., the dark region). The inhabitants disembarking from the Ark are introduced to this situation. After the Middle East is populated, the animals effectively have only two sweepstakes routes to take—southwestward to Africa or southeastward to Southeast Asia and Australia. This causes an immediate bifurcation of faunas and, among other things, explains why the tropical faunas of Africa, southeast Asia, and (later) South America have little in common.

Subsequently, (top, left) mountainous regions (such as the Urals) warm up. This is caused by the temperature inversion engendered by the atmospheric dust. A new sweepstakes route now opens up, allowing animals to migrate northward from the Middle East. Since a polar ice cap does not yet exist, the Asian Arctic is at first hospitable to these animals. Many of these continue to expand their distributions along this coast, eventually reaching North America via the Bering land bridge. Eventually the Gulf Stream becomes dominant, warming Europe and western Asia (as predicted in a nuclear winter situation: Covey *et al.* 1984). This creates yet another sweepstakes route—from the Middle East to Europe. Some of the fauna that has by now populated the Asian Arctic (and North America) also moves to Europe. This explains the faunas that occur only in Europe and North America.

Since the earlier movement of faunas between Eurasia and North America had been disjointed and subject to sweepstakes routes, it is not surprising that the faunas are so different. The Ice Age seals this situation (bottom, right). Life along the Asian and North American Arctic coasts is snuffed out, and there is no further possibility of interchange between the faunas or Eurasia and North America.

The scenario described above is an oversimplification. In reality, sweepstakes routes must have opened and re-closed repeatedly as regions of inhospitable cold changed over a time span ranging from days to decades. This caused a further vicariance of migrating animals.

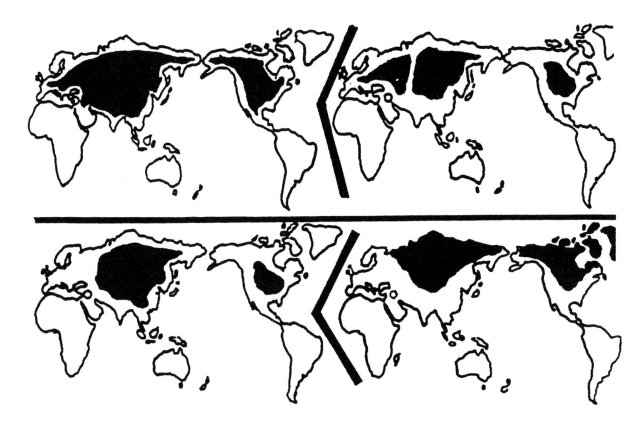

Anthropogenic Introductions

A major factor, heretofore neglected in the understanding of the spread of exotic faunas throughout remote parts of the world (i.e., relative to Ararat), is the fact that humans began a large-scale dispersal from the Middle East region only after the Tower of Babel incident (Genesis 11:78). Prior to this time, they must have been tending many of the animals that had been rapidly multiplying following their release from the Ark. As humans were forced to leave their habitations around Babel, they undoubtedly took animals with them for husbandry, game, and as a reminder of their former area of living. (For a summary of the numerous and diverse reasons for historically recent anthropogenic introductions of animals, see table 4 in Myers (1986).

These recent examples can offer only a very limited analogy to what must have taken place after the Flood. Post-Babel humans were actually in a position to bring along with them (and introduce to other continents) a much greater diversity of living things than would later be the case (when, for example, only European faunas could be brought by the post 15th century colonists to the New World). First of all, introductions into barren continents had a much greater effect on biogeography than the later introductions of living things into already-populated continents. Also, the diversity of living things in the Middle East was very great soon after the Flood. After all, first the Ark itself and then the whole Middle East region was a microcosm of the full diversity of land vertebrates that would eventually populate the entire globe. Most every group of animals initially taken from the Middle East had a good chance of being a unique faunal assemblage when introduced to distant continents.

It is important to note that introduced animals spread much more rapidly as a result of repeated anthropogenic introductions than they do through their own biological capabilities (Myers 1986). This means that, even if normal spreading tends to make faunas more homogeneous over geographic areas, anthropogenic introductions will make faunal distributions more heterogeneous at a faster rate. Also, consider the rate of population increase among Ark-released animals. If, soon after the Tower of Babel incident, the inhabitants of the Middle East knew (i.e., from advance parties) that remote areas of the earth lacked vertebrates, they had that much more motivation to take many animals with them as "they scattered" all over the globe.

Land Vertebrates with Peculiar Biogeographic Distributions

There are a number of animal groups that provide classic examples of endemic distribution. Many of these, at first, seem difficult to explain in terms of an origin from the Ark at Ararat. This work offers some novel solutions, with anthropogenic introductions being the main factor.

We have modern examples of entire faunas whose original biogeographic distributions have been completely inverted by anthropogenic introductions combined with geographically-selective extinctions. For instance, wild camels, native to North Africa and the Middle East, are now extinct there, whereas camels introduced to Australia form the largest free-living herd in the world (Myers 1986). The Middle East, originally crowded with the entire diversity of animals released from the Ark, could permanently support only a fraction of these. The rest were doomed to local extinction (or global extinction if they had no representatives beyond the Middle East). For example, the Australian marsupials are a group which was introduced there (see below) but has been long extinct in the Middle East.

Australian Marsupials

These creatures are not only highly endemic and far removed from Ararat, but also comprise a closely related group (as opposed to random assortment of unrelated exotic faunas). However, it must be remembered that the diversity of marsupials (and especially Australian ones) is exceedingly low in comparison to placentals. Only fifteen of fifty-three principal ecological niches exploited by placentals are used by any marsupial (Lee and Cockburn 1985). Furthermore, there are only seventeen families (including five extinct) of Australian marsupials in contrast to over 250 living and extinct placental families worldwide (Carroll 1988). It would have been no great difficulty for a post-Babel adventurer to have brought with himself seventeen pairs of marsupial kinds from the Middle East to Australia. Having a reminder of one's homeland is a powerful motivator for the introduction of animals (Baker 1986) and, if some of the descendants of Noah's family had grown accustomed to marsupials near their respective homes in the Middle East region, they would thus have the motivation to take marsupials with them.

I now consider possible deterministic factors in the exclusive introduction of certain marsupials to Australia. There are a number of features which nearly all Australian marsupials have in common that may have made them especially appealing for the knowledgeable traveler to have taken them along (and at the expense of placentals). Their low rates of postnatal growth (Lee and Cockburn 1985) and lesser food requirements would have made them especially suitable for long voyages, as would the near-lack of diurnal marsupials.

There is some suggestive evidence that Australian marsupials are not a naturally-occurring group but an introduced one. The thylacine, or marsupial wolf, shows close dental and pelvic resemblance to the South American borhyaenids, and evolutionists must invoke "a remarkable amount of convergent or parallel evolution" (Thomas et al. 1989) to reconcile this with DNA-based evidence that the thylacine is closer to other Australian, and not South American, marsupials. Once it is accepted that marsupials were specially created and eventually subject to anthropogenic dispersal, it is not surprising that there are astounding similarities between marsupials found on continents occurring at opposite parts of the globe. Thus, the South American Dromiciops stands out in close similarity to Australian, not South American, marsupials (Szalay 1982). Such an oddity makes sense in the light of anthropogenic dispersals of fauna: some marsupials now found only in Australia were also introduced to South America, possibly by the same crew. Indeed, the crew may have largely pre-traced the route taken by later explorers (i.e., James Cook) which would have taken them first to South America and then Australia.

It is also interesting that there are very few truly carnivorous Australian marsupials (Lee and Cockburn 1985), in contrast to the large group of carnivorous South American marsupials. It is the dingo that (apart from bats and rodents) is the only "native" Australian placental. In conventional evolutionary thought, it is claimed that the Australian marsupial fauna evolved over millions of years whereas the dingo was introduced by humans only a few thousand years ago. Accepting both the dingo and the entire Australian marsupial fauna as having been recently introduced provides us with a simple unified explanation: the ancient post-diluvian colonists evidently had preferred to bring along with them the familiar eutherian dog instead of a large group of carnivorous marsupials.

The flip side of anthropogenic introductions as a cause of the Australian marsupial fauna is that there is an explanation for otherwise surprising absences. For instance, the South American freshwater fish are completely unknown to Australia (Briggs 1987). It is not difficult to imagine why if the distributions of faunas were largely governed by the vicissitudes of anthropogenic introduction.

The Fauna of Madagascar

Next to Australia, the island of Madagascar is a striking example of a highly endemic fauna. Occurring off the east coast of Africa on a southerly maritime route from the Middle East, it is not difficult to understand why. The island was a major stopping point for colonists from the Ararat region. This not only explains the endemism of the Madagascaran biota, but also its great diversity (Mittermeier 1988). At the same time, the uniqueness of the Madagascaran fauna finds a partial explanation through African extinctions (as demonstrated, for example, by the faunas found in Madagascar and South America but not Africa: Briggs 1987).

The South American Fauna

The South American fauna contains unique groups such as the caviomorph rodents. Its avifauna is quite endemic, involving thirty-one unique families (Rich and Van Tets 1984).

Part of the South American fauna, of course, came from Eurasia via North America. This can be illustrated by those elements of the South American fauna which occur only as fossils in North America or Eurasia. Other South American forms were undoubtedly introduced by voyagers from the Middle East. Since South America is relatively close across the Atlantic on a southwesterly route from the Straits of Gibraltar, it is not surprising that it was repeatedly colonized soon after the Tower of Babel incident.

The Fauna of Mid-Oceanic Islands

For the colonization of Pacific Islands, it has been found that animals are much more capable of colonizing islands even hundreds of kilometers from a mainland than had been earlier supposed (Diamond 1987). At the same time, oceanic islands vary considerably in terms of diversity of vertebrate life and its similarity to that of the nearest continent. This can be explained by the varying successes of colonization as well as the uneven anthropogenic introductions of themselves, as they are individual counterparts to volant varieties. Flightless birds can arise from volant ancestors in a few generations (Olson 1973, Worthy 1988), making it possible for islands to have been colonized by volant birds whose recent ancestors had been released from the Ark. (This rapid "devolution" via mutations that cause loss of function and/or structure also solves the apparent problem of vestigial wings. We need not suppose that God created birds with nonfunctional wings).

Conclusions

The Creation model not only explains the distribution of living things on earth, but is also scientifically superior to the evolution model. This is because the Creation model is more parsimonious. For example, it is much simpler to explain the similarities between the Australian and certain South American marsupials in terms of anthropogenic introductions after the Flood than it is to accept their evolution, over millions of years, while the continents drifted.

Biogeographic studies can either be approached in terms of testable hypotheses or cumulative inductive evidence (Rosen 1988). This pilot study must be followed up by more detailed research into factors relevant to the spread of animals following their release from the Ark: 1) Climatic factors (i.e., the Ice Age) as a cause of sweepstakes routes operable on a transcontinental scale; 2) Anthropogenic introductions involving entire faunas of closely-related forms of life; 3) Identifiable features in Australian marsupials and Madagascaran lemurs leading to their onetime collective introductions by post-Babel humans; 4) the immediate post-Flood period and the Ararat region with its constantly-changing sweepstakes routes.

References

1. Archer, M. & G. Clayton, eds. 1984. *Vertebrate Zoogeography and Evolution in Australasia*. Australia: Hesperion Press, 1203 pp.
2. Auffenberg, W. 1974. Checklist of Fossil Land Tortoises (Testudinae). *Bulletin of the Florida State Museum* 18 (3): 121-251.
3. Baker, S. J. 1986. Irresponsible introductions and reintroductions of animals into Europe with particular reference to Britain. *International Zoo Yearbook* 24/25: 200-5.
4. Briggs, J. C. 1987. *Biogeography and Plate Tectonics*. Amsterdam: Elsevier, 204 pp.
5. Brown, J. 1983. *The Secular Ark*. Yale University Press, 273 pp.
6. Carroll, R. L. 1988 *Vertebrate Paleontology and Evolution*. New York: Freeman and Company, 698 pp.
7. Covey, C., S. H. Schneider, and S. Thompson. 1984. Global atmospheric effects of massive smoke injections from a nuclear war. *Nature* 308:21-5.
8. Diamond, J. M. 1987. How do flightless mammals colonize oceanic islands? *Nature* 327:374.
9. Jeffery, D. E. 1983. Dealing with creationism. *Evolution* 37 (5): 1097-1100.
10. Jones, A. J. 1972. Boundaries of the Min: An analysis of the Mosaic lists of clean and unclean animals. *Creation Research Society Quarterly* 9 (2): 114:23.
11. Jones, A. J. 1973 How many animals in the ark? *Creation Research Society Quarterly* 10 (2):102-8.
12. Keast, A. 1984. Contemporary ornithogeography: the Australian avifauna, its relationships and evolution, pp. 457-69, in Archer and Clayton, eds., *op. cit.*
13. Laferriere, J. E. 1989. Certainty and proof in creationist thought. *Skeptical Inquirer* 13 (2):185-8.
14. Lee, A. K. and A. Cockburn. 1985. *Evolutionary Ecology of Marsupials*. Cambridge, London: Cambridge University Press, 274 pp.
15. Lester, L. P. and R. G. Bohlin. 1984. *The Natural Limits to Biological Change*. Grand Rapids, Michigan: Zondervan, 207 pp.
16. Mittermeier, R. A. 1988. Primate diversity and the tropical forest: case studies from Brazil and Madagascar and the importance of the megadiversity countries, pp. 145-57, in E. O. Wilson, ed. *Biodiversity*. Washington, D. C.: National Academy Press, 521 pp.
17. Myers, K. 1986. Introduced vertebrates in Australia, with emphasis on the mammals, pp. 121-136, in R. H. Graves, and J. J. Burden, eds. *Ecology of Biological Invasions*. London, New York: Cambridge University Press, 166 pp.
18. Nowak, R. M. and J. L. Paradiso. 1983. *Walker's Mammals of the World*, 4th ed., vol. 1. Baltimore: John Hopkins University Press, pp. xxi-xliv.
19. Oard M. J. 1987. An ice age within the Biblical time frame, pp. 157-66 in R. Walsh, C. L. Brooks and R. S. Crowell, eds. *Proceedings of the First International Conference on Creationism*, vol. II. Pittsburgh, PA: Creation Science Fellowship, 254 pp.
20. Olson, S. L. 1973. Evolution of the rails of the south Atlantic islands (Aves: Rallidae). *Smithsonian Contributions to Zoology* 152:1-53.
21. Raup, D. M. 1982. Biogeographic extinction: a feasibility test, pp. 277-81 in L. T. Silver, and P. H. Schultz, eds. *Geological implications of impacts of large asteroids and comets on the earth*. Geological Society of American Special Paper 190, 528 pp.
22. Rich, P., and G. Van Tets. 1984. What fossil birds contribute towards an understanding of origin and development of the Australian avifauna, pp. 421-47, in Archer and Clayton, eds., *op. cit.*
23. Ronov, A. B. 1982. The earth's sedimentary shell. *International Geology Review* 24 (2):1365-1388.
24. Rosen, B. R. 1988. From fossils to earth history: applied historical biogeography, pp. 437-81, in A. A. Myers and P. S. Giller, eds. *Analytical Biogeography*. London, New York: Chapman & Hall, 578 pp.
25. Savage, D. E., and D. E. Russell. 1983. *Mammalian Paleofaunas of the World*. London, Amsterdam: Addition-Wesley, 432 pp.
26. Scherer, S. 1986. On the limits of variability, pp. 219-241, in E. H. Andrews, W. G. H. Gitt, and W. J. Ouweneel, eds. *Concepts in Creationism*, 266 pp.
27. Szalay, F. S. 1982. A new appraisal of marsupial phylogeny and classification, pp. 621-40, in M. Archer, ed. *Carnivorous Marsupials*, vol. 2. Sydney: Royal Zoological Society of New South Wales, 804 pp.
28. Thomas, R. T., W. Schaffner, WA. C. Wilson, and S. Paabo. 1989. DNA Phylogeny of the extinct marsupial wolf. *Nature* 340:465-7.
29. Van Gelder, R. G. 1977. Mammalian hybrids and generic limits. *American Museum Novitates No. 2635*, 25 pp.
30. Woodmorappe, J. 1983. A diluviological treatise on the stratigraphic separation of fossils. *Creation Research Society Quarterly* 20 (3).
31. Worthy, T. H. 1988. Loss of flight ability in the extinct New Zealand duck Euryanas finschi. *Journal of Zoology (London)* 215:619-28.

The Antediluvian Biosphere and its Capability of Supplying the Entire Fossil Record

The Antediluvian Biosphere and its Capability of Supplying the Entire Fossil Record

John Woodmorappe

Relating quantifiable elements of the fossil record (coal and oil deposits, crinoidal limestones, Meso-Cenozoic chalks, Karroo Formation vertebrates, and biogenic components of carbonates) to the number (biomass) of their biologic progenitors at the time of the Flood. Anti-creationist charges are proved specious.

Introduction

The creationist-diluvialist paradigm in its fullest development requires a young earth. With all biogenic manifestations in the earth's crust likewise confined to that time period. Certain anti-creationists (1) and professing creationists (2) have alleged certain difficulties with such a time compression. This work briefly examines these alleged difficulties, considering not only standing-crop biomass (2), but also biogenic accumulations in the 1656 years between the Creation and the Flood.

Coal and Oil Deposits

A recent study (2) attempted to relate the total mass of organic carbon in coal and oil to that of the world at the inception of the Flood. Standing-crop phytomass was assumed to be the only source of organic carbon, and 5.25×10^4 gm per sq. meter dry phytomass was the value quoted and used. It was concluded that the figures could be reconciled with a young earth only if most of the coal and oil was nonbiogenic in origin.

While considering only standing-crop phytomass, one should note that much higher values are possible. The greatest phytomass accumulations are not in the tropics (contra (2)) but in the U.S. Pacific Northwest (3). Using the factor .65 to convert live phytomass to dry matter (4), a value of 9.7×10^4 gm per sq. meter is obtained (3) for sitka spruce/western hemlock stands (Redwood forests, neglected here, can have values twice as high (3)). Log debris can provide an additional 38% dry matter. It must be stressed that none of the above cited values need approach those of the antediluvian biosphere: the maximal limit for standing-crop phytomass is unknown. Furthermore, many "virgin and primeval" forests of the type commonly used to estimate standing-crop maxima are now known (from discovered archaeological remains) to have been cleared in historic times (4).

Far more serious than underestimating standing-crop phytomass, Morton's study (2) completely neglects the contribution of peat. As will now be shown, a significant accumulation of peat dwarfs any possibly-climax forest in terms of stored organic carbon. Using a factor of .45 (5) to convert dry phytomass to organic carbon (not .18 as erroneously used by Morton (2)), Morton's value of 5.25×10^4 gm per sq. meter supplies 2.36×10^4 gm per sq. meter carbon. My alternative value quoted above (dry phytomass and dry log mass; 13.5×10^4 mg per sq. meter supplies 6.08×10^4 gm per sq. meter carbon. By contrast, a mere cubic meter of peat supplies 11.6×10^4 gm carbon (6), based on peat bulk density of 0.2 and 58% carbon content by mass. Contributions from standing-crop phytomass are henceforth neglected.

Since even a meter depth of peat supplies much organic carbon, it is worth considering plausible areal extent and thickness of antediluvian peat deposits. In 1656 years, peat deposits of 11.6 m depth result from the highest bog-plant productivities seen today (1,400 gm per sq. meter per year (7)), assuming no decay and peat density of .2 (6). Probably much higher productivities occurred in the antediluvian biosphere than even these maximal values because of favorable conditions. However, a major factor governing peat accumulations is its decomposition. The antediluvian earth was tropical, and tropical regions are inferior to temperate and boreal regions in terms of peat accumulation.

Nevertheless, tropical peats commonly occur in thicknesses exceeding 20 m (7), indicating that the high temperature of the tropics is not necessarily the limiting factor in permanent peat accumulation. While growth of aerobic bacteria (and hence increased rate of decomposition) is favored by high temperatures, stagnant water-logged soils cause local reducing conditions and favor net peat accumulation. Extensive areas of stagnant water on the antediluvian earth were favored by the inferred low topography (Psalm 104:6-9), probable high water table (Genesis 2:6), and sluggish (i.e., closed loop) hydrological cycle (Genesis 2:5).

Consider coal. If, to start our calculations, we use the previously-discussed value of 1.16×10^5 gm carbon per cu. meter peat and have the entire earth covered with 20 m of it, 1.18×10^{21} gm carbon is available. Using Morton's (2) quoted value of 1.5×10^{19} gm carbon in all the world's coal, 1.27% of the earth's surface having 20 m thick peat deposits is all that is necessary to supply the carbon. Moreover, there are smaller estimates of organic carbon in coal, but it is uncertain if the estimates refer to all coal or only to economically-usable coal. Sundquist (8) quotes values for coal as low as 5.38×10^{18} gm carbon, and only 0.46% of earth's surface with said peat deposits suffices.

Attention is now focused on oil. Consider Morton's (2) quoted value of 2.01×10^{20} gm carbon stored therein. Using the same 20 m thickness of peat, 17% of the earth's area supplies global oil deposits. (This, of course, assumes that peat is the sole source of organic carbon in oil. Marine sediments and their carbon will be considered shortly.) As with coal, Morton's (2) quoted values for oil may be much too high. Bolin (9) proposed that only 5×10^{18} gm carbon is, as a lower value, found in the world's oil, coal, and gas put together. Again, it is uncertain if these totals include disseminated occurrences. If they do, then merely 0.42% of the earth's surface with 20 m thick peat would supply the carbon.

Petroleum is thought to originate primarily from organic sludge in marine sediments rather than terrestrial peat. Shallow seas with poor circulation and reducing conditions accumulate sapropel. This could have served as an important source of organic carbon in the antediluvian earth, but it is difficult to quantify the contribution. A less concentrated but larger source of stored organic carbon is found in oceanic sediments. The present ocean floor contains 2×10^{22} gm carbon (10). If any combination of the carbon in the antediluvian oceans and that mobilized during the Flood totaled only 1% of the present oceanic amount, the high value (2) for global oil would immediately be satisfied.

This brief survey has shown that, even without nonbiogenic sources (2) of carbon, the total carbon stored in the world's coal and oil poses no problems for a young earth and global Flood.

Crinoidal Limestones

Pelmatozoan debris is a very common constituent of carbonate rocks, and it has been alleged (2, 11) that certain formations by themselves have more crinoids in them than could possibly have been alive on the entire earth at the same time. This subject must be approached by noting that only well-preserved (i.e., largely intact) crinoids found in rock must have been buried during the Flood itself. Highly fragmented crinoidal "hash" (and especially comminuted grains) could have resulted from crinoids that had lived long before the Flood. Each crinoid, when disarticulated, releases 30,000 to 50,000 plates of various sizes (12).

One must determine if certain limestones are actually crinoidal on a formation-wide scale. Clark and Stearn (cited in Morton (2)) allege that the Mission Canyon Limestone (Lower Carboniferous, western U.S.) consists of 41,000 cubic km of crinoid remains. An examination of the literature reveals that this is surely an exaggeration. Insoluble residue analyses (13) of several stratigraphic sections of Mission Canyon Limestone in Montana have revealed approximately 1% crinoid remains. A thin-section analysis of the same limestone from the same state has prompted the investigator (14) to describe fossil fragments as: ". . . ubiquitous but usually uncommon." In Wyoming, thin section analyses have shown that bioclastic debris occurs in a micritic matrix (15), while others (16) have, using whole-rock point counts, estimated at most 2% crinoid remains.

A more serious candidate for an extremely crinoidal limestone formation is the Burlington Limestone (Lower Carboniferous, central U.S.). Dott and Batten (11) have claimed that the Burlington Limestone, having a volume of 3×10^{12} m^3, consists of the remains of 2.8×10^{17} crinoid individuals. There are indeed horizons composed almost exclusively of crinoid remains (17). However, recent studies involving numerous thin-section analyses have shown an average of 50% crinoid composition for Burlington sections in central Missouri (18) and the north-central Mississippi valley (19).

Since volumes of crinoid debris generated are not limited to standing-crop of crinoids at the time of the Flood, an equation can express the total volume of crinoid debris generated in the 1656 years between the Creation and Flood:

$$V = 1656DFTA \qquad (1)$$

(V) gives the total volume of crinoid debris generated, (D) is the density of living crinoids (individuals per m^2), (F) is the volume factor (i.e., volume of debris produced by one disarticulated crinoid individual), (T) is the turnover rate (no. of generations per year), (A) is the geographic area of antediluvian crinoid growth. (V) is set at 1.5×10^{12} m^3, which is 50% of the Burlington Limestone (i.e., the crinoidal part). (D) is 300 individuals per sq. meter (20). Such densities over large areas were probable for this reason: "The single most conspicuous feature of fossil occurrences of crinoids is a strong tendency for specimens to occur in close proximity to each other" (21).

Determining a value for (F) in Equation (1) requires scrutiny. (F) equals the reciprocal of the number of crinoid individuals required to generate a cubic meter of crinoid debris. Taking Dott and Batten's (11) estimates at face value implies 93.333 individuals per m^{-3}. This value is much too high; it would mean that each crinoid produces only 10.7 cm^3 of debris. Simple volumetric computations, treating each crinoid as a cylinder 65 cm tall (the average height among some Lower Carboniferous crinoid stands (22) and one cm diameter (commonly exceeded by Burlington crinoids (18)) argue for greater debris production per crinoid individual. Indeed, Anderson's (12) estimates indicate 80 cm^3 per individual, so (F) in Equation (1) should be set at 8×10^{-5} m^3 debris per crinoid.

The turnover rate (T) is unknown for Paleozoic crinoids (20), hence estimates from extant types must be used. The turnover rate for modern echinoderms is 0.1–1.6 per year (23), with 0.2–0.3 per year the most common. Setting (T) at 0.2 per year and solving Equation (1) yields (A) equal to 189,000 km^2. This area thus supplies the Burlington Limestone with its crinoid content, and is only 2.4% of the area of the continental USA (in fact, it is less than the area of the Burlington Limestone itself). An alternative calculation, using the high turnover of 1.6 per year for (T), yields (A) equal to 23,650 km^2, which is a mere 16.5% of the area of the U.S. State of Iowa. It is evident that highly crinoidal limestones pose no problems for the young earth and the Flood. Furthermore, crinoidal limestones are not rated as highly abundant, in the Paleozoic at least (24), making it all the more unlikely that they should elsewhere be problematic for creationists.

In fact, if anything, arguments about crinoidal limestones can be turned around. It is uniformitarianism, not diluvialism, that has problems accounting for them. Boucot (24) wrote:

It is obvious that crinoidal limestone forming today is relatively rare, even in the tropical regions where crinoids tend to be more abundant and have a better opportunity for preservation than in colder waters.

In a similar vein, Carozzi and Gerber (25) wrote:

> Little is known about the Late Paleozoic ecological conditions that allowed for very large areas of coarse-grained crinoidal deposition in moderately agitated, clear, and well-aerated water. No modern analogs of these occurrences are known.

These features bespeak not only failure of uniformitarian theory and methodology, but also an antediluvian earth very different from the present one. Furthermore, the Burlington Limestone itself shows evidences of Flood deposition. Alternating layers of crinoidal debris indicate high-energy depositional events (19, 25). There are also evidences of rapid, slurry-type deposition: gravity-driven mass movements and crinoidal turbidites (18). Most intriguing of all is a shattered chert-breccia horizon in the Burlington (19). This graded-bedded horizon is attributed to a tornado-like event by Gerber (19).

Meso-Cenozoic Chalks

Chalks are a form of limestone composed primarily of micro-organisms, especially coccoliths (26). Although coccoliths are known since Jurassic (26), chalks do not become prominent until their explosive development in Late Cretaceous and (to a lesser extent) Tertiary. At present, coccoliths accumulate on the ocean floor, and Roth (27) has shown that such deposits could have accumulated in only 200 years. Returning to meso-Cenozoic chalks, it has been alleged (2) that they are a problem for Flood geology, yet no calculations had been performed.

Since it is Late Cretaceous and Tertiary that are of relevance, one must determine the total volume of chalks from those geologic periods. Such information appears to be unavailable, but one can begin with the extreme (and therefore conservative) assumption that *all* Late Cretaceous and Tertiary carbonates are chalks. Of course, they are not, and even individual formations that are mapped as chalks need not be. In a recent study involving Late Cretaceous rock, Frey and Bromley (28) wrote:

> Purer chalks crop out only in a relatively small area of Western Alabama and adjacent parts of northeastern Mississippi (and are not uniformly present even within this small area), yet the term has been used both colloquially and in a formal rock-stratigraphic sense well beyond these occurrences.

There are 17.5 million km^3 of Late Cretaceous and Tertiary limestones (29-31) and all the calculations herein are based on the said extreme assumption that they are all chalks. Existing estimates (32) for chalk accumulation rates can't be used because they rely on uniformitarian methodology that has no meaning in the creationist-diluvialist paradigm. Direct productivity estimates must be used. Using Roth's (27) calculations (which deduce 100 m thicknesses of coccoliths per 200 years), one would need 21.1 million km^2 (i.e., 4.1% of earth's surface) of coccolith-productive seas to supply the 17.5 million km^3 of coccoliths in 1656 years.

Alternate calculations, performed by the present author, follow. The larger coccoliths have diameters of 0.04 mm (36), smaller ones are more common but can reproduce at faster rates (35-36). Each 0.04 mm diameter coccolith has a volume of 3.35×10^{-14} m^3, so there are 2.99×10^{13} coccoliths per m^3. This is an exaggerated number because zero void space is assumed. In reality, coccoliths in chalks are loosely packed, as indicated by chalk's high porosity (33), and its density being commonly half that of other limestones (34).

It is difficult to model the productivity of coccoliths because coccolith concentrations in water vary in space and time by orders of magnitude. One can start by allowing each m^2 column of water to contain a conservative 3.5×10^{11} coccoliths in suspension. This situation is realized, for example, by having 35 million cells per liter (as sometimes seen today (35)) to a very conservative depth of 10 m, or a very conservative concentration of 700,000 cells per liter to depths of 500 m (commonly exceeded today by great densities of coccoliths (35)). A bidiurnal turnover rate for coccoliths can occur (36), though even greater rates can occur for smaller coccoliths (35-36). There are 1.2 million turnovers in 1656 years. At the said exaggerated value of 2.99×10^{13} coccoliths per cu. meter, a 1.4 km thick column is produced (i.e., 4.23×10^{17} coccoliths per m^2 column). Taking the said exaggerated necessity of 17.5 million km^3 of coccoliths (i.e., 5.23×10^{29} cells), a 12.5 million km^2 area (merely 2.5% of earth's area) suffices.

An alternative model of coccolith accumulation is now presented in order to reflect the fact that coccolith accumulation is not steady-state but highly episodic. There are intense blooms of coccoliths, and these can cause "white water" (36) situations because of the coccolith concentrations. Approximately 10% of earth's surface underlies, and is able to have supplied, marine Late Cretaceous and Tertiary (see Maps 32 and 34 of my work (39). If each bloom covers 10% of the earth's surface to a water depth of 500 m and generates 35 million cells per liter, 8.93×10^{26} coccoliths are spawned per bloom. Roughly 588 such blooms are needed to produce the highly exaggerated 5.23×10^{29} coccoliths required. This means one such bloom, on average, every 2.8 years in antediluvian times. Of course, these calculations are conservative even in that they assume that each massive bloom spawns *only one* generation of coccoliths.

It must be noted that neither water depths of 500 m nor concentrations of 35 million cells per liter need be limiting factors. The anti-creationist Schadewald (1) lambasted creationists by ascertaining that thermodynamic considerations prevent a much larger biomass on earth than at present. He is clearly wrong: Tappen (37) has recently noted that oceanic productivities 5–10 times greater than present could be supported by the available sunlight, and it is nutrient availability (especially nitrogen) that is the limiting factor. Present levels of solar ultraviolet radiation inhibit marine planktonic productivity (38, 40). If the antediluvian canopy screened out most or all of this injurious ultraviolet light, all the higher oceanic productivities could have been sustained.

Karroo Formation Vertebrates

The Karroo Formation (Permo-Triassic of South Africa) is said to contain the remains of 800 billion vertebrates (1). None of the references cited in (1), or elsewhere, allow a determination of how this value was calculated. Schadewald (1) has scurrilously denounced creationists as pseudoscientists for allegedly ignoring population densities. He asserted the supposed impossibility of such large simultaneously living populations (8.5 per hectare for Karroo fauna spread globally, 851 per hectare if Karroo contains 1% of all fossil vertebrates). It is easy to show the absurdity of Schadewald's arguments, and that it is he who is the pseudoscientist that has not "done his homework."

A population density of 800 per hectare results if the supposed 800 billion Karroo vertebrates are evenly spread over Africa south of the Equator (i.e., 10 million km^2 area). Compare this "impossible" density with known densities (in individuals per hectare) of some modern reptiles: 889 (1.6 kg iguanid lizards), few thousand to 110,000 (anoles and other small lizards), 548 (Colubrid snakes), 10,000 (Manchuria island pit viper), 1235 (Colorado rattlesnakes), 480 (the rhyncocephalian Tuatara), 570 (pond turtles). The first six citations are from Turner (41); the seventh is from Bury (42). It should be noted that the above-cited populations are actual habitat (as opposed to migratory) populations. Furthermore, the population densities occur over significant, even if localized, areas, and are not merely highly-provincial pockets of high population density. Although a full-fledged study of population densities of all types of animals is beyond the scope of this work, it can be seen that small areas can support large numbers of animals.

Even reptiles significantly larger than most Karroo reptiles can have surprisingly high population densities. The giant tortoise *Geochelone gigantea* (130–300 kg males; 1-1.5 m length) reaches population densities up to 160 per hectare (45) on Aldabra Island. Among dinosaurs, ceratopsians are believed to have attained densities of 28 km^{-2}, and hadrosaurids 51 km^{-2} (46). Most of the Karroo reptiles (i.e., of the *Lystrosaurus* and *Cistecephalus* zones) appear to have been primarily in the size range of 0.3–0.7 m in length (47). Compared to the reptiles cited above, the Karroo reptiles appear to most closely compare to Tuatara (.41–.66 m long; 2 kg weight (48)), and the 1.6 kg iguanid lizards. The quoted population densities for the said living reptiles (480 and 889) are within the range of that required (800 per hectare) to support the supposed 800 billion Karroo vertebrates in southern Africa.

Furthermore, Schadewald (1) is completely wrong in fingering thermodynamic considerations as limiting factors in population density. For autotrophs, sunlight is not the limiting factor in their productivity (56); nutrients commonly are. Among animals, structural complexity of the environment is a very important factor in population density (41), and low predation rates, also important (43–44), may be even more so than tropical plant productivity (43). In the antediluvian world, environments probably were more complex, and soils more nutrient-rich, than any extant environment. If man's license to eat meat after the Flood (Genesis 9:2–4) reflects wholesale increase in predator-prey ratios over those of antediluvian times, animal population densities at the time of the Flood could have been much higher than even the high values quoted above. Yet even without these additional considerations, it is evident that arguments against the creationist-diluvialist paradigm based on population densities (1) are completely fallacious. The Karroo fauna itself needs little space.

Biogenic Components of Carbonates

The biologic productivities needed to supply crinoidal limestones and chalks have already been discussed. This section considers limestones as a whole without any special consideration for type of limestone. The organic carbon in the earth's carbonate rocks has been said to be a problem (2) for creationists. The first consideration to be made is whether or not carbonates are entirely organic in origin. Grain-sized constituents of limestones are micrites (49), and most micrites are indefinite as to their origin (50). Some studies involving electron microscopy (50) indicate that many micrites are likewise of biogenic origin (i.e., especially highly-comminuted skeletal material). We can begin our calculations with the extreme assumption that all limestones are of biogenic origin, although primordial $CaCO_3$ will be briefly considered later. Highly productive marine regions today have productivities of 10^5 gm per sq. meter per year $CaCO_3$ (51). Individual types of corals can have productivities exceeding that number, while most other marine biota (green algae, red algae, molluscs, echinoderms, and forams) have productivities of 10^4 gm per sq. meter per year $CaCO_3$ (51). However, when it is remembered that productivities ten times those of present can be sustained by suitable nutrients (37), an overall $CaCO_3$ productivity of 10^6 gm per sq. meter per year can be attained.

If the entire earth had a productivity of 10^6 gm per sq. meter per year for 1656 years, 8.45×10^{23} g $CaCO_3$ would have been produced. Using a factor of 0.12 for organic carbon, 1.01×10^{23} g carbon would have been produced in carbonate rock. Consider Morton's (2) estimate of 6.42×10^{22} g carbon in earth's carbonates. Thus, 63.6% of earth's surface with said productivity would suffice to supply the carbon. But Morton's (2) values may be too high. Valiela (40) has quoted 1.83×10^{22} g carbon in limestones. Thus, 18.1% of earth's area then becomes sufficient. Even lower is Usdowski's (52) estimate of 8.4×10^{21} g carbon, for which only 8.27% of earth's area would have sufficed.

Of course, the foregoing calculations assume that 100% of carbonates have been secreted by organisms. Some disseminated as well as concentrated $CaCO_3$ may have been built-in to the antediluvian regolith since the Creation as filler material, and then have become reworked by Flood action. Calcium in soil is an important nutrient derived from $CaCO_3$, and high levels of calcium in ground water correlate with reduced incidence of dental caries (53). This nutrient role is probably another reason why the Creator most probably created some primordial $CaCO_3$. Another important source of nonbiogenic carbonate is chemical precipitation. Inorganic precipitation must have taken place alongside organic precipitation, as it does today (54). Much chemically-precipitated $CaCO_3$ must also have been generated during the Flood itself, as waters with different ions were constantly mixed.

Other Considerations: Marine Fossils

Schadewald (1) has asserted that there are far too many marine fossils in rock to have been all alive simultaneously, but presented no calculations or other evidences to support his contentions. The present author has diligently attempted, by research and contact with paleontologists, to arrive at a figure for average concentration of fossils in earth's rocks. No such information is available, and estimates are exceedingly difficult to make because concentrations of fossils vary in strata, and geographic regions, by many orders of magnitude.

Nevertheless, there certainly is no basis for Schadewald's (1) arbitrary figure of 0.1% fossil content in rock. A concentration of 0.1% implies a 10 cm^2 fossil per m^2 of outcrop face. The overwhelming majority of sedimentary rocks have far, far lower abundances of fossils, so the average for all earth's sedimentary rocks must be magnitudes lower than 0.1%. The rarity of fossils as a whole is well described by the eminent paleontologist Simpson (55):

> In spite of such exceptional examples, the great majority of fossils are found embedded in, or recently eroded from, exposures of sedimentary rocks. Yet one could dig at random in exposures of such rocks for a lifetime without encouraging a single fossil. The first discovery of an area or a stratum in which fossils of a given sort are present is often made by chance or serendipitously by someone who was looking for something else—but almost necessarily by someone who knows a fossil when he sees one. ... A paleontologist on the prowl for fossils often looks as if he were trying to find a dime that he had accidentally dropped somewhere in a hundred square miles or so of badlands.

A suggested average abundance of 0.1% for fossils is clearly ridiculous. Nevertheless, the present author plans to continue his research on fossil abundances in order to one day provide an average and then compare it with possible live biovolume at the time of the Flood.

Conclusions

It is evident that the alleged problems cited (1, 2) for scientific creationists are not problems at all. Including live Karroo biota, biogenic materials accumulated before the Flood (and then re-deposited during the Flood) is more than sufficient to supply the fossil record. In fact, many of the arguments advanced in (1) and (2) seem more devoted to the discrediting of creationism than a study of it. Nevertheless, the creationist-diluvialist paradigm only grows stronger as it passes ostensible falsification tests.

References

1. Schadewald, R. J. Six "flood" arguments creationists can't answer, in *Evolution Versus Creationism*. Oryx Press, pp. 198, 448-453.
2. Morton, G. R. 1984. The carbon problem, *Creation Research Society Quarterly*, 20(4): 212-219.
3. Franklin, J. F., and R. H. Waring. 1980. Distinctive features of the northwestern coniferous forest: development, structure, and function, in *Forests: Fresh Perspectives from Ecosystem Analysis*. Corvallis, Oregon: Oregon State University Press, p. 61.
4. Mabberly, D. J. 1983. *Tropical Rain Forest Ecology*. Glasgow, London: Blackie & Son, Ltd., pp. 10-11, 34.
5. Atjay, G. L., et al. 1979. Terrestrial primary production and phytomass, in *The Global Carbon Cycle*. New York: John Wiley & Sons, pp. 130-133.
6. Bohn, H. L. 1978. On organic soil carbon and CO_2, *Tellus* (Sweden) 30:474.
7. Clymo, R. S. 1983. Peat, in *Ecosystems of the World*, vol. 4A. Amsterdam: Elsevier, pp. 159, 196-197.
8. Sundquist, E. T. 1985. Geological perspectives on carbon dioxide and the carbon cycle, in *The Carbon Cycle and Atmospheric CO_2*. Washington, D. C.: American Geophysical Union, p. 7.
9. Bolin, B. 1983. The carbon cycle, in *The Major Biogeochemical Cycles and Their Interactions*. New York: John Wiley & Sons, p. 45.
10. Woodwell, G. M., et al. 1978. The biota and the world carbon budget. *Science* 199:143.
11. Dott, R. H., and R. L. Batten. 1971. *Evolution of the Earth*. New York: McGraw Hill, p. 307.
12. Anderson, W. I. 1983. *Geology of Iowa*. Ames, Iowa: Iowa State University Press, p. 137.
13. Sloss, L. L., and R. H. Hamblin. 1942. Stratigraphy and insoluble residues of Madison Group (Mississippian) of Montana. *American Association of Petroleum Geologists Bulletin* 26:305-330.
14. Moore, G. T. 1973. Lodgepole limestone facies in southwestern Montana. *American Association of Petroleum Geologists Bulletin* 57:1707.
15. Sando, W. J. 1967. Madison Limestone (Mississippian), Wind River, Washakie, and Owl Creek Mountains, Wyoming. *American Association of Petroleum Geologists Bulletin* 51:537.
16. Lageson, D. R. 1980. Depositional environments and diagenesis of the Madison Limestone, northern Medicine Bow Mountains, Wyoming. *Wyoming Geological Association 31st Annual Field Conference Guidebook*, p. 58.
17. Weller, J. M., and A. H. Sutton. 1940. Mississippian border of eastern interior basin. *American Association of Petroleum Geologists Bulletin* 24:794-795.
18. King, D. T. 1980. *Genetic Stratigraphy of the Mississippian System in Central Missouri*. Ph.D. Thesis. Columbia: University of Missouri, p. 121.
19. Gerber, M. S. 1975. *Carbonate Microfacies of the Burlington Crinoidal Limestone (Middle Mississippian), Western Illinois, Southeastern Iowa, and Northeastern Missouri*. Ph.D. Thesis. Urbana-Champaign: University of Illinois, p. 78.
20. Ausich, William I. 1985. Letter dated February 25. Ph.D. Thesis. Urbana-Champaign: University of Illinois, p. 197. Ausich is with the Dept. of Geology, Ohio State University (Columbus), and is a crinoid specialist.
21. Lane, N. G. Synecology. *Treatise on Invertebrate Paleontology*, vol. 1, part T.
22. Lane, N. G. 1963. The Berkeley Crinoid Collection from Crawfordsville, Indiana. *Journal of Paleontology* 37:1007.

23. Smith, S. V. 1972. Production of calcium carbonate on the mainland shelf of southern California. *Limnology and Oceanography* 17:35.
24. Boucot, A. J. 1981. *Principles of Marine Benthic Paleoecology*. New York: Academic Press, p. 103.
25. Carozzi, A. V., and M. S. Gerber. 1984. Crinoid arenite banks and crinoid wacke inertia flows: a depositional model for the Burlington Limestone (Middle Mississippian), Illinois, Iowa, and Missouri, USA, *International Congress on Carboniferous Stratigraphy and Geology* 9(3): 453.
26. Marshall, N. B. 1979. *Developments in Deep-Sea Biology*. Dorset, England: Blandford Press, p. 52.
27. Roth, A. A. 1985. Are millions of years required to produce biogenic sediments in the deep ocean? *Origins* 12(1): 52.
28. Frey, R. W., and R. G. Bromley. 1985. Ichnology of American chalks: the selma group (Upper Cretaceous), Western Alabama. *Canadian Journal of Earth Sciences* 22:802.
29. Khain, B. E., A. B. Ronov, and A. H. Balukhovskii. 1976, Cretaceous lithologic association of the continents. *International Geology Review* 18:1289.
30. Ronov, A. B., B. E. Khain, and A. H. Balukhovskii. 1979. Paleogene lithologic associations of the continents. *International Geology Review* 21:443.
31. Khain, B. E., A. B. Ronov, and A. H. Balukhovskii. 1981. Neogene lithologic associations of the continents. *International Geology Review* 23:450-451.
32. Bathurst, R. G. C. 1976. *Carbonate Sediments and Their Diagenesis*. Amsterdam: Elsevier, p. 405.
33. Mitchell, R. S. 1985. *Dictionary of Rocks*. New York: Van Nostrand Reinhold, p. 43.
34. Boynton, R. S. 1980. *Chemistry and Technology of Lime and Limestone*. New York: John Wiley, p. 23.
35. Raymont, J. E. G. 1980. *Plankton and Productivity in the Oceans*, 2nd ed., vol. 1. New York: Pergamon, pp. 251-255.
36. Sumich, J. L. 1976. *Biology of Marine Life*. Iowa: Wm. C. Brown, pp. 118, 167.
37. Tappan, H. 1982. Extinction or survival: selectivity and causes of phanerozoic crises. *Geological Society of America Special Paper 190*, p. 270.
38. Worrest, R. C. 1983. Impact of solar ultraviolet-B radiation (290-320 nm) upon marine microalgae. *Physiologia Planetarium* 58(3): 432.
39. Woodmorappe, J. 1983. A diluviological treatise on the stratigraphic separation of fossils. *Creation Research Society Quarterly* 20(3): 133-185.
40. Valiela, I. 1984. *Marine Ecological Processes*. Berlin: Springer-Verlag, pp. 43, 274.
41. Turner, F. B. 1977. The dynamics of populations of squamates, crocodilians, and rhynchocephalians, in *Biology of the Reptilia*, vol. 7, *Ecology and Behavior A*. London: Academic Press, pp. 164-263.
42. Bury, R. B. 1979. Population ecology of freshwater turtles, in *Turtles*. New York: John Wiley, pp. 586-587.
43. Schoener, T. W. 1983. Population and community ecology, in *Lizard Ecology*. Harvard University Press, p. 234.
44. Stoddart, D. R., and S. Sava. 1983. Aldabra: Island of Giant Tortoises. *Ambio* (Sweden) 12:181.
45. Grubb, P. 1971. The growth, ecology, and population structure of giant yortoises on Aldabra. *Royal Society of London Philosophical Transactions* 260B:365.
46. Beland, P., and D. A. Russell. 1978. Paleoecology of Dinosaur Provincial Park (Cretaceous), Alberta, interpreted from the distribution of articulated vertebrate remains. *Canadian Journal of Earth Sciences* 15:1020.
47. Benton, M. J. 1983. Dinosaur success in the Triassic: a non-competitive ecological model. *Quarterly Review of Biology* 58(1): 36.
48. Wood, G. L. 1982. *The Guinness Book of Animal Facts & Feats*, 3rd ed. London: Guinness Superlatives, p. 116.
49. Matthews, R. K. 1966. Genesis of recent lime mud in southern British Honduras. *Journal of Sedimentary Petrology* 36:428.
50. Lobo, C. F., and R. H. Osborne. 1973. The American Upper Ordovician Standard, XVIII: investigation of micrite in typical Cincinnatian limestones by means of scanning electron microscopy. *Journal of Sedimentary Petrology* 43(2): 478, 482.
51. Chave, K. E., *et al*. 1972. Carbonate production by coral reefs. *Marine Geology* 12:123-140.
52. Usdowski, H. E. 1968. Formation of dolomite in sediments. *Recent Developments in Carbonate Sedimentology in Central Europe*. Berlin: Springer-Verlag, p. 21.
53. Losee, F. L., *et al*. 1967. *New Zealand Trace Element Study*, in Geological Society of America Special Paper 90, p. 7.
54. Williams, H., *et al*. 1982. *Petrography*. San Francisco: W. H. Freeman & Co., p. 366.
55. Simpson, G. G. 1983. *Fossils and the History of Life*. New York: Scientific American Books, pp. 22-23.
56. Agren, G. I. 1985. Limits to plant production. *Journal of Theoretical Biology* 113:89-92.

A Diluviological Treatise on the Stratigraphic Separation of Fossils

A DILUVIOLOGICAL TREATISE ON THE STRATIGRAPHIC SEPARATION OF FOSSILS†

JOHN WOODMORAPPE*

The author and the editor have been discussing and planning this work for nearly two years. It was received 4 October 1982, and with revisions 1 June 1983.

Calculations performed on the stratigraphic separational tendencies of fossil families show that one-third of them span 3 or more geologic periods. Also, geologic periods with 4 intervening periods between them still show double-digit percentages of familial faunal similarity.

A total of over 9500 global occurrences of major index fossils have been plotted on 34 world maps for the purpose of determining superpositional tendencies. 479 juxtapositional determinations have shown that only small percentages of index fossils are juxtaposed one with another. Very rarely are more than one-third (and never more than half) of all 34 index fossils simultaneously present in any 200 mile (320 kilometer) diameter region on earth.

Flood mechanisms (pure chance, selective preservation, differential escape and hydrodynamic selectivity, and ecological zonation) are evaluated. Independent evidence is presented to demonstrate that Phanerozoic fossils were deposited under tectonically-differentiated conditions, thus justifying the concept of TABs (Tectonically-Associated Biological Provinces) as the main cause of biostratigraphic differentiation. The TAB concept is placed in an integrated study of fossil separation, and it is shown that it explains extinction trends relative to the extant biosphere. The (near) absence of pre-latest-Phanerozoic human remains is explained through low antediluvian population (primarily); preservation factors are also scrutinized.

Plan of This Article

Introduction
I. DETERMINING TRUE STRATIGRAPHIC AND SUCCESSIONAL TENDENCIES OF FOSSILS
    A. A Measurement of the Actual Stratigraphic Tendencies of Fossils
    B. The Study of Juxtapositional Tendencies of Index Fossils: A Global Geographic Approach
II. THE SEPARATION OF ORGANISMS DURING BURIAL BY THE FLOOD: PROCESSES AND MECHANISMS
    A. Indeterministic Factors leading to Stratigraphic Differentiation of Fossils
    B. Deterministic Factors leading to Stratigraphic Differentiation of Fossils: the Primacy of TABs (Tectonically-Associated Biological Provinces)
    C. The Stratigraphic Separation and Succession of Fossils: a Diluvial Synthesis
    D. Biostratigraphically-Progressive Extinctions with Respect to the Extant Biosphere: An Explanation in the Light of the TAB Concept
    E. Causes for the (Near) Absence of Pre-Pleistocene Human Fossils

Introduction

The geologic column, specifically the order of appearing and disappearing fossils, is a pivotal point in both the evolutionary-uniformitarian and Creationist-Diluvialist paradigms. Evolutionists have long cited the order of fossils as evidence for evolution, but Creationists have offered alternate explanations in terms of the Universal Deluge. Whitcomb and Morris[643] noted the role of hydrodynamic sorting as well as differential escape, while Clark[645] (and many other Creationists-Diluvialists) emphasized ecological zonation. Price[644] and Burdick[646] tended to downplay the need for any specific mechanism to account for the stratigraphic separation of fossils, pointing out that many fossils overlap large parts of the geologic column and that few different types of fossils can usually be seen to superpose at any one given locality.

This work is a rigorous study of: 1) the actual stratigraphic tendencies of fossils, 2) the actual successional tendencies of fossils, and 3) models of Flood action directly bearing on these two tendencies. The term "actual stratigraphic tendency" used herein (and throughout this work) refers to the statistical tendency of fossils to be confined within a single geologic period versus the tendency to span a large part of the geologic column. The term "actual successional tendency" means the tendency for many index fossils to be found superposed at any one given locality versus the tendency for few index fossils to be locally superposed.

There is a need to clarify the relationship between organic evolution and the geologic column because Creationists have commonly been accused of misunderstanding the relationship between the two. This issue has importance not only in relation to the Flood, but also increasingly (as pointed out by Creationists Morris and Parker[647]) in relation to the basic Creation/evolution issue. It goes without saying that evolution is based on the geologic column. McLaren[648] wrote: "All historical inference in geology comes from the positional relationships of rock and mineral bodies. Stratigraphy is a special case of this general law, and our sole knowledge of the orderly evolution of life as represented by fossils, comes from their mutual rela-

---

*John Woodmorappe has a B.A. and M.S. in Geology and a B.A. in Biology.
†This work is dedicated to the observance of the 20th anniversary of the founding of the Creation Research Society.

tions in stratified bodies. *The only proof that one fossil is younger than another lies in the relative position of the two in a sequence of rock. . . .* Hypotheses of evolution of a lineage must depend on the fact of positional relationship." (italics added)

Gingerich[649] said: "Without fossils and stratigraphic ordering, evolution itself would be little more than a speculative conjecture." (Since—as will be shown—most fossils are not superposed at any one spot, evolution *is* nothing more than a speculative conjecture.) It is also worth noting that the study of stratigraphy is not only colored by evolutionist's presuppositions, but also uniformitarian ones. This latter fact is evident in the following statement of Watson[650]: "Stratigraphy is the senior branch of historical geology . . ."

Evolutionists and uniformitarians, however, commonly claim that: 1) the geologic column is totally independent of organic evolution, and 2) that it was Creationists who had founded the geologic column. The first point is now addressed, while the second depends on definition of the word Creationists, a point to be considered later. First of all, it goes without saying that horizons of fossils (as well as lithologies themselves) have a regional character to them that enables their use in correlation (independent of any mode of origin) locally and regionally. McKerrow[651] wrote: "Some fossils can be used as a rough working basis for stratigraphy without considering them as much more than formed stones. William Smith discovered the stratigraphical application of fossils long before the publication of Darwin's *Origin of Species* in 1859; and, during the second half of the nineteenth century, palaeontologists were applying their efforts to the description of new fossils and to the establishment of a stratigraphical framework based on their new discoveries. Sedgwick, Lapworth, Murchison, and John Phillips (to put them in stratigraphic order) were all concerned with the use of fossils as indicators of geological time. Looking back, it now seems amazing that none of these early giants demonstrated much in the way of evolution in fossils."

An important distinction must be made between local and regional correlations (as exemplified by William Smith), and global correlation (as exemplified by Murchison). As one moves from local all the way to global correlation by fossils, correlations become increasingly less empirical and more conceptual. This is because there are progressively greater differences (such as lithology, local fossil succession, and overall faunal character) as one moves ever further geographically from a reference section in the type area. Accordingly, global correlation cannot rest entirely (or even primarily) upon empirically-derived superpositions but must depend upon a *conceptual* foundation linking index fossils as being time-equivalent.

The oft-repeated evolutionary-uniformitarian claim that global correlations by fossils are strictly empirical and independent of any other concept can be refuted merely by pointing to the history of geology. The Wernerians believed that basic lithologies could be correlated; hence a granite could be correlated with any other granite on earth. Note that the correlation was not purely empirical but was dependent upon the *concept of time-equivalence* and/or genetic relationships of primary lithologies. Once the concept of such equivalence among primary lithologies fell into disfavor, so did the entire Wernerian system.

An analogous situation exists for correlation by fossils, as pointed out by Price.[644] Some conceptual basis is needed for assuming a time-equivalence between fossils: this conceptual basis *is* organic evolution. Correlation by fossils has meaning *only* when they are believed to have arisen at a *definite time* and become extinct at a definite time more or less contemporaneously all over the earth.

The fact that the concept of time-equivalence of index fossils depends upon acceptance of organic evolution is proved by the following discussion concerning global correlations of Lower Cambrian cited by Cowie, et al.:[652] "Dr. W. S. McKerrow asked the authors if they considered the three Lower Cambrian zones to be satisfactory time indicators. Could these changes be due to some environmental factor like increase in depth of water? If so, the same sequence of environments might produce the same sequence of faunas at different times in different areas. In particular, would the authors state why the faunal changes between the 'non-trilobite zone' and the succeeding 'olenellid zone' should represent a time horizon rather than a change in environment?" It is evident that the stratigraphic appearance of olenellid trilobites has meaning in global correlations *only* if this appearance is the result of an isochronous *evolutionary* outburst. If this appearance is due to an ecological as opposed to evolutionary change, then there is no conceptual basis for believing that the appearance is time-equivalent all over the earth, and the mere fact that this stratigraphic appearance of olenellid trilobites is empirical in no way validates it as a time horizon for global correlations.

Attention is now focused on the question of whether or not it was Creationists who founded the geologic column. Individuals such as Cuvier and Lyell (in his earlier years) accepted special creation *only* in the organic realm, but were always evolutionistic with regard to geology. Recall that evolution is not only considered to be operative in the organic realm, but at all these five realms: 1) cosmic, 2) geologic, 3) organic, 4) organic-human, and 5) human-cultural. Total evolution repudiates all forms of Divine action and attempts to explain the origin of existence, complexity, structure, and diversity in these five realms through materialistic processes that allegedly result in innovation, usually (though not inevitably necessarily) over immense periods of time. Special Creation in these five realms not only explains mere existence, complexity, etc., in terms of Divine action, but stresses the fact that all natural changes since then have been conservative or degenerative (as opposed to innovative).

To be a consistent (or full, or true) Creationist, one must accept special creation in all five realms, and a parallel definition exists for being a consistent evolutionist. Contemporary compromising positions such as deistic evolution, theistic evolution, so-called progressive creation, the "gap" (or ruin-reconstruction) theory, and the pre-world position are thus neither consistently Creationistic nor consistently evolutionistic. These hybrid positions mix evolutionistic with Creationistic concepts and allow both special creation and evolution to split roles among (and/or within) the five realms of origins discussed previously.

Cuvier, Lyell, and other originators of the geologic column also held to hybrid positions, so it is fallacious to say that they were really Creationists. Fossils were used for global correlation and special creation was used as the basis for their alleged time equivalence, but this whole notion of multiple repopulations is purely the result of special creation (in the organic realm only) being unequally yoked with evolution (in the geologic realm).

It has been already demonstrated that global correlation by fossils requires some concept of time-equivalence of fossils to be operative. While Cuvier and Lyell used special creation as the concept for time-equivalence, their hybrid position of multiple creations of life over immense amounts of time has been long since repudiated (in fact, Lyell himself became a total evolutionist towards the end of his life). Nowadays, it is the concept of organic evolution which provides the basis for alleged time-equivalence of index fossils. Since the distorted concept of special creation used by the originators of the geologic column was never truly Creationistic, and organic evolution has long since become the conceptual basis for time-equivalence of index fossils, modern Creationists can justifiably point out that organic evolution *is* the basis for the geologic column.

I. DETERMINING TRUE STRATIGRAPHIC AND SUCCESSIONAL TENDENCIES OF FOSSILS

A. A Measurement of the Actual Stratigraphic Tendencies of Fossils

It is a well-known fact that not all fossils are believed to have time significance, and many range through several geologic periods. In evaluating tendencies for fossils to be stratigraphically differentiated, one must first evaluate the credibility of the taxonomy involved. In my work[653] on cephalopods, I had advocated that fossil species and genera (as well as their stratigraphic ranges) not be recognized. It was extensively documented that; 1) fossil species and genera are highly subjective—even to such an extent that the number of fossil species and genera identified in a given collection often varies by more than a factor of two, 2) there is an artificially high diversity of short-range taxa, 3) the taxonomy is deliberately biased to produce short-ranged "species" and "genera," 4) the same taxa are given different names in different stratigraphic positions.

Some additional evidence is now presented. In setting up Lyellian curves for progressive extinction of marine faunas in Late Tertiary (stated to be circa 20 million years ago), a large scatter of 4 to 5 million years was noted by Stanley, et al.,[654] who wrote: "It is possible that at least part of the apparent disparity between the gastropods and bivalves is an *artifact of taxonomy.*" (italics added). If fossil species that often (or usually) have actual living representatives are subjective, how much more so ancient forms with no living representatives even at higher taxa! Elsewhere, Chaloner and Lacey[655] wrote: "It is the nature of palaeontology that as knowledge of material increases, particularly from a wide range of localities, concepts of generic limits change. This makes any attempt to collate records from all over the world, involving data published over a considerable amount of years, considerably vulnerable." Though they were speaking of biogeographic differentiation, the same applies for stratigraphy. The (implied) proclivity to multiply taxonomic names at different stratigraphic horizons is evident in this use of foraminifers with respect to Cretaceous stages, described by Bartenstein and Bolli[656]: "A Middle to Upper Albian assignment in Rumania by Costea (1974) is stratigraphically so young that the species determination should be checked again." In another situation, Windle[657] advocated that Carboniferous spores had been reworked into the Triassic; the spores erroneously having been given two different names by others depending on which geologic period they were in although they were nearly identical. Ethington and Schumacher[658] wrote: "We are reluctant to extend its range downward into Middle Ordovician rocks without evidence of its concurrence in Upper Ordovician strata as well as in rocks representing almost the entire Silurian System." They wanted to invoke different names for look-alike conodonts just because they were in different strata! (so-called homeomorphy).

The family level of taxonomy was taken as the basic unit for calculating the overlap of all fossils with respect to the geologic column, and the results are shown in Figure 1. The data for fossil families came from the volume by Harland.[659] He listed 2,617 fossil families (with a small admixture of slightly higher and lower taxons where necessary for approximate equivalence). They were shown as lines spanning part or all of the geologic column. The present author manually counted all 2,617 lines in terms of how many geologic periods they span and the results were thus graphed in Figure 1. Although fossil genera are not recognized as valid entities, they were included in Figure 1 be-

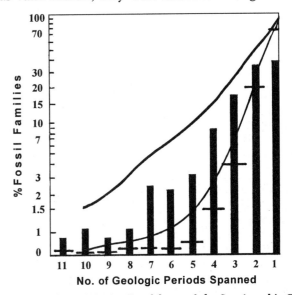

Figure 1. A Quantitative Breakdown of the Stratigraphic Overlap of Fossils with Respect to the Geologic Column. The horizontal line segments and thin cumulative-frequency curve refer to fossil genera; the thick vertical bars and thick cumulative frequency curve refer to fossil families.

cause they were already available from the work of Raup[660] in direct numerical form (in contrast to the stratigraphic lines for fossil families). The total number of fossil genera is 19,805.

Caution must be used in interpreting the data because of the following reason given by Cutbill and Funnel[661] concerning all such quantitative manipulations of bulk taxonomic data: "Moreover, we are not at all convinced that there is any real equivalence in rank even between nominally equivalent taxa." Thus one is in the proverbial situation of mixing apples with oranges. Nevertheless, the data in Figure 1 do give an idea of the degree of stratigraphic overlap of fossils. It can be seen that fossils are highly differentiated stratigraphically, but on the other hand there is significant overlap of many geologic periods. One-third of all fossil families span 3 or more of the 10 geologic periods (the present is listed in Figure 1 as an 11th geologic period). The number of all fossil families spanning the entire geologic column, while a very small minority, is still measureable on the percentage scale. At the same time, only one-third of all fossil families are stratigraphically confined to only one geologic period. The net result of the data shown in Figure 1 is that, while the Creationist-Diluvialist must account for the stratigraphic differentiation of fossils, the evolutionist-uniformitarian must resort to special pleading in using fossils as time markers because of the fact that he must ignore many fossils that span a large portion of the geologic column.

It must be realized that even the stratigraphic confinement of families is self-fulfilling to a considerable extent because circular reasoning plays a major role in biostratigraphy and because most index fossils do not actually overlie one another; both points are extensively discussed in subsequent chapters. Even taking Figure 1 just at face value, one must note many reasons for shifting both the histograms and curves leftwards (towards increasing stratigraphic overlap of fossils). First of all, the subjectivities discussed in conjunction with the rejection of fossil species and genera apply to a certain extent to fossil families. Koch[662] showed that there is an artificially low diversity of long-range taxa because they, having little or no stratigraphical utility, have not been as well studied as short-range taxa. He concluded: "The published fossil record has significant bias in favor of common and biostratigraphically important taxa . . ." Simultaneously, there is an artificially high diversity of short-range taxa caused by taxonomic oversplitting by stratigraphers. This was amply demonstrated in this author's work[653] on the cephalopods. Elsewhere, in a study of Archaeocyathids, Sepkoski[663] noted an "excess of families" probably caused by their "biostratigraphical value."

Even when taxa are accepted as valid there is a noteworthy trend for stratigraphic ranges to increase with further collecting. As a matter of historical interest, Kielan-Jaworska[664] wrote: "Not until 1925 were remains of the placental mammals found in pre-Tertiary deposits, specifically from the Cretaceous." There are numerous recent instances of significant stratigraphic-range increases, and some of these are summarized by the Creationist Lubenow[665] and also by the present author in his work on cephalopods,[653] his first Anthology,[666] and his second Anthology;[667] not to mention over 200 stratigraphically-anomalous fossils tabulated in his second Anthology[667] (and explained away by uniformitarians as being reworked). In just the last few years there have been interesting developments in the area of extention of stratigraphic ranges. Shu said:[668] "It is still necessary to explain why so many Paleozoic plants persisted into earliest Triassic time in South China." Bengston[669] wrote: "In all the investigated characteristics, *Atractosella* is indistinguishable from a modern soft coral of the family Alcyoniidae. It is interpreted as the earliest known representative of the octocoral order Alcyonacea, extending the range of this group from the Lower Jurassic to the Lower Silurian." Collins and Rudkin[670] reported a find of barnacles that extends their range downward from "the Upper Silurian to the Middle Cambrian" and they also noted that it is ". . . a barnacle of such modern aspect."

Such stratigraphic-range extensions are not exceptional, and one can never be certain that a sufficiently large number of stratigraphic observations of a taxon have been made, for the following reason given by Crick:[198] "Chances of fragmenting the fossil record and truncating stratigraphic ranges are increased if small geographic areas are sampled." Cutbill and Funnel[661] wrote: ". . . collection failure usually tends to produce bunched and shortened ranges." In his work on fossil genera (here graphed in Figure 1), Raup[660] said: "If the early members of a genus are not preserved . . . then that genus may be placed in a later cohort in ignorance of its earlier history. Another general effect of non-preservation is to truncate geologic ranges . . ." There is thus no way of knowing whether the stratigraphic ranges of taxa shown in Figure 1 are reasonably final. At the same time, there is evidence presented by Raup[671] and by Simpson[672] that most fossil lower taxa have already been discovered, so it is unlikely that future discoveries of short-range taxa will statistically offset the continual increase of stratigraphic ranges exhibited by currently-known taxa.

The factor of circular reasoning will be discussed in a later chapter; but some of its employment is mentioned here because of its bearing in terms of the artificial exaggeration of the number of short-range taxa in relation to long-range ones; again justifying a leftward shift in Figure 1. Cutbill and Funnel[661] wrote: "It seems to us that the number of taxa shown in our figures as commencing or ending their ranges at major Era or System boundaries may well be influenced at least in part by preconceptions on the part of systematists on the limiting effect of these boundaries." The consequences of such circular reasoning were well described by Maheshwari:[673] "Fossils are relied on to provide an indication of geologic age; if age is accepted as a criterion for taxonomic distinction, a perfect circularity of reasoning results that would nullify one of the important purposes of paleontologic work." When considering the implications of Figure 1, one should keep in mind that there is a significantly greater share of long-range taxa than shown because of the factors just discussed.

Stratigraphic Separation of Fossils

	Ordovician	Silurian	Devonian	Carboniferous	Permian	Triassic	Jurassic	Cretaceous	Tertiary	Recent
Recent —										80,* 89
Tertiary —									70, 51	60,* 82
Cretaceous —								83, 52	64, 29	57, 49
Jurassic —							67, 49	61, 27	50, 16	48, 30
Triassic —						52,* 65	40, 36	37, 21	32, 13	31, 24
Permian —					72, 71	39,* 48	31, 27	29, 16	25, 10	25, 19
Carboniferous —				58, 54	42, 38	24,* 28	18, 15	18, 9.1	15, 5.8	15, 10
Devonian —			85, 59	52, 33	39, 25	26, 20	19, 11	19, 6.9	16, 4.2	15, 7.5
Silurian —		70, 68	61, 41	39, 25	31, 19	21, 16	17, 9.3	17, 5.8	14, 3.5	11, 6.1
Ordovician —	55, 20	37, 13	36, 8.7	27, 6.1	24, 5.4	20, 5.6	18, 3.7	18, 2.3	17, 1.5	15, 2.5
Cambrian —										

Table 1. The Tendencies for Sharing of Fossil Families Between Geologic Periods. The numbers are percentages of fossil families reciprocally shared between Geologic Periods. For example, 29% of all families that cross at least part of the Carboniferous also cross at least part of the Cretaceous; 16% of all families crossing Cretaceous also cross Carboniferous. The asterisks (*) indicate exceptional situations in which a younger geologic period has more families in common with an older period than vice versa.

Figure 1 gives only degree of stratigraphic overlap but does not specify the actual geologic periods where the taxa occur. Table 1 was constructed to show quantitatively how fossil families spread stratigraphically across geologic periods with respect to specific geologic periods versus each other. Each geologic period has a definite number of fossil families that cross it stratigraphically; irrespective of whether they originate, terminate, or totally span the period. Two percentages are given in Table 1: the former is the total number of families in common between two periods divided by the total number of families crossing the older geologic period (times 100), the latter is the total number of families in common with the two periods divided by the total number of families crossing the younger geologic period (times 100).

The raw data are from the Harland[659] volume, and the percentages were computed by manually reducing all 2,617 families to a numerical abundance relative to all possible stratigraphic ranges: Cambrian-Cambrian, Cambrian-Ordovician. . . . Cambrian-Recent, Ordovician-Ordovician, . . . Ordovician-Recent. . . . . Recent-Recent. The total number of possible Phanerozoic ranges is 66 (= 11 + 10 + 9 + . . . 1). An example is now presented to show how the entries in Table 1 were computed: Silurian vs. Triassic. The first term is given by: S+T+(100x)/(S+T+ plus S+T−) whereas the latter term is given by: S−T+(100x)/(S+T+ plus S−T+). The S and T stand for Silurian and Triassic, whereas a (+) sign to the right of the letter denotes that a family crosses it and the opposite is the case for a (−) sign. Thus S+T+ means families common to both Silurian and Triassic. S+T+ equals: Σ(Cambrian-Triassic+ . . . Cambrian-Recent) + (Ordovician-Triassic+ . . . Ordovician-Recent) + (Silurian-Triassic+ . . . Silurian-Recent). S+T− means families crossing Silurian but not Triassic and equals: Σ(Cambrian-Silurian+ . . . Cambrian-Permian) + (Ordovician-Silurian+ . . . Ordovician-Permian) + (Silurian-Silurian+ . . . Silurian-Permian). S−T+ denotes families not present in Silurian but present in Triassic and equals: Σ(Devonian-Triassic+ . . . Devonian-Recent) + (Carboniferous-Triassic+ . . . Carboniferous-Recent) + (Permian-Triassic+ . . . Permian-Recent) + (Triassic-Triassic+ . . . Triassic-Recent).

Probably the most interesting result of Table 1 is the fact that, in all but 5 out of 55 cross-comparisons, the per cent value at left is greater than the right. This means that, in all but the few exceptional cases, older geologic periods have more of their families in common with younger periods than younger ones have their families in common with older ones. In other words, the main trend going stratigraphically upwards is not as much the disappearance of old forms as the addition of brand new forms. If this trend in Table 1 had been shown as a Venn Diagram, the older geologic period would be represented with a smaller circle than the younger and thus the area of overlap of the circles would be a smaller share of the larger circle than of the smaller.

The trend shown in Table 1 may be primarily an artifact of biostratigraphic methods, and reveal in its own way how the fossil record is artificially made to appear more stratigraphically differentiated than it really is. This takes place because stratigraphic conflicts are resolved by allowing old taxa stratigraphically to range into younger strata in preference to allowing younger taxa to range downward into the older strata. Thus stratigraphic differentiation is made to appear more compelling by having as many groups as possible not appearing until late in the geologic column. A concrete example of this was provided by Karamlov,[674] who observed an anomalous stratigraphic coexistence of Riphean-Cambrian algae with Devonian Brachiopods, Corals, and Crinoids. He commented: "Since it is quite impossible for the host strata here to be Riphean or Riphean-Cambrian age, the conclusion that the range of the above forms is limited to the Riphean and early Cambrian can be queried." Karamlov did not even consider the possibility that the conflict could be resolved by allowing the Devonian forms to range downward into the

Riphean-Cambrian! How the trends seen in both Table 1 and Figure 1 relate to the Flood will be discussed in a later chapter, taking into consideration both stratigraphic differentiation and overlap.

B. The Study of Juxtapositional Tendencies of Index Fossils: A Global Geographic Approach

Whenever one considers biostratigraphic differentiation (Figure 1, Table 1), one is saddled with the tacit assumption that the fossils actually overlie each other on earth. When the Diluvialist is challenged by the question, for example, why Cambrian Trilobites never are in stratigraphic coexistence with Tertiary mammals, the question has meaning only if Cambrian trilobites and Tertiary mammals have a chance ever to have become mixed; a situation true only if the two fossils actually overlie one another somewhere on earth. Even after it is shown that there are such locations, the fact that they are few in number makes biostratigraphic differentiation largely vacuous or at least highly amenable to nonevolutionary, nontemporal explanations.

Table 2. Summary of data used in constructing Maps 1-34. The "No. of Localities" refers to the number of fossil localities plotted on a given map; whereas the "Data-Base References" denotes the reference nos. of the articles used in compiling global fossil occurrence data for each map.

Map No.	Age and Fossil	Some Prominent Representatives of Fauna/Flora	No. of Localities	Data-Base Reference No.
1	Precambrian Miscellanea	*Eosphaera, Kababekia, Stratifera, Eostion, Eomycetopsis, Conophyton*	250	2-28, 549-51, 731-4
2	Cambrian Trilobites	*Olenellus, Paradoxides, Redlichia, Conaspis, Geragnostus*	354	29-58, 110, 552-3, 731-2, 735-6
3	Cambrian Archaeocyathids	*Archaeocyathus, Zonocyathus, Aldanocyathus, Coscinocyathus, Radiocyathus*	174	4, 30-1, 33, 36-8, 59-70, 555-7, 732, 736-7
4	Ordovician Trilobites	*Selenopeltis, Chasmops, Bathyurellus, Illaenus, Cyclopyge, Symphysurina*	482	31-2, 34, 71-128, 134-6, 175, 203, 341, 552, 554, 570, 572
5	Ordovician Graptolites	*Dictyonema, Tetragraptus, Clonograptus, Climacograptus, Nemagraptus*	319	31-2, 35, 97-109, 111, 122, 126, 130, 132-3, 137-62, 175, 203, 558-61, 571, 738
6	Ordovician Brachiopods	*Spirigerina, Zygospira, Platystrophia, Leptaena, Christiania*	262	31-2, 34, 98-117, 122-4, 130, 163-8, 175, 203, 400, 570-2, 739-41
7	Ordovician Conodonts	*Cordylodus, Periodon, Amorphognathus, Belodina, Pygodus, Oistodus*	279	31, 53, 104-6, 123-33, 169-89, 562, 570, 572, 742
8	Ordovician Nautiloids	*Ellesmeroceras, Endoceras, Discoceras, Actinoceras, Tarphyceras, Baltoceras*	218	31, 34-5, 101-8, 113-22, 127-9, 131, 139, 190-203, 563-5, 570-2, 742-3
9	Siluro-Ordovician Echinoderms	*Mitrocystella, Pemphocystis, Pisocrinus, Petalocrinus, Scyphocrinites*	205	31, 175, 204-12, 226, 228, 566, 570, 741
10	Silurian Brachiopods	*Stricklandia, Atrypoidea, Pentamerus, Eocoelia, Clarkeia*	303	31, 98, 106, 180, 209-28, 567-9, 740
11	Siluro-Devonian Graptolites	*Glyptograptus, Linograptus, Monograptus hercynicus, M. turriculatus, M. thomasi*	287	31-2, 98, 137-8, 175, 210-1, 228-36, 272, 568-70
12	Siluro-Devonian Fish	*Bothriolepis, Cephalaspis, Psammosteus, Thelodus, Logania*	188	175, 237-56, 270, 573-6, 744-50
13	Siluro-Devonian Trilobites	*Acernaspis, Dalmanites, Encrinurus, Warburgella, Acastella, Coniproetus*	314	31, 140, 175, 209-11, 217, 226-8, 257-75, 568-70, 577-80, 746, 751-3
14	Devonian Floras	*Zosterophyllum, Callixylon, Dawsonites, Phacophyton, Cooksonia*	137	32, 140, 272, 276-89, 578, 581-4, 746, 754-9
15	Devonian Ammonoids	*Cabrieroceras, Imitoceras, Manticoceras, Foordites, Cheiloceras*	161	200-1, 217, 270-2, 290-7, 570, 577-8, 585-8, 751, 759-63
16	Devonian Coelenterates	*Heliophyllum, Moravophyllum, Hexagonaria, Salairophyllum, Favosites*	305	31, 175, 270-2, 298-306, 577, 589

STRATIGRAPHIC SEPARATION OF FOSSILS

Map No.	Age and Fossil	Some Prominent Representatives of Fauna/Flora	No. of Localities	Data-Base Reference No.
17	Devonian Brachiopods	*Isorthis, Stringocephalus, Howellella, Strophochonetes, Basilicorhynchus*	307	140, 175, 217, 227, 270-2, 297, 307-18, 570-1, 578, 590-4, 740, 746, 748, 751-3, 759-60, 764-6
18	Carboniferous Ammonoids	*Orthoceras, Protocanites, Eoasianites, Muensteroceras, Reticuloceras*	183	140, 175, 200, 217, 230, 305, 319-40, 462, 570, 587, 595-603, 767-9
19	Carboniferous Fusulinaceans	*Wedekindellina, Eostaffella, Triticites, Schubertella, Beedeina, Eofusulina*	238	324, 341-50, 601, 604-10, 770
20	Permo-Carboniferous Floras	*Lepidodendron, Cordaites, Glossopteris, Pecopteris, Sphenophyllum*	505	175, 285, 351-64, 741, 754, 771
21	Permo-Carboniferous Corals	*Kueichouphyllum, Cyathaxonia, Syringopora, Parawentzellella, Waagenophyllum*	262	301, 325, 365-77
22	Permian Fusulinaceans	*Veerbeekina, Neoschwagerina, Codonofusiella, Reichelina, Palaeofusulina*	236	348-50, 375, 378-86, 570, 612, 772
23	Permian Brachiopods	*Crytospirifer, Meekella, Richthofenia, Urushthenia, Linoproductus*	356	325, 354, 375, 387-93, 611, 741, 773
24	Permian Ammonoids	*Xenodiscus, Cibolites, Uraloceras, Timorites, Cyclolobus*	145	140, 217, 305, 325, 375, 394-410, 602-3, 611-8, 741, 774
25	Permian Ectoprocts	*Fenestella, Hexagonella, Streblascopora, Polypora, Fistulipora*	147	325, 375, 411-4, 570, 612, 775
26	Permo-Triassic Reptiles	*Dimetrodon, Cynognathus, Rutiodon, Lystrosaurus, Kannemeyeria, Mesosaurus*	255	200, 415-38, 619-23, 776-9
27	Triassic Fish	*Semionotus, Redfieldius, Palaeolimnadiopsis, Boreosomus, Dictopyge*	147	200, 417-20, 439-50, 623-30, 780-3
28	Triassic Ammonoids	*Owenites, Tirolites, Cochloceras, Proavites, Otoceras*	157	200, 341, 401, 442, 451-60, 631-3, 784-5
29	Triassic-Jurassic Floras	*Sagenopteris, Araucaria, Nilssoniopteris, Gingko, Marattia, Cycadeoidea*	244	461-71, 733, 754, 786-9
30	Jurassic Ammonoids-Belemnites	*Ameoboceras, Phylloceras, Epipeltoceras, Amaltheus, Cylindroteuthis*	440	472-81, 638, 741, 790-5
31	Jurassic-Cretaceous Dinosaurs	*Ankylosaurus, Iguanodon, Brachiosaurus, Stegosaurus, Pteranodon, Titanosaurus*	188	438, 482-500, 634-7, 796-9
32	Cretaceous Ammonoids-Belemnites	*Turrilites, Protexanites, Hoplites, Clioscaphites, Actinocamax*	499	480, 501-4, 63, 794-5, 800-3
33	Tertiary Mammals	*Unitatherium, Coryphodon, Bathyopsoides, Eumys, Hipparion*	535	505-28, 640-2, 804-6
34	Tertiary Foraminifers	*Globigerina, Globorotalia, Lepidoclina, Cycloclypeus, Fabiania, Miogypsina*	478	529-48, 807
			TOTAL: 9,560	

The only method for determining how index fossils of different geologic periods actually overlie one another is to construct locality maps for each type of fossil and then superimpose such maps, e.g., over a light table, to determine superpositions of fossils. This has been done (Table 2, Maps 1-34). Data were meticulously gathered from hundreds of sources; the sources including individual (particularly recent) fossil discoveries, sources yielding mapped fossiliferous regions or provinces, and sources already providing global fossil occurrence data. All 34 fossil occurrence maps had all three types of sources utilized; but the preponderance of references for Paleozoic fossils (Table 2) reflects the fact that fewer comprehensive fossil-occurrence sources had been available for that part of the geologic column. All data were pre-screened for accuracy; authors who used many small and overlapping symbols on either regional or world maps were used, because usage of many small symbols indicated that the cited authors were concerned with accurate representation of fossil-bearing localities, while overlapping symbols indicate that the symbols represent true localities and thus are not merely crude schematic representations of fossil occurrence. A latitude-longitude grid map was made (shown as Map 35 for any reader wishing to make a transparency of it to determine the

Maps 1-37. These maps, on the next few pages, show where the fossils have actually been found, as is explained in the text. Note that the small region shown at the middle right is that part of Antarctica for which data are available.

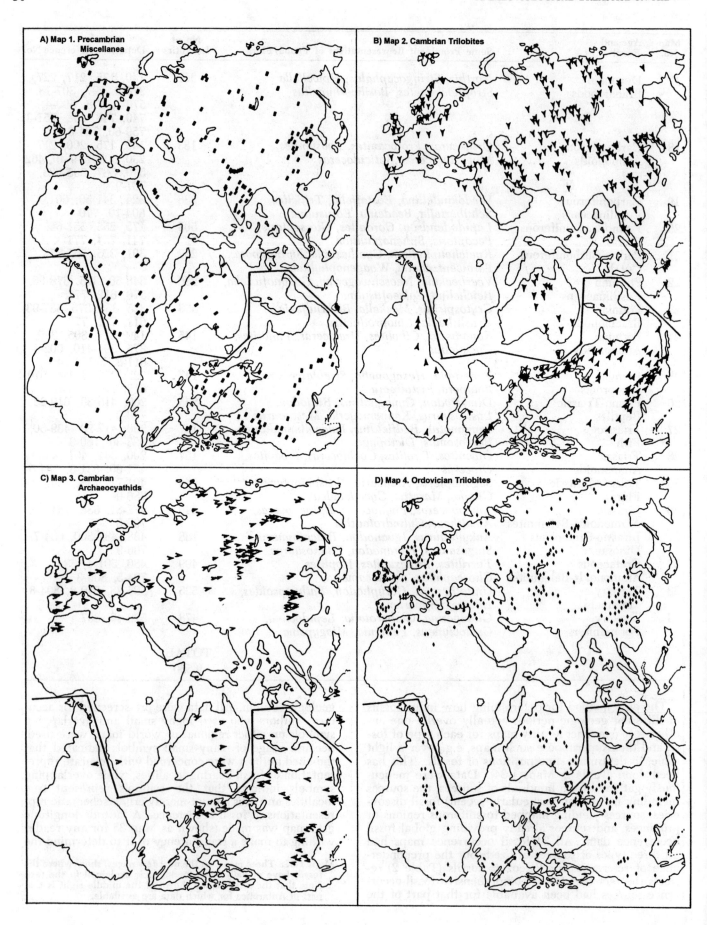

Stratigraphic Separation of Fossils

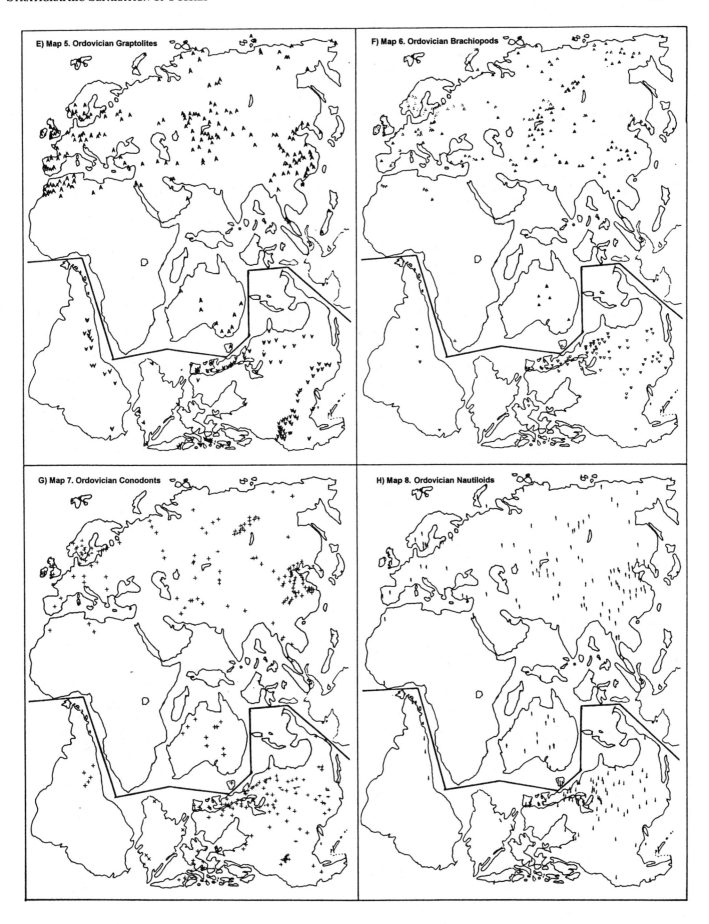

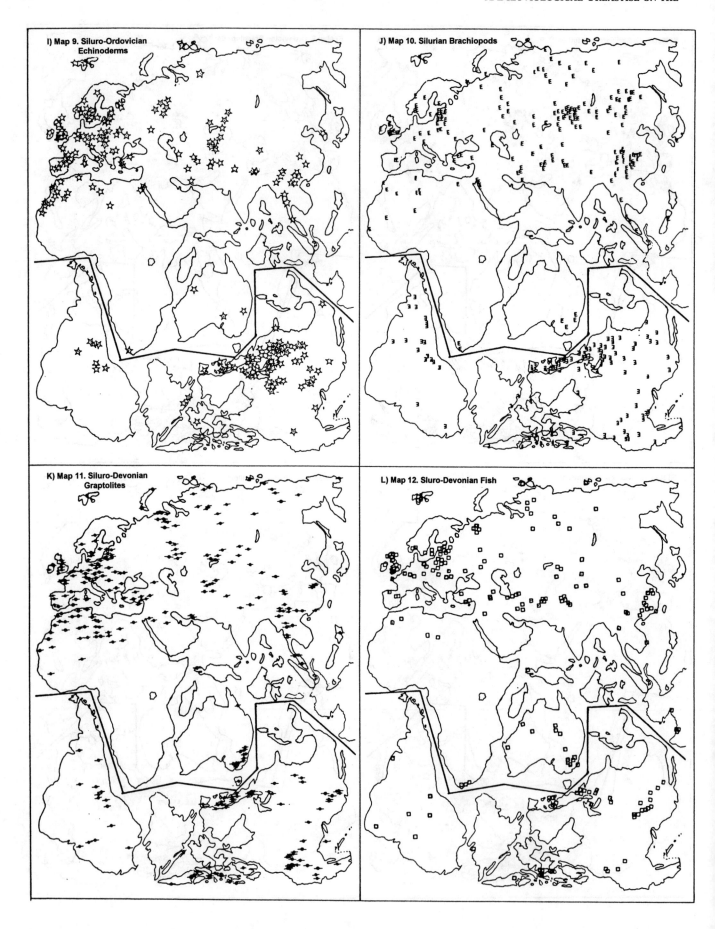

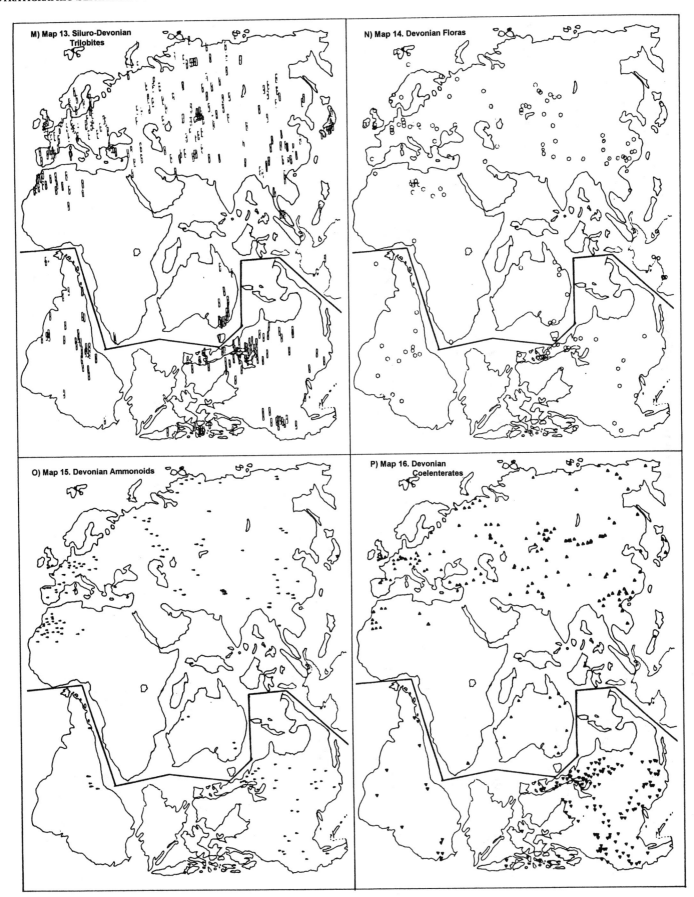

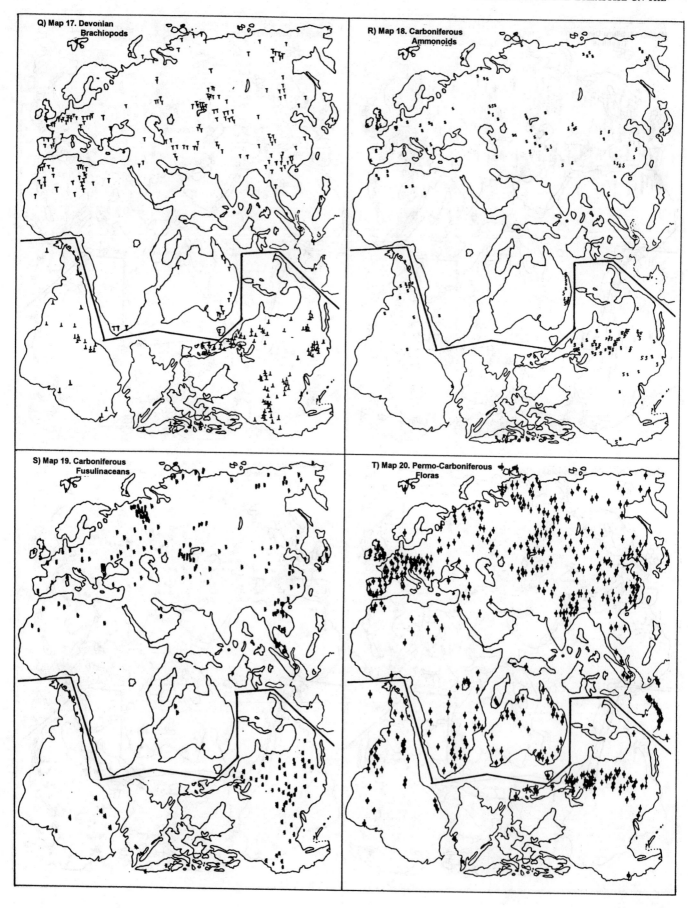

STRATIGRAPHIC SEPARATION OF FOSSILS

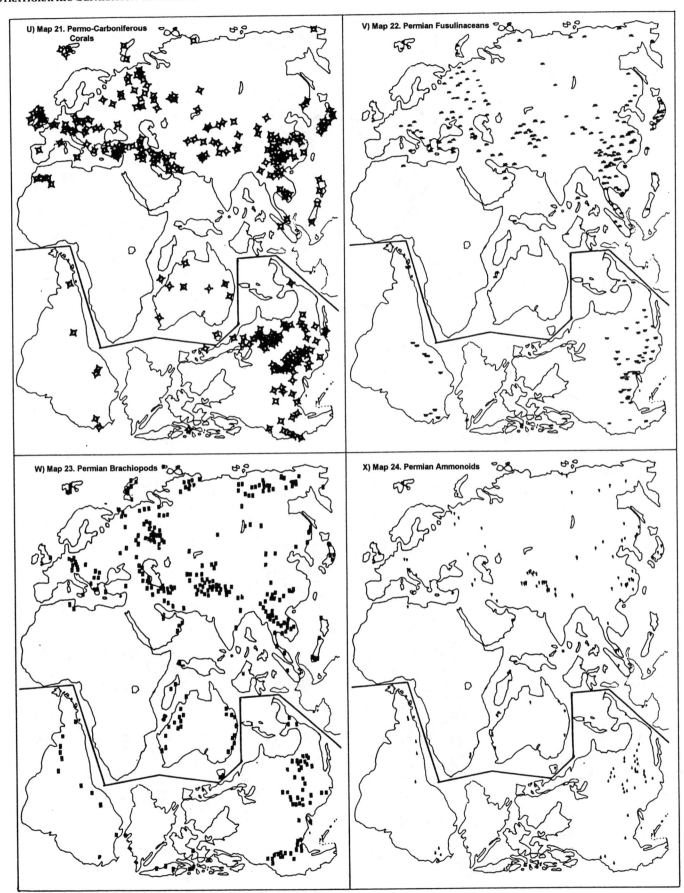

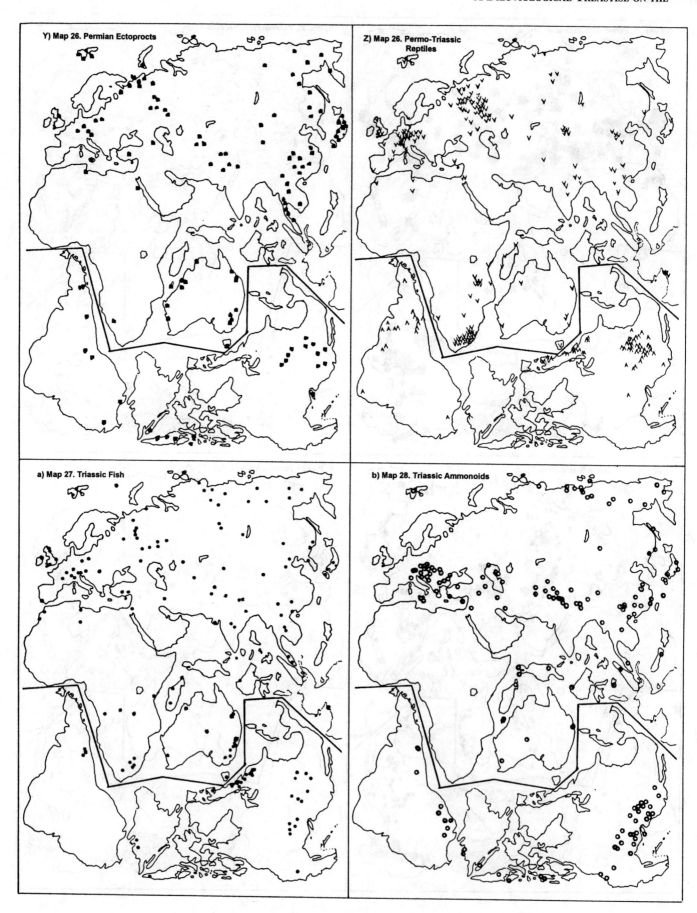

STRATIGRAPHIC SEPARATION OF FOSSILS

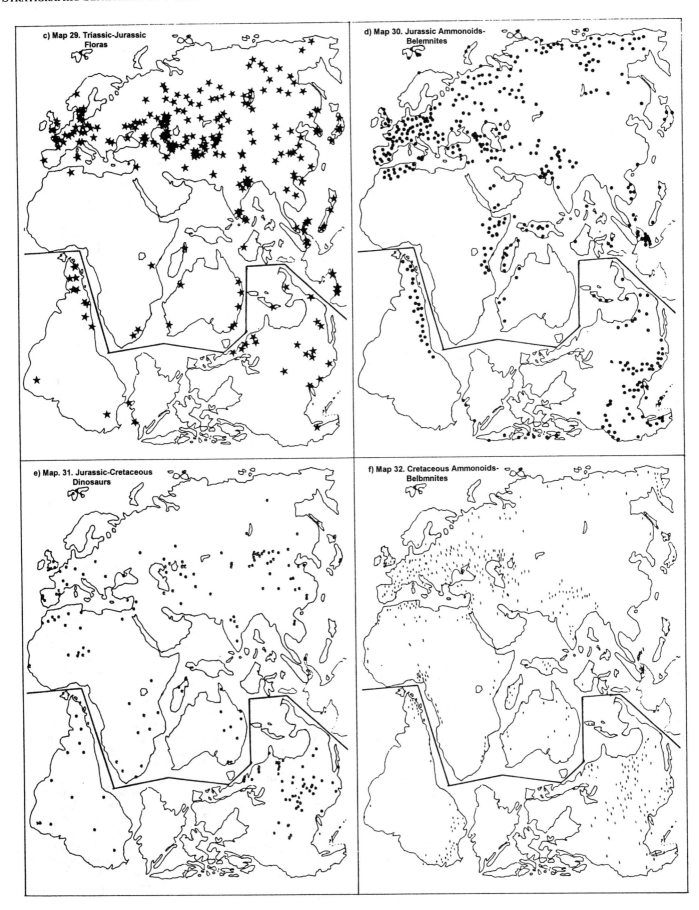

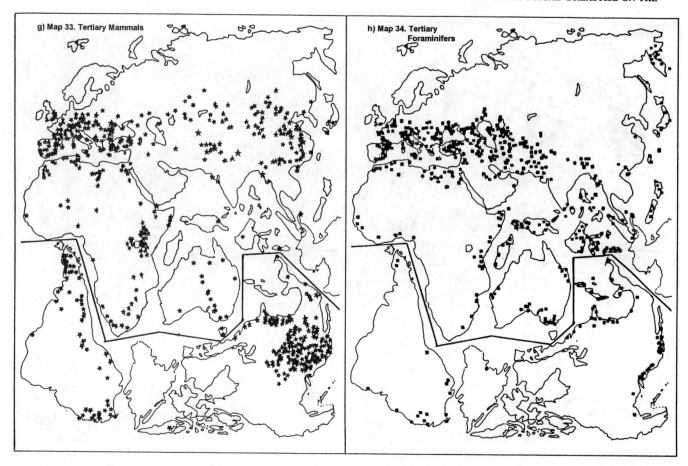

g) Map 33. Tertiary Mammals

h) Map 34. Tertiary Foraminifers

exact location of any fossil or juxtaposition) in order to plot fossil localities accurately. The Winkel's Triple Projection was used as the world base map in order to synchronize all data with the world maps already available from the *Atlas of Palaeobiogeography*.[1] However, the continents were repositioned to eliminate ocean space so that continents could be shown at the largest possible scale. (Each map originally covered an entire page when made and when juxtapositional determinations were performed, but the maps had to be drastically reduced here as a result of space limitations. A Winkel Map in its natural form can be seen ahead in Figure 8).

It is necessary to define what is meant by juxtaposition. Obviously fossils that can be seen superposed in an outcrop face or drill core are juxtaposed; but fossiliferous beds (and sedimentary strata generally) have a regional character to them, so fossils occurring several tens of kilometers apart but in different strata levels must also be recognized as superposed (provided that the region is tectonically uncomplicated). However, such extrapolation can not be extended much beyond this. In speaking of the discovery of Jurassic Corals on Sakhalin Island, Krasnov and Savitskiy[675] wrote: "The age of this series was thought to be late Paleozoic, but this dating was based only on its lithological similarity to the Paleozoic of Japan and the continental part of the Far East." Clearly, lithological similarity did not agree with the stratigraphic order of fossils. At the other end of the spectrum, the character of fossil assemblages changes over short distances. In discussing fossil succession, Harper[676] commented: "Owing to facie changes, the principle is best restricted, where possible, to individual sites where superpositional order can be seen in outcrop or when it is obvious as in a borehole in a structurally uncomplicated area."

Since strata have a (however justifiably limited) regional character to them, the level of resolution of Maps 1-34 is sufficient for juxtapositional determinations to have been made (bear in mind that there must be allowed a margin of error of several tens of kilometers for each fossil locality plotted. Incidentally, the fossil-occurrence symbols on all maps are unequal in significance. One symbol may denote a locality where a solitary questionable fragment was found, while another may indicate a cluster of outcrops yielding thousands of specimens of wide taxonomic diversity). While the level of resolution of fossil localities in Maps 1-34 is several tens of kilometers, it is worthwhile to consider juxtapositions of index fossils on a regional level—so that fossil occurrences only several kilometers apart could be resolved. Map 36 was constructed especially for this purpose; showing fossil localities in Utah-Nevada and the British Isles. The fossils and the respective reference numbers for sources of data are: Cambrian Trilobites (42, 677-9), Silurian Brachiopods and Graptolites (211, 220, 680), Lower Carboniferous Corals (365, 370, 681-2), Jurassic Ammonites (683-6).

It is difficult to say which method (world Maps 1-34 or the regional one-Map 36) offers a "truer" picture of fossil juxtaposition. The high resolution of Map 36 ob-

Stratigraphic Separation of Fossils

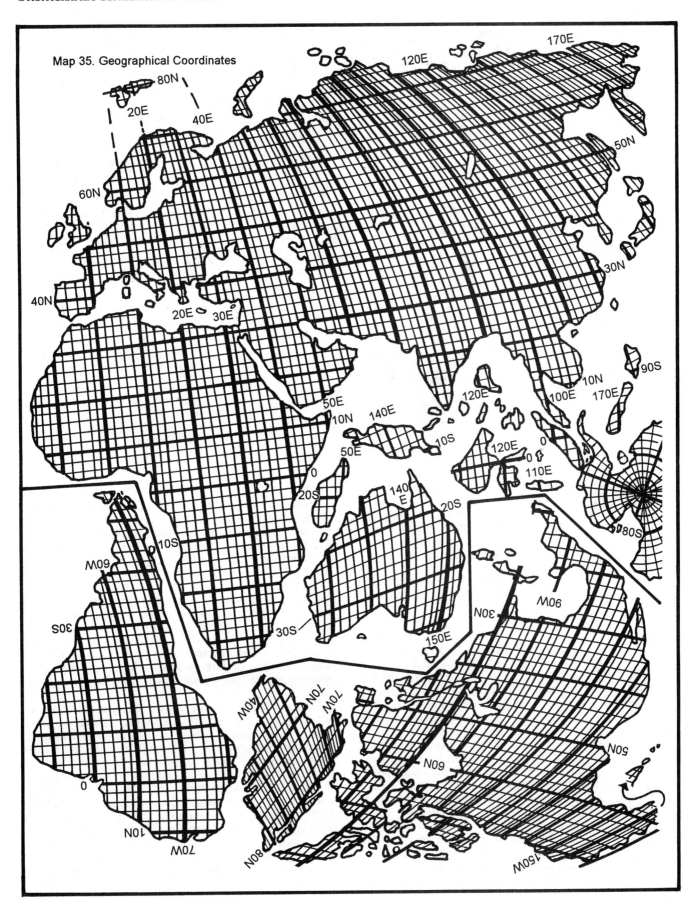

Map 35. Geographical Coordinates

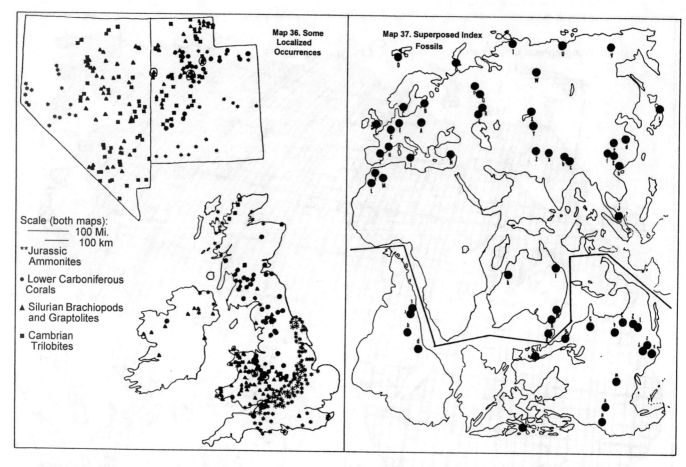

Map 36. Some Localized Occurrences

Map 37. Superposed Index Fossils

Scale (both maps): 100 Mi. / 100 km
** Jurassic Ammonites
• Lower Carboniferous Corals
▲ Silurian Brachiopods and Graptolites
■ Cambrian Trilobites

viously offers great detail, but because of this, detail is more vulnerable to outcrop availability bias and the exaggeration of other local factors. However, Map 36 shows that very seldom are 3 of the 4 cited fossils within a few tens of miles of each other (encircled areas show those locations). Juxtapositions of two fossils at a time are more common, but this is offset by the fact that all fossils shown are actually a single regional occurrence and so every individual fossil-locality manifestation on Map 36 should not really count as a separate candidate for juxtaposition.

It thus appears best to judge juxtapositions on a global scale. Table 3 has been drafted to show the results of superposing Maps 1-34 against each other. There are 479 cross-comparisons: every fossil versus every other that belongs to another geologic period. It can be seen that only small percentages of all localities of any given fossil overlie, or are overlain by, any other single fossil of another geologic period. Thus fossils of different geologic periods invariably tend to shun each other geographically, and this in itself may be taken as *prima facie* evidence that all fossils are ecological and/or biogeographic equivalents of each other—negating all concepts of evolution, geologic periods, and geologic time. To the Diluviologist, this tendency of any two different-"age" fossils to be geographically incompatible not only allows an understanding of fossils in light of the Universal Deluge, but also makes mechanisms of fossil separation (discussed extensively in the next two chapters), for the juxtapositions that do occur, workable without any need of unrealistic efficiency on their part.

From Table 3 it is evident that fossils which are closer in biostratigraphic "age" tend to have more geographic juxtapositions in common with each other. But the apparent significance of this is offset by the fact that fossils of close biostratigraphic age (Figure 1, Table 1) have a considerable number of other fossils in common. Viewed in the opposite direction, the smallness of the number of fossil families in common between geologic periods at opposite ends of the geologic column is made vacuous by the very small number of opportunities for those fossils ever to have had a chance to become mixed during the Flood (note the preponderance of bar symbols—denoting very small percentages of juxtaposition—at the lower right of Table 3).

Whereas Table 3 only shows the juxtapositions of two fossils at a time, Table 4 shows regions on earth where many index fossils are possibly juxtaposed. "Possibly juxtaposed" is used here because the circles shown on Map 37 cover large areas (they have a diameter of over 200 miles or 320 kilometers): the largeness of the encircled areas being made necessary to allow a large margin of propagated error resulting from multiple juxtapositions. The numbers of juxtaposed fossils portrayed in Table 4 are thus an exaggeration. The 59 regions of juxtaposition shown are those where at least 7 of the 34 index fossils occur in the same encircled area though not all 7-occurrence localities have been shown due to space limitations. Note that there are only singular instances on earth where over 10 of the 34 index fossils are possibly juxtaposed, and no case at all where half of all index fossils are possibly juxtaposed.

There does not appear to be any trend for individual fossils to be exceptionally commonly juxtaposed or non-juxtaposed with others. A positive correlation exists between the number of fossils present on a given map and the commonness of that fossil's presence among the 59 biostratigraphic columns of Table 4. In Table 3, fossils which have relatively few numbers of occurrence have either exceptionally high or exceptionally low rates of juxtapositions with other fossils, indicating the somewhat erratic effects of relatively small numbers of occurrences. The number of occurrences of a given fossil (Table 2) are primarily a function of their abundance in their respective ancient faunas. For example, ammonoids are rare constituents of Paleozoic marine faunas, but very abundant constituents of Mesozoic marine faunas.

II. THE SEPARATION OF ORGANISMS DURING BURIAL BY THE FLOOD: PROCESSES AND MECHANISMS

A. Indeterministic Factors Leading to Stratigraphic Differentiation of Fossils

This work thus far measured the degree of stratigraphic differentiation and local succession of fossils; the remainder, commencing with this section, concerns itself with causes of these phenomena in terms of the Flood. Let it be noted, first of all, that there is nothing particularly "natural" about an evolutionary-uniformitarian explanation for the stratigraphic differentiation of fossils. In describing interpretations of faunal lists from two fossil populations, Raup and Crick[687] wrote: "If two lists have no taxa in common, it can be assumed that *something* was different. The possible causes vary from ecological differences (marine vs. fresh water; shallow vs. deep water, etc.) to temporal differences (complete evolutionary turnover) to biogeographic differences (provinciality, separation by geographic barriers, etc.)." (italics theirs) It is thus clear that, even within the evolutionary-uniformitarian paradigm, evolutionary turnover is only one of several potential lines of evidence for interpretation of differentiated fossils, so there is nothing exotic about the Creationist-Diluvialist Paradigm considering non-evolutionary, non-temporal explanations for fossil separation.

The most mundane cause for stratigraphic separation of fossils is pure chance. It would be odd indeed if, even with all other causes eliminated, organisms buried by the Flood were equally present at all stratigraphic horizons. At the same time, the fact that most index fossils do not actually overlie each other (Table 3) allows chance to have a significant role in generating biostratigraphic differentiation.

Figure 2 has been drawn to illustrate this principle. Note that there are few cases where fossils N, P, and/or S occur in the same stratigraphic section. The only possible combinations (two at a time, with or without overlap) are: N/P and P/N, N/S and S/N, P/S and S/P. If there were many mutual juxtapositions of these fossils, then all six combinations would occur and hence there would be no global biostratigraphic differentiation. But since actual juxtapositions are few, one or more of these six combinations may never occur; solely by chance. This follows from a well-known principle in statistics that artifactual (i.e., apparently significant) trends can occur when the population sampled is sufficiently small. For example, it would be highly significant if 500 out of 600 coin tosses were "heads" but not so if 5 out of 6 were "heads" because of the small population in the latter case. The Founder Effect in Population Genetics is another example.

Applying this principle to biostratigraphy, one should note that in Case 1 of Figure 2 the combination N/S never occurs. In Case 2, the Flood is hypothetically allowed to happen all over again. This time, it is the combination P/S that never occurs. Keeping in mind that index fossils shun each other geographically (Table 3), one can see a directly comparable situation with Figure 2; in both cases there are few opportunities for any two index fossils of different "ages" to mix with each other, so many non-mixings can occur by chance. Since in Case 1 S/N is the only way fossils S and N are stratigraphically related to each other (since N/S never occurs, by chance, due to the rarity of instances where fossils N and S occur in the same location), then uniformitarians imagine that S and N are index fossils relative to each other and thus delineate different spans of geologic time. The time horizon has been drawn in to show how the sections are time-correlated according to this "knowledge." The boundary is firm (and thus drawn as a solid line where both N and S occur in the same section) but indeterminate (as in stratigraphic section no. 4) when neither occur. Where one of the two index fossils is present, then the boundary is capable of being placed under S or above N but not with exactness. In Case 2, it is S and P that are index fossils and it is N that has no time significance.

The principle of the origin of biostratigraphic differentiation by chance can be extended to multitudes of fossils, in contrast to the mere 3 shown in Figure 2. Again, the fact that most index fossils are geographically incompatible relative to each other (Table 3, Figure 2) makes it possible. In such a group of fossils, some stratigraphic combinations will fail to occur (by chance); these will be the source of index fossils. Other combinations will occur, and these fossils will be rejected as index fossils. Since adjacent geologic periods have a majority (or very large minority) of families stratigraphically common to each other (Figure 1, Table 1), the principle of chance by itself may be sufficient to account for the biostratigraphic differentiation of any three adjacent geologic periods. The geologic column and its fossil population can be therefore broken down into four principle divisions; these divisions needing deterministic factors to account for their biostratigraphic differentiation relative to each other. Such deterministic factors are the topic of the succeeding chapter.

A major ramification of the origin of biostratigraphic differentiation by chance is the fact that many (if not most) stratigraphic occurrences of index fossils are solitary (see occurrence nos. 4, 5, 11, 17, and 19). Once fossils are elevated to index fossil status, their time-stratigraphic confinement becomes largely circular and therefore self-fulfilling. Creationists have commonly pointed out the circular reasoning in the use of index fossils but, in view of the fact that evolutionists-uniformitarians have commonly sought to deny this fact, it is worthwhile to provide additional evidence. Potapenko and Stukalina[688] wrote: "The crinoids found here rule out a Precambrian or Cambrian age for the host limestone because no reliably identified primitive crinoids have ever been found in Paleozoic rocks older than Early Ordovician." The use of crinoids in ruling out a Precambrian-Cambrian age is justified by the "fact" that such crinoids have never been found in Precambrian-Cambrian rocks, and then the circle of reasoning closes by claiming that rocks are not Precambrian-Cambrian solely because they contain such crinoids! In being uncertain whether to date a certain lithology as Precambrian or Cambrian until a trilobite was found, Yochelson and Stump said:[689] "The trilobite fragment precludes a Precambrian age." In another situation, Skehan, et al.,[690] wrote: "Middle Cambrian trilobites of Acado-Baltic affinities have been found in southern Narragansett Bay, Rhode Island, in phyllites previously mapped as part of the Pennsylvanian stratigraphy of the Narragansett Basin." The claim that certain trilobites are confined to the Cambrian begs the question because rocks are dated as Cambrian (and not some other geologic period) often solely because they contain such trilobites. Many other examples could be given.

Table 3. The Actual Geographic Compatibilities and Incompatibilities of Index Fossils: A Quantitative Tabulation. Each row-column intersection shows the juxtapositional tendencies of two index fossils relative to each other. There are 479 possible different-"age" juxtapositional combinations of the 34 index fossils; all of these are shown. The symbols indicate percentages of fossils juxtaposed based on the number of juxtapositions divided by the total number of fossils given in Table 2. The star denotes percentages over 10%, blank space indicates percentages between 5% and 10%, and vertical bar indicates percentages under 5%. The symbol (or blank) at the left refers to the "older" fossil; at the right to the "younger" one. For example, over 10% of the 157 Triassic ammonoid localities (Map 28, Table 2) are overlain by Tertiary Foraminifers, but less than 5% of the 478 Tertiary Foraminifer localities (Map 34, Table 2) overlie Triassic ammonoids. (Overall, very few of the percentages over 10% are greatly in excess of that figure.)

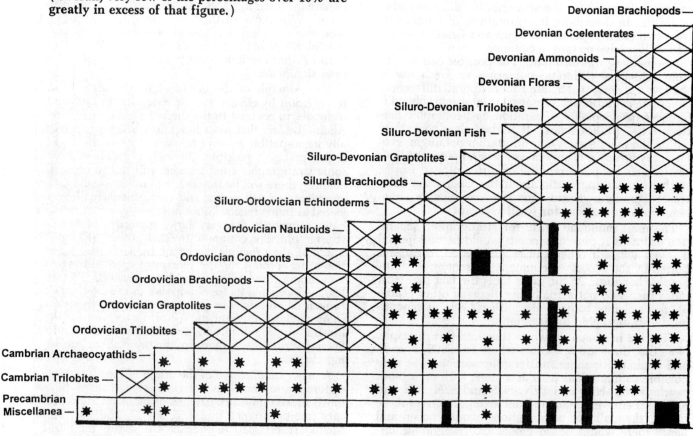

STRATIGRAPHIC SEPARATION OF FOSSILS

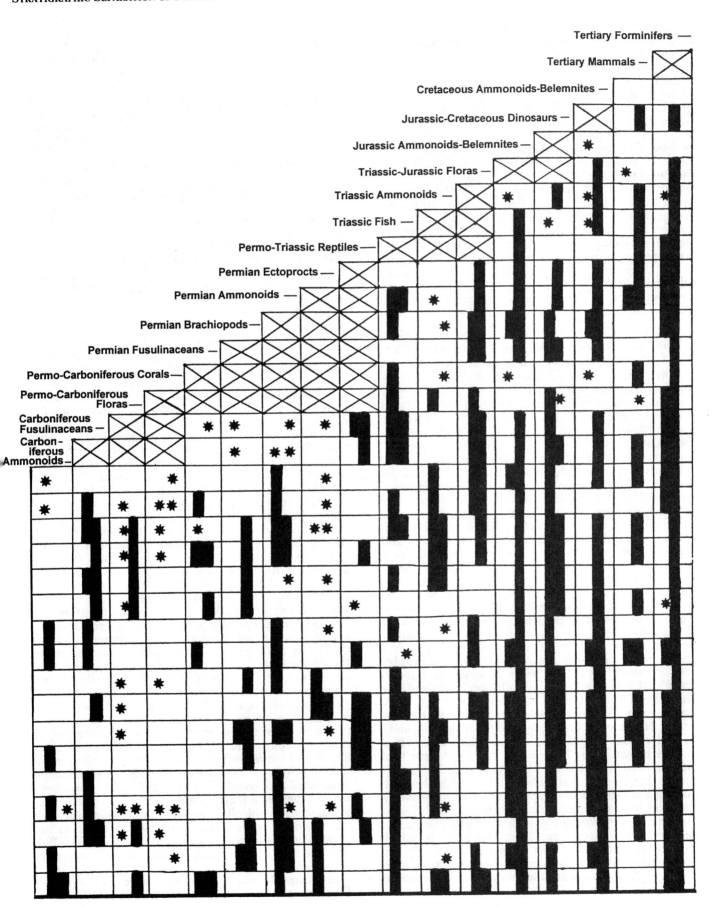

Still another major factor in the origin of biostratigraphic differentiation by chance is preservation bias. Collier[691] wrote: "When similar living faunas are preserved in the fossil record, they become much less similar, due to incomplete habitat representation and small sample size." This factor, applied to biostratigraphy, can be visualized by referring to Figure 2 (Case 2). In stratigraphic section 5, only P is found but there may have been N and/or S also originally present that had not been preserved. If any occurrence of N or an occurrence of S above the P had in fact been preserved, it would have made no difference on the special S/P relationship and on the use of S and P as index fossils. But if there had originally been a mixture of S with the P or else an instance where P overlay S in section 5, then the fortuitous non-preserving circumstance with respect to S in section 5 would have spared the whole S/P relationship. Thus, in Figure 2, there were many instances where there were additional fossils in given sections that had not been preserved and at times this non-preservation eliminated would-be biostratigraphic conflicts.

B. Deterministic Factors leading to Stratigraphic Differentiation of Fossils: The Primacy of TAB's (*Tectonically-Associated Biological Provinces*)

This section considers how the physical, ecologic, and biogeographic properties of organisms led to their biostratigraphic differentiation; emphasis being placed upon possible connections between biogeographic realms of antediluvian organisms and their tectono-sedimentary environment. These factors are complementary to the indeterministic ones discussed in the previous chapter and also make considerable use of the fact that there is considerable net biostratigraphic overlap (Figure 1, Table 1) and that index fossils tend not to superpose (Tables 3 and 4, Figure 2).

Creationists have cited deterministic factors which lead to a stratigraphic differentiation of fossils, and these factors have their greatest realization because index fossils shun each other geographically (Table 3, Figure 2). Consider hydrodynamic sorting and differential escape, proposed by Whitcomb and Morris,[643] in the light of the information in Figure 2, Case 1. Suppose that, when in the same geographic area, fossil S has 70% chance of being buried later than fossil N due to sorting and/or differential escape. The fact that S and N so rarely coexist geographically enables the 30% tendency of N/S never to occur. In Case 2, the Flood is allowed to happen all over again and this time the 30% situation of N/S does occur and so N and S are not made into index fossils relative to each other. However, in Case 2, hydrodynamic sorting and/or differential escape cause a burial bias where P is buried before S, say 80% of the time, the same bias having been thwarted by the 20% chance in Case 1. It can thus be seen that factors such as hydrodynamic sorting and differential escape do not have to be overly efficient in order to generate biostratigraphic differentiation. This overcomes objections about the turbulence of Floodwater: since index fossils rarely superpose relative to each other (Table 3, Figure 2) the sorting, etc., need work consistently only a few times relative to any two organisms in order for them to be buried in consistent biostratigraphic order.

The factor of ecological zonation (discussed by Clark[645] and also applied to fossil cephalopods by the present author[692]) probably is more efficient in generating biostratigraphic differentiation than passive sorting or differential escape. Nevertheless, ecological zonation also does not need to be highly efficient to generate biostratigraphic differentiation for the same reason as was discussed in the preceding paragraph for hydrodynamic sorting and differential escape. Examining Figure 2, it can be seen that (in Case 1) N could be in a lower habitat than S. The combination S/N would then be generated through ecological zonation if: N was benthic while S was pelagic, N was either benthic or pelagic while S was planktonic (both situations in marine ecology were discussed in my work[692] on cephalopod ecological zonation), N lived on lower ground while S lived on higher ground. Again, the fact that S and N rarely coexist geographically means that ecological zonation needs to work consistently only several times for the S/N biostratigraphic relationship to be established. The process for S and N is applicable to any other situation where index fossils are members of different ecological habitats.

Ecological zonation is in many instances so prominent that it not only plays a major role in total biostratigraphic differentiation, but also causes biotal incompatibilities *within* geologic periods. Many fossils are rejected as index fossils because they are facies fossils—fossils restricted to some particular lithology or well-defined sedimentological circumstance. Yet even index fossils show by their predominance in certain lithologies that they were ecologically-controlled and hence capable of flourishing only in certain environments. For instance, Nelson[693] cited: ". . . the extreme rarity of graptolites in limestone." In view of this fact of ecological dependence of even the most ideal index fossils, there is no *a priori* reason why the role of ecological zonation cannot be extended beyond faunal differences within (alleged) time-horizons to differences between (alleged) time-horizons (that is, between different geologic periods).

Thus far, the factors discussed (hydrodynamic sorting, differential escape, preservation bias, and ecological zonation) are well known to informed Creationists. The present author now proposes a whole new mechanism to account for biostratigraphic differentiation of fossils. It is based on the fact that sedimentation in the Phanerozoic record is strongly influenced by tectonics, and at the same time on the fact that fossil organisms are not only ecologically zoned but also biogeographically zoned. If tectonics and biogeographic zonation are linked (see Figure 3) then biogeographic provinces must be superposed in a consistent manner, thus resulting in biostratigraphic separation of fossils. The Flood model herein proposed that envisions such linkage is termed *the concept of TAB's (Tectonically-Associated Biological Provinces)*, and will be discussed later. There is a major trend of changes in tectonics going stratigraphically upward in the Phanerozoic and this trend (to be discussed later) may be taken as independent evidence for the existence of TAB's. But first the role of biogeography in the fossil record is described.

It is common when considering modern examples of biogeography erroneously to think of it only in terms of climatic differences (tropical plants versus the high-latitude pines of Canada and Siberia) or continental differences (the marsupials native to Australia and South America versus placentals elsewhere). Many factors, in actuality, cause biogeographic zonation and such zones need not cover large areas. Looking at paleobiogeographic examples, even within the context of geologic periods, bears this out. For instance, the *Tuvaella* fauna[223] is a distinctive Silurian brachiopod biogeographic zone, and it is restricted to only Mongolia and adjacent parts of the USSR and China. One need only consult the *Atlas of Palaeobiogeography*[1] to see how fossil organisms of all geologic periods are divided up into paleobiogeographic provinces. Thus, the evolutionist-uniformitarian will note (to give another example) that Ordovician trilobites differ markedly in different places on earth and ascribe such differences to paleobiogeographic provinces (such as the bathyurid province, remopleuridid province, etc.).[72-3] At the same time, he will note differences between Ordovician and Silurian trilobites and ascribe such differences to evolution and geologic time. The Creationist-Diluvialist can reject such a dualism and view the same fundamental biogeographic processes that cause faunal differences *within* Ordovician trilobites to be the basic cause of differences between Ordovician and Silurian trilobites. In fact, biogeographic differences between marine faunas ascribed to the same geologic periods are so pronounced that Sheehan[694] proposed that there is a major breakdown in any comparison between extant marine communities and ancient ones. Since biogeographic differentiation *within* geologic periods is so considerable, there is nothing farfetched about the Creationists-Diluvialists' use of the same basic mechanism (when developed in the TAB model) to explain faunal differences *between* geologic periods.

It is worthwhile to make a distinction between ecological zones and biogeographic zones. Taylor and Forester[695] point out that biogeographic zones (which, as they note, can also be termed faunal provinces or biofacies) may be ecologically controlled; hence the definitions, strictly speaking, overlap. The working definitions used in this work are as follows: The term ecological zonation refers to organisms that are mutually proximate but do not live together because they occupy different habitats or have different environmental tolerances. The term biogeographic zonation refers to organisms that are geographically separated, irrespective of whether or not they occupy the same ecological niche. The term biome would apply to organisms that are both ecologically different (such as those possessing different climatic tolerances) and biogeographically zoned. When organisms are members of the same ecological niche but biogeographically zoned, then they could live together were it not for their geographical separation and any geographic barriers that enforce it.

A contrast is now made between the evolutionary-uniformitarian and Creationist-Diluvialist paradigms with respect to the origins of ecological zonation and biogeographic zonation. In the evolutionary-uniformitarian paradigm, ecological zonation is caused by organisms evolving to match their environment: biogeographic zonation is caused by organisms evolving in a distinct geographic area and being imprisoned in that (or somewhat larger) area by geographic barriers. The Creationist-Diluvialist, not confined by the implicitly atheistic presuppositions of the evolutionist-uniformitarian, is free to explore possible Divine causes in the origins of ecological as well as biogeographic zonation. In attempting to "think God's thoughts after Him," it is worthwhile to note the fact that both ecological and biogeographic zonations are means by which a higher diversity and number of organisms can be supported on earth. There is also less conflict for space and for food when organisms are ecologically partitioned or geographically separated. Thus God may have created ecological and biogeographic zones in order to be able to Create a far wider variety of organisms than would have been the case had He Created only one ecological niche or only one global biogeographic zone. His actions with respect to His New Creation (the Church) may help clarify His Creative actions with respect to the Old Creation (the natural world). We are told that: "Now there are varieties of gifts, but the same Spirit." (1 Corinthians 12:4 NASB). Just as the Spirit gives different gifts so that believers can occupy different "spiritual niches," so also God Created different organisms suited for their respective ecological niches. (Ecological zonation may itself have Scriptural basis—see Isaiah 41:19-20). Biogeographic zonation may find its analogy within the New Creation in the form of geographic separation of ministries (such as in the geographically differentiated preaching of the Gospel—Romans 15:20). Just as the Spirit can multiply the number of ministries if each has definite geographic boundaries, so God can Create more organisms if each is subject to biogeographic partitions.

Having reflected upon teleological considerations with respect to ecology and biogeography, attention is now focused upon how biogeographic provinces could have been linked with tectonics and thus have been the primary source of biostratigraphic differentiation (the TAB concept—see Figures 3 and 4). Note that in Figure 3 biogeographic provinces repeat themselves; on the land surface of the world each biological province may occupy an area with a transverse distance across it of up to a few hundred miles, the same province appearing again several hundred miles away. But, according to the TAB concept, the same biogeographic province is linked with the same tendency for tectonic downwarp irrespective of where it occurs on earth. Thus, note that in Figure 3 the biogeographic province symbolized by solid rectangles is always linked with the areas on earth having the greatest tendency for tectonic downwarp. Actually, there is greater biogeographic differentiation than tectonic differentiation (as evidenced by the previously-discussed fact that even biotas *within* geologic periods exhibit biogeographic differentiation). This is shown in Figure 3 in the form of asterisk-type stars sharing the same tectonic proclivity (the next to greatest) with ovals. Most other symbols in Figure 3 show the same effect. It is thus not special pleading to invoke the TAB concept as the major causative factor for total biostratigraphic differentiation in view of the fact that there would actually be more biogeographic differentiation than tectonic.

Table 4. The Global Successional Tendencies of Index Fossils. The Table shows which of the 34 index fossils can be seen superposed at the localities shown on Map 37: the localities shown being those where the greatest number of index fossils can be seen to superpose. Blacked-out rectangles denote absence of that given fossil at that given locality.

Stratigraphic Separation of Fossils

One may wonder if there is any independent evidence for such a proposed linkage between tectonics and biogeography, and also if there are plausible reasons to account for such a linkage. The answer to both question is yes; and so the concept of TAB's can be independently justified. Studies on modern marine biogeography described by Taylor and Forester,[695] and Crick,[696] have shown that oceanic current patterns result in biogeographic differentiation (a differentiation that is also biomic in character), temperature of the water being the major factor. Salinity is another. Crick added that water-temperature patterns of oceanic current flow can be controlled by submarine topography. This concept can be applied to antediluvian epicontinental seas. Suppose that the entire Phanerozoic is divided into four divisions—I, II, III, and IV: these are simultaneously four antediluvian biogeographic provinces and also are approximately equivalent to Eras. Thus, no. I (see Figure 4) represents the biogeographic province (roughly corresponding to Lower Paleozoic in biotal content) that is associated with the regions on earth showing the greatest tectonic proclivity. Returning to the discussion concerning Crick,[696] one can see that, in this instance, the no. I biogeographic province could have had uniform temperature of water, and that this temperature could have been regulated by submarine topography. Submarine topography could have, in turn, been a reflection of the tectonic stability of the region. This is one possible causal connection between biogeography and tectonics. Another temperature-based biomic situation could result from the fountains of the deep (Genesis 7:11) having been partly geothermal springs and underground rivers. The temperature and number of such springs in a region could have depended upon the tectonic proclivity of the region; more numerous and hotter springs being generated in regions of greatest tectonic proclivity because of numerous deep fissures (no. I of Figure 4), the biotic contents of the associated biological province living at a high temperature environment due to the number and temperature of the geothermal springs. Note that in the examples discussed thus far (marine currents and geothermal springs), the temperature of the water was the causal factor of the TAB's and also the factor linking the tectonic proclivity with the biogeographic provinces. Still another connecting factor between biogeography and tectonics could have been the chemistry (eH − pH nutrients, trace elements) of the seabed and seawater (in marine regions of biogeographic provinces) and the chemistry of soils (in land regions of same). Geothermal springs, once again, could have been the causal connection between tectonic proclivity and chemistry (and thus biogeography).

An example from modern ecology where biogeographic distribution is regulated by chemistry is discussed by Parker and Toots:[697] "Proboscidians are highly advanced in the evolution of their dentition but are primitive in their sodium metabolism. Because of the latter fact, distribution of elephants in modern Africa is closely correlated with high environmental sodium levels (Weir 1972), and elephants are known to depend on food that is particularly rich in sodium."

TAB's could have also been Created without any ecological (biomic) character to them; the tectonic proclivities being part of the structure of geographic barriers designed to prevent significant migration of organisms from one biogeographic province to another. (It should be emphasized that TAB's did not arise from trial-and-error migrations but were present since the Creation and were based on teleological design.)

Independent evidence that there is a connection between biostratigraphic segments of the Phanerozoic and tectonics of sedimentation is now presented. When the present author conceived the TAB process, he predicted that, if it is valid, then the lower part of the Phanerozoic column should contain evidences of sedimentation under more tectonically-active conditions than the upper part of the same. *After* making such a prediction, it was discovered that such is indeed the case. First of all, note that in Table 3, Lower Paleozoic fossils have a greater tendency to superpose among themselves than is the case for fossils of adjacent geologic periods higher in the geologic column (this is clearly manifested by the concentration of star symbols in the left half of Table 3 and simultaneous negative concentration of bars). This indicates that Lower Paleozoic fossils have been deposited in smaller geographic areas than was the case for higher fossils, indicating that tectonic downwarp was greatest for Lower Paleozoic and thus forced the concentration of those fossils into geographically restricted areas. Other lines of evidence, presented by other authors, bears this out. Note that geosynclines are regions of greater tectonic activity (downwarp followed by uplift) than platforms. Ronov[698] did extensive calculations on the areas and volumes of Phanerozoic rock with respect to geologic periods. He showed (his Figure 32) that geosynclines occupy 40% of the area of oldest geologic periods but less than 20% of the recent ones. The present author performed calculations on the data presented by Ronov[699] in another article. It was determined that the Mesozoic-Cenozoic (roughly corresponding with TAB's III and IV) contains 57.4% of the total volume of Phanerozoic platform sediments but only 41.3% of the total volume of Phanerozoic geosynclinal sediments. The ratio (by volumes) of geosynclinal to platformal sediments (taking the Phanerozoic as a whole) is 2.4. A clear trend is evident between the Eras (and sub-Eras). In the Lower Paleozoic (TAB I), the ratio is 3.1; in the Upper Paleozoic (TAB II), it is 3.4; in the Mesozoic (TAB III) it is only 1.8; and in the Cenozoic (TAB IV), it is but 1.6. The trends discussed in this paragraph demonstrate that there is a tectonic trend going stratigraphically upward in the geologic column: this trend provides independent evidence for the TAB concept.

Some ramifications of the TAB concept should be discussed; both biogeography and tectonics being considered. The reason why the Phanerozoic was divided into four lateral equivalents (the biogeographic provinces I through IV) was because (as demonstrated in the earlier chapter on indeterministic factors) geologic periods adjacent to each other share a great many families between them (Figure 1, Table 1), so that there are really only four (not more nor less) groupings of Phanerozoic faunas. This approximately corresponds to Eras and Sub-Eras in terms of a "natural" division of the Phanerozoic. Furnish, et al.,[615] wrote:

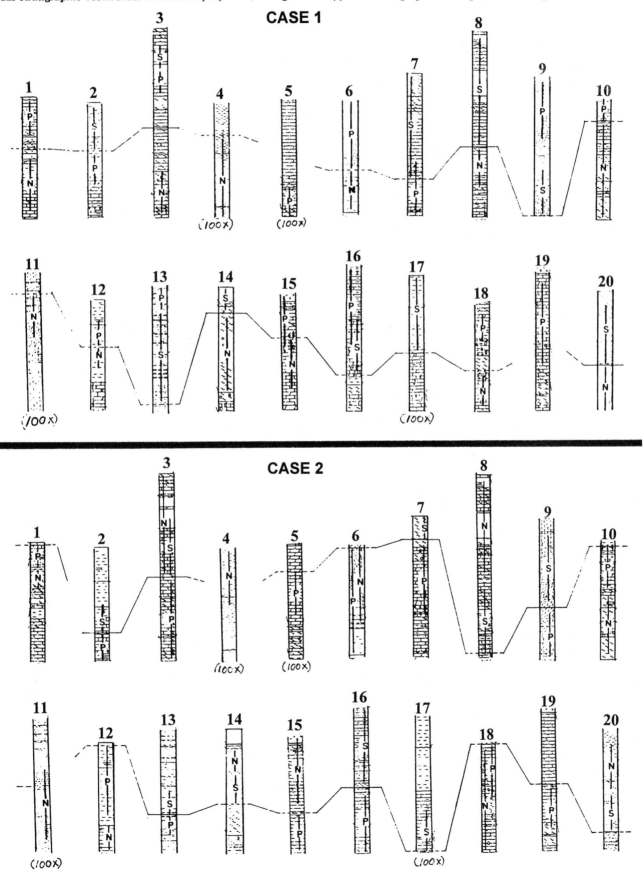

Figure 2. A schematic representation of the effects of pure chance on stratigraphic ranges. Note that Stratigraphic Sections 4, 5, 11, and 17 are numerically weighted 100 times, so there are actually 416 different sections shown per case. Relatively rare mutual stratigraphic occurrences of fossils P, N, and /or S generate *apparent* stratigraphic incompatibilities. Explained in Text.

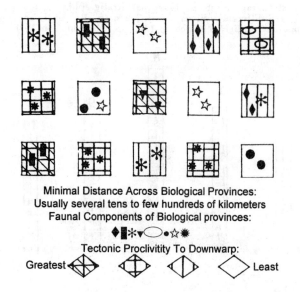

Figure 3. The Approximate Geographic Congruence of Antediluvian Biogeographic Provinces with Differentiated Tectonically-Prone Regions. The squares symbolize different areas on earth, and show given organisms associated with areas having the same tendency to downwarp. Explained in text.

"Since the first half of the 19th century, it has become apparent that assemblages of fossil organisms can be grouped as 'ancient,' 'medial,' or 'recent' in overall aspect. These groupings formed the basis for Phillips' definition (1) of the geological eras: Palaeozoic, Mesozoic, and Kainozoic. Even casual comparison reveals basic differences between marine invertebate faunas of those three eras, with the era boundaries representing intervals of faunal crisis." In considering tectonic proclivities, it should be pointed out that the tectonic proclivities became operable upon the application of great subterranean stress, irrespective of the source of the stress. Just as a dark fabric will get hotter than a similar white fabric regardless of the source of heat (be it sunlight, ordinary fire, or a nuclear explosion), so also note that the TAB process will have worked regardless of whether the stress was imposed by direct Divine will, passive Divine will (such as undirected flow of Divine energy), or providentially-timed naturalistic causes (such as providentially-timed release of earth-interior stress that had been built-in into the earth since Creation Week, gravitational stresses from a passing celestial body, or a bolide impact). Thus it should be clear that the TAB concept can be incorporated into the theories of other Creationists-Diluvialists, since it does not matter what the overall causal factor of the Flood was: the TAB process operated regardless of cause. However, it should be noted that the exact process by which earth-interior stress becomes manifested as crustal tectonic movements is not yet clarified. Hobson and Tiratsov[700] cited: "Tectonic forces, the nature of which are still only partly understood . . ." This partial knowledge applies to both uniformitarian and Diluvialist understandings of tectonic action: uniformitarians are certainly not deterred by this fact from proposing models of tectonic action, so neither should the Diluvialist be discouraged from accepting TAB's. Just as the TAB concept operated regardless of the ultimate cause of earth-interior stress, so also it operated regardless of the mechanism by which earth-interior stresses get converted into crustal tectonics. The lack of knowledge of cytological and biochemical genetics in the 19th century did not prevent Gregor Mendel from proposing his laws of inheritance; neither should the overall partial knowledge of tectonics hinder the Diluvialist. The TAB process finds its greatest realization with respect to tectonics by the clear geological evidences of tectonic motion. Stokes[701] cites troughs, grabens, downwarps, depressions, rifts, and "pull apart" structures (not to mention orogens and geosynclines) and concludes: "As a matter of fact, negative features should be even more common than positive ones. The forces that cause uplift must work against gravity, while those causing depression work with it."

TAB's are not equivalently distributed over the earth. Thus, the reason why ocean floors are almost exclusively Mesozoic and Cenozoic is because they were exclusively overlain by TAB's II and IV. (The percentage of all 15 possible combinations of TAB successions with respect to earth's land surface will be discussed later in conjunction with Figure 6.) TABs have been described in terms of Phanerozoic biotas, but Precambrian biotas can be assigned to TAB I; one need realize that the sum of Precambrian biotas is minute in comparison with the number and diversity of Phanerozoic biotas, and that Precambrian biotas can be attached to TAB I because they commonly range into Lower Paleozoic and because (as previously mentioned in conjunction with trilobites) the Cambrian-Precambrian boundary is circularly defined. Dott,[702] in an anti-Creationist article, completely misses the mark when he asserts that the Noachian Deluge would need to have deposited all Precambrian, along with Phanerozoic, sedimentary rock. Only a vanishingly small percentage of Precambrian is fossiliferous and therefore must be post-Creation Week. Even if this were not so, Late Precambrian (Riphean and Vendian, which together are ascribed to the time span 1600 m.y. to 600 m.y. ago and in which the vast bulk of Precambrian biotas are concentrated), volumetrically occupy only 16% of the total combined volume of Riphean-Vendian-Phanerozoic sediment, according to Ronov.[699] When intensely metamorphosed equivalents are included, the figure rises to 27%, but again only a very small percentage of even this volume is fossiliferous and therefore must be Diluvial in origin.

The Lower Paleozoic (TAB I) contains almost exclusively marine fossils, and this indicates that the biological province was exclusively marine. All other geologic periods, while containing land faunas and floras, are still dominated by marine biotas. This indicates that TAB's II, III, and IV contained both marine and terrigenous regions. When a particular regional TAB was land in entirety, then the geologic periods deposited had only land faunas. For example, the Gondawana Formations of South Africa were primarily TAB III of nearly entirely terrigenous geography. Many TAB's, by contrast, contain alternations of marine and nonmarine biotas, indicating that the particular regional manifestation of a given TAB si-

Figure 4. The Generation of Biostratigraphically-Differentiated Strata through the Operation of TAB's (Tectonically-Associated Biological Provinces). The four quadrants depict four representative depositional regions during the Flood (from anywhere on earth). The denotation for Tectonic Proclivity is identical to that in Fig. 3. The biogeographic components of the areas are shown by the numerals I-IV (corresponding approximately to a fourfold division of Phanerozoic life) replacing the geometric symbols of life shown in Figure 3. The numbers in ovals show the type of local biostratigraphic succession generated; the age possible successions are shown on the bottom of the diagram. The arrows show how Floodwater deterministically flows across regions, irrespective of whether it is progressional, intra-Flood, or recessional. Explained in text.

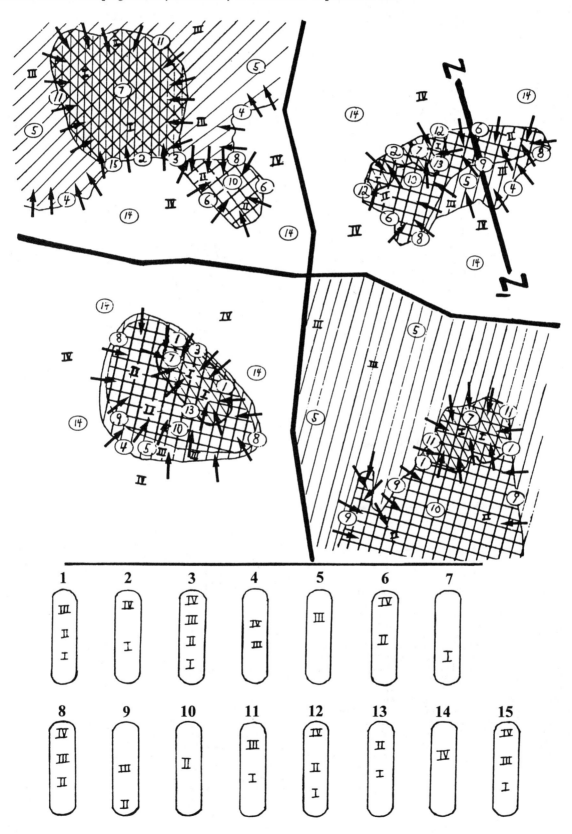

multaneously contained both land and marine areas. Every regional manifestation of a TAB is independent from any other regional manifestation in terms of sedimentology and most sediment is regional in origin, so there are no major global volumetric tendencies with respect to primary lithologies. Thus, Ronov[699] showed that every geologic period has sandstone, shale, carbonate, etc., in percentages that fluctuate considerably going from one geologic period to the next, but without any major volumetric trends across the whole Phanerozoic.

TAB's have thus far been discussed in terms of their biogeographic and tectonic components, as well as in terms of their implications and ramifications. The following discussion concerns their *modus operandi* with respect to the Flood (see Figures 4 and 5). Note that Floodwater (and its transported sedimentary particles and organisms) always flows from an area of lesser tectonic proclivity to a higher one. This is because an area of greater tectonic proclivity always downwarps before an area of lesser tectonic proclivity. Thus the sequence of TAB's: IV/III/II/I is always preserved in that relative order no matter how many of the four TAB's are actually present in a given area. Note that in the regions illustrated in Figure 4, stratigraphic successions of multiple TAB's are generated only at and near junctions of TAB's. In geographic centers of large representatives of specific TAB's only biotic members of that same TAB are superposed. This is shown in Figure 4 under stratigraphic succession numbers 5, 7, 10, and 14. On the earth, such singular successions are seen in the form of thick geosynclinal deposits containing a few mutually-adjacent geologic periods (for example, very thick Lower Paleozoic Caledonian geosynclinal accumulations) and also platform deposits with singular geologic periods represented.

In order to get an idea of how TAB's would be geographically proportioned (wide TAB representatives which would have given singular successions versus small and narrow representatives sharing boundaries with other similar TAB's and hence yielding multiple-TAB successions), the earth's land surface was divided into the 15 possible TAB successions shown in Figure 4 (the 16th possibility being a region having none of the four TAB's present there). The raw data came from Table 1 of the author's previous work[703] on the nonexistence of the evolutionary-uniformitarian geologic column. The results, shown in Figure 6, indicate that over half the earth's land surface has 2 or fewer of the 4 TAB's superposed at any one locality. This indicates that the dominant mode of sedimentation during the Flood involved little tendency for TAB constituents to be transported much beyond their boundaries; hence less than half of earth's land surface has more than 2 locally superposed TAB's. When oceanic data are included (only TAB's III and/or IV), this tendency is increased to such an extent that only about 15% of the earth's entire surface has more than 2 locally superposed TAB's. It is interesting to note (from Maps 11-15 of the author's previous work[703]) that regions of greatest completeness of the geologic column are also regions of greatest sedimentary thickness (that

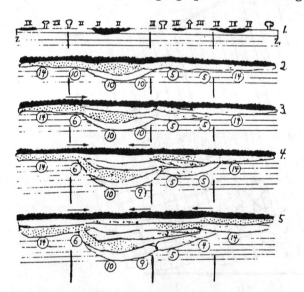

Figure 5. A cross-section showing how biotic members of TAB's become superposed throughout the course of the Flood. The section dissected is shown by the line segment Z-Z, in the upper-right quadrant of Fig. 4. The denotation for Roman numerals and for numbers in ovals is identical to that of Figure 4 (except that the situation in Figure 4 is in plan (areal) view whereas this figure is in cross-section). No. 1 (at right) refers to the antediluvian section: the projections symbolize trees whereas the black symbolizes antediluvian epicontinental seas. The thin, continuous horizontal lines indicate antediluvian regolith; the thick horizontal line segments indicate boundaries of TAB's and their respective tectonic proclivities. Nos. 2-5 refer to progressive stages of Flood deposition; the black band denotes the surficial cover of Floodwater whereas the thin arrows indicate the net direction of Floodwater and its sediment transport in response to the sequential downwarp of TAB's. The vertical scale is approximately 1 centimeter to a few thousand meters; the horizontal is 1 centimeter to a few tens of kilometers. Explained in text.

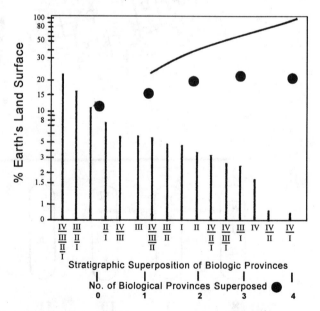

Figure 6. A quantitative breakdown showing tendencies of superposition of TAB's over the earth's land surface. The superposed Roman numerals show every possible combination of TAB's; listed from most to least frequent going rightward together with the per cent of earth's land surface each covers. The black circles and cumulative frequency curve refer to the absolute number of TAB's and their per cent occurrence. Explained in text.

is, the geosynclines). Map 37 also indicates this in its own way: areas of most superposed index fossils tend to be geosynclines. Both these trends are logically explicable in terms of TAB's; in fact, they may be yet another independent evidence for the existence of TAB's. Regions that have most or all TAB's locally superposed will naturally tend to have the greatest thickness of accumulated sediment because they have the most sources of sediment (multiple TAB's) all delivering their sediment into those regions.

It is worthwhile to consider how TAB's operate during the Deluge in terms of a specific representative region: this is shown in Figure 5. Note that only TAB's II, III, and IV are present in the cross-section pictured. Stage 1 is the antediluvian situation. Stage 2 the earliest part of the Flood, erosion and deposition taking place thus far only within TAB's (shown as successions 5, 10, and 14 in both Figures 4 and 5). The Flood progresses on to Stage 3, and TAB II begins to downwarp. As a result, the Flood transports (as indicated by arrow) sediment and organisms from TAB IV on top of the sediment already deposited within TAB II; hence the succession no. 6 (IV/II) is generated. In Stage 4, the downwarp of TAB II continues and now the Floodwater (again shown by arrow) transports sediment and organisms from TAB III unto the previously-deposited sediment of TAB II; thus succession no. 9 (III/II) is produced. (There is a greater differential between IV and II than between III and II, so succession no. 6 starts to form somewhat before succession no. 9.) Finally, TAB III begins to downwarp (Stage 5) so sediment from IV becomes transported on top of III and succession no. 4 (IV/III) is generated. The sequential downwarp of TAB's is largely independent of the hydrologic stages of the Flood (encroaching, prevailing, and recessional). Thus the covering of Floodwater (shown as black film in Figure 5) is already prevailing on land in Stages 2-5. (While TAB I starts downwarping, it may be occurring just as Floodwaters encroach on land, but it need not be.)

The TAB concept firmly rebuts the objection to the Flood that great thicknesses of sediment could not be laid down in one global Flood. Since sedimentation is primarily controlled by tectonics, and sequentially-downwarping TAB's operate throughout the duration of the Flood, there is constant impetus for mass Floodwater transport of sediment from one region to another. Thus the caricature of the Flood being merely a passive rise and fall of ocean levels (with then only relatively small extent of erosion and deposition—certainly not tens of thousands of feet) is shown to be just that. (Not to mention the probable fact that antediluvian regolith was unconsolidated and thick; hence easy to erode.) The anti-Creationist Milne[704] did a farcical misrepresentation of the Flood by claiming that, if it had occurred, then organisms in the fossil record would be uniform and identical to those organisms that are extant. Such a straw-man situation would have had validity only if the Flood had been merely a passive rise of water (like a river flood but global in extent). Once the geologic record is consulted and uniformitarian preconceptions are dispelled, the record in all its catastrophic implications testifies to the fact of the Deluge; the TAB concept showing how the paleontological details of the fossil record can be understood in terms of the Flood. In fact, a Diluvian interpretation (such as the TAB Concept espoused in this work) is scientifically superior to the evolutionary-uniformitarian view because the Diluvian interpretation explains fossil succession with less multiplication of concepts and hypotheses, and less special pleading, than the evolutionary-uniformitarian interpretation. A concrete example of this, where Occam's Razor favors the Diluvian interpretation, is shown by biostratigraphic differentiation itself. The uniformitarian must resort to special pleading in citing some fossils as being index fossils, while disregarding long-range forms. The Diluvialist can consistently explain both in the TAB concept, and need not imaginatively ascribe time properties to selected fossils. The Diluvian interpretation of the fossil record is simpler and more direct because it does not proliferate grandiose unobserved processes (such as organic evolution) but offers a more mundane cause for biostratigraphic differentiation. Also, the TAB concept is a more unifying cause than evolution and geologic time, because the former is a single-shot cause while the latter is a proliferation of (imagined) causes and (imagined) processes. The Diluvialist can use ecology, biogeography, etc. to explain fossil differentiation, while the uniformitarian must not only utilize those causes but also invoke evolution with geologic time. Thus the uniformitarian position multiplies hypotheses to a greater extent than the Diluvian position, which again is why Occam's Razor favors the Diluvian position and makes it more scientific. Why invoke evolution and geologic time in addition to ecology and biogeography as causes of biostratigraphic differentiation when the latter two causes are sufficient?

C. The Stratigraphic Separation and Succession of Fossils: a Diluvial Synthesis

This chapter unites all the factors involved in the stratigraphic separation of fossils; that is, all the factors discussed in the previous two chapters. The TAB process is the dominant factor in fossil separation and all other factors supplement it.

Figure 7 has been drafted in order to illustrate how the TAB process causes organisms to be restricted to specific stratigraphic intervals. For illustrative purposes, Figure 7 has been constructed with the stipulation that sedimentation rates are identical and that all four TAB's are locally present; the stipulations only temporarily held as real so that the full scope of TAB operations can be clarified. Thus, organisms A, E, and I are TAB I organisms and hence tend to have their stratigraphic ranges confined to the lowest quarter thickness of sediment. Organisms B, F, and J are TAB II organisms and tend to be injected into the second quarter thickness (from bottom) of sediment, etc. However, organisms are not absolutely restricted to the stratigraphic interval "belonging" to their respective TAB's because biogeographic boundaries are usually gradational and because the TAB mechanism is statistically-not absolutely-efficient. Nor need it be; since most index fossils do not actually superpose locally (Figure 2, Table 3) the few cases where any two fossils A through L superpose determine which be-

Figure 7. The generation of biostratigraphic differentiation (primarily through TAB's; with an interplay (both constructive and destructive) of other relevant factors discussed in text). The 21 occurrences each of fossils A-L, shown clustered together, actually are from diverse locations from all over the world. The four horizons per fossil denote quarter-thicknesses of rate-normalized sediment accumulations over the entire earth; each TAB deterministically injecting its sedimentary and biotic constituents into one of four quarter thicknesses. Explained in text.

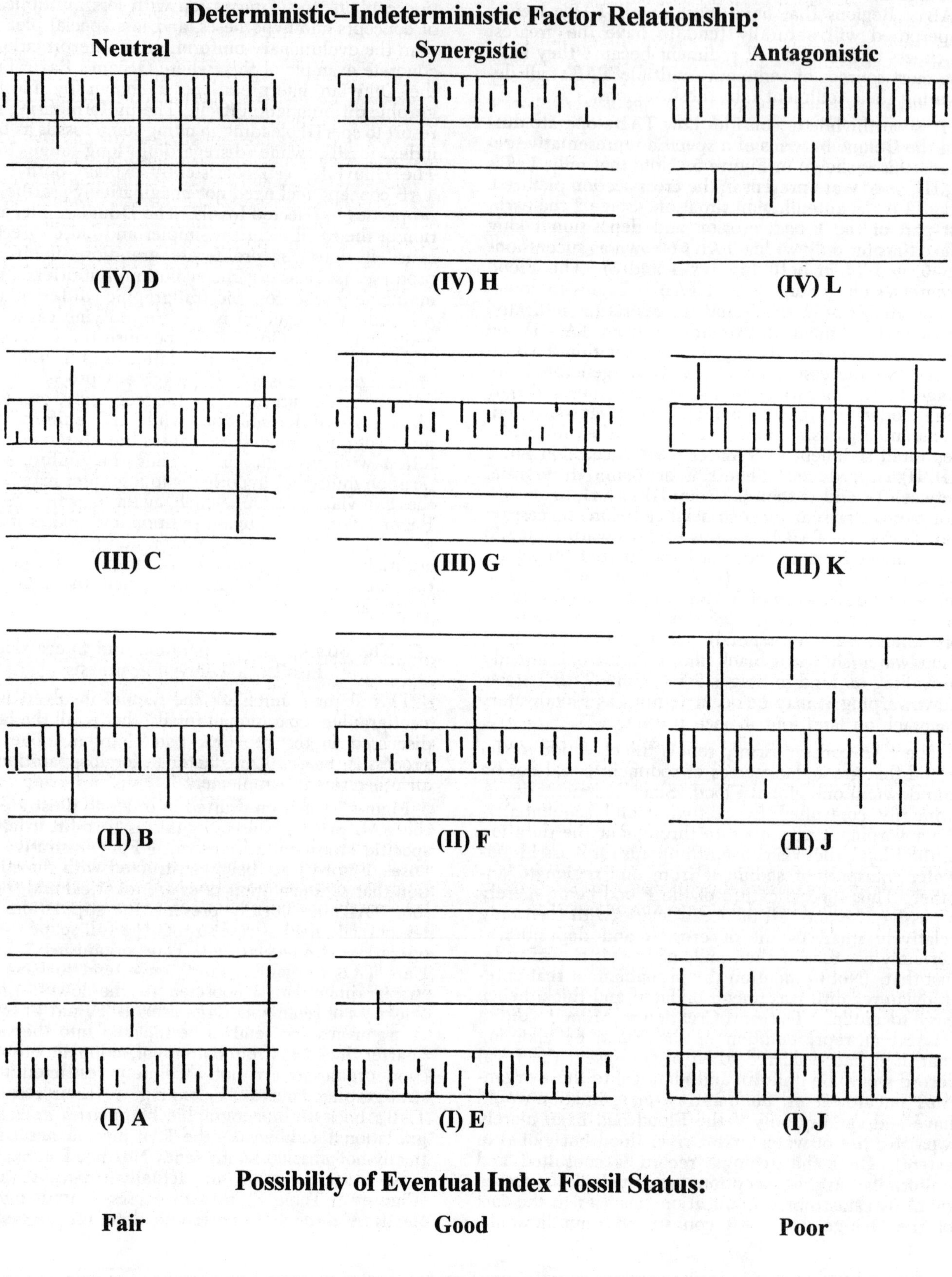

Deterministic–Indeterministic Factor Relationship:

Neutral   Synergistic   Antagonistic

(IV) D   (IV) H   (IV) L

(III) C   (III) G   (III) K

(II) B   (II) F   (II) J

(I) A   (I) E   (I) J

Possibility of Eventual Index Fossil Status:

Fair   Good   Poor

come index fossils—as will be discussed later in conjunction with Figure 8.

In the real world, neither sedimentation rates are constant nor are usually all four TAB's locally present. Suppose all four TAB's are present but the amount of sediment is the greatest for TAB II: TAB II organisms and sediment will be still deposited after TAB I organisms and sediment and before TAB III organisms and sediment, but the stratigraphic interval of TAB II will be greater than one quarter of the total thickness of total sediment deposited locally. Since TAB's provide both organisms and sediment, their respective absences deprive a given region of both. Thus, if a region only contains TAB's II and IV, one will not usually find organically-blank thicknesses of sediment where I and III would have been; only IV superposed over II with either a paraconformity or angular unconformity between them, depending on local tectonic dynamics.

The effects of the TAB process alone on the stratigraphic partitioning of organisms is shown under the "Neutral" column in Figure 7. All other relevant factors (pure chance, preservation bias, hydrodynamic sorting, differential escape, ecological zonation) acted either neutrally, synergistically, or antagonistically with respect to TAB's. For example, if an organism was benthic (and hence tending to be buried at the lowest stratigraphic interval) but it belonged to TAB III or IV (which would have tended to have it buried at high stratigraphic intervals) then ecological zonation and TAB's were antagonistic in that situation. This would have produced an organism as K or L, with many stratigraphic occurrences smeared over several reference quarter-thicknesses or translocated outside the expected quarter-thicknesses. However, even in such cases, the dominance of the TAB process in determining what stratigraphic interval an organism was deposited guaranteed that organisms will still have tended to be confined to stratigraphic intervals "belonging" to that TAB. Where multiple effects on the burial of organisms were synergistic, then the more consistent restriction of such an organism to a particular stratigraphic interval will have increased its chances of never being found overlapping with a like organism confined to a different stratigraphic interval—and thus both being index fossils. Note also from Figure 7 that, when effects are synergistic with TAB's, not only are organisms almost totally prevented from deviating beyond their TAB's quarter-thickness of sediment or from being translocated beyond the thickness interval, but even within each quarter thickness the organisms occupy shortened stratigraphic intervals.

Keeping in mind that every individual stratigraphic occurrence (each of the 252) shown in Figure 7 is geographically discrete, one should note that each occurrence can be juxtoposed with any other one. Figure 8 has been drafted to show how index fossils are concocted from juxtapositions of the discrete occurrences portrayed in Figure 7. Each of the 252 occurrences is denoted by a letter-numeral: for example, E7 means the 7th occurrence (from left; out of the 21 possible) of fossil E. The ordered pairs denote specific juxtapositions: for example, (E7, K12) denotes a juxtaposition of stratigraphic occurrence E7 with stratigraphic occurrence K12. Compatible juxtapositions are ones where the stratigraphic occurrences involved overlap; incompatible juxtapositions are ones where they do not. It is obvious that incompatible juxtapositions (and index fossils) are generated most frequently from fossils E, F, G, and H, the organisms most rigidly restricted to stratigraphic intervals of rock.

The concoction of index fossils is an interplay of actual TAB-generated stratigraphic restrictions of organisms (Figure 7) with the limited number of opportunities for fossils to be juxtaposed (Table 3, Figure 2). The fact that there are few chances for any two fossils to juxtapose means that stratigraphic mixtures and overlaps take place only when certain combinations of TAB-generated stratigraphic ranges simultaneously occur. For instance, consider fossils F and H from Figure 7. The only possible way that the juxtaposition of F and H could result in a stratigraphic overlap would be for the juxtaposition (F11, H7) to take place. The chance of an occurrence of F to be F11 is 1 in 21, and the same probability exists for an occurrence of H to be H7. Thus, the probability that a juxtaposition of F and H will result is thus only 1 in 441. But since there are only a handful of places on earth where fossils F and H occur juxtaposed, there are few chances for that 1 in 441 combination to have ever come up. Thus, if there are 10 locations on earth where F and H juxtapose, the chances are only (10)(1/441) that F and H coexist stratigraphically. The same principle of limited opportunities for juxtaposition governs all fossils. Thus, in Figure 8, the limited opportunities (six shown) for juxtaposition of any combinations of two fossils is schematically illustrated by six lines emanating from representative places on earth where juxtapositions may have actually occurred.

Uniformitarians take the results of juxtapositional situations and imaginatively ascribe time-stratigraphic significance upon them. Seeing that the combinatons EC, IH, IC, HE, and CH (see Fgure 8) are consistently incompatible, they imagine that each fossil denotes a time horizon relative to the other fossil and that such (imagined) time-horizons can be correlated with other such (imagined) time-horizons. Thus (referring to the right side of Figure 8) E is always stratigraphically below C, and C is always stratigraphically below H: the time-relationship E-C-H (going chronostratigraphically from earliest to latest) is concocted. Other fossils are rejected as index fossils because they are found to be partly or totally compatible stratigraphically with each other. J is found to be compatible with both C and E (no relationship with H nor I is shown developed in Figure 8). I is found to be incompatible (and below) H, but the (imagined) time-relationship is refined by noting that I is also incompatible (and below) C. However, I is found to be compatible with E, so it is regarded as being time-equivalent with E but not ranging stratigraphically higher (and earlier) than E because, like E, fossil I is incompatible with (and stratigraphically below) C. In summary, fossils E, C, and H are the main index fossils while I is an auxiliary one; J is dismissed as an index fossil and considered to be a long-ranging form. The same line of thinking discussed in conjunction with Figure 8 applies to actual index fossils. In conclusion, the TAB process pro-

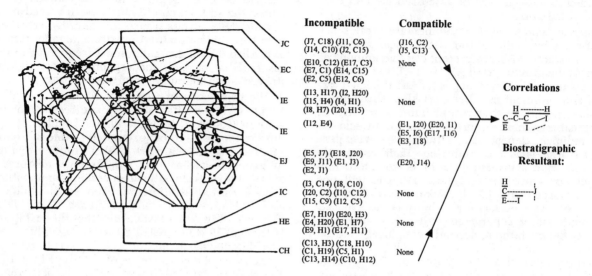

Figure 8. The selection of index fossils: a synthesis. The letters denote the same fossils as they do in Fig. 7, with the numbers indicating the numbered stratigraphic occurrences. The ordered pairs indicate which types of occurrences (of Fig. 7) participate in the juxtapositions. Incompatible combinations are those where stratigraphic occurrences do not overlap each other; only these can be index fossils. How the fossils are placed in an (imagined) time-stratigraphic relationship is illustrated on the right. Explained in text.

vides the main source of actual biostratigraphic differentiation, while the general non-superposition of fossils adds a large element of imagination to the whole concept of such differentiation.

D. Biostratigraphically-Progressive Extinctions with Respect to the Extant Biosphere: An Explanation in Light of the TAB Concept

In previous chapters the focus of attention was biostratigraphic differentiation; the why of fossils being different from one rock horizon to another. This chapter considers why there is progressive extinction; why the lower one goes into the geologic column, the fewer are the taxons still extant (and, conversely, why the present biosphere has the greatest number of taxons in common with the most recent geologic periods and the least number common with the most remote ones). This trend is quantitatively illustrated in Table 1; the rightmost column showing the percentage of families shared between Recent and the different Phanerozoic geologic periods. The periods of the Lower Paleozoic have only 11-15% of their families extant while the figure rises to 60-80% for the most recent geologic periods. An even sharper trend is shown for the number of families in the present biosphere in common with geologic periods; only 2.5% of all presently-living families are to be found in Cambrian.

The same basic TAB process that explains biostratigraphic differentiation also explains stratigraphically-progressive extinction with respect to the present biosphere. But before it is explained how it does let it first be noted that there is nothing intrinsically natural about the evolutionistic-uniformitarian explanation for progressive extinction. In fact, it is difficult for the evolutionist-uniformitarian to explain what causes extinctions! Benton[705] recently wrote: "Many hundreds of pages have been written about how the dinosaurs became extinct without our being any the wiser." Both concepts of gradual extinction of groups (such as by climatic change) and catastrophes within the context of geologic time (such as by a bolide impact at the end of the Cretaceous) encounter the difficulty of explaining how they could be sufficiently efficient and global in extent to totally obliterate taxa from off the earth. The Flood provides the best overall explanation for extinction becaues it is simultaneously global in extent and pervasive in effects (in contrast to the bolide impact whose effects on the biosphere would have been considerable but not as pervasive on a global scale). Moreover, the Diluvian interpretation is scientifically superior to a uniformitarian one because while the uniformitarian explanation must invoke multiple causes for extinctions throughout the geologic column, the Diluvian interpretation offers a single unified explanation. Occam's Razor thus favors it.

The initial understanding of how the Flood caused extinctions is the realization that the vast majority of organisms living at the time of the Flood were killed by it. The TAB process governed how many organisms of each antediluvian biogeographical zone (I, II, III, and IV) survived. The deeper the burial of a group of antediluvian organisms (irrespective of whether they were buried nearly *in situ* or transported significant distances) the less the probability that any organisms (or their eggs, larvae, seeds, spores, etc.) survived in the residual Floodwater and hence were available to repopulate the post-Diluvian earth. The sediment carried by the Floodwater not only acted as an entombing and filtering agent with respect to organisms, but also served to suffocate marine organisms. Where the burial was shallow, it meant that large amounts of sediment had not been suspended in the Floodwater and/or the period of deposition was not prolonged. In such a situation, it was more probable that some organisms had been spared from the entombing action of sediment or had not been filtered out of the Floodwater by descending sedimentary particles.

Figure 9 illustrates how depth of burial (and hence probability of survival, and thus ultimately the probability of not becoming extinct after the Flood) was controlled by the TAB process. Note that almost all of the representative stratigraphic sections (from all over the world) laid down in TAB I are thick, so this is indicative of very few organisms from that antediluvian biogeographic province having survived the Flood (this is shown in Figure 9 by the single line denoting minimal contribution to the postdiluvian biosphere). By contrast, the TAB IV sections are nearly all thin, so relatively many organisms from that biogeographic province survived the Flood (this is shown by the many lines fanning out from the TAB IV stratigraphic sections). This differential survival of TAB faunas and floras is the key to progressive extinctions relative to the contemporary biosphere. Note that the antediluvian biosphere (Figure 9, top right) had a considerable presence of all four biogeographic-province biotas (in terms of both population and low-taxon diversity). The center circle illustrates the effects of differential survival of TAB biotas: the organisms immediately surviving the Flood were numerically and taxonomically impoverished in TAB I and II constituents, and the largest share of survivors were from TAB IV.

As organisms began to repopulate the earth after the Flood, the lopsided representation of antediluvian biogeographic provinces became even more lopsided as organisms that had lived in separate biogeographic provinces before the Flood now coexisted and were now in direct competition. The organisms from the lower TAB's, being at a numerical disadvantage, were much more likely to be driven to extinction in the competition against the numerically-abundant higher-TAB biotas than this latter group. Numerical abundance was not the only extinction-biasing factor: the fact that the higher-TAB organisms were more taxonomically abundant gave them a reproductive advantage over lower-TAB organisms because the antediluvian ecological "webs" of higher TAB's were more likely to be nearly intact than was the case for lower TAB's. The end result is that the contemporary biosphere (Figure 9, lower right) is overwhelmingly dominated by TAB IV organisms: in biostratigraphic terms, the contemporary biosphere thus has much in common with Cenozoic but very little with Lower Paleozoic.

The discussion in the last few paragraphs was concerned with organisms that directly experienced the Flood; i.e., were not on the Ark. The present author follows Jones[706] in accepting ". . . (1) all birds, (2) all land-dwelling reptiles and mammals, (3) possibly some of the more terrestrial amphibia . . ." as having been the only animals on the Ark. Attention is now focused on the question of why the Ark-inhabiting organisms show the same progressive extinction with respect to the contemporary biosphere as do organisms that experienced the Flood. It has just been shown how the TAB process itself accounts for progressive extinction of organisms that went through the Flood. But the animals on the Ark were ecologically dependent on organisms that went through the Flood! One manifestation of this was the food chain. The Paleozoic reptiles ate TAB II vegetation, the dinosaurs primarily TAB III vegetation, and mammals were designed to subsist on TAB IV vegetation. Since vegetation, being outside the Ark, was subject to immediate differential extinction, the animals released from the Ark were subject to differential extinction which necessarily parallelled that of the vegetation they were dependent upon. Thus, the main reason why mammals survived at the expense of dinosaurs, "primitive" reptiles, etc., was because mammals had such a great reproductive advantage due to the overwhelming predominance of TAB IV vegetation soon after the Flood. The preferential relationship with humans of mammals probably was also a significant factor. It is concluded by reiterating that differential extinction was statistical, not absolute, which is why there are "living fossils" still extant (Table 1 shows measurable percentages of families from the Lower Paleozoic still extant) and why—conversely—many types of "advanced" mammals are extinct.

E. Causes for the (Near) Absence of Pre-Pleistocene Human Fossils

One major ramification of biostratigraphic differentiation is the (near) fact that humans do not appear until the very top of the geologic column. In a sense this should not be surprising to the Creationist-Diluvialist in view of the fact that man is totally different in capabilities and manner of life from all other organisms.

First of all, there are significant reports of pre-Pleistocene human remains pointed out by Creationists (for example, Jochmans[707]), and there are good reasons to suspect that there are many more that are not recognized — unintentionally and intentionally. Non-recognition of human fossils is unintentional if fragmentary skeletal remains are erroneously ascribed to some other vertebrate. As far-fetched as this may appear, it actually frequently happens. Walker[708] said: "Sometimes mistakes occur and, since the specialist is usually unfamiliar with groups other than his own, he may not recognize the mistake. In this way crocodile femora have been described as hominoid clavicles (Le Gros Clark and Leakey 1951), lateral toes of *Hipparion* called *Australopithecus* clavicles (Bone 1955), crocodile naviculocuboids called *Palaeopropithecus* capitates (Sera 1935), and so on. It is no accident, perhaps, that mistakes occur most frequently in human and primate paleontology, because every scrap is seen as important and the anatomists are sometimes unfamiliar with other orders of mammals, let alone other classes of vertebrates. As a general rule, the smaller the fragment, the greater the chances of mistaken identity. The chances are especially high, it seems, if the bone to be examined is presented together with a set of similar bones . . . Apart from bones finding their way to the wrong specialist, it may also be true that a large number of unrecognized primate fossils still remain in museum collections. This is because the time involved in sorting through boxes of bone scraps and unidentified fragments is too long for most visiting scientists to expend." This process operates even more so down in the geologic column, where there is the widely-held evolutionary-uniformitarian belief that there are no human remains there. For example, a

Figure 9. A schematic illustration of the effect of TAB's on progressive extinction trends in the fossil record. The stratigraphic columns at left are representative of thicknesses that biotic members of TAB's are buried under; decreasing in trend going from I to IV. The width of each right triangle is directly proportional to the number of relatively thin stratigraphic sections, whereas the height is directly proportional to the thickness of such strata. The lines terminating at the arrow are units of biological survivors' representation—the organisms (primarily eggs, larvae, spores, seeds, etc.) which survived the straining action of sediments in Floodwater and were thus available to repopulate the postdiluvian world. The circles at right depict changes in the global biosphere since the Creation; the sections of the circle being proportional to the taxonomic diversity of each of the respective Biological Provinces with the amount of numeral figures proportional to the numbers of individuals extant. Explained in text.

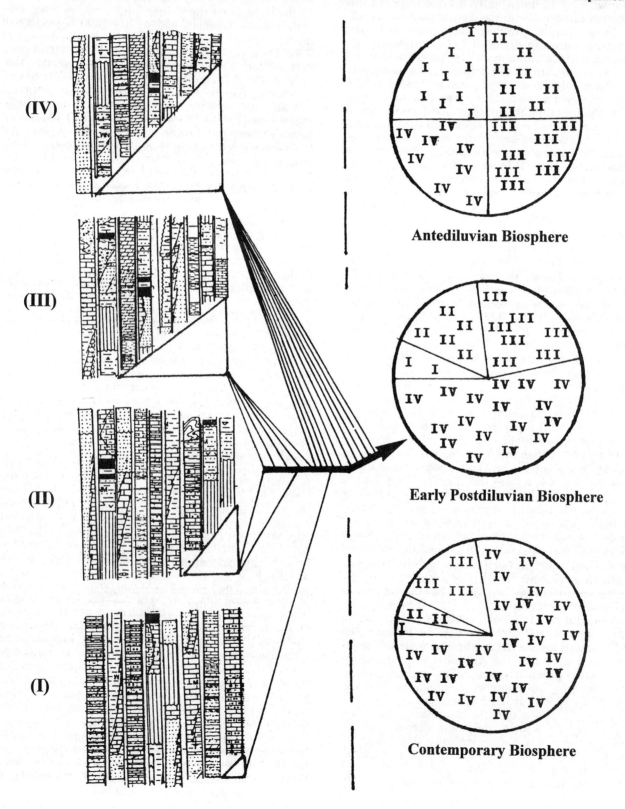

specialist in Paleozoic tetrapods will have no training in human paleontology and will misidentify human skeletal fragments found in a Paleozoic bone assemblage as belonging to some vertebrate of that (alleged) time. The same holds true for Mesozoic, early Cenozoic, and Middle Cenozoic bone assemblages.

One must face the rather unpopular question of intentional nonrecognition of pre-Pleistocene human remains. Is one justified in suspecting that discoveries of human remains low in the geologic column are deliberately ignored or discounted? The likely answer can be found by considering the uniformitarian reaction to fossil finds that appear much earlier in the geologic column than had been previously accepted. Consider the "fact" that angiosperms do not appear until Late Cretaceous, and how a much earlier candidate was treated. Daghlian[709] wrote: "The Triassic age rather than morphological considerations appears to be the main obstacle to accepting *Sanmiguella* as a possible angiosperm." Evolutionary-uniformitarian preconceptions are vividly evident. In similar fashion is the example of belemnites, discussed in the author's work on cephalopods.[710] Reports of Devonian belemnites were admittedly "ignored or discounted" for nearly a century because of a widely-held preconception that there were no belemnites until the Triassic. If angiosperms and belemnites could be subject to such stratigraphic preconceptions, how many magnitudes more so would early Phanerozoic human remains! *If* such ancient human remains were ever recognized as valid by uniformitarians, they would probably be taken as evidence for, of all things, time travel! Lipps[711] wrote: "It is more likely that anomalies produced by time travel would be 'anachronisms' in historical or fossil records—for example, the fossilized remains of a modern human in Jurassic rock strata or descriptions of nuclear weapons in ancient literature. Should errata like these be found, they would constitute evidence that time travel is possible and will be developed."

Even having considered pre-Pleistocene human remains to be rare, one should note that such a situation is not unusual nor problematic when one considers it in the context of the whole fossil record. In other words, there are many extant organisms with a nonexistent (or poor) fossil record—not only organisms (for example, worms) lacking hard parts but also vertebrates. For instance, concerning certain modern amphibians, Carroll[712] wrote: "There are approximately 34 genera and 160 species of living Apoda. None has a fossil record. A single vertebra from the Upper Paleocene of Brazil is the only known fossil." Other examples include monotremes and marsupials: the latter has only 12 known Pre-Pleistocene fossil specimens.[713]

The clearest, least complex, and most probable single explanation for the near nonexistence of lower Phanerozoic human fossils is low antediluvian human population. Whitcomb and Morris[730] proposed a population of 1 billion, resulting from 6 surviving-reproducing offspring per generation. The population shrinks to 10 million if the number drops only slightly to 4.6.

The evidence derived from the extreme sinfulness of the antediluvians, as will now be shown, provides an independent basis for concluding that there were relatively few humans (perhaps only several million worldwide) as candidates for fossilization at the time of the Flood. It must be remembered, first of all, that plants and animals existed since Creation week as populations, whereas humans began with a single pair (Adam and Eve). The almost universal depravity of the antedeluvians guaranteed low fecundity. Anywhere from a large minority to a majority of the antediluvian population undoubtedly engaged in homosexual, zoophiliac, or pedophiliac contact. Even heterosexual contact was promiscuous, causing rampant venereal diseases and thereby damaging reproductive organs beyond use in childbearing and making it likely that any children born to parents whose organs had not been irreversibly damaged would themselves be born diseased and die shortly after birth. If a certain apocryphal tradition cited by von Wellnitz[714] is historically accurate, then low fecundity was also caused by the widespread use of contraceptives by women interested only in beauty.

The second major cause of low antediluvian population was the high murder rate (Genesis 6:11). A large fraction (perhaps majority) of babies and young children of every generation died through infanticide and child sacrifice to idols. Both children and adults were subject to the gross disrespect of life so characteristic of florid depravity. It would have mattered little that the longevity of antediluvians was measured in centuries if the vast majority of people were murdered long before that age. Lest it be considered that low fecundity and high murder rates discussed here are an exaggeration, let it be noted that the merciful God would have withheld the Flood had there been any significant percentage of people not totally depraved (just as He would have spared Sodom and Gomorrah if only 10 (relatively) righteous people were found— Genesis 18:32).

The net effect of the great evils of the antediluvians was a low population, and this caused a parallel situation with preservable implements. However, just because metal-work existed among the antediluvians (Genesis 4:22) does not mean that it was common among the populace. If there were sharp distinctions between antediluvian social classes, then probably only the upper classes had significant metal and ceramic implements.

Suppose that the antediluvian population totaled 10 million, and (for purpose of discussion) the number of preservable implements balanced out the number of skeletons that had not been preserved. According to Ronov,[699] there are 700 million cubic kilometers of Phanerozoic rock in the earth's crust. If randomly distributed, the antediluvian anthropogenic remains (bones and implements) would occur at a rate of one specimen per 70 cubic kilometers of rock. It takes little reflection to appreciate the vanishing probability of such remains ever being discovered. (If skeletons were disarticulated and the fragments scattered, this would increase the absolute number of individual remains. But this would be more than offset by the fact, known in studies of taphonomy,[715] that fragments are much more likely to go unnoticed than complete skeletons.) In reality, fossils are very inhomogeneously distributed in rock. However, the rarity of antediluvian anthropogenic remains overcomes any apparent increase in chance of discovery caused by exceptional concentration of remains. Suppose that the 10 million antediluvian anthropogenic remains, instead of being randomly scattered over the 700 million cubic kilometers of Phanerozoic rock, were concentrated in only 1 million cubic kilometers. Occurring at a rate (in the special rocks) of 10 specimens per cubic kilometer, the chances of discovery would still be quite small; and the chance that any cubic kilometer is one of the special ones would be only 1 in 700. It is thus improbable that the 1 million cubic kms. would simultaneously be those of many outcrops and in situations attracting particular investigative interest. In fact, paleontological interest is probably being inadvertently diverted away from any special humaniferous rocks. One of the sampling biases reviewed by Signor[716] is that of paleontologists' interest: paleontologists (and not to mention other professional—and amateur—collectors) tend to study rocks which are highly fossiliferous. But since humans were rare and humans probably did not live near regions of great animal population, then areas highly fossiliferous in these animal vertebrates (even more so those of concentrated marine invertebrates) are very negatively concentrated in human remains. This makes it all the more unlikely that the exemplary 1-in-700 million cubic kilometers has attracted any significant collectors' interest.

The present author believes that the smallness of the antediluvian human population is more than sufficient in itself as an explanation for the near absence of pre-Pleistocene human remains. Yet there are still other significant factors tending to greatly reduce the number of human fossils. Shotwell[717] wrote: "Forms which are nearly always rare or missing in fossil mammalian faunas, irrespective of their probable abundance in the area, are those with volant or arboreal habits. This characteristic has hindered the study of such groups as bats, primates, and flying squirrels. Their usual small size and fragility does not seem to be the important factor since insectivores and small rodents are not uncommon in quarries." Bishop[718] wrote: ". . . primates are normally 'shy' candidates for fossilization." Since humans lived away from regions of deposition (those first covered by the sediment of the Flood), their remains are more likely to have rotted away before having much chance to be buried. Figure 10 has been drafted to illustrate how antediluvian human communities were probably distributed—in a way that would have minimized the number of human bones eventually preserved. Every student of ancient history knows that most ancient civilizations were situated near rivers. This was probably even more so in the antediluvian world, because the dense forestation (and impenetrable grasslands) probably made rivers the only practical mode of long-distance transport and trade.

A very major clarification about the sedimentology of rivers is necessary in order to differentiate between rivers under ordinary (local flood) conditions and those at the time of the Flood. Rivers are normally areas of burial and preservation of vertebrates, under local flood conditions, because sediment is deposited on the floodplain or delta, entombing organisms. But in the global Flood situation, rivers were quickly elevated to flood stage from rain runoff and the runoff from subterranean springs. They were maintained at this flood stage at great intensity and for a prolonged period of time (relative to any local river flood), enabling the river and its adjacent floodplain to be entirely erosional along the entire length of the river. Under these conditions, humans living near the rivers were not entombed on the floodplains but were flushed out into the open ocean (Stage 2 of Figure 10). Since the rivers under these extremely erosional conditions did not become depositional until they entered the oceans, man-made implements and the human corpses that had sunk did not get deposited until they reached the mouths of the rivers. Any such remains were concentrated into small volumes of sediment (at the prodeltas) and—as discussed previously—overall few remains concentrated into small volumes of sediment means very low probability of discovery. Yet it is even plausible that the volumes of prodeltaic sediment were metamorphosed beyond recognition, obliterating all their anthropogenic remains. This is because there occur at the center of mountain ranges evidences of ancient oceanic crust (ophiolites): in plate-tectonics models, mountains are believed to be largely the result of ancient oceans that had been compressed into relatively small linear areas. Irrespective of whether or not there was continental drift during or after the Flood, narrow antediluvian oceans became compressed to form extant ophiolite-containing orogens. Any prodeltaic remains from antediluvian rivers (as shown in Figure 10) thus were probably metamorphosed as a result of being associated with oceanic crust.

Humans flushed down the antediluvian rivers (Stage 2, Figure 10) usually did not even get deposited in the prodeltaic sediments at the mouths of these rivers, but instead tended to float on out into the ocean where they decomposed or else were devoured by predators or scavengers. This follows from the fact that both living mammals[719] and their corpses[720] tend to float on water (the latter buoyed up by the gases of decomposition). Since the antediluvian topography was probably low, only a relatively small water-level elevation of the antediluvian rivers was required to wash away the antediluvian human communities downstream. This has important ramifications. It took much less time for the antediluvian rivers to reach flash flood stage and wash out the human communities into the ocean (Stage 2, Figure 10) than it took for the ocean bottoms to be uplifted and the continental interiors to be submerged (Stage 3, Figure 10). Thus appreciable flooding of the continents, initiation of the TAB process, and actual burial of organisms found in the Phanerozoic fossil record (Stage 3) all did not begin until after humans had been flushed out into the oceans. The flatness of the topography meant that the antediluvian human communities could have been flooded in a matter of hours, denying the antediluvians time to flee their homes near the rivers.

Once human corpses were out at sea, they could have decomposed in a very few weeks—this fact based

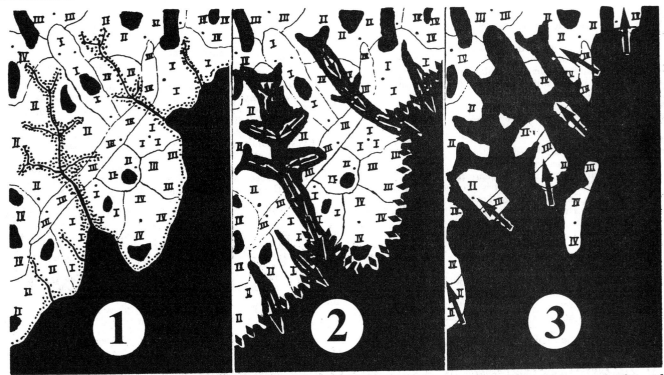

Figure 10. Antediluvian human communities and the near non-preservation of human remains in the pre-Pleistocene fossil record. Roman numerals and their enclosures depict TABs, dots indicate units of antediluvian human population, thick sinuous and dendritic lines denote rivers, the black denotes sea. No. 1 refers to geography before the Flood, No. 2 to the same area at the earliest part of the Flood (diamonds indicate tsunamis battering the coast, arrows indicate rivers swollen at flash flood stage), and No. 3 to Flood proper with its encroachment of seawater on continents (shown by arrows) and initiation of the TAB process. Explained in text.

on taphonomic studies[721] on floating mammal corpses. The skeleton could disarticulate before all the flesh was gone.[722] Flume experiments[723] have shown that complete crania are the easiest transported of all skeletal components by water currents. Thus the great preservability of teeth was offset by their tendency to be scattered; teeth gradually falling out of floating skulls as the alveoli decayed. (It should be pointed out that humans who had died before the Flood would not have much of a candidacy for fossilization because—as anthropological studies[724] have shown—perishable buried anthropogenic remains decay completely within 20 years at most.)

Another major factor limiting the number of fossil humans was diagenesis, but its exact role is not yet known. Behrensmeyer and Hill[725] wrote: "There are too many variables, too many unknowns, and a general lack of understanding of how bones become fossils." However, the selectivity of diagenetic conditions for preservation of buried bone is evident in the following statements of Hill:[726] "Not all environments will favour a bone's ultimate preservation, and in this many factors are involved . . . Many Miocene hominoid localities in East Africa are associated with carbonatitic volcanics. Analysis show that the chemical composition of such rocks is similar to that of bone, producing a stable environment for fossilization. Similar work is needed to determine what chemical conditions are necessary for fossilization, and which of the whole range of possible palaeoenvironments might have possessed them."

It is evident that if most antediluvian humans lived in areas whose diagenesis following Flood burial was not suitable for fossilization of even deeply buried bone, then this factor alone could account for the near absence of pre-Pleistocene humans. Organic acids help weather bone,[727] and modern taphonomy[728] suggests that alkaline conditions favor bone preservation. Still another important factor in diagenesis and fossilization is eH. Positive eH (oxidizing conditions) favors prompt decomposition of not only flesh but also bone.[729] Thus fluvial regions where antediluvians lived (Figure 10) may have generated sediment too oxidized for the final preservation of any human bones that had managed to get buried. By contrast, reducing conditions probably prevailed in the poorly-circulated, poorly-ventilated shallow antediluvian seas, facilitating the preservation of the endless number of Phanerozoic marine fossils once the seas were Flooded (Stage 3, Figure 10). Likewise, land areas away from rivers (where few humans lived—according to the model proposed in Figure 10) were water-logged and therefore reducing, thus facilitating the preservation of Phanerozoic land biotas. Yet it must be remembered that low antediluvian human population itself accounts for the paucity of pre-Pleistocene human remains.

References

AC — Academic Press, New York, London
AF — Palaeontologica Africana
AG — American Association of Petroleum Geologists Bulletin
AJ — American Journal of Science
AM — Geological Society of America Memoir
AP — Bulletin d'le Academie Polonnaise des Sciences
AS — Geology and Paleontology of Southeast Asia
AZ — Proceedings of the Geologists' Association
BB — Bibliography and Index of Geology

BF — Bulletin Societe Geologique de France
BI — Geobios
BM — BMR Journal of Australian Geology and Geophysics
BO — Palaeobotanist
BP — Bulletins of American Paleontology
BR — Bulletin of the British Museum (Natural History) Geology section
CA — Geological Association of Canada Special Paper
CB — Canada Geological Survey Bulletin
CE — Canadian Journal of Earth Sciences
CH — Alcheringa
CO — Scottish Journal of Geology
CP — Bulletin of Canadian Petroleum Geology
CR — Creation Research Society Quarterly
DE — Soviet Academy of Sciences Doklady: Earth Science Section (English-language Translations)
DO — Dowden, Hutchison, and Ross Publishing Co., Stroudsburg, Pennsylvania
EC — Ecology
EL — Elsevier Scientific Publishing Company, Amsterdam, London
FI — Fieldiana: Geology
FR — Memoir hors serie Societe geologique de France
FS — Fossils and Strata
GA — Geological Society of America Abstracts with Programs
GB — Geological Society of America Bulletin
GE — Geology
GL — Journal of the Geological Society of London
GM — Geological Magazine
GP — Geological Society of America Special Paper
GU — Journal of the Geological Society of Australia
HE — Eclogae Geologicae Helvetiae
HO — Journal of the Faculty of Science of the University of Hokkaido
IG — International Geology Review
IN — Journal of the Geological Society of India
JA — Neues Jahrbuch fur Geologie und Palaontologie
JJ — Japanese Journal of Geography and Geology
JP — Journal of Paleontology
JW — John Wiley and Sons, New York
KA — University of Kansas Paleontological Contribution
LE — Lethaia
LI — (Liverpool-Manchester) Geological Journal
ME — Ameghiniana
MM — Geological Society of America Memoir
NA — Nature
ND — Proceedings of the Indian Geologists' Association
NJ — Memoirs of Nanjing Institute of Geology and Palaeontology, Academia Sinica
NO — Norsk Geologisk Tiddskrift
OL — Palaeontologia Polonica
PA — Palaeontology
PB — Paleobiology
PE — Journal of Petroleum Geology
PH — Palaeontographica
PJ — Paleontological Journal (translation from Russian)
PL — Palaeontological Association of London Special Papers in Palaeontology
PO — Acta Palaeontologica Polonica
PP — Palaeogeography, Palaeoclimatology, Palaeoecology
PV — Palaeovertebrata
PZ — Palaeontologisch Zeitschrift
RB — Revista Brasileira de Geosciencas
RU — Geologisch Rundschau
SB — Akademiia Nauk CCCP Doklady, Sibirskoe Otdelenie
SC — Science
SB — Sedimentary Geology
SI — Acta Paleontologica Sinica
SJ — Palaeontological Society of Japan Special Paper
SL — Senckenbergaia Lethaia
SN — Scientia Sinica
SP — Springer-Verlag Publishing Company, Berlin, Heidelberg
SV — Soviet Geology and Geophysics (translations from Russian)
TE — Tectonophysics
TO — Journal of the Faculty of Science of the University of Tokyo
TP — Treatise on Invertebrate Paleontology
TR — Trudy Instituta Paleontologicheskovo Akademii Nauk CCCP
UB — United States Geological Survey Bulletin
US — University of Chicago Press, Chicago and London
UG — United States Geological Survey Professional Paper
UK — Akademiia Nauk CCCP Doklady: Seriya Geologicheskaya
UN — Gronlands Geologiske Undersogelse Rapport
VO — Voprosy Mikropaleontologii
WI — George Allen and Unwin Publishing Company, London
YG — Proceedings of the Yorkshire Geological Society
YO — University of Wyoming Contributions to Geology

1. Hallam A. (ed.). 1973. *Atlas of Palaeobiogeography*. EL
2. Murray G., Kaczor M. J., and R. E. McArthur. 1980. Indigenous Precambrian Petroleum Revisited. *AG* 64:1685
3. Halin G. and A. D. Pflug. 1980. Ein Neuer Medusen-Fund aus dem Jung-Prakambrium von Zentral Iran. *SL* 60:449
4. Paley, I. P. and Z. A. Zhuravleva. 1979. New data on the structure of the Kernlin sutural zone (Mongolia). *IG* 21:709
5. Peel, J. S. and K. Sechen. 1979. A second fossil occurrence from the Precambrian Shield of southern west Greenland. *UN* 91: 99
6. Binder, P. L. and M. M. Bokhaii. 1979. Chitinozoan-like microfossil of a late Precambrian dolostone from Saudi Arabia. *GE* 8:70
7. Gunia, T. 1981. The First Discovery of Precambrian Microflora in Paragneisses of the Sowie Gory Mountains, Sudetes. *AP* XXIX(2)131
8. Horodyski, R. J. and J. A. Donaldson. 1980. Microfossils from the Middle Proterozoic Dismal Lakes Group, Arctic Canada. *RE* 11:126
9. Thi, P. T. 1978. Stratigraphy and Petrology of the Precambrian Formations of Vietnam. *SV* 19:31-37
10. Hofmann, H. J. 1981. First record of a Late Proterozoic faunal assemblage in the North American Cordillera. *LE* 14:303
11. Lindiao, C., Z. Huimm, X. Yasheng, and M. Gnogan. 1981. On the Upper Precambrian (Sinian Subratem) in China. *RE* 15:207-28
12. Schopf, J. W., *et al.* 1977. Six New Stromatolitic microbiotas from the Proterozoic of the Soviet Union. *RE* 4: 271-82
13. Shenfil, V. Yu. 1978. Algae in Precambrian Deposits of the Yenisey Ridge Region. *DE* 240:224
14. Nyberg, A. V. and J. W. Schopf. 1981. Precambrian Microbiota from the Min'yen Formation, Southern Ural Mountains, U.S.S.R. *Palynology* 5:221
15. Muir, M. D. 1978. Alcheringa news item. *CH* 2:310
16. Fedonkin, M. H. 1980. New Precambrian Coelenterata in the North of the Russian Platform. *PJ* 14:1
17. Mendelson, C. V. and J. W. Schopf. 1982. Proterozoic Microfossils from the Sukhuya Tunguska, Shorikha and Yudoma Formations of the Siberian Platform, USSR. *JP* 56:44
18. Martin, A., Nisbet, E. G. and M. J. Bickle. 1980. Archaean Stromatolites of the Belingwe Greenstone Belt, Zimbabwe (Rhodesia). *RE* 13:338
19. Ruiji, C. and Z. Uenjie. 1981. Sequence of Precambrian Stromatolite Assemblages in North China. *SI* 20:516
20. Yun, Z. 1981. Proterozoic Stromatolite Microfloras of the Gaoyuzhuang Formation (Early Sinian:Riphean) Hebei, China. *JP* 55:486
21. Shirmon, A. E. and A. Horowitz. 1972. Precambrian Organic Microfossils from Sinai. *Pollen et Spores* XIV(3)333
22. Lo, S. S. C. 1980. Microbial Fossils from the Lower Yudoma Suite, Earliest Phanerozoic, Eastern Siberia. *RE* 13:111
23. Zhongying, Z. 1981. A New Oscillatoria-like filament microfossil from the Sinian (late Precambrian) of western Hubei Province, China. *GM* 13:202
24. Cloud, P. E., *et al.* 1979. Earliest Phanerozoic or Latest Proterozoic Fossils from the Arabian Shield. *RE* 10:75
25. Shixing, Z. 1982. An Outline of Studies on the Precambrian Stromatolites of China. *RE* 18:369

26. Raher, P. K. and M. V. A. Sastry. 1982. Stromatolites and Precambrian Stratigraphy of India. *RE* :294
27. Cheng-Hua, D. 1982. Precambrian algal megafossils *Churaria* and *Tawuia* in some areas of eastern China. *CH* 6:58
28. Buck, F. 1980. Stromatolite and Ooid Deposits within the fluvial and Lacustrine Sediments of the Precambrian Ventersdorp Supergroup of South Africa. *RE* 12:312
29. Burrett, C. and R. Richardson. 1980. Trilobite Biogeography and Cambrian Tectonic Models. *TE* 63:163
30. Cowie, J. W. 1971. Lower Cambrian Faunal Provinces (*in* Middlemiss, F. A., Rawson, P. E., and G. Newall. 1971. *Faunal Provinces in Space and Time.* LI Special Issue, No. 4) p. 32-3
31. Wolfart, R. 1981. Lower Palaeozoic Rocks of the Middle East (*in* Holland, C. A., ed. 1981. *Lower Palaeozoic of the Middle East, Eastern and Southern Africa, and Antarctica.* JW.) pp. 1-130
32. Klitzsch, E. 1981. Lower Palaeozoic Rocks of Libya, Egypt, and Sudan (*in* Holland. 1981. *ibid.*) pp. 131-63
33. Kabankov, V. Ya, Shasnurina, I. T., and N. A. Shishkin. 1973. New Data on the Lower Cambrian Stratigraphy of the Kolyma Uplift in the Northeastern USSR. *DE* 210:41
34. Piyasin, S. 1980. Tentative Correlation of the Lower Paleozoic Stratigraphy of Western Part of Southern Shan State, Burma and Northwestern Through Peninsular of Thailand. *AS* 21:19-25
35. Dean, W. T. 1973. Cambrian and Ordovician correlation and trilobite distribution in Turkey. *FS* 4:355
36. Theokritoff, G. 1979. Early Cambrian provincialism and biogeographic boundaries in the North Atlantic Region. *LE* 12:282
37. Ke-xing, Y. and Z. Sen-gui. 1981. Lower Cambrian Archaeocyathid Assemblages of Central and Southwestern China (*in* Teichert, C., Lu, L., and C. Pei-ji. 1981. *Paleontology in China, 1979.* GP 187) P. 41
38. Kobayashi, T. 1967. Stratigraphy of the Chosen Group in Korea and South Manchuria. *TO* 16:431, 472, 499
39. Palmer, A. R. 1969. Cambrian Trilobite Distributions in North America and their bearing on Cambrian Palaeogeography of Newfoundland (*in* Kay, M. 1969. *North Atlantic-Geology and Continental Drift.* AG Memoir 12) p. 141
40. Wentang, Z. and Y. Jin-liang. 1981. Trilobites from the Hsuchuang Formation (Lower Middle Cambrian) in western marginal parts of the North China Platform (*in* Teichert *et al.* 1981. *op. cit.*) p. 161-9
41. Zhao-ding, Z. and J. Li-Fu. 1981. An Early Cambrian trilobite faunule from Yeshan Luhe District, Jiangus (*in* Teichert *et al.* 1981. *op. cit.*) p. 154
42. Palmer, A. R. 1971. The Cambrian of the Great Basin and Adjacent Areas, Western United States (*in* Holland, C. H. 1971. *Cambrian of the New World.* JW), p. 4
43. Korobov, M. N. and T. V Yarkauskus 1980. First Trilobite Find In the Middle Cambrian of the Baltic Region (Lithuania). *DE* 253:138
44. Palmer, A. R. 1972. Problems in Cambrian Biogeography. 24th *CO*, Sec. 7, pp. 312-4.
45. Palmer, A. R. 1973. Cambrian Trilobites (*in* Hallam. 1973. *op. cit.*) p. 4
46. Palmer, A. R. 1977. Biostratigraphy of the Cambrian System—A Progress Report. *RV* 5:14
47. Palmer, A. R. 1982. Fossils of Dresbachian and Franconian (Cambrian) Age from the Subsurface of West-Central Indiana. *Indiana Geological Survey Special Report* 29, 12 p.
48. Palmer, A. R. and J. S. Peel. 1979. New Cambrian faunas from Peary Land, eastern North Greenland. *UN* 91:30
49. Henningsmoen, G. 1956. The Cambrian of Norway. 20th *CO*, p. 46
50. Taylor, M. E. 1976. Indigenous and Redeposited Trilobites from Late Cambrian Basinal Environments of Central Nevada. *JP* 50:670-6
51. Shah, S. K., Raina, B. K., and M. L. Razdan. 1980. Redlichid Fauna from the Cambrian of Kashmir. *IN* 21:511
52. Jell, P. A. 1975. Australian Middle Cambrian Eodiscoids with a Review of the Superfamily. *PH* A150:1-97.
53. Tai-Xiang, A. 1981. Recent progress in Cambrian and Ordovician conodont biostratigraphy of China (*in* Teichert *et al.* 1981. *op. cit.*) p. 210
54. Taylor, M. E. 1977. Late Cambrian of Western North America: Trilobite Biofacies, Environmental Significance, and Biostratigraphic Implications (*in* Kauffman, E. G. and J. E. Hazel. 1977. *Concepts and Methods of Biostratigraphy.* DO), p. 402
55. Fortey, R. A. and W. A. Rushton. 1976. *Chelidonocephalus* Trilobite Fauna From the Cambrian of Iran. *BR* 27(4)322
56. Bell, W. C., Feniak, O. W., and V. E. Kurtz. 1952. Trilobites of the Franconia Formation. Southeast Minnesota. *JP* 26:178
57. Ahlberg, P. 1981. Ptychopariid Trilobites in the Lower Cambrian of Scandinavia. (*in* Taylor, M. E. 1981. *Short Papers for the Second International Symposium on the Cambrian System.* US Geological Survey Open-File Report 81-743), p. 6
58. Korobob, M. H. 1973. Trilobity Semiestva Concoryphidae i ich znacheniye dlya stratigrafii Kembriiskich otlozhenii. *UK* 211 (transactions), p. 119 (in Russian)
59. Rozanov, A. Yu and F. Debrenne. 1974. Age of Archaeocyathid Assemblages. *AJ* 274:834
60. Hill, D. 1972. Archaeocyatha. *TP* E40-E41
61. Holland, C. H. and B. H. Sturt. 1970. On the Occurrence of Archaeocyathids in the Caledonian Meatmorphic Rocks of Soroy and their Stratigraphic Significance *NO* 50:343
62. Shengzhe, G. 1981. Lower Cambrian Archaeocyathids from the Central Part of Da Hinggan Ling. *SI* 20:63
63. Kobluk, D. R. 1982. First record of *Labyrinthies soranfi* Kobluk from the southern Appalachians, Lower Cambrian Shady Dolomite, Virginia. *CE* 19:1094
64. Zhantkov, T. M. and N. V. Polyanskiy. 1972. Stratigraphy of the Basal Part of the Sequence in the Chingiz-Tarbagatay Maganticlinorium. *DE* 204:26
65. Nitecki, M. H. 1967. Bibliographical Index of North American Archaeocyathids. *FI* 17(2)111
66. Kruse, P. D. and P. W. West. 1980. Archaeocyatha of the Amadeus and Georgina Basins. *BM* 5:166
67. Zhambyn, B. 1971. The Wendian and Lower Cambrian of Northern Mongolia. *IG* 13:916
68. Ke-Xing, Y. and Z. Sen-giu. 1980. Lower Cambrian archaeocyatha of Central and Southwestern China. *SI* 19(5)391
69. Palmer, A. R. and A. Yu. Rozanov. 1976. Archaeocyatha from New Jersey: Evidence for an intra-Cambrian unconformity in the north-central Appalachians. *GE* 14:714
70. James, N. P. and F. Debrenne. 1980. First regular archaeocyaths from northern Appalachians, Forteau Formation, western Newfoundland. *CE* 17:1609-10
71. Whittington, H. B. 1956. Presidential Address: Phylogeny and Distribution of Ordovician Trilobites. *JP* 40:703, 8
72. Whittington, H. B. and C. P. Hughes. 1972. Ordovician Geography and Faunal Provinces Deduced from Trilobite Distribution. *Royal Society of London Philosophical Transactions*, Ser. B, p. 247
73. Whittington, H. B. 1973. Ordovician Trilobites (*in* Hallan. 1973. *op. cit.*) pp. 14-15
74. Ross, R. J. 1975. Early Paleozoic trilobites, sedimentary facies, lithospheric plates, and ocean currents. *FS* 4:312
75. Apollonov, S. K. 1974. *Ashgillskie Trilobity Kazakhstana.* Nauka, Moskva, P. 4 (in Russian)
76. Anderston, R., Bridges, P. H., Leeder, M. R., and B. W. Sellwood. 1979. *A Dynamic Stratigraphy of the British Isles* WI, p. 34
77. Tian-rong, Z. 1981. New Materials of Early Tremadocian Trilobites from Sandu and Pu'an, Guizhou. *SI* 20:246
78. Shaw, F. C. and R. A. Fortey. 1977. Middle Ordovician facies and trilobite faunas in North America. *GM* 114(6) 441
79. Ludvigsen, R. 1976. New cheirurinid trilobites from the lower Whittaker Formation (Ordovician), southern McKenzie Mountains. *CE* 13:948
80. Fortey, R. A. 1973. Early Ordovician trilobite communities. *FS* 4:331
81. Wilson, J. L. 1957. Geography of Olenid Trilobite Distribution and Its Influence on Cambro-Ordovician Correlation. *AJ* 255:321-40
82. Hong-jun, G. and D. Ji-ye. 1979. Cambrian and Early Ordovician Trilobites from Northeast Hebei and West Liaoning. *SI* 17(4)458
83. Winder, C. G. 1960. Paleoecological Interpretation of Middle Ordovician Stratigraphy in Southern Ontario, Can-

ada. 21st CO, pt. 7, pp. 18-27
84. Nakhorosheva, L. D. 1976. Ordovician Bryozoa of the Soviet Arctic (*in* Bassett, M. G., ed. 1976. *The Ordovician System.* University of Wales Press), p. 576
85. Whittington, H. B. 1968. *Cryptolithus* (Trilobita): Specific Characters and Occurrence in Ordovician of Eastern North America. *JP* 42:704
86. Kobayashi, T. and T. Hamada. 1970. A Cyclopygid-bearing Ordovician Faunule discovered in Malaya with a Note on the *Cyclopykidae*. *AS* 8:8
87. Landing, E. and C. R. Barnes. 1981. Conodonts from the Cape Clay Formation (Lower Ordovician), southern Devon Island, Arctic Archipelago. *CE* 18:1609-10
88. Ross, R. J. 1957. Ordovician Fossils from Wells in Williston Basin, Eastern Montana. *UB* 1021-M, p. 440
89. Ludvigsen, R. 1978. Middle Ordovician Trilobite Biofacies, Southern Mackenzie Mountains (*in* Stelck, C. R., and B. D. E. Chatterton. 1978. Western and Arctic Canadian Biostratigraphy. *CA* 18), p. 3
90. Taylor, M. E. 1973. Biogeographic Significance of Some Cambrian and Ordovician Trilobites from Eastern New York State. *GA* 5:226-7
91. Fortey, R. A. 1976. Correlation of Shelly and Graptolitic Early Ordovician Successions Based on the Sequence in Spitzbergen (*in* Bassett. 1976. *op. cit.*), p. 268
92. Berry, W. B. N. 1972. Early Ordovician bathyurid province lithofacies, their relationship to a proto-Atlantic Ocean and correlations. *LE* 5:73
93. Ross, R. J. and J. K. Ingham. 1970. Distribution of the Toquima-Table Head (Middle Ordovician Whiterock) Faunal Realm, Northern Hemisphere. *GB* 81:394, 8
94. Ormiston, A. R. and R. J. Ross. 1979. *Monorakos* in the Ordovician of Alaska and its Zoogeographic Significance (*in* Boucot, A. J. and J. Gray. 1979. *Historical Biogeography, Plate Tectonics, and the Changing Environment.* Oregon State University Press), p. 53
95. Ross, R. J. 1958. Trilobites in a Pillow-Lava of the Ordovician Valmy Formation, Nevada. *JP* 32:570
96. Ross, R. J. 1970. Ordovician Brachiopods, Trilobites, and Stratigraphy of Eastern and Central Nevada. *UG* 639:2-40
97. Kobayashi, T. and R. Hamada. 1978. Upper Ordovician Trilobites from the Langkawi Islands, Malaysia. *AS* 19:1-29
98. Cocks, L. R. M. and R. A. Fortey. 1982. Faunal evidence for oceanic separation in the Palaeozoic of Britain. *GL* 139:469-73
99. Koren', T. N., *et al.* 1979. New Evidence on Graptolite Successions Across the Ordovician-Silurian Boundary in the Asian Part of the USSR. *PO* 24(1)126
100. Walters, M., Lesperance, P. J., and C. Hubert. 1982. The biostratigraphy of the Nicolet River Formation in Quebec and intra-North American correlations in Middle and Upper Ordovician strata. *CE* 19:571-88
101. Romano, M. 1982. The Ordovician biostratigraphy of Portugal—A Review with new data and re-appraisal. *LI* 17:89-110
102. Nikitin, I. F. 1972. *Ordovik Kazakhstana*. Chast' 1: Stratigrafiia. Izdatelstov Nauka, Kazahskoye SSR, Alma Ata. 239 p. (in Russian)
103. Obut, A. M. 1977. Stratigrafiya i Fauna Ordovika i Silura Chukotskovo Polyostrova. *SB* 351, 222 p. (in Russian)
104. Cooper, B. J. 1981. Early Ordovician Conodonts from the Horn Valley Siltstone, Central Australia. *PA* 24:148
105. Sheng, S. F. 1980. *The Ordovician System in China*. International Union of Geological Sciences Pub. 1, 7 p., charts.
106. Rozman, Kh. S. and Ch. Minzin. 1980. Stratigraphy of the Ordovician Ashgillian Stage of Western Mongolia. *IG* 22:578
107. Kobayashi, T. 1960. The Ordovician of Korea and its Relation to the other Ordovician Territories. 21st *CO*, sec. 7, pp.34-44
108. Sokolov, B. S., *et al.* 1960. Stratigraphy, Correlation and Paleogeography of the Ordovician Deposits of the USSR. 21st *CO*, sec. 7, pp. 4-58
109. Jaanusson, V. 1979. Ordovician. *TP* A139-A155
110. Antsygin, N. YA., Varganov, V. G., and V. A. Nasedkina. 1970. Upper Cambrian and Lower Ordovician of the Orsk District of the Urals. *DE* 193:27-8
111. Severzina, L. G. 1978. Fauna i Biostratigrafiia Verkhnovo Ordovika i Silura Altae-Sayanskoi Oblasti. *SB* 405, p. 4 (in Russian)
112. Kanizin, A. B. 1977. Razrez Ordovika i Silura Reki Moyero. *SB* 303, p. 14 (in Russian)
113. Nikiforova, O. I. 1968. *A Guide to the Geological Excursion on Silurian and Lower Devonian Deposits of Pedolia (Middle Dniestr River).* Leningrad. pp 8-9
114. Melnikova, L. M. 1976. Late Ordovician Ostracods of the Bolshaya Nirunda River. *PJ* 10:459
115. Rozman, Kh. S. 1967. Ordovician Biostratigraphy of the Sette Daban Range (South Verkhoyansk Region). *DE* 184:33
116. Bialy, B. I. Biostratigrafiia Pogranichiyh Otlozhenii Nizhnego i Srednovo Ordovika Na Yugo Sibirskoi Platformy. *SB* 372, p. 64 (in Russian)
117. Nikolskii, F. V. and A. A. Podvesko 1980. Characteristics of the Formation of Deposits of the Krivolutskii Formation of the Middle Ordovician in the Lena-Tunguska Interfluvial Area. *SV* 21:33-4
118. Lofgren, A. 1978. Arenigian and Llandvernian conodonts from Jamtland, northern Sweden. *FS* 13:6
119. Roomusoks, A. 1960. Stratigraphy and Paleogeography of the Ordovician in Estonia. 21st *CO*, sec. 7, p. 60
120. Apollonov, M. K. 1975. Ordovician trilobite assemblages of Kazakhstan. *FS* 4:375-80
121. Owens, R. M. 1973. Ordovician Proetidate from Scandinavia. *NO* 53:150, 175
122. Volkova, K. A., Latpov Yu., i K. R. Wuiznikova. 1978. Ordovik i Silur Yuzhnovo Verkhoyana. *SB* 381 (in Russian)
123. Nazorov, B. B. 1977. A New Radiolarian Family from the Ordovician of Kazakhstan. *PJ* 11:166
124. Kanygin, A. V., Moskalenko, T. A., and A. G. Yadrenkina. 1980. On the Lower to Middle Ordovician Boundary Deposits of the Siberian Platforms. *SV* 21:11
125. Stock, C. W. 1981. *Cliefdenella alaskaensis* N.S.P. (Stromatoporoidea) from the Middle/Upper Ordovician of Central Alaska. *JP* 55:998
126. Stouge, S. and J. S. Peel. 1979. Ordovician conodonts from the Precambrian Shield of southern West Greenland. *UN* 91:106
127. Robison, R. A. and J. Pantoja-Alos. 1968. Tremadocian Trilobites from the Nochixtlan region, Oaxaca, Mexico. *JP* 42:768-9
128. Melnikova. 1979. Some Early Ordovician Ostracodes of the Southern Urals. *PJ* 13:71
129. Flower, R. H. 1979. A New raedemannocerid cephalopod from the Ordovician of western North Greenland. *UN* 91:93
130. Ermikov, V. D., *et al.* 1979. The Tremadocian of Northern Gornyi Altai. *SV* 20:15-25
131. Li-wen, X. and L. Cai-gen. 1981. The Cambrian-Ordovician Boundary in China (*in* Taylor, M. E., ed. 1981. *op. cit.*), p. 242-3
132. Acenolaza, F. G. 1976. The Ordovician System in Argentina and Bolivia (*in* Bassett. 1976. *op. cit.*), p. 481
133. Chugaeva, M. N. 1976. Ordovician in the North-Eastern USSR (*in* Bassett. 1976. *op. cit.*), p. 286-7
134. Benedetto, J. L. 1977. Una Nueva Fauna De Trilobites Tremadocianos de La Provincia De Jujuy (Sierra de Cajas), Argentina. *ME* 14:186
135. Koroleva, M. N. 1978. New Ordovician Harpidae (Trilobita) of North Kazakhstan. *PJ* 12:215
136. Baldis, B. A. and G. Blasco. 1974. Trilobites Ordovicos de la Comarca de Jachal Precordillera Argentino. *ME* 11:71
137. Bulman, O. M. B. 1964. Lower Palaeozoic plankton. *GL* 120:470-3
138. Bulman, O. M. B. 1971. Graptolite faunal distribution (*in* Middlemiss, *et al.* 1971. *op. cit.*), pp. 49-56
139. Nelson, S. 1963. Ordovician Paleontology of the Northern Hudson Bay Lowland. *AM* 90, pp. 63-80
140. Pomerol, Ch. and C. I. Babin. 1977. *Stratigraphie et Paleogeographie Precambrian et Paleozoique.* DOIN Editeiurs, Paris, pp. 201-440
141. Kilpatrick, D. J. and P. D. Fleming. 1980. Lower Ordovician sediments in the Wagga Trough: discovery of early Bendigonian graptolites near Eskdale, north Victoria. *GU*
142. Twenhofel, W. H. ed. 1954. Correlation of the Ordovician Formations of North America. *GB* 65:247-98
143. Bergstrom, S. M. and R. A. Cooper. 1973. *Didymograptus bifidus* and the trans-Atlantic correlation. *LE* 6:331

144. Jiantino, Yu. and F. Yiting. 1981. *Arienigraptus*, A New Graptolite Genus from the Ningkuo Formation (Lower Ordovician) of South China. SI 20:32
145. Jackson, D. E. 1969. Ordovician Graptolites in Lands Bordering North Atlantic and Arctic Oceans (*in* Kay, M. 1969. *op. cit.*), p. 505
146. Acenolaza, F. G., Gorustovich, S., and J. Solis. 1976. El Ordovicio Del Rio La Alumbierra, Departmento Tinogasta, Provincia de Catamarca. ME 13:269
147. Oradovaskaya, M. M. 1970. Ordovician and Silurian Stratigraphy of the Chukotska Peninsula. DE 191:36
148. Rosen, R. N. 1979. Permo-Triassic Boundary of Fars-Persian Gulf area of Iran. JP 53:92
149. Shiding, J. 1981. Some New Graptolites from the Ningkuo Formation (Lower Ordovician) of Zhejiang. SI 20:69
150. Skevington, D. 1974. Controls Influencing the Composition and Distribution of Ordovician Faunal Provinces (*in* Rickards, R. B., Jackson, D. E., and C. P. Hughes. 1974. *Graptolite Studies in Honour of O. M. B. Bulman.* PL 13), p. 62
151. Moskalenko, T. A. 1972. Conodonts from Llandoverian Deposits of the Siberian Platform. DE 204:236
152. Peterson, M. S., Rigby, J. K., and L. F. Hintze. 1980. *Historical geology of North America*, 2nd edition. Brown Publ. Co., Dubuque, Iowa, p. 57
153. Jinding, L., *et al.* 1980. Discovery of Ordovician Graptolites from Wufeng of Yougan, Fujian Province. SI 19 (6)512
154. Yao-kun, L. 1980. *Ordosograptus* — A New Graptolite Genus and Its Affinities. SI 19(6)479
155. Riva, J. 1974. Late Ordovician Spinose Climacograptids from the Pacific and Atlantic Faunal Provinces (*in* Rickards, *et al.* 1974. *op. cit.*), p. 109
156. Miller, H. 1979. Das Gundegebirge der Anden im Choros-Archipel, Region Aisen, Chile. RU 68:449
157. Sobolevaskaya, R. F. 1971. New Ordovician Graptolites of the Omulev Mountains. PJ 5:76-9
158. Kobayashi, T. 1968. Stratigraphy of the Chosen Group in Korea and South Manchuria. TO, sec. 2, Vol. 17:260
159. Gang, W. 1981. On the Discovery of New Graptolites from the Tungtze Formation (Lower Ordovician) in Gulin of Sichuan. SI 20:351
160. Alikhova, T. N. 1976. Principal problems of the stratigraphy of the Ordovician system. IG 18:896-9
161. Degarden, J. M. 1979. Decouverte du genre *Phyllograptus* (Graptolites) dans l' Ordovicien des Pyrenees Atlantiques Consequences Stratigraphiques. BI 12(2)321
162. Jackson, D. E. 1966. Graptolitic Facies of the Canadian Cordillera and Arctic Archipelago: A Review. CP 14:470
163. Jaanusson, V. 1973. Ordovician Articulate Brachiopods. (*in* Hallam. 1973. *op. cit.*), pp. 20-25
164. Williams, A. 1973. Distribution of Brachiopod Assemblages in Relation to Ordovician Palaeogeography (*in* Hughes, N. F. 1973. *Organisms and Continents Through Time.* PL 12), p. 178, 182
165. Potter, A. W. and A. J. Boucot. 1971. Ashgillian Late Ordovician Brachiopods from the Eastern Klamath Mountains of Northern California. GA 3:180-1
166. Rozman, Kh. S., *et al.* 1970. Biostratigrafiia Vyerkhnovo Ordovika Severo-Vostoka CCCP. UK 205:246-7
167. Hill, D., Playford, G., and J. T. Woods. 1969. *Ordovician and Silurian of Queensland.* Queensland Palaeontographical Society, Brisbane, p. 2
168. Biernat, G. 1973. Ordovician Inarticulate Brachiopods from Poland and Estonia. OL 28:16
169. Barnes, C. R., Rexroad, C. B., and J. F. Miller. 1972. Lower Paleozoic Conodont Provincialism. GP 141:167-73
170. Bergstrom, S. M. 1973. Ordovician Conodonts (*in* Hallam. 1973. *op. cit.*), p. 50-4
171. Lindstrom, M. 1976. Conodont Palaeogeography of the Ordovician (*in* Bassett, M. G. ed. 1976. *op. cit.*), pp. 503-17
172. Votaw, R. 1979. Upper Ordovician conodonts from the Upper Peninsula of Michigan. GA 11(5)259
173. Bergstrom, S. M. 1977. Early Paleozoic Conodont Biostratigraphy in the Atlantic Borderlands (*in* F. M. Swain. 1977. *Stratigraphic Micropaleontology of Atlantic Basin and Borderlands.* EL), p. 89
174. Bergstrom, S. M. 1981. Personal Communication.
175. Bellini, F. and D. Massa. 1980. A Stratigraphic Contribution to the Palaeozoic of the Southern Basins of Libya (*in* Salem, M J. and M. T. Busreevil. 1980. *The Geology of Libya.* Academic Press, London, NY), p.3-on.
176. Abaimova, G. P. 1971. New Early Ordovician Conodonts from the Southeastern Part of the Siberian Platform. PJ 5:486-90
177. Gastil, R. G. and R. H. Miller. 1981. Lower Palaeozoic strata on the Pacific Plate of North America. NA 292:828
178. Gerasimova, N. A., Dubinina, S. V., and N. I. Zardrashvile. 1977. Age of the Siliceous Elastic Complex of the Atasu Anticlinorium, Central Kazakhstan. DE 235:39-41
179. Kurtz, V. F. and J. F. Miller. 1981. Early Ordovician Conodont Faunas from Central East Greenland. GA 13 (6)285
180. ———. 1978. *The Geology of New Zealand.* New Zealand Geological Survey. Vol. I, p. 66
181. Gnoli, M. and E. Serpagli. 1980. The Problematic Microorganisms *Nivia* in the Lower Ordovician of Precordilleran Argentina and its Paleogeographic Significance. JP 54:1246
182. Talent, J. A. 1981. Palaeontology and Stratigraphy in India: Retrospect and Prospects. IN 22:454
183. Moskalenko, T. A. 1976. Environmental Effects on the Distribution of Ordovician conodonts of the western Siberian Platform (*in* Barnes, C. R. 1976. *Conodont Paleoecology.* CA 15), p. 60
184. Tipnis, R. S., Chatterton, B. D. E., and R. Ludvigsen. 1978. Ordovician Conodont Biostratigraphy of the Southern District of Mackenzie, Canada (*in* Stelck and Chatterton. 1978. *op. cit.*), p. 41
185. Moskalenko, T. A. 1976. Unique Conodontophorid Finds in the Ordovician Deposits of the Irkutsk Amphitheater. DE 229:232
186. Abaimova, G. P. and E. P. Markov. 1977. Pyerviye Nahodki Conodontov Nizhneordoviskich zony *Kordylodus* na yuge Sibirskoi Platformy. SB 372:87 (in Russian)
187. Novikova, M. Z., Ryazantsev, A. V., and S. V. Dubinina. 1978. Age of the Akdym Series of the Yerementau-Niyaz Anticlinorium, Central Kazakhstan. DE 241:57
188. Nasedkina, V. A. and V. N. Puchkov. 1972. Srednoordovikskie Konodonty Severa Urala i ikh stratigraficheskoe Znacheniye. UK (*Uralskiy tsentr.*), vip. 145:5-24 (in Russian)
189. Kushnareva, T. I. and N. B. Rasskazowa. 1978. The Ordovician of the Pechora syneclise. IG 20:700-1
190. Crick, R. E. 1978. *Ordovician Nautiloid Biogeography: A Probabalistic Multivariate Analysis.* Unpublished Ph.D. Thesis, University of Rochester, p. 12
191. Crick, R. E. 1980. Integration of paleobiogeography and Paleogeography: Evidence from Arenigian Nautiloid Biogeography. JP 54:1224
192. Dun-lin, Q. 1980. Ordovician Cephalopods from Wuwei of Anhui and their Stratigraphical Significance. SI 18 (4)260
193. Ross, R. J. 1965. Early Ordovician trilobites from the Seward Peninsula, Alaska. JP 39
194. Xi-ping. 1981. Early Ordovician Nautiloids from Quingshuihe, Nei Monggol (Inner Mongolia) and Pianguan, Shanxi Province. SI 20(4)361
195. Stait, B. 1980. *Gouldoceras* N. Gin (Cephalopoda, Nautiloidea) and a Revision of *Hecatoceras*, Teichert and Glenister, from the Ordovician of Tasmania, Australia. JP 54:1114
196. Zhong-fa, L. 1981. Ordovician Cephalopods from Hunjiang Region of Jilin and Northern Neimongol. SI 20 (5)398
197. Miller, A. K., Youngquist, W., and C. Collinson. 1954. Ordovician Cephalopod Fauna of Baffin Island. AM 62:4-16
198. Crick, R. E. 1981. Diversity and evolutionary rates of Cambro-Ordovician nautiloids. PB 7:225
199. Flower, R. H. 1976. Ordovician Cephalopod Faunas and their Role in Correlation (*in* Bassett. 1976. *op. cit.*), pp. 529-42
200. Antze, Mu., *et al.* 1973. Stratigraphy of the Mount Jolmo Lungma Region in Southern Tibet, China. SN 16(1) 96-on.
201. Zhuravleva, F. A. 1978. Some Mongolian Early and Middle Paleozoic Cephalopods. PJ 12:485
202. Flower, R. H. 1963. New Ordovician Ascocerida. JP 37:69-82
203. Sheehan, P. M. 1979. Swedish Late Ordovician Marine

204. Paul, C. R. C. 1976. Palaeogeography of Primitive Echinoderms in the Ordovician (*in* Bassett. 1976. *op. cit.*), pp. 564-71
205. Derstler, K. 1979. Biogeography of the Stylophoran Carpoids (Echinodermata) (*in* Boucot and Gray. 1979. *op. cit.*), pp. 100-2
206. Rozhkov. C. V. 1981. Morskiye Lilii Nadsemeistva Pisocrinacea. *TR* 192:47 (in Russian)
207. Jobson, L. and C. R. C. Peel. 1979. *Compagiginus fenestratus*, a new Lower Ordovician inadunate crinoid from North Greenland. *UN* 91:72
208. Witzke, B. J., Frest, T. J., and H. L. Strimple. 1979. Biogeography of the Silurian-Lower Devonian Echinoderms (*in* Boucot and Gray. 1979. *op. cit.*), pp. 125-6
209. Berry, W. B. N. and A. J. Boucot. 1973. Correlation of the African Silurian Rocks. *GP* 147 all pages, maps.
210. Berry, W. B. N. and A. J. Boucot. 1972. Correlation of the South American Silurian Rocks. *GP* 133, 59 p., maps.
211. Berry, W. B. N. and A. J. Boucot. 1970 Correlation of the North American Silurian Rocks. *GP* 102, 289 p., maps.
212. Rust, F. C. 1981. Lower Palaeozoic Rocks of Southern Africa (*in* Holland. 1981. *op. cit.*), p. 181
213. Ziegler, A. M., et al. 1977. Silurian Continental Distributions, Paleogeography, Climatology, and Biogeography. *TE* 40:39
214. Ivanovskii, A. B. and N. P. Kul'kov. 1975. Biogeographic Zoning of Silurian Deposits. *SV* 17:36-9
215. Yu, W., Jia-yu, R., and Y. Xue-chang. 1980. The Genus *Atrypoidea* (Brachiopoda) of Southwest China and Its Stratigraphical Significance. *SI* 19(2)103
216. Cocks, L. R. M. and W. S. McKerrow. 1973. Brachiopod Distributions and Faunal Provinces in the Silurian and Lower Devonian (*in* Hughes. 1973. *op. cit.*), 295-9
217. Minato, M. and M. Kato. 1965. Waagenophyllidate. *HO* 12:31
218. Wang, Yu., et al. 1981. Stratigraphic distribution of brachiopoda in China (*in* Teichert, et al. 1981. *op. cit.*), p. 99
219. Baoyu, L. 1980. *The Silurian System of China*. Institute of Geology—Chinese Academy of Geological Science, Peking, pp. 2-5
220. Ziegler, A. M., Rickards, R. B., and W. S. McKerrow. 1974. Correlation of the Silurian Rocks of the British Isles. *GP* 154:9-13
221. Berry, W. B. N. and A. J. Boucot. 1972. Correlation of the Southeast Asian and Near Eastern Silurian Rocks. *GP* 137:14-61, maps.
222. Copper, P. 1977. The late Silurian brachiopod genus *Atrypoidea*. *Geologiska Foreningens i Stockholm Forhandlingar*, 99:11
223. Zhengsu, Y. 1981. On the Geological and Geographic Distribution of *Tuvaella* with Reference to its Habitat. *SI* 20(6)570-1
224. Jia-yu, R. and Z. Zi-xin. 1982. A Southward extension of the Silurian Tuvaella brachiopod fauna. *LE* 15:142
225. Cocks, L. R. M. 1972. The Origin of the Silurian *Clarkeia* Shelly Fauna of South America and its Extension to West Africa. *PA* 15:625
226. Shevchenko, V. I., et al. 1977. New Data on Silurian sediments of the Volga Region near Volgograd. *DE* 233:125
227. Lanbacher, G., Boucot, A. J., and J. Gray. 1982. Additions to Silurian Stratigraphy, Lithofacies, Biogeography, and Paleontology of Bolivia and Southern Peru. *JP* 56:1138-70
228. Kaljo, D. L. 1978. On the Bathymetric Distribution of Graptolites. *PO* 23(4)524
229. Berry, W. B. N. 1973. Silurian-Early Devonian Graptolites (*in* Hallam. 1973. *op. cit.*), p. 83
230. Bordet, P., Colchen, M., and L. Fort. 1972. Some Features of the Geology of the Anapuina Range, Nepal Himalaya (*in* Jhingran, A. G. and K. S. Paldiya. 1973. *Himalayan Geology*, Delhi)
231. Lenz, A. C. 1979. Llandoverian Graptolite Zonation in the Northern Canadian Cordillera. *PO* 24:139
232. Furon, R. 1972. *Elements de Paleoclimatologie*. Librairie Vuibert, Paris, p. 92
233. Koren', T. N. 1973. The Silurian and Lower Devonian Graptolite-bearing strata in the USSR (a review). *GM* 110:3
234. Berry, W. B. N. and V. J. Gupta. 1966. Monograptids from the Kashmir Himalayas. *JP* 40:1339
235. Jaeger, H. 1979. Devonian Graptolithina (*in* House, M. R., Scruton, C. T., and M. G. Bassett. 1979. *The Devonian System. PL* 23), p. 337
236. Amatov, V. A., et al. 1970. *Basic Features of the Paleozoic Stratigraphy and tectonics of the Mongolian Peoples Republic*. Peoples Republic Translation, Vol. 1, p. 31-2
237. Tarlo, L. B. H. 1964. Psammosteiformes (agnatha)—A Review with descriptions of New Material From the Lower Devonian of Poland. *OL* 13:10
238. Halstead, L. B. and S. Turner. 1973. Silurian and Devonian Ostracoderms (*in* Hallam. 1973. *op. cit.*), pp. 68-75
239. Turner, S., Jones, P. J., and J. J. Draper. 1981. Early Devonian thelodonts (agnatha) from the Toko Syncline, western Queensland, and a review of other Australian Discoveries. *BM* 6:52
240. Tankard, A. J., et al. 1982. *Crustal Evolution of Southern Africa. SP*, pp. 353-6
241. Koemmelbein, K. 1968. Devonian of the Amazonas Basin, (*in* Oswald, D. H. 1968. *International Symposium on the Devonian System*, Calgary) Vol. 2, p. 201-8
242. Cross, W. 1973. Kleinschuppen, Flossenstacheln, und Zaline om Fischen Aus Europaischen und Nordamerik aurscer Bonebeds Des Devons. *PH* A142:51-155
243. Gupta, V. J. and Ph. Janvier. 1979. Late Devonian Vertebrate Remains from Western Himalayas (Himachal Pradesh, India). *ND* 12:161
244. Gupta, V. J. and P. Janvier. 1981. Remarks on an Osteolipid Fish from the Devonian of Zanskar, Ladakh. *ND* 14:80
245. Blieck, A. 1982. Les Grandes Signes De La Biographie des Heterostraces du Silurien Superieur-Devonien Inferieur dans le domane nord-Atlantique. *PP* 38:286
246. Kiang, P. 1981. Devonian Antiarch Biostratigraphy of China. *GM* 118:70
247. Guscherkov, V. A. 1968. A Find of Placoderms in the Northern Tien Shan. *DE* 179:53-4
248. Denison, R. H. 1968. Early Devonian Lungfishes from Wyoming, Utah, and Idaho. *FI* 17(4)353-413
249. Janvier, P. 1977. Descriptions des Restes D'Elasmobrances (Pisces) Du Devonien Moyen de Bolivia. *PV* 7-IV, pp. 127-32
250. Gupta, V. J. and S. Turner. 1973. Oldest Indian Fish. *GM* 110:483
251. Gupta, V. J. and R. A. Denison. 1966. Devonian Fishes from Kashmir, India. *NA* 211:177
252. Tikhiy, V. N. and M. S. Stanichkova. 1973. Age of the Kazanla Suite and the beginning of Devonian Sedimentation on the Russian Platform. *DE* 210:94
253. Blieck, H., et al. 1980. A New Vertebrate locality in the Eifelian of the Khush-Yeilagh Formation, Eastern Alborz, Iran. *PV* 9-V:1-4
254. Bernacsek, G. M. 1977. A Lungfish Cranium from the Middle Devonian of the Yukon Territory, Canada. *PH* A157:176
255. Gaikusha, M. P., Yengoyoan, M. A., Oganesyan, D. Va., and S. S. Sakrasyan. 1971 A Find of Fish Remains in the Upper Devonian of Armenia *DE* 196:77
256. Janvier, P. 1980. Osteolipid Remains from the Middle East (*in* Panchen, A. L. 1980. *The Terrestrial Environment and The Origin of Land Vertebrates. AC*), p. 224
257. Schrank, E. 1977. Zur Palaeobiogeographie silurischer Trilobiten. *JA* 155(1)110-23
258. Chlupac, I. 1975. The distribution of phacopid trilobites in space and time. *FS* 4:400-5
259. Eldredge, N. and A. R. Ormiston. 1979. Biogeography of Silurian and Devonian Trilobites of the Malvinokaffric Realm (*in* Boucot and Gray. 1979. *op. cit.*), pp. 147-69
260. Baldis, B. A., et al. 1976. Trilobites Silurico-Devonicos de la Sierra De Zapla (Nordeste de Argentina). *ME* 13:185
261. Talent, J. A., et al. 1975. Correlation of the Silurian Rocks of Australia, New Zealand, and New Guinea. *GP* 150, 108 p.
262. Li-wen, X. 1981. Some Late Devonian trilobites of China (*in* Teichert. 1981. *op. cit.*), p. 183
263. Ormiston, A. R. 1967. Lower and Middle Devonian Trilobites of the Canadian Arctic Island. *CB* 153:8

264. Kobayashi, T. and T. Hamada. 1974. Silurian Trilobites of Japan and Adjacent Regions. *PJ* 18, 155 p.
265. Kobayashi, T. and T. Hamada. 1977. Devonian Trilobites of Japan and Adjacent Regions. *PJ* 20, 202 p.
266. Ormiston, A. R. 1975. Siegenian trilobite zoogeography in Arctic North America. *FS* 4:391
267. Alberti, G. K. B. von. 1969. Trilobiten des jungeren Siluriums sowie des Unter und Mitteldevons. *Abhandlungen der senckenbergischen Naturforschenden Gesselschaft* 520:34
268. Lane, P. D. and A. J. Thomas. 1978. Silurian trilobites from Northeast Queensland and the classification of effaced trilobites. *GM* 115:351
269. Ormiston, A. R. 1977. Trilobites and the Silurian-Devonian Boundary (*in* Martinsson, A. 1977. *The Silurian-Devonian Boundary*. E. Schwiezerbeutsche Verlagsbuchhandblung, Stuttgart), p. 324
270. Menner, Vl. V., Krylova, A. K., Kolodeznikov, K. Ye., and G. S. Fradkin. 1970 Correlation of the Middle Devonian of the Siberian Platform *DE* 193:106-9
271. Anderson, M. M., Boucot, A. J., and J. G. Johnson. 1969. Eifelian Brachiopods from Padaukpin, northern Shan States, Burma. *BR* 18(4)109
272. Shih-pu, Y., Kiang, P., and H. Hung-fei. 1981. The Devonian System in China. *GM* 118:113-38
273. Talent, J. A. 1972. Provincialism and Early Devonian Faunas. *GU* 19:88
274. Durkoop, A., Mensink, H., and G. Plodowski. 1968. Devonian of Central and Western Afghanistan and Southern Iran (*in* Oswald. 1968. *op. cit.*), V1, pp. 540-3
275. Alberti, G. K. B. von. 1981. Zur Biostratigraphie und Fauna (Tentaculiten, Trilobiten, Graptolithen) des Unter-, und Mittel-Devons von Benzueg (Becken von Bechar, SW-Algerien) *JA Monatschefte* 1981(11)643
276. Edwards, D. 1973. Devonian Floras (*in* Hallam. 1973. *op. cit.*), pp. 106-10
277. Chaloner, W. G., Mensah, M. K., and M. D. Crane. 1974. Non-Vascular Land Plants from the Devonian of Ghana. *PA* 17:933
278. Alvarez, R. C. 1981. Devonian Plants from Hornachos (Badajoz), Spain. *BO* 28:15
279. Zdebska, D. 1982. A New Zosterophyll from the Lower Devonian of Poland. *PA* 25:247-63
280. Chaloner, W. G. and Z. Sheerin. 1979. Devonian Macrofloras (*in* House, et al. 1979. *op. cit.*), p. 159
281. Lejal-Nicol. 1975. Sur Une Nouvelle Flora A Lycophytes De Devonien Inferieur de la Libye. *PH* B151:53
282. Hueber, F. M. 1971. Early Devonian Land Plants from Bathurst Island, District of Franklin. *Kansas Geological Survey Paper* 17-28, p. 1
283. Irwina, A. et Y. Lemoigne. 1979. Sur la Presence du *Callixylon newberryi* (Dawson) Elkins et. Wieland 1814 en Kazakhstan (URSS) au Devonien Superieur. *PH* 170B:1
284. Andrews, H. N., Gensel, P. G., and W. H. Forbes. 1974. An Apparently Heterosporous Plant from the Middle Devonian of New Brunswick. *PA* 17:338
285. Lacey, W. S. 1975. Some Problems of "Mixed" Floras in the Permian of Gondwanaland (*in* Campbell, K. S. W. 1975. *Gondwana Geology*, Canberra), p. 129
286. Churkin, M., Jr., et al. 1969. Lower Devonian Land Plants from Graptolitic Shale in South-Eastern Alaska. *PA* 12:560
287. Zahkarova, T. V. 1981. On the Systematic Position of the Species "Psilophyton" goldschmidti from the Lower Devonian of Eurasia. *PJ* 15(3)109-on.
288. Grierson, J. D. 1976. Sedersia Complexa (Lycopsida, Middle Devonian) Its Anatomy, and the Interpretation of Pyrite Petrifactions. *American Journal of Botany* 63:1187
289. Cornet, B., Phillips, T. L., and H. N. Andrews. 1976. The Morphology and Variation in *Rhacophyton Ceratangium* from the Upper Devonian and Its Bearing on Frond Evolution. *PH* 158B:107-21
290. House, M. R. 1973. Devonian Goniatites (*in* Hallam. 1973. *op. cit.*), p. 101-2
291. House, M. R. 1980. Early Ammonoids in Space and Time (*in* House, M. R. and J. R. Senior. 1980. *The Ammonoids. AC*), p. 365
292. House, M. R. 1982. Written Communication.
293. Bogoslovsky, B. I. 1980. Early Devonian Ammonoids of the Zeravshan Range. *PJ* 14:54
294. Ameta, S. S. and K. Gaur. 1980. New Fossil Find from the Muth Quartzite, Pen Valley, Lahaul and Spiti District, Himachal, India. *ND* 13(1)73
295. House, M. R. 1978. Devonian ammonoids from the Appalachians and their bearing on International zonation and Correlation (*in* House. 1978. *op. cit.*), p. 30
296. House, M. R. and R. B. Blodgett 1982 The Devonian goniatite genera *Pinacites* and *Foodites* from Alaska. *CE* 19:1873
297. Yuen, Weng C. and W. Zigler. 1981. Middle Devonian conodonts from Xequitu Q1., Inner Mongolia Autonomous Region, China. *SL* 62:127
298. Oliver, W. A. 1976. Presidential Address: Biogeography of Devonian Rugose Corals. *JP* 50(3)369-71
299. Oliver, W. A. 1977. Biogeography of Late Silurian and Devonian Rugose Corals. *PP* 22:85-135
300. Chang-Ming, Y. and K. Guo-dun. 1980. Rugose Corals from Devonian Ertang Formation of Central Guangxi. *SI* 19(3)180
301. Hill, D. 1957. Presidential Address: The Sequence and Distribution of Upper Palaeozoic Coral Faunas. *Australian Journal of Science* 119(3A)42-61
302. Sharkova, T. T. 1980. Rifogennie Postroiki Rannevo Devona Yuzhnoi Mongolii (*in* Sokolov, B. C. 1980. *Koraly i Rifi Fanerozoi CCCP*, Nauka, Moskva), p. 93 (in Russian)
303. Pedder, A. E. H. 1982. *Chostophyllum*, A New Genus of Charactophyllid Corals from the Middle Devonian of Western Canada. *JP* 56:579-80
304. Flugel, E. 1975. Fossile Hydrozoen-Kenntnisstand und Probleme. *PZ* 49(4)385-9
305. Minato, M., et al. 1965. *The Geologic Development of Japanese Islands*, pp. 39-84
306. Oliver, W. A. 1980. Corals in the Malvinokaffric Realm. *Munsterche Forschungen Zur Geologie und Palaeontologie*, Heft 52:17
307. Boucot, A. J., Johnson, J. G., and J. A. Talent. 1969. Early Devonian Brachiopod Zoogeography. *GP* 119, pp. 14-23
308. Boucot, A. J., Johnson, J. G., and W. Struve. 1966. *Stringocephalus*, Ontogeny and Distribution. *JP* 40:1358
309. Hamada, T. 1971. Early Devonian Brachiopods from the Lesser Khingan District of Northeast China. *SJ* 15:4
310. Sartenaer, P. 1969. Late Upper Devonian (Famenian) Rhynchonellid Brachiopods from Western Canada. *CB* 169:12
311. Tyazheva, A. P. and P. A. Zhavoronkove. 1972. *Korali i Brakiopody Pogranichnich Otlozhenii Siluria i Nizhnevo Devona Zapadnovo Sklona Yuzhnovo Urala*. Nauka, Moskva, p. 3 (in Russian)
312. Walmsley, V. G. and A. J. Boucot. 1975. The Phylogeny, Taxonomy, and Biogeography of Silurian to Mid Devonian Isorthinae (Brachiopoda). *PH* 148A:49
313. Thanh, T. D. 1980. The Stratigraphy of the Devonian Deposits in Vietnam. *SV* 17:34-5
314. Yan, Z. 1981. Early Devonian Brachiopods from Zhusilghairhan Reg. Western Neumongol. *SI* 20(5)391
315. Boucot, A. J., Isaacson, P. E., and G. Leubacher. 1980. An Early Devonian Eastern Americas Realm Faunule From the Coast of Southern Peru. *JP* 54:361
316. Gatsianova, P. T. 1975. Brakiopody Rannevo i Srednovo Devona Altae-Sayanskoi Oblasti. *SB* 248:6 (in Russian)
317. Brice, D. 1977. Biostratigraphie du Devonien d'Afghanistan. *FR* 8:267-76
318. Garcia-Alcalde, J. L. and P. R. Racheboeuf. 1975. Donnes paleobiologiques Strophochonetinae de Devonien et palaeobiogeographiques sur quelques d'Espagne et du Massif Armorican. *LE* 8(4)337
319. Hodson, F. and W. H. C. Ramsbottom. 1973. The Distribution of Lower Carboniferous Goniatite Faunas in Relation to Suggested Continental Reconstructions for the Period (*in* Hughes. 1973. *op. cit.*), p. 323
320. Rocha-Campos, A. C., De Carvalho, R. G., and A. J. Amos. 1977. A Carboniferous (Gondwana) Fauna from Subandean Bolivia. *RB* 7:291
321. Closs, D. 1967. Orthocone Cephalopods from the Upper Carboniferous of Argentina and Uruguay. *ME* 5(3)123-on.
322. Ruzina, L. F. 1980. Sayoorskiye Ammonoidei. *TR* 181, Ris. 1 (in Russian)

323. Wang, M. 1981. Carboniferous Ammonoids from Eastern Xinjiang. SI 20(5)480
324. Gandin, V. G. 1971. The First Continuous Upper Paleozoic Sequence Found in the Northeastern USSR. DE 200:35
325. Amos, A. J. 1979. *Guia Paleontologica Argentina*, Buenos Aires, p. 13-on.
326. Gless, M. J. M., et al. 1980. Pre-Permian Depositional Environments around the Brabant Massif in Belgium, the Netherlands, and Germany. SG 27(1)69
327. Ruzhencev, Ye. 1974. Late Carboniferous of the Russian platform and Cisuralia. PJ 8(3)313-22
328. Higgins, A. C. and C. H. T. Wagner-Gentries. 1982. Conodonts, Goniatites, and the Biostratigraphy of the earlier Carboniferous from the Cantabrian Mountains, Spain. PA 25:313-50
329. Kuzina, L. F. 1971. New and Little Known Early Visean (Silurian) Ammonoids. PJ 5:34-8
330. Thomas, D. H. 1928. An Upper Carboniferous fauna from the Amotape Mountains, north western Peru. GM 65:146-52
331. Ruzhentsev, V. Ye. and M. F. Gogoslovskaya. 1975. The Family Reticuloceratidae and Related Taxa. PJ 9(1)49-58
332. Sturgeon, M. T., Windle, D. L., Mapes, R. H., and R. D. Hoare. 1982. New and Revised Taxa of Pennsylvanian Cephalopods in Ohio and West Virginia. JP 56:1453-79
333. Nassichuk, W. W. 1975. Carboniferous Ammonoids and Stratigraphy of the Canadian Arctic Archipelago. CB 237:3
334. Semichatova, S. V., et al. 1979. The Bashkirian Stage as a Global Stratigraphic Unit (in Wagner, R. H., Higgins, A. C., and S. V. Meyen. 1979. *The Carboniferous of the USSR.* YG Occasional Publication 4), p. 106
335. Martin, H., Walliser, O. H., and N. Wilczewski. 1970. A Goniatite from the Glaciomarine Dwyka Beds near Schlip, South West Africa (in _____. 1970. *2nd Gondwana Symposium Proceedings*), p. 625
336. Riccardi, A. C. and N. Sabattini. 1975. Cephalopoda from the Carboniferous of Argentina. PA 18:119
337. Ruzhentsev, B. E. and M. F. Bogushevski. 1978. Namyoorski Etap i Evolutsyi Ammonoidey. TR 167:22-6 (in Russian)
338. Brown, D. A., K. S. W. Campbell, and J. Roberts. 1964. A Visean Cephalopod Fauna from New South Wales. PA 7:682-3
339. Solonina, R. V. and A. A. Gieke. 1977. New Data on the Upper Carboniferous in the Northern Kharaulakh Area Soviet Arctic Maritime Region). SV 18:11-on
340. Miller, A. K. and H. F. Garner. 1953. Upper Carboniferous Goniatites from Argentina. JP 27:823
341. Runnegar, B. 1979 Marine Fossil Invertebrates of Gondwanaland: Palaeogeographic Implications (in _____. 1979. *IV International Gondwana Symposium*, 1977, Calcutta, India), pp. 155-6
342. Ross, C. A. 1973. Carboniferous Foraminiferida (in Hallam. 973. op. cit.), p. 127
343. Sando, W. J., Mamet, B. L., and J T Dutro 1969. Carboniferous Megafaunal and Microfaunal Zonation in the Northern Cordillera of the United States. UG 613E, p. E2
344. Ozawa, T. 1976. Late Visean *Eostafella* (Fusulininean Foram) from West Malaysia. AS 17:120
345. Li, Land, G. Feng. 1980. Late Carboniferous brachiopods from Yanji of Jilin, Northeast China. SI 17(6)490
346. Metcalfe, I. 1980. Palaeontology and Age of the Panching Limestone, Pahang, West Malaysia. AS 21:13
347. Rogozov, Yu. G., et al. 1971. Moscovian Stage of North-Central Chukotka. DE 197:56
348. Thompson, M. L. 1967. American Fusulinacean Faunas Containing Elements from other Continents (in Teichert, C. and E. L. Yochelson. 1967. *Essays in Paleontology and Stratigraphy.* University of Kansas Press.), p. 107
349. Ross, A. 1967. Development of Fusulinid (Foraminiferida) Faunal Realms. JP 41(6)1342-3
350. Douglas, R. C. 1977. The Development of Fusulinid Biostratigraphy. (in Kauffman and Hazel. 1977. op. cit.), p. 473
351. Meyen, S. V. 1970. On the Origin and Relationship of the Main Carboniferous and Permian Floras (in _____. 1970. op. cit.), p. 553
352. Chaloner, W. G. and S. V. Meyen. 1973 Carboniferous and Permian Floras of the Northern Continents (in Hallam 1973. op. cit.), pp. 170-8
353. Plumstead, E. P. 1973. The Late Palaeozoic *Glossopteris* Flora (in Hallam. 1973. op. cit.), p. 192
354. Ageev, K. S., et al. 1981. Permian Deposits in Severnaya Zemlya SV 22(3)135-8
355. Ziegler, A. M., et al. 1981. Paleozoic Biogeography and Climatology (in Niklas, K. J. 1981. *Paleobotany, Paleoecology, and Evolution.* Praeger Publishers, NY); Vol. 2, pp. 236-59
356. Asama, K. 1976. *Gigantopteris* flora in Southeast Asia and its Phytopaleogeographic Significance. AS 17:204
357. Jennings, J. R. 1980. Fossil Plants from the Fountain Formation (Pennsylvanian) of Colorado. JP 54:149
358. El-Khayal, A. A., Chaloner, W. G., and C. R. Hill. 1980. Palaeozoic Plants from Saudi Arabia. NA 285:33
359. Lele, K. M. 1974. Late Palaeozoic and Triassic Floras of India and Their Relation to the Floras of Northern and Southern Hemispheres. BO 23(2)106
360. Gorelova, S. G. 1978. The Flora and Stratigraphy of the Coal-Bearing Carboniferous of Middle Siberia. PH 165B:54
361. Pfefferkorn, A. W. and W. H. Gillespie. 1980. Biostratigraphy and Biogeography of Plant Compression Fossils in the Pennsylvanian of North America (in Dilcher, D. L. and T. N. Taylor. 1980. *Biostratigraphy of Fossil Plants.* DO), p. 95
362. Phillips, T. L. 1980. Stratigraphic and Geographic Occurrences of Permineralized Coal—Swamp Plants—Upper Carboniferous of North America and Europe (in Dilcher and Taylor. 1980. ibid.), pp. 32.3
363. Hsu, J. 1976. On the Palaeobotanical Evidence for Continental Drift and Himalayan Uplift. BO 25:135
364. Meyerhoff, A. A. 1978. Petroleum in Tibet and the India-Asia Suture(?) Zone. PE 1:110
365. Ting-Yang, H. Ya. 1954. Climate and Relative Positions of the Continents During the Lower Carboniferous. *Acta Geologica Taiwanica* 6:5-8
366. Hill, D. 1948. The Distribution and Sequence of Carboniferous Coral Faunas. GM LXXXV(3)126-7
367. Hill, D. 1973. Lower Carboniferous Corals (in Hallam. 1973. op. cit.), p. 135
368. Sayutina, T. A. 1973. Nizhnekammyenougulniye Koraly Severnevo Urala. TR 140:9
369. Nelson, S. J. 1982. New Pennsylvanian syringoporid coral from Kamloops area, British Columbia. CE 19:376-7
370. Sando, W. J. 1980. The Paleoecology of Mississippian Corals in the Western Conterminous United States. PO 25:620
371. Scruton, C. T. 1973. Palaeozoic Coral Faunas from Venezuela. II. Devonian and Carboniferous Corals from the Sierra De Perija. BR 23(4)225
372. Khoa, N. D. 1977. Carboniferous Rugosa and Heterocorallia from boreholes in the Lublin Region (Poland). PO 22(4)301-3
373. Minato and Kato. 1965. op. cit. (ref. 217), pp. 40-54
374. Rowett, C. L. 1975 Stratigraphic Distribution of Permian Corals in Alaska. UG 823D, pp. D64-D72
375. Hoover, P. R. 1981. Paleontology, Taphonomy, and Paleoecology of the Palmanto Formation (Permian of Venezuela). BP 80(313)7-10
376. Wilson, E. C. 1974. Bibliographic Index of North American Permian Rugose and Tabulate Coral Species. JP 48:598-604
377. Rowett, C. L. 1972. Paleogeography of Early Permian Waagenophyllid and Durhaminid Corals. *Pacific Geology*
378. Gobbett, D. J. 1967. Palaeozoogeography of the Verbeekinidae (Permian Foraminifera) (in Adams, C. G. and D. Ager. 1967. *Aspects of Tethyan Biogeography.* Systematicists Association (of London) Publication No. 7, p. 79
379. Gobbett, D. J. 1973. Permian Fusulinacea (in Hallam. 1973. op. cit.), pp. 153-6
380. Ross, C. A. 1979. Evolution of Fusulinacea (Protozoa) in Late Palaeozoic Space and Time (in Boucot and Gray. 1979. op. cit.), p. 225
381. Bostwick, D. A. and M. K. Nestell. 1967 Permian Tethyan Fusulinid Faunas of the Northwestern United States (in Adams and Ager 1967. op. cit.), p. 95
382. Yancey, T. E. 1975. Permian Marine Biotic Provinces in North America. JP 49:763

383. Kalmikova, M. A. 1975. Znachenie Fuzulinid v. Rasshifrovkie Paleografii Asselskovo Vyeka Rannei Permi. *VO* 18:126 (in Russian)
384. Toriyama, R. 1973. Upper Permian Fusulininan Zones (*in* Logan, A. and L. V. Hills. 1973. *The Permian and Triassic Systems and Their Mutual Boundary.* CP Memoir 2), pp. 498-9
385. Stow, D. A. V. 1975. New Fusulinid Evidence for the Permian Age of the Palaeozoic rocks of Hydra, Greece. *GM* 112:72
386. Waterhouse, J. B. 1976. *World Correlations for Permian Marine Faunas.* University of Queensland Paper of Department of Geology 7(2)167
387. Stehli, F. G. 1957. Possible Permian Climatic Zonation and Its Implications. *AJ* 255:611-14
388. Termier, H. W., Termier, G. et. D. Vachard. 1977. Monographie Paleontologique des Affleurements Permiens De Djebel Tebaga (Sud Tunisien) *PH* 156A:16
389. Nakamura, K. 1979. Additional Occurrences of Urushtenoidea (Brachiopoda) from the Permian of Asia. *HO,* Ser. IV 18:224
390. Waterhouse, J. B. and G. B. Bonham-Carter. 1972. Permian Paleolatitudes Judged from Brachiopod Diversities. 24th *CO*, Sec. 7, pp. 354-9
391. Stehli, F. G. 1973. Permian Brachiopods (*in* Hallam. 1973. *op. cit.*), pp. 145-6
392. Nakamura, N. and F. Golshani. 1981. Notes on the Permian Brachiopod Genus *Crytospirifer*. *HO*, Ser. IV 20:69
393. Archbold, N. W., *et al.* 1982. Indonesian Permian brachiopod fauna and Gondwana-South East Asia relationships. *NA* 296:
394. Glenister, B. F. 1981. Written Communication.
395. Waterhouse, J. B. and V. J. Gupta 1977 Permian Faunal Zones and Correlations of the Himalayas *ND* 10(2)1-19
396. Waterhouse, J. B. 1972. The evolution, correlation, significance of the Permian ammonoid family Cyclolobidae. *LE* 5:268
397. Spinosa C., Furnish, W. M. and B. F. Glenister. 1975. The Xenodiscidate, Permian Ceratitoid Ammonoids. *JP* 49:240
398. Leven, E. Ya. 1982. The Permian Yakhtashian Stage: its basis, characteristics, and correlation. *IG* 24(8)945-54
399. Xi-lou. 1981. Early Permian Cephalopods from Northwest Gansu and Western Nei Monggol. *SI* 21(6)499
400. Valdiya, K. S. and V. J. Gupta. 1972. A Contribution to the Geology of Northeastern Kumuan, with Special Reference to the Hercynian Gap in Tethys Himalaya (*in* Jhingran and Valdiya. 1972. *op. cit.*), 1-34
401. Vachard D. and C. Montenat. 1981. Biostratigraphie, Micropaleontologie, et Paleogeographie du Permian de la Region de Tezak (Montagnes Centrales D'Afghanistan). *PH* 178B:4-5
402. Yegorov, A. Yu. and V. S. Andreyev. 1982. Structure of Permian deposits of Northern Verkhoyansk. *IG* 24(8) 979-87
403. Teichert, C. and M. Rilett. 1974. Revision of Permian Ecca Series Cephalopods, Natal, South Africa. *KA* 68:1
404. Frest, T. J., Glenister, B. F. and W. M. Furnish. 1981. Pennsylvanian-Permian Chedoceratacean Ammonoid Families Maumitidae and Pseudonaloritidae. *JP* (Supplement) for Vol. 55, pp. 31-45
405. Glenister, B. F. and W. M. Furnish. 1961. The Permian Ammonoids of Australia. *JP* 35:676-7
406. Kamen, Kaye M. 1978. Permian to Teritary Faunas and Paleogeography: Somalia, Kenya, Tanzania, Mozambique, Madagascar, South Africa. *PE* 1:82, 95
407. Ruzhencev, V. Ye. 1976. Late Permian Ammonoids from the Soviet Far East. *PJ* 10:279
408. McClure. 1980. Permo-Carboniferous glaciation in the Arabian Pennisula. *GB* 91:708-9
409. Kulikov, M. V., Pavlov, A. M., and V. N. Rostovtsev. 1973. A Find of Goniatites in the Lower Kazanian in the Northern Part of the Russian Platform. *DE* 211:112
410. Glenister, B. F., Nassichuk, W. W., and W. W. Furnish. 1979. Essay Review: Ammonoid Successions in the Permian of China. *GM* 116:232-8
411. Ross, J. R. P. 1979. Permian Ectoprocts in Space and Time (*in* Boucot and Gray. 1979. *op. cit.*), p. 265
412. Ross, J. R. P. 1981. Written Communication.
413. Ross, J. R. P. 1978. Biogeography of Permian Ectoproct Bryozoa *PA* 21:341-on.
414. Sakagami, S. 1976. Paleobiogeography of Permian Bryozoa on the Basis of the Thai-Malayan District. *AS* 17:157
415. Charig, A. J. 1971. Faunal Provinces on Land: evidence based on the distribution of fossil tetrapods with especial reference to the reptiles of the Permian and Mesozoic (*in* Middlemiss, *et al.* 1971. *op. cit.*), p. 118
416. Anderson, J. M. and A. R. T. Cruickshank. 1978. The Biostratigraphy of the Permian and the Triassic. Part 5. A Review of the Classification and Distribution of Permo-Triassic Tetrapods. *AF* 21:23, charts.
417. Olson, E. C. 1957. Catalogue of Localities of Permian and Triassic Terrestrial Vertebrates of the Territories of the USSR. *JG* 65:212-25
418. Cox, C. B. and D. G. Smith. 1973. Triassic Vertebrate faunas of Svalbard. *GM* 110:408
419. Ingavat, R. and P. Janvier. 1981. *Cyclotosaurus Posthumus* Fraas (Capitosauridae, Stereospondyli) from the Huai Hih Lat Formation (Upper Triassic) Northeastern Thailand. *BI* 14:712
420. Olsen, E. P., McCune, A. R., and K. S. Thomson. 1982. Correlation of the early Mesozoic Newark Supergroup by vertebrates, principally fishes *AJ* 282:35-9
421. Cox, C. B. 1973. Triassic Tetrapods (*in* Hallam. 1973. *op. cit.*), p. 217
422. Brown, D. A. 1967. Some Problems of Distribution of Late Palaeozoic and Triassic Terrestrial Vertebrates. 40th *ANZAAZ. Australian National University: Geology Department Publication* No. 105A, pp. 28-37
423. Leonardi, G. 1980. *Isochnothium* sp. Pista de un Gigantesco Teconte na Formacao antenor Navarro (Triassico). Sousa, Paraiba, Brasil. *RB* 10(3)186.
424. Ai-lin, S. and H. Lian-hai. 1981. *Hazhenia*, a New Genus of Scalaposauria. *SI* 20(4)310
425. Cei, R. I. and J. Gargiula. 1977. Icnites de Tetrapodos Permicos Del Sin De Mondozo. *ME* 14:127
426. Barbarera, M. C., Correia, N. D. R., and J. J. Aumond. 1980. Contribuicao a Estrata e Bioestratigrafia do Grupa Passa Dois Na Serra Do Cadead o (Nordeste do Parana, Brasil). *RB* 10(3)268
427. Reeside, J. B., *et al.* 1957. Correlation of the Triassic Formations of North America Exclusive of Canada. *GB* 68:1451-1514
428. Haubold, H. Die Tetrapodenfahrten des Buntsandsteins in der Deutsche Demokratischen Republik und in West deutschland und ihie Equivalente in der gesamten Trias. *Palaontologische Abhandlungen* A IV(3)409
429. Stehli, F. G. 1964. Permian Zoogeography and Its Bearing on Climate (*in* Cloud, P. 1970. *Adventures in Earth History.* W. H. Freeman and Sons, San Francisco), p. 828
430. Kitching, J. W. 1977. *The Distribution of the Karroo Vertebrate Fauna.* University of Witwatersrand, Johannesburg, enclosed map.
431. Chudinov, P. K. 1965. New Facts about the Fauna of the Upper Permian of the USSR. *JG* 73:118
432. Ai-lin, S. 1973. Permo-Triassic of Sinkiang. *SN* 16(1) 152-3
433. Harris, J. M. and R. L. Carroll. 1977. *Kenyasaurus*, A New Eosuchian Reptile from the Early Triassic of Kenya *JP* 51:139
434. Bonaparte, J. F. 1978. *El Mesozoico de America Del Sur y Sus Pterapodos.* Tucaman, Argentina, pp. 176-95
435. Sigogneair-Russel, D. and A. L. Sun. 1981. A Brief Review of Chinese Synapsids. *BI* 14(2)276-7
436. Panchen, A. L. 1970. *Handbuch de Palaoherpetologie* Teil 5a., p. 70
437. Anderson, H. M. and J. M. Anderson. 1979. A Preliminary Review of the Uppermost Permian, Triassic, and Lowermost Jurassic of Gondwanaland. *AF* 13:9-22, enclosures.
438. Techter, D. 1972. Introduction (*in* Glut, D. F. 1972. *The Dinosaur Dictionary* Citadel Press, New Jersey), p. 6
439. Schaeffer, B. 1970. Mesozoic Fishes and Climate (*in* Yochelson, E. L. 1970. *Proceedings of the North American Paleontological Convention*, Vol. I), pp. 378-83
440. Minikh, A. V. 1981. *Saurichthys* Species from the Triassic of the USSR. *PJ* 1981(1)81-on.
441. Kobayashi, T. 1975. Upper Triassic Estherids in Thailand and the Conchostracan Development in Asia in the Mesozoic. *AS* 16:57-60
442. Miska, R. C., Sahni, A., and N. Chhabra. 1973. Triassic Conodonts and Fish Remains from Niti Pass, Kumuan

Himalaya (*in* Jhingran and Paldiya. 1973. *op. cit.*), p. 148
443. Vorobyeva, E. I. 1973. Age of the Fish Fauna of the Vilyuy Syneclise. *DE* 213:212
444. Cosgrif, J. 1974. Lower Triassic Temnospondyli of Tasmania. *GP* 149:129
445. Lovovskiy, V. R. and M. H. 1974. The first discovery of labyrinthodont in Lower Triassic sediments in Mangyshlack. *IG* 16:611
446. Schaeffer, B. and M. Mangus. 1976. An Early Triassic Fish Assemblage from British Columbia. *American Museum of Natural History Bulletin* 156:519-63
447. Forey, P. and B. G. Gardiner. 1973. A New Dictyopygid from the Cave Sandstone of Lesotho, Southern Africa. *AF* 15:29-31
448. Lu yi, J. 1981. Late Triassic Lamellibranchs from Datong of Qinghai, Northwest China. *SI* 20(6)584
449. Hutchinson, P. 1975. Two Triassic Fish from South Africa and Australia, with comments on the Evolution of the Chondrostei. *PA* 18(3)613
450. Uppal, S., Sahni, A. and V. J. Gupta. 1981. New Fish Locality from the Permian of Northern Pirpanjal Flank, Kashmir. *ND* 14(2)155
451. Kummel, B. Lower Triassic (Scythian) Molluscs (*in* Hallam. 1973. *op. cit.*), pp. 228-9
452. Wiedmann, J. 1973. Upper Triassic Heteromorph Ammonites (*in* Hallam. 1973. *op. cit.*), p. 244
453. Ishibashi, T. 1975. Triassic Ammonites from Indonesia Malaysia. *AS* 16:46-7
454. Brookfield, M. E. and G. E. G. Westermann. 1982. Mesozoic Ammonites from the Spong Valley, Zanskar, Northwest India. *IN* 23:1
455. Loughman, D. L. and A. Hallam. 1982. A Facies Analysis of the Pucara Group (Norianto Toanaan Carbonates, Organic-Rich Shale and Phosphate) of Central and North Peru. *SG* 32:163
456. Buriy, I. V. and N. K. Zharnikova. 1981. Ammonoids from the *Triolites* Zone in the Southern Primorye Region. *PJ* 15(3)58
457. Qing-ge, H. 1980. Discovery of the Late Anisian *Peraceratites Trinodosus* Fauna (Ammonoidea) from Doilung deqen, Tibet and Its Significance. *SI* 19(5)345
458. Bychkov, Yu. M. and A. D. Chekhov. 1979. Find of Tethys Ammonoids of Triassic Age in the Koryak Mountains. *DE* 245:55
459. Vavilov, M. N. 1978. Some Anisian Ammonoids of Northern Siberia. *PJ* 12:331
460. Zonenshayn, L. P., Kiparosova, L. D., and T. M. Okuneva. 1971. First Find of Marine Triassic Sediments in Mongolia. *DE* 199:32
461. Barnard, P. D. W. 1973. Mesozoic Floras (*in* Hughes. 1973. *op. cit.*), p. 178.182
462. Xing-xue, L. 1981. Thirty Years of Paleobotany in China (*in* Teichert. 1981. *op. cit.*), pp. 22-3
463. Wesley, A. 1973. Jurassic Plants (*in* Hallam. 1973. *op. cit.*), p 329
464. Ash, S. 1982. Occurrence of the Controversial Plant Fossil *Sanmiguella CF. S. Lewisi* Brown in the Upper Triassic of Utah. *JP* 56:752
465. Ash, S. R. 1980. Upper Triassic Floral Zones of North America (in Dilcher and Taylor. 1980. *op. cit.*)
466. Krassilov, V. A. 1981. Changes of Mesozoic Vegetation and the Extinction of Dinosaurs. *PP* 34:213
467. Ge, S. 1981. Discovery of Dipteridaceae from the Upper Triassic of Eastern Jilin. *SI* 20(5)467
468. Kobayashi, T. and T. Hamada. 1974. Non-marine Mesozoic Formations and Fossils in Thailand and Malaysia. *AS* 15:209
469. Vachrameev, V. A. 1972. Mesozoic Floras of the Southern Hemisphere and Their Relationship to the Floras of the Northern Continents. *PJ* 6:412
470. Buriy, I. V. and N. K. Zharikova. 1981. Plant-bearing Strata of the Ladinian Stage (Middle Triassic) in South Primorye. *IG* 23:32
471. Vigran, J. O. 1970. Fragments of a Middle Jurassic Flora from Northern Tronddag, Norway. *NO* 50(3)193
472. Stevens, C. R. 1963. Faunal Realms in Jurassic and Cretaceous Belemnites. *GM* 100:484
473. Frebold, H. 1975. The Jurassic Faunas of the Canadian Arctic. *CB* 243:1-5
474. Yun-zhu. 1980. Studies on Lower Jurassic Ammonites from Kaiping-Enping Area, Guangdong. *SI* 19(2)76
475. Sato, T. 1975. Marine Jurassic Formations and Faunas in Southeast Asia and New Guinea. *AS* 15:156
476. Stevens, G. R. 1973. Jurassic Belemnites (*in* Hallam. 1973. *op. cit.*), pp. 261-4
477. Howarth, M. K. 1973. Lower Jurassic (Pliensbachian and Toarcian) Ammonites (*in* Hallam. 1973. *op. cit.*), pp. 277-80
478. Dietl, G. 1973. Middle Jurassic (Dogger) Heteromorph Ammonites (*in* Hallam. 1973. *op. cit.*), p. 284
479. Cariou, E. 1973. Ammonites of the Callovian and Oxfordian (*in* Hallam. 1973. *op. cit.*), p. 292-3
480. Wiedmann, J. 1973. Ancyloceratina (Ammonoidea) at the Jurassic/Cretaceous Boundary (*in* Hallam. 1973. *op. cit.*), p. 310
481. Aliev, M. M. 1980. Stratigraphic position and geographic range of *Inoceramus azerbaid janensis*. *IG* 22:806
482. Maryanska, T. 1977. Ankylosauridae (Dinosauria) from Mongolia (*in* Kielan-Jaworska, Z. 1977. Results of the Polish-Mongolian Palaeontological Expeditions—Part VII. *OL* No. 37), p. 87
483. Monteillet, J., Lanparient, J. R., et e. P. Taquet. 1982. Un Pterosaurien geant dans le Cretace superiare de Paki. *Comptus Rendus des Sciences de L'Academie des Sciences.* Serie II, Tome 295, p. 409
484. Sochava, A. V. 1969. Dinosaur Eggs from the Upper Cretaceous of the Gobi Desert. *PJ* 3:518
485. Reyment, R. A. 1981. Colombia (*in* Reyment, R. A. and P. Bengston. 1981. *Aspects of Mid-Cretaceous Regional Geology.* *AC*), pp. 186-9
486. Charig, A. J. 1973. Jurassic and Cretaceous Dinosaurs (*in* Hallam. 1973. *op. cit.*), p. 486
487. ———. 1982. *Britannica Science and Future Library.* VI, p. 176 (Map by Ralph Stobart)
488. Morris, W. J. 1981. A New Species of hadrosaurian dinosaur from the Upper Cretaceous of Baja California—*Lambeosaurus laticaudus.* *JP* 55:453
489. Buffetaut, E. 1981. A plesiosaur vertebra from the Chichali Formation (Late Jurassic to Early Cretaceous) of Pakistan. *JA Monatschefte* 1981, heft 6, p. 334
490. ———. 1977. Dinosaurs found in Tibet. *New Scientist* 73(1038)329
491. Galton, P. M. and J. A. Jensen. 1975. *Hypsilon* and *Iguanodon* from the Lower Cretaceous of North America. *NA* 257:668
492. Dodson, P. and A. K. Behrensmayer. 1980. Taphonomy and paleoecology of the dinosaur beds of the Jurassic Morrison Formation. *PB* 6:209
493. Horner, J. R. 1979. Upper Cretaceous Dinosaurs from the Bearpaw Shale (Marine) of South-Central Montana with a checklist of Upper Cretaceous Dinosaur remains from Marine Sediments in North America. *JP* 53:514-5
494. Carpenter, K. 1982. Baby dinosaurs from the Late Cretaceous Lance and Hell Creek Formations and a description of a new species of Theropod. *YO* 20:125
495. Galton, P. M. and J. A. Jensen. 1978. Remains of Ornithopod Dinosaurs from the Lower Cretaceous of North America. *Brigham Young University Studies in Geology* 25(3)1-10
496. Taquet, P. 1977. Les Decouvertes recentes et Dinosaures du Jurassique et du Cretace en Afrique au Roche et Moyen-Orient et en Inde. *FR* No. 8, pp. 327-8
497. Molnar, R. E., Flannery, T. F., and T. A. V. Rich. 1981. An Allosaurid Theropod dinosaur from the Early Cretaceous of Victoria, Australia *CH* 5:141-2
498. Thulborn, R. A. and A. Warren. 1980. Early Jurassic plesiosaurs from Australia. *NA* 285:224
499. Buffetaut, E. 1981. Elements pour une historie paleobiographique du Sud-Est Asiatique: l'apport des Vertebres Fossiles Continentaux. *BF* 23(6)589
500. Molnar, R. E. and R. A. Thulborn. 1980. First Pterosaur from Australia. *NA* 288:362
501. Matsumoto, T. 1973. Late Cretaceous Ammonoidea (*in* Hallam. 1973. *op. cit.*), pp. 422-7
502. Stevens, G. R. 1973. Cretaceous Belemnites (*in* Hallam. 1973 *op. cit.*), 387-90
503. Pergament, M. A. 1981. Pacific Region of the USSR (*in* Reyment and Bengston. 1981. *op. cit.*), pp. 69-102
504. Kennedy, W. J. and W. A. Cobban 1977 The Role of Ammonites in Biostratigraphy *PL* 316-20

505. Kurten, B. 1973. Early Tertiary Land Mammals (*in* Hallam. 1973. *op. cit.*), pp. 438-40
506. Ming-zhen, Z. 1981. Vertebrate paleontology in China, 1949-1979. (*in* Teichert. 1981. *op. cit.*), pp. 17-8
507. Macfadden, B. J. 1980. The Miocene Horse *Hipparion* from North America and from the Type Locality in Southern France PA 23(3)631
508. Woodburne, M. O. and R. L. Bernor. 1980. On Superspecific Groups of Some Old World Hipparionine Horses. JP 54:1320
509. Davies, M. 1975 (Updated Edition). *Tertiary Faunas*. WI, pp. 375-8
510. Patterson, B. and Pascual. 1972. The Fossil Mammal Fauna of South America (*in* Keast, A., Erk, F. C., and B. Glass. 1972. *Evolution, Mammals, and Southern Continents*. State University of New York Press, Albany), pp. 250-2
511. Keast, A. 1972. Australian Mammals: Zoogeography and Evolution (*in* Keast, et al. 1972. *ibid*), p. 225
512. Cooke, H. B. S. 1972. The Fossil Mammal Fauna of Africa (*in* Keast, et al. 1972. *ibid.*), p. 97
513. Min-chen, C. and Z. Jia-Jian. 1980. The Mammal-Bearing Early Tertiary Horizons of China. *Museum of Paleontology (University of California; Berkeley)* 32
514. Sahni, A., *et al.* 1981. Vertebrates from the Subeth formation and comments on the biogeography of Indian subcontinent during the early Paleogene. BF (1981)(7) 23(6)690
515. Savage, R. J. G. 1967. Early Miocene Mammal Faunas of the Tethyan Region (*in* Adams and Ager. 1967. *op. cit.*), p. 273
516. Cooke, H. B. S. 1978. Africa: The Physical Setting (*in* Haglio, V. J. and H. B. S. Cooke 1978 *Evolution of African Mammals*, Harvard University Press, Cambridge), p. 24
517. Storer, J. E. 1978. Tertiary Sands and Gravels in Saskatchewan and Alberta: Correlation of Mammalian Faunas (*in* Stelck and Chatterton. 1978. *op. cit.*), p. 97
518. Archer, M. and A. Bartholomai. 1978. Tertiary Mammals of Australia: A Synoptic Review CH 2(1)3
519. Dashzeveg, D. 1980. New Pantodonts from the Eocene of Mongolia. PJ 14(2)97
520. Kennedy, G. E. 1980. *Paleoanthrolopogy*. McGraw Hill, p. 138
521. Goldsmith, N. F., *et al.* 1982. Ctenodactid rodents in the Miocene Negev fauna of Israel. NA 296:645
522. Lytshev, G. F. 1978. A New Early Oligocene Beaver of the Genus *Agnatocastor* from Kazakhstan. PJ 12:542
523. Brandy, L. D. 1981. Rongeues Muroides Du Neogene Superieur D'Afghanistan. Evolution, Biogeography, Correlations. PV 11(4)136
524. Dashzeveg, D. and M. C. McKenna. 1977. Tarsoid Primate from the Early Tertiary of the Mongolian People's Republic. PO 22:122
525. Tang, X. and Z. Ming-zhen. 1965. The Vertebrate-Bearing Early Tertiary of South China: A Review. IG 7:1349
526. Belyaeva, E. I. 1968. USSR (Tertiary) Faunas (*in* Orlov, Yu. A. 1968. *Fundamentals of Paleontology*. Vol. 13), p. 54
527. Barry, J. C., Lindsay, E. H., and L. L. Jacobs. 1982. A Biostratigraphic Zonation of the Middle and Upper Siwaliks of the Potwar Pleateau of Northern Pakistan PP 37:99
528. Yongsheng, T. 1982. Chinese Dintatheres and the Evolution of the Dinocerata. JP (Abstracts, North American Paleontological Convention III), Supplement for 1982.
529. Hottinger, L. 1973. Selected Paleogene Larger Foraminiferida (*in* Hallam. 1973. *op. cit.*), pp. 448-51
530. Adams, C. G. 1973. Some Tertiary Foraminiferida (*in* Hallam. 1973. *ibid.*), pp. 456-66
531. Rau, W. W. 1981. Pacific Northwest Tertiary benthic Foraminiferal biostratigraphic framework—An Overview. GP 184:69
532. Samanta, B. K. 1982. *Fabiania* Silvestii (Foraminiferida) from India, with notes on its global distribution. GM 119:260
533. Serova, M. Ya. 1978. Planktonoviye Foraminifery Paleogena i Neogena Severniy Chasti tikookeanskoi Provincii. VO 21:172-5
534. Henderson, G., Rosenkrantz, A., and E. J. Schiener. 1976. Cretaceous-Tertiary sedimentary rocks of West Greenland (*in* Escher, A. and W. S. Watt. 1976. *Geology of Greenland*, Copenhagen), p. 341
535. Poore, R. Z. 1980. Age and Correlation of California Paleogene Benthic Foraminiferal Stages. UG 1162-C, p. CR
536. Chang-Min, Y. and W. Hui-ji. 1981. Some Tube-like Fossils from the Early Tertiary of Northern Jiangsu. SI 20(5)415
537. Turner, D. L. 1970. Pacific Coast Miocene Foraminiferal Stages (*in* Bandy, O. L. ed. 1970. *Radiometric Dating and Paleontological Zonation*. GP 124:94-5
538. Yi-chun, H. and Z. Xue-lu. 1980. Early Tertiary Foraminiferida from the Kaslu Basin of Xinjiang. SI 19(2)167
539. Saperson, E. and M. Jahal. 1980. Biostratigraphy of the Anomalinidate and Cibicidate in the Soviet Tethyan Paleogene. *Micropaleontology* 26(4)393
540. Siesser, W. G. 1982. Cretaceous Calcareous Nannoplankton in South Africa. JP 56:335-50
541. Yu-lin, L. and L. Cai-hua. 1981. A New Echinoid with Sexual Dimorphism from Late Tertiary Deposits of Beibuwan, Guangxi. SI 20(5)484
542. Chaprioniere, G. C. H. 1981. Influence of Plate Tectonics on the distribution of Late Palaeogene to Early Neogene larger foraminiferids in the Australasian region. PP 31:306-9
543. Zinsmeister, W. J. 1981. Middle to Late Eocene Invertebrate Fauna from the San Julian Formation at Punta Casamayor, Santa Cruz Province, Southern Argentina. JP 55:1090
544. Varhatova, N. N., *et al.* 1979. Paleobiografiya i Paleotemperatura Eotsenskih Morei Yevrazii Ustanovleniye po Nummulitidam. VO 22:76, 93
545. Yi-chun, H. 1981. Thirty years of micropaleontological research in China (*in* Teichert, *et al.* 1981), pp. 12-14
546. Yan-hao, *et al.* 1981. Invertebrate paleontology in China (1949-79). (*in* Teichert, *et al.* 1981. *op. cit.*), pp 6
547. Stainforth, R. M. and J. L. Lamb. 1981. An Evaluation of Planktonic Foraminiferal Zonation of the Oligocene. KA Paper 104, p. 3
548. Shi-lan, Z. 1982. Neogene Calcareous Nannofossils from the Huanglui Formation of the Yinggehai Basin, South China Sea. SI 21(2)200
549. Amard, B. 1983. Decouverte de microfossiles dans le Proterozoique metamorphique de l'adrardes Iforas (Mali). CR *Academie des Sciences Paris* Serie II, Tome 296(1)85
550. Bekker, Yu. R. 1980. A New Locality with Fossil Fauna of the Ediacara type in the Urals. DE 254:236
551. Cheng-Hua, D. 1982. Late Precambrian algal megafossils *Chuaria* and *Tawaia* in some areas of eastern China. CH 6:58
552. Andrawis, S. F., *et al.* 1983. Lower Paleozoic Trilobites from subsurface rocks of the Western Desert, Egypt. JA *Monatschefte* 1983(2)65-6
553. Zhenhun, S. 1982. Late Lower Cambrian Trilobites from Southern Dahongshan Region, Hubei. SI 21(3)307
554. Zhi-yi, Z. and Z. Zhi-qiang. 1982. An Ashgillian (Rawtheyan) Trilobite Faunule from Ejin Qi, Nei Monggol (Inner Mongolia). SI 21:669
555. Matthews, S. C. and V. V. Missarzhevsky. 1975. Small shelly fossils of late Pre and early Cambrian age: a review of recent work. GL 131:290
556. Maithy, P. K. and S. Gupta. 1981. Archaeocyatha from the Vindhyan Supergroup of India. BB 46(11)2802
557. Krylov, *et al.* 1981. Middle Reaches of the Aldan River (*in* Raaben, M. E. 1981. *The Tommotian Stage and the Cambrian Lower Boundary Problem*. National Science Foundation), p. 17
558. Jordan, M. 1979. The Stratigraphic Significance of Graptolites in Romania. PO 24:122
559. Kobayashi, T. and H. Igo. 1965. On the Occurrence of Graptolitic Shales in North Thailand. JJ 36:37
560. Ji-jin, L. 1982. Graptolites from the Jeanling Formation (M. Ordovician) of Yaxian, Hainan Island. SI 21(2)205
561. Tsai, D. T. 1982. The Graptolitic Zonal Scale of the Pacific Ocean Province. SV 23(4)9
562. Wang, C. Y. and W. Zeigler. 1983. Conodonten aus Tibet. JA *Monatscheft* 1983(2) p. 70
563. Ulrich, E. O., *et al.* 1944. Ozarkian and Canadian Cephalopods. GP 58
564. Flower, R. and C. Teichert. 1957. The Cephalopod Order Discosorida. KA Article 6, p. 33

565. Ingavat, R., Maunlek, S., and C. Vdomratin. 1975. On the Discoveries of Permian fusulinids and Ordovician Cephalopods of Ben Rai, West Thailand. *BB* 43(1)785
566. Kaplan, A. A. 1917. Silurian-Devonian Boundary Beds of Central Kazakhstan. *DE* 199(1-6)45-7
567. Xue-chang, Y. and R. Jia-yu. 1982. Brachiopods from the Upper Xiushan Formation (Silurian) in the Sichuan-Guishou-Hunan-Hubei Border Region. *SI* 21(4)432, 419
568. Yu-nan, N., et al. 1982. The Silurian Rocks of West Yunnan. *SI* 21:119
569. _____. 1982. Silur Sibirskoi Platformy. *SB* 508 (in Russian)
570. Hagen, D. and E. Kemper. 1976. Geology of the Thong Pha Plurm Area (Kancharaburi Province, West Thailand). *Geologisches Jahrbuch* B21, pp. 53-91
571. Benton, M. J. 1982. Dictyodora and associated trace fossils from the Palaeozoic of Thuringia. *LE* 15:115-32
572. Draper, J. J. 1980. Rusophycus (Early Ordovician ichnofossil) from the Mithaka Formation, Georgina Basin. *BM* 5:57-61
573. Barcellos, M. T. 1979. Scales and Teeth of fish of the Budo facies, Itarare Subgroup, Rio Grande de Sul. *BB* 43(4)1458.
574. Babin, et al. 1976. The Schistes de Porsiguen Formation (Upper Devonian) of Brest Roads, Armorican Massif. *BB* 46(2)383
575. Shitao, W., et al. 1980. The discovery of Silurian agnathans and Pisces from Chaoxian County, Anhui Province and its stratigraphic significance. *BB* 46(7)
576. Young, G. C. 1982. Devonian Sharks from South-Eastern Australia and Antarctica. *PA* 25(4)818
577. Vicente, J. C. 1975. Essai d'organisation palaeogeographique et structural du Paleozoique des Andes Meridionales. *RU* 64(2)358
578. Isaacson, P. E. 1974. First South American Occurrence of *Globithyris* Its Ecological and Age significance in the Malvinokaffric Realm. *JP* 48:779
579. Tian-rui, L. 1982. Trilobite Fauna from the Fentou Formation (Middle Silurian) of Nanjing and its Geologic Age. *SI* 21(4)455
580. Kobayashi, T. and T. Hamada. 1960. A New Proetid from Perlis, Malaysia. *JJ* 37:89
581. Chibukova, Ye. V. 1974. Floral Assemblages of Devonian-Carboniferous Boundary Beds and the age of the Solonchatka Creek Flora. *DE* 215:68
582. Xing-xue, L. and W. Hong-feng. 1982 On the Occurrence of Late Devonian Plants from Mt. Longmenshan, North Sichuan. *SI* 21(1)94
583. Tripathi, C. 1980. Note on the Report of diamiotite and plant remains from Salari area, Kaming district, Arunachal Pradesh. *BB* 47(1)170
584. Cai, C. and H-J. Schweitzer. 1983. Uber *Zosterophyllum yunnanicum* Hsu Aus dem Unterdevon Sud Chinas. *PH* B185:2
585. Yochelson, E. 1983. A Devonian aptychas (Cephalopoda) from Alabama. *JP* 59:124
586. Suarez-Riglos, M. 1979. Review of the Bolivian Devonian based on a new goniatite from Huamampampa Formation in Campo Redendo, Chuquisaca. *BB* 45:2991
587. Yiping, R. 1981. Devonian and Earliest Carboniferous Ammonoids from Guangxi and Guizhou. *NJ* 15:139
588. Luo-zhao, L. 1982. *Gansuceras*, a New Devonian Cephalopoda from Gansu. *SI* 21(3)342
589. Yong-yi, C. 1982. Some Tabulate Corals from Late Middle Devonian in Baijinksham District of Dahinganling. *SI* 21(4)479
590. Boucot, A. J., Brunton, C. H. C., and J. N. Theron. 1983. Implications for the age of the South African Devonian rocks in which *Tropidoleptus* (Brachiopoda) has been found. *GM* 120(1)51-3
591. Zheng-xiang, T. 1982. Earliest Devonian Brachiopod Fauna from Ruergai, Sichuan and Diebu, Gansu. *SI* 21(3)337
592. Johnson, J. G. 1970. Great Basin Lower Devonian Brachiopoda. *MM* 121, p. 4-9
593. Cooper, G. A. and J. T. Dutro. 1982. Devonian Brachiopods of New Mexico. *BP* 82, 83, no. 315, p. 6
594. Isaacson, P. E. and D. G. Perry. 1977. Biogeography of *Tropidoleptus* (Brachiopoda, Orthida during the Devonian, and Morphological Considerations. *JP* 51:1109
595. Yiping, R. 1981. Carboniferous Ammonoid faunas from Qixu in Nandan of Guangxi. *NJ* 15:226
596. Kochanssky-Devide, V., et al. 1980. Carboniferous of Northwest Yugoslavia. *BB* 44(4)1406
597. Young, J. A. 1942. Pennsylvanian Scaphopoda and Cephalopoda from New Mexico. 16:121-5
598. Wilson, R. M. 1980. A goniatite from the Mill Hill Maine Band, Lower Limestone Group of East Fife. *SG* 16:33
599. Ming-qian, W. 1981. Carboniferous Ammonoids from Eastern Xinjiang. *SI* 20(5)480
600. Wang-shi, W. and Z. Cai-lin. 1982. Early Carboniferous Corals in the Ammonoid Facies from Barkol, Xinjiang. *SI* 21(2)151
601. De-you, W. 1982. Some Marine Pelecypods from the Middle Carboniferous in Gushi of Henan. *SI* 21(4)46
602. Mapes, R. H. 1979. Carboniferous and Permian Bactritoidea (Cephalopoda) in North America. *KA Article* 4:3-4
603. Miller, A. K. and W. Yongquist. 1942. American Permian Nautiloids. *MM* 41
604. Vdorenko, M. V. 1975. Zoogeograficheskoe Rannemraronirovaniye Evraziiskoi Oblasti Karbonie (Vizeiski Vyek) po danym Foraminifer. *VO* 18:22-5 (in Russian)
605. Lipina, O. A. 1973. Zonalnaya Stratigrafiya i Paleobiogeografiya turne po Foraminiferam. *VO* 16:11-25 (in Russian)
606. Dumont, J. F. and M. Lys. 1973. Description of an autochthonous Carboniferous Sequence, Coller Bolgesi, Egrider. *BB* 42(1)545
607. Hamadat. 1964. Two Carboniferous Brachiopods from Loeiy, Thailand. *JJ* 35(1)6
608. Konovalova, M. V. 1975. Nekatoriye Paleobiogeograficheskoe i Paleoecologicheskoe ossobenosti postdnekammenougolnich i Rannepermskich Foraminifer Timano-Cechorskoi oblasti. *VO* 18:148 (in Russian)
609. Jiang-xiu, H. 1982. Middle and Upper Carboniferous Fusulinids from the Nadanhada Range, Heilongjiang Province. *SI* 21(3)
610. Marfenkova, M. M. 1975. Paleogeografiya i Foraminifery Rannevo Karbona Chu-Betrak-Dalinskovo Baseina (Yuzhniy Kazakhstan). *VO* 18:63 (in Russian)
611. Ivanovskiy, A. B. and G. S. Kropacheva. 1980. A Discovery of *Pseudofavosites* (tabulata) in the Permian rocks of the Soviet Far East. *DE* 252:189
612. Barthvon, W. 1972. Das Permokarbon Bie Zudiniz (Bolivia) und eine Ubersedt des Jung palaezoikums in Zentral Teil der Andem. *RU* 61:249-65
613. Sokratov, B. G. 1983. Oldest Triassic strata and the Permian-Triassic boundary in the Caucasus and Middle East. *IG* 25(4)488
614. Xi-luo, L. 1982. Some Early Permian Ammonoids from Jilin and Nei Monggol. *SI* 21(6)5
615. Furnish, W. M., et al. 1973. Permian Ammonoid *Cyclolobus* from the Zerwan Formation, Guryul Ravine, Kashmir. *SC* 180:188
616. Zhuo-quan and M. Jun-wen. 1982. Early Lower Permian Ammonoids from Yichuan, Jiangxi. *SI* 21(3)288
617. Miller, A. K. and S. G. Unklesbay. 1974. Permian Nautiloids from Western United States. *JP* 16:111-17
618. Dowson, D. T. 1978. A Permian aulocerid from Saraburi, central Thailand. *BB* 43(1)446
619. Chung-duen, Y. 1979. A nothosaur from Lu-hsi County, Yunnan Province. *BB* 43(4)1521
620. Demathieu, G. et M. Wiedram. 1982. Les empreiutes de pas de reptiles dans le Trias du Vieux Emossen (Finhaur, Valais, Suisse). *HE* 75(3)722
621. Bermudo, Melendez D. 1980. Discovery of a "protoavian" reptile from the Triassic of the Prudes Mountain Range, Taragona. *BB* 44(4)1868.
622. Zhong-jian, Y. 1980. A New Late Permian fauna from Jiyuan County, Henan Province. *BB* 44(4)1869.
623. Beltan, L., et al. 1979. A new marine fish and placodont reptile fauna of Ladinian age from Southwestern Turkey. *JA Monatschefte* 1979(5)258
624. Selozneva, H. A. 1982. Triassic Fish Finds in the Franz Jozef Land Archipelago. *PJ* 16(2)131
625. Martin, M. 1980. Diptheronotus Gibbosus (Actinopterygi, Chondrostie) Nouveu Colobodortide du Trias Superieur Continental Marocain. *BI* 13(3)445
626. Stipanicic, P. N. 1980. The Triassic of the Rio de Los Patos Valley, San Juan. *BB* 44:1781

627. Russell, D. and D. Russel. 1977. Preliminary Results of paleontologic exploration in the Triassic of Algarve, Portugal. BB 45(2)2660
628. Vollrath, A. 1977. A discovery of *Semionotus* (Pisces, Ganoidea) in the Stuben Sandstone near Winnenden. BB 42(2)2109
629. Johnson, G. D. 1980. *Xenacanhtodii* (Chondrichthees) from the Tecovas Formation (Late Triassic) of West Texas. JP 54(5)925
630. Nairn, A. E. M. 1978. Northern and Eastern Africa (*in* Moullade M. and A. E. M. Nairn. 1978. *The Phanerozoic Geology of the World.* A. Mesozoic) El., p. 356-8
631. ———. 1982. Bio-i Lithostratigrafiya Triasa Sibiri. SB 462 (in Russian)
632. Podstolski, R. 1980. Cephalopods from the Gorazde Beds (lower Muschelkalk) quarried at Gorazde near Opole. BB 45(2)2446
633. Turculet, I. 1980. Norian fauna of the Ciungi Klippe, Rarau Bucovine, eastern Romanian Carpathians. BB 47 (2)377
634. Shilin, P. V. and Yu. V. Suslov. 1982. A Hadrosaur from the Northeastern Aral Region. PJ 16(1)132
635. Sanz, J. L. 1982. A Sauropod Dinosaur from the Lower Cretaceous of Galve (Province of Teruel, Spain). BI 15(6)944
636. Zeng, D. and Z. Jin-jian. 1980. Fossil Dinosaur eggs from the eastern part of the Dongting Basin, Hunan Province. BB 44(2)2667
637. Zhao, Z. 1979. The discovery and significance of a new type of dinosaur eggs and a dinosaur footprint from Neixiang County, Henan Province. BB 45(2)3448
638. Smith, P. L. 1983. The Pliensbachian ammonite *Dayriceras dayiceroides* and Early Jurassic paleogeography. CE 20(1)89
639. Hirano, H. 1982. Cretaceous biostratigraphy and ammonites in Hokkaido. *Proceedings of the Geologists' Association* 93(2)213
640. Shotwell, J. A. 1961. Late Tertiary Biogeography of Horses in the Northern Great Basin. JP 35:205-8
641. West, M. R. M. 1979. Apparent prolonged evolutionary stasis in the Middle Eocene hoofed mammal *Hyopsodus.* PB 5(3)253
642. Olson, S. L. and Y. Hasegawa. 1979. Fossil Counterparts of Giant Penguins from the North Pacific. SC 206:688-9
643. Whitcomb, J. C. and H. M. Morris. 1961. *The Genesis Flood.* Presbyterian and Reformed Pub. Co., pp. 273-5
644. Price, G. M. 1923. *The New Geology.* Pacific Press, California
645. Clark, H. W. 1968. *Fossils, Flood, and Fire.* Outdoor Pictures, California
646. Burdick, C. L. 1976. What about the zonation theory? CR 13(1)37-8
647. Morris, H. M. and G. E. Parker. 1982. *What is Creation Science?* Creation-Life Pub., San Diego, p. 212
648. McLaren, D. J. 1977. The Silurian-Devonian Boundary Committee: A Final Report (*in* Martinson, A. 1977. *The Silurian-Devonian Boundary.* E. Schweizerbutsche, Stuttgart) 27-8
649. ———. 1982. Response by Phillip Gingerich. JP 56:831
650. Watson, J. V. 1982. Single-system geology after Arkell. NA 295:268
651. McKerrow, W. S. 1971. Palaeontological prospects—the use of fossils in stratigraphy. GL 127:455-6
652. Cowie, J. W., et al. 1972. A Correlation of Cambrian rocks in the British Isles. GL *Special Report No. 2*, p. 40
653. Woodmorappe, J. 1978. The Cephalopods in the Creation and the Universal Deluge. CR 15(2)100-3
654. Stanley, S. M., Andicott, W. O., and K. Chinzie. 1980. Lyellian curves in paleontology; Possibilities and Limitations. GE 8:424
655. Chaloner, W. G. and W. S. Lacey. 1973. The Distribution of Late Palaeozoic Floras (*in* Hughes. 1973. *op. cit.*), p. 274
656. Bartenstein, H. and H. M. Bolli. 1977. The Foraminifera in the Lower Cretaceous of Trinidad, West Indies. HE 70(2)552
657. Windle, T. M. F. 1979. Reworked Carboniferous Spores: An Example from the Lower Jurassic of Northeast Scotland. *Review of Paleobotany and palynology* 27:174-on. Chapter under "Unrecognized Reworking."
658. Ethington, R. L. and D. Schumacher. 1969. Conodonts of the Copenhagen Formation (Middle Ordovician) in Central Nevada. JP 43:478
659. Harland, W. B. (ed.). 1967. *The Fossil Record: a symposium with documentation,* Geological Society of London/Palaeontological Association, 827 p.
660. Raup, D. M. 1978. Cohort analysis of generic survivorship. PB 4:3-13
661. Cutbill, J. L. and B. M. Funnel. 1967. Numerical Analysis of *The Fossil Record* (*in* Harland. 1967. *op. cit.*), p. 793-4
662. Koch, C. F. 1978. Bias in the published fossil record. PB 4(3)367
663. Sepkoski, J. J. 1979. A kinetic model of Phanerozoic taxonomic diversity A. Early Phanerozoic families and multiple equilibria. PB 5(3)230
664. Kielan-Jaworska, Z. 1975. Late Cretaceous Mammals and Dinosaurs from the Gobi Desert. *American Scientist* 3:150
665. Lubenow, M. L. 1980. Significant fossil discoveries since 1958: Creationism Confirmed. CR 17(3)148-60
666. Woodmorappe, J. 1980. An Anthology of Matters Significant to Creationism and Diluviology: Report 1. CR 16:11-2
667. Woodmorappe, J. 1982. An Anthology of Matters Significant to Creationism and Diluviology: Report 2. CR 18:219 (stratigraphic-range extensions) and 209-16 (alleged reworking)
668. Shu, O. 1982. Upper Permian and Lower Triassic palynomorphs from eastern Yunnan, China. CE 19:79
669. Bengston, S. 1981. *Atractosella,* a Silurian alcyonacean Octocoral. JP 55:281
670. Collins, D. and A. M. Rudkin. 1981. *Priscansermarinus Bernetti,* a Probable Lepadomorph Barnacle from the Middle Cambrian Burgess Shale of British Columbia. JP 55:1014
671. Raup, D. M. 1976. Species diversity in the Phanerozoic: a tabulation. PB 2:288
672. Simpson, G. G. 1980. *Why and How: Some Problems and Methods of Historical Biology.* Pergamon Press, Oxford, New York, p. 6
673. Maheshwari, H. K. 1972. Permian Wood from Antarctica and Revision of Some Lower Gondwana Wood Taxa. PH B138:5
674. Karamlov, V. B. 1967. Discoveries of Ancient Onealites and Catagraphs in Paleozoic Sediments of the Shanton Islands. DE 175:85-7
675. Krasnov, Ye. V. and V. O. Savitskiy. 1973. Upper Jurassic Coral Reefs of Sakhalin and the Hypothesis of Drift of the Japanese Islands. DE 209:53
676. Harper, C. W. 1981. Inferring Succession of Fossils in Time: The Need for a Quantitative and Statistical Approach. JP 55:442
677. Stubblefield, C. J. 1956. Cambrian Palaeogeography in Britain. *20th CO*, vi, p. 6
678. Brand, P. J. 1965. New Lower Cambrian Fossil Localities in Northwest Scotland. SG 1:285-6
679. Cowie, et al. 1971. op. cit., entire volume.
680. Sheehan, P. M. 1980. Paleogeography and Marine Communities of the Silurian Carbonate Shelf in Utah and Nevada (*in* Fouch, T D., et al. 1980. *Paleozoic Paleogeography of the West-Central United States.* SEPM Rocky Mt. Section) p. 20
681. Gutschick, R. C., et al. 1980. Mississippian Shelf Margin and Carbonate Platform from Montana to Nevada. (*In* Fouch. 1980. op. cit.)
682. Mudds, J. R. 1981. Discovery of the Carboniferous Coral *Dorlodotia* in Northern England. YG 43:331-40
683. Stanley, K. O. 1971. Tectonic and Sedimentologic History of Lower Jurassic Sunrise and Dunlop Formations, West-Central Nevada. AG 55(3)455
684. Cope, J. C. W., et al. 1980. A Correlation of Jurassic Rocks in the British Isles. GL *Special Paper 15,* both volumes.
685. Imlay, R. W. 1980. Jurassic Paleobiogeography of the Conterminous United States in Its Continental Setting. UG 1062:43
686. Howarth, M. K. 1980. The Toarcian Age of the Upper Part of the Marlstone Rock Bed of England. PA 23:638
687. Raup, D. M. and R. E. Crick. 1979. Measurement of Faunal Similarity in Paleontology. JB 53:1213
688. Potapenko, Yu. Ya. and G. A. Stukalina. 1971. First Find

of Fossils in the Metamorphic Complex of the Main Caucasus Range. *DE* 198:134
689. Yochelson, E. L. and E. Stump. 1977. Discovery of early Cambrian Fossils at Taylor Nunatak, Antarctica. *JP* 51: 873
690. Skehan, J. W., Murray, D. P., and A. R. Palmer. 1972. p. 694
691. Collier, K. O. 1981. An Empirical Estimate of Preservation Bias in Paleobiogeography. *GA* 13(7)429
692. Woodmorappe. 1978. (ref. 653; *op. cit.*), pp. 104-9
693. Nelson, S. J. 1981 Stratigraphic Position of Ordovician Oil Shales, Southhampton Island, Northwest Territories, Canada *AG* 65:1174
694. Sheehan, D. M. 1982. The Dichotomy of Mid-Paleozoic Community Evolution-Epicontinental Seas vs. Open Seas. *GA* 5(14)288
695. Taylor, M. E. and R. M. Forester. 1979. Distributional model for marine isopod crustaceans and its bearing on early Paleozoic paleozoogeography and continental drift. *GB* 90:405-13
696. Crick, R. E. 1980. Integration of Paleobiogeography and Paleogeography: Evidence from Arenigian Nautiloid Biogeography. *JP* 54:1220-1
697. Parker, R. B. and H. Toots. 1980. Trace Elements in Bones are Paleobiological Indicators (*in* Behrensmeyer, A. K. and A. P. Hill. 1980. *Fossils in the Making. UC*, p. 206
698. Ronov, A. B. 1982 The Earth's sedimentary shell (quantitative patterns of its structure, compositions, and evolution. *IG* 24(12)1381 (Part II)
699. Ronov, A. B. 1982. The Earth's sedimentary shell (quantitative patterns of its structure, compositions, and evolution. *IG* 24(11)1321-39 (Part I)
700. Hobson, G. D. and E. N. Tiratsov. 1981. *Introduction to Petroleum Geology*, 2nd Edition. Gulf Pub. Co. Houston, Texas, p. 271
701. Stokes, W. L. 1983. Diastrophy—a Word Whose Time Has Come. *Journal of Geological Education*. 31:35
702. Dott, R. H. 1982. The Challenge of Scientific Creationism. *JP* 56:268
703. Woodmorappe, J. 1981. The Essential Nonexistence of the Evolutionary-Uniformitarian Geologic Column: A Quantitative Assessment. *CR* 18(1)58-67
704. Milne, D. H. 1981. How to Debate with Creationists—and "Win." *American Biology Teacher*. 43(5)238, his figure 1b
705. Benton, M. J. 1983. Large-scale replacements in the history of life. *NA* 302(3)17
706. Jones, A. J. 1973. How Many Animals in The Ark? *CR* 10:102-8
707. Jochmans, J. R. 1978. *Strange Relics from the Depths of Earth*. Bible-Science Organisation.
708. Walker, A. C. 1980. Functional Anatomy and Taphonomy (*in* Behrensmeyer and Hill. 1980. *op. cit.* Ref. 197), p. 182-3
709. Daghlian, C. P. 1981. A Review of the Fossil Record of Monocotyledons. *Botanical Review* 47(4)523
710. Woodmorappe. 1978. *op. cit.* (Ref. 653), p. 101
711. Lipps. 1983. (Letter) *Science Digest*, February 1983, p 9
712. Carrol, R. L. 1977. Patterns of Amphibian Evolution: an extended Example of the incompleteness of the Fossil Record (*in* Hallam, A. 1977. *Patterns of Evolution in the Fossil Record. EL*), p. 422
713. Simpson. 1980. *op. cit.*, (Ref. 672), p. 252-5
714. von Wellnitz, M. 1979. Noah and the Flood: The Apocryphal Traditions. *CR* 16(1)44
715. Behrensmeyer, A. and R. E. D. Boaz. 1980. The Recent Bones of Amboseli Park, Kenya, in Relation to East African Paleoecology (*in* Behrensmeyer and Hill. 1980. *op. cit.*, Ref. 697), p. 80
716. Signor, P. W. III. 1982. Species richness in the Phanerozoic: Compensating for sampling bias. *GE* 10:626
717. Shotwell, J. A. 1955. An Approach to the Paleoecology of Mammals. *EC* 36:328
718. Bishop, W. W. 1980. Paleogeomorphology and Continental Taphonomy (*in* Behrensmeyer and Hill. 1980. *op. cit.*), p. 32
719. Wall, W. P. 1983. The Correlation Between High Limb-Bone Density and Aquatic Habits in Recent Mammals. *JP* 57:197
720. Schafer, W. 1972 (translated). *Ecology and Palaeoecology of Marine Environments. UC*, p. 20
721. Hill, A. P. 1980. Early Postmortem Damage to the Remains of Some Contemporary East African Mammals (*in* Behrensmeyer and Hill. 1980. *op. cit.*), p. 134
722. Hill, A. 1979. Disarticulation and scattering of mammal Skeletons. *PB* 5(3)262
723. Boaz, N. T. and A. K. Behrensmeyer. 1976. Hominid Taphonomy: Transport of Human Skeletal Parts in an Artificial Fluviatile Environment. *American Journal of Physical Anthropology* 45:59
724. Gifford, D. P. and A. K. Behrensmeyer. 1977. Observed Formation and Burial of a Recent Human Occupation Site in Kenya. *Quaternary Research* 8:245-66
725. Behrensmeyer, A. K. and A. P. Hill. 1980. Introduction (*in* Behrensmeyer and Hill. 1980. *op. cit.* Ref. 697), p. 1
726. Hill, A. 1978. Taphonomical background to fossil man-problems in palaeoecology (*in* Bishop, W. W. 1978. *Geological Background to Fossil Man*. Scottish Academic Press), p. 98
727. Behrensmeyer, A. K. 1978. Taphonomic and ecologic information from bone weathering. *PB* 4(2)150-62
728. Behrensmeyer, A. K., et al. 1979. New perspectives in vertebrate paleoecology from a recent bone assemblage. *PB* 5(1)14
729. Rolfe, W. D. I. and D. W. Brett. 1969. Fossilization Processes (*in* Eglinton G. and M. T J Murphy. 1969. *Organic Geochemistry. SP*), p. 226
730. Whitcomb & Morris. 1961. *op. cit.* p. 26
731. Cooper, et al. 1982. Late Precambrian and Cambrian Fossils from Northern Victoria Land and their Stratigraphic Implications (*in* Craddock, C. 1982. *Antarctic Geoscience*. University of Wisconsin Press), p. 630
732. Clarkson, P. D., Hughes, C. P., and M. R. A. Thomson. 1979. Geological Significance of a Middle Cambrian fauna from Antarctica *NA* 279:791
733. Dalziel, I. W. D. 1982. The Early (Pre-Middle Jurassic) History of Scotia Arc Region: A Review and Progress Report (*in* Craddock. 1982. *op. cit.*), pp. 114-5
734. Ke-xing, Y. and Z. Sen-gui. 1983. Discovery of the *Tommotia* Fauna in Southwest China *SI* 22(1)40
735. Stumm, E. C. 1956. Upper Cambrian Trilobites from Michigan. *Contributions of the Museum of Paleontology, University of Michigan*. XIII(4)95
736. Oosthuizen, R. D. 1981. An Attempt to Determine the Provenance of the Southern Dwyka from Palaeontological Evidence. *AF* 24:27-8
737. Sen-Gui, Z. 1983. Early Cambrian Archoeocyathids from Kuruktag, Xinjiang. *SI* 22(1)18
738. Yen-hao, L. 1976. Ordovician Biostratigraphy and Palaeogeography of China. *NJ* 7, p. 23
739. Alexander, R. R. 1975. Phenotypic Lability of the Brachiopod *Rafinesquina Alternata* (Ordovician) and Its Correlation with the Sedimentological Regime. *JP* 49:608
740. Sheehan, P. M. 1975. Lower Devonian Brachiopods from the Solis Limestone, Chihuahua, Mexico. *JP* 49:445
741. Bassi, U. K., Chopra S. and B. M. Datta. 1983. A New Phanerozoic Basin in Kinnaur, Himachal Himalayo. *IN* 24:281-90
742. Crick, R. E. and C. Teichert. 1983. Ordovician endocerid genus *Anthocerass* its occurrence and morphology. *CH* 7(2)155
743. Cai-gen, L. 1982. Ordovician Cephalopods from Xainza, Xizang (Tibert). *SI* 21(5)558
744. Karatajute-Talimaa, V. N. 1978. Silurskiye i Devonskiye Telodonty CCCP i shpitzbergena. Mosklas, Vilnius, 334 p. (in Russian)
745. Blieck, A., et al. 1982. Vertebres du Devonien superieur d'Afghanistan. *Bull du Museum National d'Historie Naturell* 4e serie T, p. 5
746. Boucot, A. J., Doumani, G. A., and G. F. Webers. 1967. Devonian of Antarctica (*in* Oswald, D. H 1967 *op. cit.*), p. 640-1
747. Young, G. G. and J. D. Gorter. 1981. A new fish fauna of Middle Devonian age from the Taemas/Wee Jasper region of New South Wales. *Australia, Bureau of Geology and Mineral Resources*, p. 86
748. Carozzi, A. V. 1979. Petroleum Geology in the Paleozoic Clastics of the Middle Amazon Basin, Brazil. *PE* 2(1) 66-7

749. Denison, R. H. 1966. *Cardipeltis* An Early Devonian Agnathan of the Order Heterostraci. FI 16(4)89-92
750. Long, J. A. 1983. New Bothriolepid Fish from the Late Devonian of Victoria, Australia. PA 26:296
751. Stumm, E. C. 1953. Lower Middle Devonian Proetid Trilobites from Michigan, Southwestern Ontario, and Northern Ohio. *Contributions of the Museum of Paleontology, University of Michigan* XI(2)15
752. Carls, P. and R. Lages. 1983. Givetium und Ober-Devon in der Ostlichen Iberischen Ketten (Spanien). *Zietschrift der Deutchen Geologischen Gesselschaft* 134(1)120-3
753. Rzhonsnitskaya, M. A. 1978. *Yarusnoye Raschlenenie Nizhnevo Devona Tiho'okeanskoi oblasti na Teritorii CCCP.* Moskva, Hedra, p. 10 (in Russian)
754. Plumstead, E. P. 1964. Palaeobotany of Antarctica (*in* Adie, R. J. 1964. *Antarctic Geology*. JW), p. 638
755. Jing-zhi, et al. 1979. *Advances in the Carboniferous Biostratigraphy of China.* Academia Sinica, Nanjing China, p. 4
756. Tanner, W. R. 1983. A New Species of *Gosslingia* (Zosterophyllophyton) from the Lower Devonian Beartooth Butte Fm. of Northern Wyoming (*in* Mamet, B. and M. J. Copeland. 1983. *Third North American Paleontological Convention Proceedings*) V. 1, p. 541
757. Singh, G., Maithy, P. K., and M. N. Bose. 1982. Upper Palaeozoic Flora of Kashmir Himalaya. BO 30(2)194
758. Grindley, G. W., et al. 1980. A Mid-Late Devonian Flora from the Rupert Coast, Marie Byrd Land, West Antarctica. *Journal of the Royal Society of New Zealand* 10(3)271-2
759. Bogoslovskiy, B. I., Poslavskaya, I. A., and O. Ye. Belyayev. 1982. Frasnian ammonoids from Central Kazakstan. PJ 16(3)32
760. Gupta, V. J. and H. K. Erben. 1983. A Late Devonian ammonoid faunule from Himachal Pradesh, India. PZ 57(1/2), p. 93
761. Anderson, T., et al. 1974. Geology of a Late Devonian Fossil Locality in the Sierra Buttes Fm., Dugan Pond, Sierra City Quadrangle, California. GA 6(3)139
762. Petersen, M. S. and W. L. Stokes. 1983. A clymenid ammonoid from the Pinyon Peak Limestone of Utah. JP 57:717
763. Yan, Z. and R. Yiping. 1983. Discovery of a Devonian ammonoid Species from Ejin Banner of Western Inner Mongolia. SI 22
764. Boucot, A. J., Massa, D., and D. G. Perry. 1983. Stratigraphy, Biogeography, and Taxonomy of Some Lower and Middle Devonian Brachiopod-Bearing Beds of Libya and Northern Niger. PH 180A, p. 95
765. Kropachev, A. P., et al. 1982. The Fammenian Deposits in the North of the Sette-Daban Range. SV 23(5)117
766. Miller, H. 1982. Geologic Comparison between the Antarctic Peninsula and Southern South America (*in* Craddock. 1982. *op. cit.*), p. 127
767. Yin, T. H. 1935. *Upper Palaeozoic Ammonoids of China.* Peiping, pp. 1-6
768. Yi-ping, R. 1979. *On the Occurrence and Stratigraphical Significance of the Carboniferous Ammonoid Faunas in Nandan of Guangxi.* Academia Sinica, Nanjing, China, pp. 3-7
769. Campbell, K. S. W., Brown, D. A., and A. R. Coleman. 1983. Ammonoids and the correlation of the Lower Carboniferous rocks of eastern Australia. CH 7(2)76
770. Lin, R. 1983. Fusulinacean Fauna from the Quanwangtou Limestone (Early Upper Carboniferous) in Jiawang Coal Field, Northern Jiangsu. SI 22(2)179
771. Chandra, A. and A. K. Srivastava. 1982. Plant Fossils from the Talchir and Coal-Bearing Formations of South Rewa Gondwana Basin, India and their Biostratigraphical Significance. BO 30(2)143-67
772. Benita, M. and D. L. Jones. 1983. Tectonic and Paleobiologic Significance of Permian Radiolarian Distribution in Circum-Pacific Region. AG 67(3)522
773. Archbold, N. W. 1983. Permian Marine invertebrate provinces of the Gondwanan Realm. CH 7(1)61-70
774. Nassichuk, W. W. 1970. Permian ammonoids from Devon and Melville Islands, Canadian Arctic Archipelago. JP 41:77
775. Kobayashi, I., et al. 1982. Discovery of Carboniferous-Permian bryozoans at Ikadomari in Sado Island. BB 47(7)1792
776. Olson, E. C. 1967. Early Permian Vertebrates. *Oklahoma Geological Survey* Circular 74
777. Clemens, W. A., et al. 1979. Where, When, and What (*in* Lillegraven, J. A., et al. 1979 *Mesozoic Mammals.* University of California Press), p. 14
778. Hammer, W. R. and J. W. Cosgriff. 1981. *Myosaurus Gracilis*, an Anomodont Reptile from the Lower Triassic of Antarctica and South Africa. JP 55:411
779. Colbert, E. H. 1982. Mesozoic Vertebrates of Antarctica (*in* Cradock. 1982. *op. cit.*), p. 620-1
780. Monod, O., et al. 1983. Decouverte de Dipneustes Triasianes (Ceratodontiformes, Dipnoi) dans la Formation de Cenger ("Arkoses Rouges") da Touras Lycien (Turquie Occidentale) BI :167
781. Banks, M. R. 1978. Correlation Chart for the Triassic System of Australia. *Australia: Bureau of Mineral Resources, Geology, and Geophysics* Bulletin 156C
782. Dziewa, T. J. 1980. Note on a Dipnoan Fish from the Triassic of Antarctica. JP 54:488-9
783. Thulborn, R. A. 1983. A Mammal-like reptile from Australia. NA 303:330
784. Pie-xia, G. 1982. On the Occurrence of Late Lower Triassic Ammonoids from Anhui and Jiangsu. SI 21(5)567
785. Yi-gang, W. 1983. Ammonoids from Falang Formation (Ladinian-E. Carnian) of Southeast Guizhou, China. SI 22(2)160
786. Mahashwari, H. K. 1982. Mesozoic Plant Fossils from the Himalayas—A Critique. BO 30(3)243
787. Bose, M. N., et al. 1982. *Pachypteris Haburensis* N.S.P. and Other Plant Fossils from the Pariwar Formation. BO 30(1)1-2
788. Ballance, P. F. and W. A. Watters. 1971. The Mawson Diamictat and the Carapace Nunatak, Victoria Land, Antarctica. *New Zealand Journal of Geology and Geophysics* 14(3) 521
789. Askin, R. H. and D. H. Elliot. 1982. Geologic Implication of recycled Permian and Triassic palynomorph in Tertiary rocks of Seymour Island, Antarctic Peninsula. GE 10:547-51
790. Behrendt, J. and T. S. Laudon. 1964. Cretaceous Fossils Collected at Johnson Nunatak, Antarctica. SC 143:353-4
791. Quilty, P. G. 1970. Jurassic Ammonites from Ellsworth Land, Antarctica. JP 44:110
792. ———. 1982. Atlas Bespozvonochnikh Pozdnemyelovih Morei Prikaspiiskoi Upadini. TR 187:4, 193-228
793. Davoudzadeh, M. and K. Schmidt. 1983. Contribution to the Paleogeography and Tectonics of the Middle and Upper Jurassic of Irah. JAAbh. 166(3)331
794. Thomson, M. R. A. 1980. Late Jurassic ammonite faunas from the Latady Formation, Orville Coast. *Antarctic Journal of the United States* 15(5)28-30
795. Thomson, M. R. A. 1982. A comparison of the ammonite faunas of the Antarctic Peninsula. GL 139:763-70
796. Carpenter, K. 1982. The Oldest Late Cretaceous Dinosaurs in North America? *Mississippi Geology* 3(2)13
797. Bose, M. N., Kutty, T. S., and H. K. Mahashwari. 1982. Plant Fossils from the Gangapur Formation BO 30(2)122
798. Pei-ji, C. 1983. A Survey of the non-marine Cretaceous in China. *Cretaceous Research* 4:128
799. Wellenhofer, P., et al. 1983. A Pterosaurian from the Lower Cretaceous of Brazil. PZ 57(1/2)149
800. Birkenmayer, K., et al. 1983. Cretaceous and Tertiary fossils in glaciomarine strata at Cape Melville, Antarctica. NA 303:56
801. Zaborski, P. M. 1982. Campanian and Maastrichtean sphenodiscid ammonites from southern Nigeria. BR 36(4)304
802. Farquharson, G. W. 1982. Late Mesozoic sedimentation in the Northern Antarctic Peninsula and its relationship to the southern Andes. GL 139:721-2
803. Dong-li, S. and Z. Bing-gao. 1983. Aspects of the Marine Cretaceous of China. *Cretaceous Research* 4:147
804. Lillegraven, J. A. and A. R. Tabrum. 1983. A new species of Centetodon (Mammalia, Insectivora, Geolabididae) from southwestern Montana. *University of Wyoming Contributions to Geology*. 22(1)58
805. Savage, D. E. and D. E. Russell. 1983. *Mammalian Paleofaunas of the World.* Addison-Wesley Pub. Co., 432 p.
806. Woodburne, M. O. and W. J. Zinsmeister. 1982. Fossil Land Mammal from Antarctica. SC 218:284
807. Zinsmeister, W. J. and H. H. Camacho. 1982. Late Eocene Molluscan Fauna of LaMeseta Formation of Seymour Island, Antarctic Peninsula. (*in* Craddock. 1982. *op. cit.*), p. 300-1

An Anthology of Matters Significant to Creationism and Diluviology: Report 2

AN ANTHOLOGY OF MATTERS SIGNIFICANT TO CREATIONISM AND DILUVIOLOGY: REPORT 2

John Woodmorappe[*]

Received 20 February, 1981

This report is not about one specific topic, but is a collection of miscellaneous findings conveying a diverse body of information of interest to Creationists and Diluvialists. It is thus a natural sequel to the author's first anthology.[1]

Highlights of points concerning biological evolution include: 1) fallacies in claims of life from non-life, 2) lack of a proven driving mechanism; 3) the problem of "living fossils, 4) fundamental biologic phenomena not explained by evolution.

The section on "ancient reefs" further shows that: 1) ancient "reefs" lack a reef network, 2) these deposits were cemented inorganically, 3) growth orientation is no proof of growth in situ over immense periods of time.

Previous Creationists' observations about "overthrusts", such as lack of gouge, and perfectly conformable "thrust" contacts, are confirmed. Over two hundred cases of anomalous fossils are tabulated; and it is shown that such fossils typically do not show morphological evidence of the "reworking" which has been invoked to explain them.

A final section on uniformitarianism notes evidence to show that thick igneous and metamorphic rocks have formed and cooled quickly; and illustrates the blinding influence of uniformitarianism.

Plan of this Article

I. BIOLOGICAL EVOLUTION: SCIENTIFIC AND PHILOSOPHICAL CONSIDERATIONS.
II. FALLACIOUS CLAIMS OF "ANCIENT REEFS" IN THE GEOLOGIC RECORD.
III. UNIFORMITARIAN CONFIRMATIONS OF CREATIONISTS' OBSERVATIONS CONCERNING (ALLEGED) OVERTHRUSTS.
IV. SOME EXAMPLES OF "REWORKING" RATIONALIZATIONS FOR ANOMALOUS FOSSILS
V. RATES OF GEOLOGIC PROCESSES AND UNIFORMITARIANISM: SCIENTIFIC AND PHILOSOPHICAL CONSIDERATIONS.

I. BIOLOGICAL EVOLUTION: SCIENTIFIC AND PHILOSOPHICAL CONSIDERATIONS

1. Incongruity of Claiming Evolution to be a Fact.

In a recent comprehensive work on evolution, Darlington, a zoologist based at Harvard, wrote: "Different minds will require different 'degrees of cogency', but I think that most persons who look at the evidence for themselves, and are not prevented by religious or political prejudices (i.e., by judgements before the evidence) will accept evolution as a fact".[3]

Publishing in 1969 (long before the Creationist movement had become widely known and many evolutionists had resorted, in retaliation, to dogmatic proclamations of evolution as fact) Savage[51] wrote: "No serious biologist today doubts the fact of evolution, the development of all living organisms from previously existing types under the control of evolutionary processes ... We do not need to list evidences demonstrating the fact of evolution any more than we need to demonstrate the existence of mountain ranges", (sic).

By contrast, Davies[4] wrote: "No hypothesis is ever proven, only mathematicians prove things. In science, we can only ask whether or not a hypothesis seems to correspond with the real world, as best we perceive it ...To take another hoary, but still politically active example, the debate about the theory of evolution would generate less heat if some of its proponents did not claim that the theory is a proven fact."

Comment: Even if evolution was very strongly supported by evidence, evolutionists should still have no right to proclaim evolution to be fact or attempt to browbeat students and readers of their works to accept evolution as fact.

As is, all the premises of evolution completely break down under close examination, so " ... most persons who look at the evidence for themselves and are not prevented by religious or political prejudices" should agree that evolution does not "correspond to the real world."

2. Utter Baselessness of all Evolutionistic Origin-of-Life Hypotheses

Brownlow[2] wrote: "Special conditions *may have* been required for the next step, the combination of biomonomers into the structurally complicated biopolymers, such as proteins. Only relatively simple biopolymers have been formed in laboratory experiments, and none of the extremely complex polymers of living organisms has been synthesized. It *seems* probable that fairly special (but not necessarily unusual) conditions were required for the evolution of biopolymers. For instance, this evolution *may have* taken place in isolated ponds where the necessary biomonomers were concentrated by evaporation and a chemical catalyst was present to make certain reactions occur efficiently. On the other hand, this evolution *may have* occurred in the oceans, where the clay minerals *could have* served as concentrators and as catalysts. We know that clay minerals have chemically active surfaces and interact with organic molecules. Laboratory research has shown that clay minerals can bring together different organic molecules and can stabilize amino acids. All this is, however, *pure speculation*. We *know very little* about the formation of biopolymers on the earth by non-

[*] John Woodmorappe has an M.S. in Geology, and a B.A. in both Geology and Biology.

biological processes. The next step, the formation of a living thing, is also *not understood* in terms of chemical processes. This step marks the actual origin of life and was followed by biological evolution." (italics added)

Comment: These statements demonstrate once again all evolutionistic claims of abiogenesis are sheer fantasy.

3. Origin of Metazoans Entirely Conjectural

Valentine[52] wrote: "Just which of the many possible early metazoan forms was the actual primitive metazoan may never be known; all metazoan lineages of this grade are extinct. Fossils are unlikely to occur and would not be conclusive if found. There is no shortage of *speculative* reconstructions, however, the most famous being the blastaea-gastraea pathway envisioned to Haeckel (1874)." (italics added)

Comment: After the (supposed) origin of life from nonlife, the next evolutionary step imagined is the origin of metazoans and increase of biotal complexity and diversity prior to the origin of the phyla. It has been even said the 80% of molecules-to-man evolution had already been completed before the origin of the phyla. But just as the evolutionary origin of life from nonlife is fantasy and wishful thinking, so also is the evolution of metazoans.

4. Lack of Proven Driving Mechanism for Evolution

Ruse[53] wrote: "Evolution is, to put it simply, the result of natural selection working on *random* mutations." (italics added)

In speaking of famous modern examples of supposed evolution at work (drug-resistant viruses, the peppered moth *Biston betularia*, etc.), Ruse[54] wrote: "The experiments are designed to tell one about the theory—this they do. They are not aimed primarily at the actual reconstruction of the history of life, and for this reason should not be blamed when they do not tell us about it."

Hallam[55] said: "Certainly, elucidation of evolutionary *mechanisms* must remain the province of geneticists and ecologists, but these scientists are denied the invaluable time dimension which allows us to investigate evolutionary *patterns* in a comprehensive and meaningful way. Such topics as crucial to a full understanding of evolution as the nature of diversity, change through time, rates of origination and extinction, progressive colonization of and adaptation to ecological niches, convergence and parallelism, paedomorphosis, size change, *the temporal aspect of speciation and origination of new higher taxa*, and radiation and extinction in relation to changing macro-environments, are decidedly the realm of the paleontologist." (italics added for phrase: "... the temporal aspect of speciation and origination of new higher taxa ...". Other italics his)

Ruse[56] wrote: "However, although geneticists know of some mutations which cause fairly drastic changes, they have entirely failed to discover the kind of macromutations required by the saltation theory—the kind of mutation which would take a group of organisms from one order to another. Moreover, the large-effect mutations which are known are usually just those mutations which are the most crippling to their carrier ... Of course, one might argue that the failure to find the right kind of macromutations does not necessarily prove their non-existence, but, like unicorns, there is a difference between saying that *logically* they might exist and that it is *reasonable* to suppose that they exist." (italics his)

Comment: The first statements (ref. 53, 54) of Ruse and the statements of Hallam concern evolution by gradual accumulation of selected micromutations, whereas the final statement of Ruse (ref. 56) concerns evolution via drastic mutations in relatively short periods of time.

Ruse (ref. 54) claims that such examples as the peppered moth are *bona fide* examples of and proofs of evolution, yet acknowledges that they in no way demonstrate that actual evolution of new living forms occurs. The statements of Hallam make it clear that all claims of real evolution (origin of new forms) appeal to paleontology. Knowledge from genetics, population biology, ecology, etc. in no way proves that macro-evolution is taking place. The claim that there *is* such a process as macroevolution must always appeal to the fossil record to imagine that it *did* happen. One can never prove from modern living things that macroevolution is happening or even *exists*.

Some newer concepts of evolution as "saltatory evolution," "hopeful monsters," "punctuated equilibria," etc., rely to a large extent on supposed macromutations. The final statement of Ruse (ref. 56) makes it clear that there is not one iota of evidence for these macromutations as sources for evolution.

5. Non-Preservation of Hard Parts of Ancestral Forms No Excuse for the "Cambrian Explosion"

Frey[5] wrote: "A popular theme in organic evolution holds, in essence, that the 'sudden' appearance of fossils in Cambrian rocks reflects the acquisition by animals of hard parts capable of being preserved, not the rapid diversification of organisms themselves during earliest Cambrian time. If true, one should expect to find diverse assemblages of trace fossils in Precambrian rocks, which is not the case. Studies ... show that trace fossils reflect an explosion in diversity and complexity in Cambrian rocks that is comparable to that of body fossils ... thus ruling out the above theme as a simple explanation for impoverished Precambrian biotas."

Comment: Any ancestors of the phyla appearing suddenly at the basal Cambrian would leave a record of traces even if the ancestors lacked hard parts to be preserved as body fossils. The fact that traces also have a "Cambrian explosion" means that the only rationalization that evolutionism can invoke for the abrupt appearance of the phyla (and many lower taxonomic categories) in the Cambrian is the convenient claim that evolution was so rapid that it left no fossil record at that point.

6. Illogical Reasoning in Appeal to the Incompleteness of the Fossil Record as a Rationalization for Absent Transitions

Darlington[6] wrote: "Many gaps and ambiguities occur in the fossil record and are stressed by critics, but (as Darwin noted) they are expected. Fossilization is and must be rare and chancy ... The fossil record *must* be fragmentary. It would almost be more logical to criticize the record, not because it is incomplete, but because it is better than it ought to be." (italics his)

144 pages later, Darlington[7] wrote: "Nevertheless, in spite of being fragmentary and biased, the fossil record gives us a surprisingly good view—almost a magical one—of the course of evolution at least of higher plants and animals."

Comment: Evolutionists can appeal to an incomplete fossil record as a rationalization for absent transitions and then turn around, contradict themselves, and point out that the fossil record is actually very rich.

Creationists Anderson and Coffin[8] noted that further collecting of fossils sharpens, not closes, the gaps in supposed evolutionary fossil sequences. More and more evolutionists are recognizing this, and the "punctuated equilibrium" concept has been invented for this purpose (see below).

7. Incompleteness of Fossil Record No Excuse for Absent Transitions

Waterhouse[9] wrote: "Darwin set aside most of the fossil evidence for evolution with the proposal that it was massively incomplete. But there were polemic rather than scientific reasons for this attitude because he insisted on gradualistic evolution which most fossils did not substantiate. But the fossil record can no longer be set aside as woefully incomplete. More than 100 years of study demand instead that the gradualistic concept be reassessed."

Comment: Because many evolutionists no longer believe that missing transitions can be explained away by appeals to non-preservation, they have proposed the "punctuated equilibrium" concept, where it is conveniently supposed that evolution of new taxonomic categories was so rapid that it left no fossil record at that point."

8. "Missing Links" Substantiating (Alleged) Human Evolution No Longer Expected to be Ever Discovered

Laporte[57] recently wrote: "Ironically, as the hominid fossil record improves, arguments for phyletic gradualism lose their force, particularly the old chestnut that 'if the record were better, we'd see transitional change occurring'."

Comment: The "punctuated equilibrium" concept is applied to human evolution. How often human evolution is presented as fact to the gullible public, yet now some evolutionists not only admit the lack of transitions in human evolution, but do not even pretend a hope that they will ever be found!

9. Human Evolution: Welter of Contradictions, Imaginations, and Ill-Defined Taxons

In an excellent up-to-date review article on human evolution, Cronin *et. al.*[138] wrote: "However, not all palaeoanthropologists are convinced that *H. erectus* is ancestral to *H. sapiens*." They then described 4 mutually-contradictory widely-held views of human evolution. The *H. habilis-H. erectus-H. sapiens* lineage was accepted and held common to the 4 views. One view held *Australopithecus africanus* and *A. afarensis* both ancestral to the aforementioned *Homo* line, another held the former but not the latter to be ancestral, another held the latter but not the former to be ancestral, and still another view held neither form to be ancestral to genus *Homo*.

Cronin *et. al.*[139] also said: "Specimens often quoted as displaying 'intermediate' or 'mosaic' characters between *H. erectus* and *H. sapiens* include Broken Hill and Omo (Kibish) in Africa, Ngandong and Arago, Vertesszollos and Petralona in Europe. Other recently discovered fossils which may belong to this intermediate category are the ..." "... present concepts of variation within the species *H. erectus* and *H. sapiens* need to be re-examined."

Cronin *et. al.*[140] wrote: "Second, while the K-Ar dating of tuffs at Laetoli and Hadar remain to be confirmed by other dating techniques, such as palaeomagnetism, there are preliminary faunal indications that Hadar and Laetoli may be closer in time than the absolute dates would suggest."

Comment: The highly imaginative character of presumed human evolution is once again demonstrated by the highly contradictory views of how this supposed evolution took place (Cronin *et. al.*, ref. 138).

The claims of Cronin *et. al.* that the *H. habilis-H. erectus-H. sapiens* lineage is transition-filled (gradual cranial-size and body-weight increase, p. 118) is betrayed by the fact that these "species" of *Homo* are quite amorphous. The fact that many forms are "mosaic" and "intermediate" (ref. 139 cited above) between *H. erectus* and *H. sapiens* need not be evolutionary but simply indicative of the ill-definition, amorphousness, and artificiality of these "species." My article on cephalopods[141] described similar phony transition-filled evolution in artificial genera. Cronin's statement (ref. 140) shows once again the selective acceptance of age-dating results.

10. "Living Fossils" Admittedly are not Explained by Evolution

Paul[58] wrote: "Nevertheless, examples are known among vertebrates (*Sphenodon*, *Latemeria*), invertebrates (*Neopilina*, *Platasterias*), and plants (*Ginkgo*, *Araucaria*) and are thus too widespread to be ignored. Clearly there is some pervasive effect which allows the survival of small groups for long periods, but what it is remains unknown. The occurrence of 'living fossils' is partly due to our better understanding of the living world compared with the fossil record. Small groups have a relatively low preservation potential, but this only explains why small groups are not preserved, not why they manage to survive for long periods."

Comment: Creationists have long been calling attention to "living fossils" such as the coelacanth *Latimeria* and the maiden-hair tree *Ginkgo*. *Platasterias* is an extant echinoderm that has persisted unchanged since the Ordovician, with last fossil appearance in the Devonian. Paul admits that there is no evolutionary explanation as to why some forms persist for hundreds of millions of years after allied forms either evolved into something else or became extinct. It is easily explained by the discarding of evolution and geologic time via the Creationist-Diluvialist Paradigm.

11. Many Fundamental Biologic Phenomena Not Easily Reducible to Evolutionistic Explanations

Van Heyningen[59] said: "Why the tetanus and botulinus bacilli should produce these immensely potent toxins is a problem of great philosophical and practical interest. Diphtheria toxin and most other bacterial toxins attach and break down the tissues of the animal infected by the parent organism. In doing so, they assist the bacteria in the invasion because the bacteria grow well in disintegrating tissues. The tetanus and botulinus toxins, however, do not attack animal tissues generally. It does not appear to be of any survival value to the tetanus and botulinus bacilli to produce toxins that not only confine their action to nerve tissue but also, as far as can be seen, cause no damage even in this tissue. Yet on evolutionary ground it is hardly conceivable that the bacilli should produce the toxins unless they have some survival value."

Comment: This is yet another example of biologic phenomena that do not appear to have any survival value to the organism and hence do not support evolution and its "survival of the fittest" dictum.

12. Persistence of Belief in the Biogenic Law among Evolutionists

Darlington[10] wrote: "In simple principle ontongeny does recapitulate phylogeny, but the recapitulation is at best incomplete and often also imprecise and complexly modified by omission of stages, distortion of sequences, or premature termination..."

Comment: Because of the inconsistencies noted above, the claim that ontogeny recapitulates (supposed) phylogeny is superficial at best.

13. Fallacies of Evolutionistic Attempts to Circumvent Probabalistic Arguments

Darlington[11] wrote: "To apply this concept to, for example, man: it was probably inevitable that an intelligent organism in some ways like man would evolve on earth. The probability that this organism would be man as he is, in all details, approaches zero. But the possibilities approach infinity, and one of these possibilities was (practically) sure to occur, and man is in fact the one."

In responding to the classic Monkeys-Typing argument, Darlington[12] said: "This is the kind of situation that obtains in evolution: the source of energy (analogous to the monkey) is the molecular and chemical energy of atoms and molecules, the 'living letters' are genes and their components, mutants, and combinations; and the additions have in fact been greater than the erasures, for otherwise evolution would not have occurred. Under these conditions, the probability that evolution will produce any pre-designated organism approaches zero, but the probability that some organism comparable to, say, man, in organization, complexity, and intelligence may approach certainty."

Comment: First of all, this argument begs the question. It is known that something *had* to evolve because evolution does take place, and evolution is known to take place because something *had* to evolve.

It has not been demonstrated (much less proved) that any living system *could* evolve, let alone that *some* sort of living system *had* to evolve. To say that complex living things are here because *some* complex living system *had* to appear is folly and presumption.

14. Evolutionistic View of the Basic Nature of Man: An Animal

Heinlein[13] wrote: "Man is an unspecialized animal. His body, except for his enormous brain case, is primitive. He can't dig; he can't run very fast; he can't fly. But he can eat anything and he can stay alive where a goat would starve, a lizard would fry, a bird freeze. Instead of special adaptations, he has general adaptability."

Clebsch[14] wrote: "Then Darwin's epoch making *Origin of the Species by Means of Natural Selection* (of 1859) and *The Descent of Man* (of 1871) made fables of the notion of human uniqueness and the story of the animals in Noah's ark. Put crudely, humanity bore the image of the ape-*creatio in imago simii*." (italics his)

Comment: Despite efforts by many evolutionists to downplay it, evolution clearly teaches that man is mere animal. It is fallacious for "theistic evolutionists" to imagine that man became the image and likeness of God (Gen. 1:27) at a certain point in his (alleged) evolution, for the simple reason that evolution does not recognize any qualitative difference between animals and man.

15. Evolutionistic Advocacy of a Cruel and Violent Nature

Darlington[15] wrote: "The first point is that selfishness and violence are inherent in us, inherited from our remotest animal ancestors. They are not peculiar to man. Some biologists have tried to persuade themselves that cooperation rather than competition is the rule in nature, and that violence (a form of competition) is unnatural or secondary, but they are mistaken. Nature is, conspicuously, red in tooth and claw, and I do not see how naturalists who look carefully around them can doubt it... Violence is, then, natural to man, a product of evolution."

Comment: Evolution is therefore not only scientifically fallacious but also immoral. It reinforces the delusion that violence, etc. are natural and inevitable rather than the products of man's fallen sinful nature of his own responsibility. The fact that most evolutionists do not practice nor advocate violence is of no importance; reinforcing sinful delusions on a matter of such profound implication as the fundamental nature of man can do only harm, as spawning evil socio-political philosophies (see below).

16. Evolution and the Origin of Evil Socio-Political Philosophies

Concerning Darwin, Littell[16] wrote: "He proposed that natural selection governs the evolution of forms of life; with the fittest surviving. The latter proposition became the basis of several schools of politics and social philosophy, including both laissez-faire economics and Nazism. The former displaced the view of man as a fallen angel and replaced it with man conceived as risen animal."

Hoffman[17] wrote: "Hitler believed in struggle as a Darwinian principle of human life that forced every people to try to dominate all others; without struggle they would rot and perish ... Even in his own defeat in April 1945 Hitler expressed his faith in the survival of the stronger and declared the Slavic peoples to have proven themselves the stronger."

Comment: These statements illustrate once again that many evil socio-political philosophies were and are based on evolution. It is interesting to note that Hitler's committment to evolution was stronger than his committment to Pan-Germanism.

While evolutionists deny any legitimate association with, or responsibility for, Nazi ideology and practice, their position reinforces delusions (see no. 15 above). Since the evolution tree is evil, it can only, sooner or later, in one form or another, give rise to evil fruit (Matt. 7:17-18). Some other evil fruits are discussed in sections below.

17. Evolution and Nihilism

Darlington[18] wrote: "For example, we may say 'wings are to fly with', and this seems to imply purpose, but to most evolutionists it means only that the function of wings is flight and that flight gives wings the selective advantage that results in their evolution. This implies cause but not purpose. *Most evolutionists see no purpose in evolution even when they use language that seems to imply it.* I shall try not to use teleological language and (if I use it inadvertently) shall never intend to imply purpose by it." (italics added)

Comment: The strongly anti-teleological position of evolution is in diametric contrast to the Creationist position and its recognition of purposeful Divine design in nature (Psalm 139:14, Romans 1:20). Evolution could not possibly have been "God's method of Creation" as "theistic evolutionists" imagine because evolution vehemently rejects all notions of purposeful design. Evolution is thus nihilistic, as there truly is no purpose in anything other than simply that combination of matter which has survived.

18. Evolution and Monistic Principles

Clebsch[19] said: "Most of the century's scientists, including Darwin, sensed the tension between the conclusions their data implied and their personal attachments to the traditional sense of human uniqueness. Even Haeckel wrote a book (in 1892) reconciling religion and science under monistic philosophy."

Comment: Theism and evolution are not "reconciled" under monism for the simple reason that monism is irreconcilable with theism. The growing popularity of Eastern mysticism among western universities may be partly the result of evolutionary concepts.

19. Impossibility of Reconciling Theistic Religion with Evolution

Darlington[20] wrote: "Holism and emergent evolution are only parts of a continuing and, I think, *continually unsuccessful* effort of some evolutionists to reconcile evolution, mystic ideas, and religions. Teilhard de Chardin was perhaps the best-known evolutionist who was still trying to reconcile them until his death a few years ago-after which the Vatican condemned his works." (italics added)

Comment: Any "reconciliation" of theism and evolution is doomed to utter failure because, while attempting to satisfy both, it inevitably ends up satisfying neither. Evolution and theism can't be combined because evolution is thoroughly and decisively atheistic. (see below)

20. Evolution: Atheism and Materialism

Darlington[21] wrote: "The outstanding evolutionary mystery now is how matter has originated and evolved, why it has taken its present form in the universe and on the earth, and why it is capable of forming itself into complex living sets of molecules. This capability is *inherent in matter as we know it*, in its organization and energy.

"*It is a fundamental evolutionary generalization that no external agent imposes life on matter.* Matter takes the forms it does because it has the *inherent* capacity to do so ... This is one of the most remarkable and mysterious facts about our universe: that matter exists that has the capacity to form *itself* into the most complex patterns of life. By this I do *not* mean to suggest the existence of a vital force or entelechy or universal intelligence, but just to state an *attribute of matter* as represented by the atoms and molecules we know ... We do not solve the mystery by using our inadequate brains to invent mystic explanations." (italics added, except for *not*)

Comment: Evolution is clearly atheistic and materialistic. "Theistic evolution" is totally fallacious because: 1) evolution does not tolerate the notion of God having either started the process, nor directed it, nor intervened in any part of it. Nor could God have "used" evolution because evolution does not require Divine consent. There is no place for God in evolution *whatsoever*, while *everything* in evolution is purely the result of matter in motion. Nor could God have even created the natural laws that supposedly make evolution happen because even natural laws are inherent in matter. It is high time that believers stop attempting to compromise with evolution through fallacies such as "theistic evolution", recognize that evolution is unabashedly atheistic, and reject evolution *in toto* because it has no scientific basis.

How interesting that Darlington can simultaneously maintain that evolution is fact (see No. 1, this section) and at the same time acknowledge that how and why things (supposedly) evolve at all are an "outstanding mystery." How interesting also that how and and why evolution (supposedly) occurs is an "outstanding mystery" yet it is something asserted to be demonstrably "inherent in matter." "Professing to be wise, they became fools ..." (Romans 1:22)

II. FALLACIOUS CLAIMS OF "ANCIENT REEFS" IN THE GEOLOGIC RECORD

1. Introduction: Varying Uniformitarian Opinions Concerning "Ancient Reefs"

Shaver[22] wrote: "Perhaps as many of these (studies) have directly or indirectly cast doubt on the existence of true reefs of Silurian age in the Midwest as have supported the idea... Indeed, the graduate students at Indiana University who have joined with me to present the information for... (this guidebook) do not all agree that 'reef' is wisely applied to the Silurian structures so often called 'reefs.'"

Comment: Even within uniformitarianism there is controversy as to whether or not certain carbonate lithologies are "ancient reefs." This entire section is a sequel to a similar section in the author's first Anthology,[23] where evidence was presented that "ancient reefs" were not reefs and could be understood in terms of the Creationist-Diluvialist paradigm.

2. Conjectural and Imaginative Character of Uniformitarian Reef Models Ascribed to Ancient Rock

Mountjoy[43] wrote: "The Alberta Basin is one of the best known Paleozoic reef provinces in the world, especially in terms of the amount of geological data available particularly in the subsurface. It has been extensively studied and various depositional and diagenetic models have been published... The models are based on conjecture, unproven ideas, and the interpretations of limited factual observations germinated and enhanced by imagination."

Comment: Mountjoy recommends that further studies be done to strengthen reef interpretations, but, in view of the fact that these lithologies have already been extensively studied, one might realize that the conjecture actually reflects the fallacious uniformitarian ancient-reef premises.

3. Some "Ancient Reefs" Merely Artifacts of Erosion

After describing effects of subaerial erosion in modern carbonates of the Persian Gulf, Shinn[24] wrote: "Subaerial sculpturing might explain the steep-sided 'cores' and flank beds seen in some ancient 'reefs' and 'bioherms'."

In an ancient example, Squires[25] wrote: "The pseudo-reefs do resemble true reefs in that they seem to lack clearly defined bedding in certain cliff-face exposures."

In another ancient example (this one from a classic locality) Twenhofel[30] wrote: "The Schlern dolomitic limestones of South Tyrol were originally interpreted as a great barrier reef system... with bordering strata dipping steeply from the margins. Ogilvie-Gordon questioned this origin of the steep dips and considered these features as due to faulting and erosion."

Comment: Much "reef geometry" may be apparent rather than real. That which is real is not proof of reef (see No. 12 below)

4. Capriciously-Dipping "Reef Flank" Strata

Carozzi and Hunt[26] wrote: "This reversal of dip is not an uncommon phenomenon for Niagaran reefs and is believed by Cummings and Shrock (1926, 1928) to result from the settling of the heavy reef masses into the underlying sediments."

Comment: Frequently, the "reef flank" strata fail to dip away from the "reef core." The real reason is that these are not ancient reefs but Flood deposits, and the capricious dip reflects variability in Flood currents. This is especially realized because of the fact that ascriptions of "reef core" and "reef flank" are themselves quite arbitrary (see below).

5. Vague Facies Distinction Between "Reef Core" and "Reef Flank"

In describing the Silurian Thornton Reef of Chicago, McGovney[27] reinterpreted much "reef core" as being truly "reef flank" when bedding was discovered in it. He concluded: "The reef flank facies is about 95% of the preserved reef." Even this he considered an arbitrary distinction, contended that there is really no "reef core" as such, and suggested that the deposit was not a shoal-water wave-battered reef but a carbonate mound.

In speaking of the Silurian "reefs" of Iowa, Hinman[28] wrote: "The core and flanks are petrographically nearly identical, but structurally and paleontologically dissimilar... The material of the flank beds is lithologically quite similar to that of the core and can be distinguished from it solely on the basis of the stratification of the flank material. The flank beds are also fossiliferous..."

In reporting on the Silurian "reefs" of Ohio, Kahle[29] wrote: "Although fossils tend to be concentrated within bioherms, they are locally equally abundant in rocks between bioherms."

Comment: It is clear that there is, at best, tenuous justification for attributing rock to the "reef core" and "reef flank" facies. Once again, reefs have been read into the rock record instead of out of it.

6. Conspicuous Absence of Framework in "Ancient Reefs"

Twenhofel[31] wrote: "Walther's definition may hold for some reefs, but few reefs known in the geologic column known to the writer contain many branching corals."

Kahle[29] wrote: "Except for bryozoa, fossils within bioherms typically *do not touch one another* and do not appear to have formed a rigid framework." (italics added)

Lane[32] wrote: "A variety of sedimentary deposits that contain abundant crinoid ossicles, at places completely disarticulated or coherent in the form of calcyes and crowns, are widespread in Paleozoic rocks. A number of such deposits have been called reefs or bioherms even though commonly there is little evidence that they were raised above the surrounding sea floor at the time they were being formed, that they had rigid organic framework, or that they were wave-resistant."

Concerning the Carboniferous "reefs" of England, Anderton *et. al.*[44] wrote: "Problems still remain concerning the origin of the mudbanks, due to the rarity of framebuilders."

Stanley[45] said: "Unlike modern reefs, the North American examples were all small biostromal buildups that never developed in the high-energy surf zones ... These findings acquire special significance when compared with thick Triassic sequences in Germany, Austria, and Italy which have long been regarded as classic reefs. Studies in the Northern and Southern Alps have shown that extensive coral framework is absent."
"Although Middle Triassic sequences in the Dolomites have been referred to as reefs, they have little reef framework and corals are minor constituents.[46]"

Klovan[33] wrote: "The dearth of three-dimensional framework in many ancient reefs has led to a gradual change in the conceptual model of the organic reef from that of a 'reef wall' to that of a thin, discontinuous rim."

Comment: These statements provide further confirmation of the lack of framework in "ancient reefs." This lack is precisely because these were not reefs and consquently organisms did not intergrow and bind lime muds accumulating over immense periods of time. These are Flood deposits, with material and organisms washed in. Even fossils appearing in growth position are not proof of *in situ* reef development (see Nos. 9 and 10 below). Furthermore, the lime muds could have been cemented inorganically and not by organisms over immense periods of time (see No. 11 below).

The Triassic deposits discussed by Stanley provide an interesting sidelight. They are probably the deposits cited by Giordano Bruno (1548?-1600) who insisted that there could have been no Flood. Instead of reacting with dogmatic appeals and with censure, churchmen should have examined these rocks and thereby refuted Bruno's claims of reef origin for these deposits.

7. Evidence Against Reef Origins from Dearth of Predation

Twenhofel[34] pointed out that ancient reefs are well preserved in contrast of modern reefs.

Comment: This lack of predation in "ancient reefs" is because there are Flood deposits and not reefs. Consequently, there was insufficient time for appreciable predation to occur on these deposits.

8. Major differences in Scale between Modern and Ancient "Reefs"

Twenhofel[34] pointed out that the Great Barrier Reef is 1000 miles long and (with the exception of the Silurian reefs of the Michigan Basin): "Few ancient reefs approach this dimension..." He noted that the modern reef in Maratoea is 1400 ft. thick and (except for some Permian algal reefs): "No ancient reef made by coral as an important contributor is known to have anything like a comparable thickness."

By contrast, in speaking of relatively minor differences between modern and ancient "reefs", Klovan[35] wrote: "Therefore, although similar controls are likely to affect recent and ancient reefs, precise analogues between the two can seldom be drawn."

Comment: This major scale difference between modern and ancient "reefs" can be understood in terms of the ancient deposits not being reefs but (much thinner) Flood deposits. Lack of precise analogies is another consequence.

9. Many claims of In-Situ Reef Organisms based upon Conjecture

Philcox[36] wrote: "The corals in the thickets *appear* to be *more or less* in their position of growth, as they are self-supporting." (italics added)

Ager[37] claimed that 87.5%-97.5% of corals and stromatoporoids in the Chicago-area Silurian reefs are in growth position. In a review of Ager's book, Manten[38] disagreed, writing: "In reefs, comparable to those in the English Wenlock, around 81% of the coral colonies are found in position of growth, this percentage decreasing to only 46-58 for reefs formed in relatively shallow water."

Comment: The statement of Philcox has an air of vagueness about it. Indeed, many claims in the literature about reef organisms occurring in growth position are based on conjecture rather than careful investigation. Furthermore, the contradictory opinions of Ager and Manten reveal a subjectivity in judging what is and what isn't in growth position. Even those reef organisms found definitely in growth position do not prove *in situ* reef growth over immense periods of time (see below).

10. Reef Organisms in Growth Position No Proof for Reef Growth

Ager[37] wrote: "A more serious objection, even when the subjectivity of such observations has been reduced by counting, is that the same effect could be brought about by mechanical means. It may be argued that a hemispherical body such as shown in Figure 5.5. would tend to come to rest in this position of greatest stability. Clearly, we need to consider other evidence as well, such as the nature of enclosing sediments."

Concerning the Silurian "reefs" of Iowa, USA, Philcox[47] wrote: The majority of flat-based favositid colonies lie upright, but since this is their most stable orientation, they may have rolled into this position. In the absence of other evidence they should be regarded as ambiguously situated. It is frequently claimed in the literature that a given upright colony is in its growth position, implying *location*, when the evidence really only shows the colony to be in its growth *orientation*." Furthermore: "... rolled conical or cylindrical colonies could come to rest in an upright position if supported by projections in the substrate or by other colonies." Philcox then suggested that *in situ* growth could be substantiated if the contact of the colony with the substrate could be seen, but then acknowledged that a transported colony could bring the contact along with it (as an uprooted tree can bring undisturbed soil layers among its roots). Even after suggesting that relatively rare instances of intergrown, branched colonies crossing different horizons demonstrate *in situ* growth, he would only say the the situation offers: "... a greater chance that they have been buried where they grew.[47]"

In speaking of how to recognize *in situ* fossil forms as a whole, Fursich[48] recently wrote: "Recognition of epifaunal species preserved in life position is difficult where the life position is a hydrodynamically stable position and thus could also have resulted from transport."

The Soviet paleoecologist Hekker[39] cited: ". . . the occasionally preserved life groups."

Comment: Clearly *in situ* ancient reefs are not proven by the existence of reef organisms in growth position. Since organisms occurring in growth position are very rare overall (statement of Hekker) it may be a matter of chance that the Flood currents sometimes deposited a large percentage of fossils in growth position locally. Such occurrences were facilated by the stability tendencies pointed out. Usually, the Flood would deposit organisms and lime mud with low concentration of fossils and very few of them in growth position, and this would be attributed to a non-reef carbonate shelf environment. Occasionally, however, the Flood deposited a high concentration of fossils and these turned out in growth position; such deposits being attributed to ancient reefs by uniformitarians.

11. Inorganic Cementation of "Ancient Reef" Deposits

In speaking of the Chicago-area Silurian "reefs", Pray[40] wrote: "I believe that submarine cementation played a role of equal or greater importance than organic binding in the construction of the Thornton Reef Complex."

Neumann[49] said: "Ancient mounds, in contrast to the modern lithoherms, appear to have accumulated largely from submarine cementation of products of *in situ* origin."

Comment: Lime muds need not have been bound by reef organisms over immense periods of time. Just as in clastic sedimentary rocks, inorganic cementation can account for the cementation of these deposits. Even a mound-shaped geometry of the "reef" deposit is not proof that it was a reef and consequently the organisms in it had bound it. (see below)

12. "Reef" Geometry No Proof for a Deposit Being a Reef

Keith[41] wrote: "This reef and others in the region, together with their associated carbonate-shelf deposits, fit into a well-expressed orthogonal tectonic pattern controlled by larger-scale basement features."

Lane[32] wrote: "The specific geometry developed in a deposit that contains abundant crinoid debris is thought to be most importantly controlled by two factors—the rate of sedimentation and the strength of bottom currents."

Comment: Many forms of geometry of "reef" deposits can be explained by tectonics during sedimentation. The statement of Lane indicates that geometry may be purely the result of sedimentological factors. "Ancient reefs" may be dune-type Flood deposits. (see below)

13. "Ancient Reefs" Actually Dune Deposits

Anstey and Pachut[42] wrote: "Longitudinal Ophiurids (star dunes) result from radially converging longitudinal vortices forming a collective updraft on the dune appex."

Many "ancient reefs" (especially of the Lower Paleozoic) are composed mainly of crinoids and crinoidal debris. Jenkyns[50] proposed that crinoidal lenses are a type of sand wave deposited in a pelagic seamount: a giant ripple mark of tidal origin oriented perpendicular to current direction.

Comment: During the Flood, the waters deposited organisms and sediment into variously-shaped dunes, accounting for the "mound"-shaped "reef" geometry of these deposits.

III. UNIFORMITARIAN CONFIRMATIONS OF CREATIONISTS' OBSERVATIONS CONCERNING (ALLEGED) OVERTHRUSTS

1. Introduction to "Overthrusts"

Gretener[60] wrote: "The following observations seem to have universal validity: 1. The contact is usually sharp and unimpressive in view of the great amount of displacement. 2. Structures which have been named 'tongues' appear to be common. They are features where material from the overridden sequence is seemingly injected as a tongue into the base of the overthrust plate. 3. Secondary (splay) thrusts are common. 4. Coalescence of tongues may produce pseudo-boudins. 5. Minor folding and faulting can usually be observed in both the thrust plate and the underlying rocks. The intensity of such deformations is normally comparatively weak, at least in view of the large displacements these thrust plates have undergone . . .[6] Late deformations, particularly by normal faulting, are present in many thrust plates. They should be recognized for what they are: post-thrusting features completely unrelated to the emplacement of the thrust plates."

Comment: Creationists and Diluvialists have long noted instances of strata resting in "wrong" order, and how overthrusts have been claimed because of this inverse order. The works of Burdick,[61] Burdick and Slusher,[62] and many others have all pointed out that: 1) the contact between "overriding" and "overriden" plate is faint and lacks slickensides, gouge material, etc., 2) the "overriding" plate shows little deformation, and 3) there is an overall paucity of evidences pointing to extensive tectonic motion supportive of such alleged thrusting.

The statements of the uniformitarian Gretener confirm these points. Such confirmation is especially useful because a uniformitarian could not be accused of making observations partial to Creationism. The following entries elaborate upon the major points and provide an example from the Soviet Union.

2. Lack of Evidences of Motion Along "Thrust" Contacts

Gretener[63] wrote: "However, modern mapping has led to many places where the thrusts themselves can be observed. Invariably the descriptions use such words as: 'knife sharp', 'drawn with a knife edge', etc. Figure 6 shows the McConnell thrust as exposed in Mount Yamnuska (detail, middle of Figure 1). It is indeed possible to place one's hand on the surface separating the Middle Cambrian Eldon Formation from the Upper Cretaceous Belly River Formation . . . The fact that this boundary is so sharp becomes more impressive when one considers that each point above the fault surface in Figure

6 has moved a minimum distance of 15 km. (10 mi).''

In generalizing on all "overthrusts", Gretener[60] added: "gouge material is essentially absent or, if present, is very thin (Brock, 1973; Engelder, 1974). Be it as it may, the idea of easy gliding, so persistent in the geologicial literature, is certainly supported by even a cursory examination of any thrust belt."

Gretener[64] said: "Different lithological units, usually with stratigraphic separation measured in kilometers, are in juxtaposition along a sharp contact, *often no more impressive than a mere bedding plane.*" (italics added)

Rezvoy[65] said: "*Fully conformable* with the general structure of the Silurian deposits is the underlying complete section of Devonian or Carboniferous terrigenous Totubay suite." ". . . the thrust plates, where they can actually be observed, show no traces of large displacements along them.[66] (italics added)

Comment: The descriptions of Rezvoy (refs 65, 66) concern a situation in the Tien Shan Mountains (located near the USSR/northwest-China border) where Silurian rests upon Devonian and Carboniferous. He claims that the inversion is not from a true overthrust but from the Silurian strata having been tectonically squeezed out of position and re-implaced upon the presumably younger strata. The example of Gretener (ref. 63) refers to Mount Yamnuska (located near Calgary, Alberta, Canada) where Cambrian rests on top of Cretaceous.

It is credulous to believe that thrust contacts involving miles of rock layers would result in a contact lacking gouge and capable of being straddled with one's hand. The real reason why these "thrust" contacts are "knifesharp", "conformable," and "no more impressive than a mere bedding plane" is precisely because they *are* bedding planes. The contacts are thus sedimentary and not tectonic; strata have been deposited in "wrong" order and the entire evolutionary-uniformitarian geologic column is shown to be false.

Incidentally, the situation described by Rezvoy illustrates once again the circular reasoning of using index fossils. The age of Silurian was attributed to the lithology on top of Devonian and Carboniferous as a result of a single graptolite occurrence. Certain such graptolites are known to be Silurian because they only occur in Silurian rock, and rocks are dated as Silurian solely because they contain such graptolites.

3. Paucity of Overall Deformation in Strata Involved in "Thrusting"

Gretener[63] wrote: "While the sharp nature of the faults has been emphasized by many authors . . . , one should not forget that the rocks above and below a thrust may show considerable deformation for distances of some hundreds of feet. *Still the thrust plates as a whole have remained intact and are in no way comparable to the jumbled masses of landslides.* Thus, the presumption underlying this paper—that we are dealing with *virtually undistorted plates* moving over the undisturbed sequence along sharp planes—is basically correct." (italics added)

Comment: It is credulous to believe that lithologies involved in thrusting would not at least show massive disruption and deformation. The fact that there is so little evidence of that is another evidence that these are not overthrusts but sedimentary contacts.

Even features such as "tongues," "false boudins", etc., (see item 1, Gretener[60]) can be explained by sedimentary instead of tectonic processes. It is common for submarine erosion to result in tongues of material from underlying beds being dragged into the overlying beds, according to Ager.[67]

4. Associated Tectonism No Proof for Overthrust

Gretener[68] wrote: "Most thrust plates have been subjected to post-thrusting deformation of one kind or another. One must beware of and separate such observations and not falsely ascribe them to the thrusting movement."

Comment: Evidences of tectonism are to be expected in tectonically complex regions. One wonders if the tectonic evidences associated with the alleged overthrusts are not also prevalent in regions where no overthrusts are claimed. Presumed evidences for overthrust can easily be claimed because of the uniformitarian preconception of a thrust being present where strata occurs in "wrong" order.

IV. SOME EXAMPLES OF "REWORKING" RATIONALIZATIONS FOR ANOMALOUS FOSSILS

1. Introduction to "Reworking"

Fossils frequently occur where they are not "supposed" to. It is then claimed that either the fauna or flora have lived longer than previously known (simple extention of stratigraphic range) or that the fossil has been reworked. In "reworking," it is claimed that the fossil has beeen eroded away from a much older host rock and has thus been incorporated into a rock of more recent age. The reciprocal situation is "downwash," where it is claimed that an organism has been washed down into rock much older than the time it lived and has become fossilized there.

Table 1 is a compilation of both situations. The entries are examples only and do not represent a comprehensive literature search. Nor does the table include examples previously discussed by the author in his first anthology[69] or the Creationist-Diluvialist and evolutionary-uniformitarian references cited therein.

As later items of this section will demonstrate, "reworking" is very often (if not usually) not justified by any independent lines of evidence such as state of preservation of the fossil. Claims of "reworking" are thus invoked solely because of the "improper" stratigraphic occurrence of the fossil. The numerous instances of fossils occurring in "wrong" strata is thus yet another line of evidence against the validity of the evolutionary-uniformitarian geologic age system.

Comment: Some of the entries in Table 1 have special significance.

The entries under references 84 and 96 involve the presence of Phanerozoic pollen in the Precambrian. The occurrence of Jurassic pollen in the Precambrian Ukrainian Shield was explained away by Krassilov[84] through claiming that a Jurassic weathering episode in the Shield caused a contamination with the pollen of that

Table 1. This table is a compilation of over 200 published instances of anomalously-occurring fossils. At left is the kind of fossil(s) involved followed by the accepted age of occurrence, the instance of anomalous occurrence, its locality, and the reference.

Instances were tabulated as separate entries when they involved: 1) different fossils (usually), 2) different "proper" ages, 3) different anomalous ages, or 4) different geographical locations (different nations, provinces, sedimentary basins, etc.)

The instances entered under references 70, 73, 77, 78, 88, 100, and 150 are subaqueous occurrences (deep marine) offshore marine, or lacustrine): most others involve sedimentary lithologies on land. References 70, 17, 72, 82, 84, 89, 96, 147, 148, 153, and 160 are "downwash," "infiltration," or supposed "contamination." Most other entries are "reworking" of fossils into younger-age beds.

A few of the entries in this table are claimed by the cited authors to be possible stratigraphic-range extentions rather than necessarily "reworking". This table does not include fossils of clearly secondary position (such as fossils within constituents of conglomerates and the very many instances of Phanerozoic fossils in Pleistocene glacial till).

For further discussion of the entries in Table 1, see section VI, No. 1.

Type of Fossil:	"Proper" Age:	Found In:	Location	Reference
Foraminifers	Pleistocene	Tertiary	Atlantic Ocean	70
Pollen	Tertary	Quaternary	Bolkow, Poland	146
Pollen	Tertiary	Cretaceous	East Netherlands	147
Mammal bones	Tertiary (late)	Tertiary (early)	Kazakhstan, USSR	148
Algae (?)	Tertiary	Cretaceous	France	71
Foraminifers	Tertiary	Cretaceous	Nigeria	72
Nannoplankton	Tertiary (lower)	Tertiary (higher)	Tabqa, Syria	164
Spores	Tertiary (recent)	Permian	Southland, New Zealand	165
Spores	Tertiary	Pleistocene	North Caspia, USSR	166
Foraminifers	Tertiary	Cretaceous	North Italy	167
Pollen	Tertiary	Pleistocene	Yenisei, USSR	168
Palm wood	Tertiary	Jurassic	Utah, USA	169
Foraminifers	Tertiary (early)	Tertiary (medial)	Trinidad, West Indies	206
Foraminifers	Tertiary (early)	Tertiary (medial)	Victoria, Australia	207
Foraminifers	Tertiary (early)	Tertiary (medial)	Hungary	208
Nannoplankton	Tertiary (early)	Tertiary (medial)	Romania	209
Spores	Tertiary (recent)	Jurassic	Louisiana-Texas, USA	210
Mammal bones	Tertiary (early)	Tertiary (medial)	Northeast Siberia, USSR	211
Foraminifers	Cretaceous	Tertiary	Ionian Sea	212
Foraminifers	Cretaceous	Tertiary	West Israel	213
Coccoliths	Cretaceous	Tertiary	Crimea, USSR	214
Spores	Cretaceous	Tertiary	Kazan, USSR	215
Pollen	Cretaceous	Tertiary	British Columbia, Canada	216
Pollen	Cretaceous	Tertiary	Arctic Canada	216
Pollen	Cretaceous	Tertiary	Fushun, China	216
Pollen	Cretaceous	Tertiary	Shandong, China	216
Pollen	Cretaceous	Tertiary	South Coastal China	216
Pollen	Cretaceous	Tertiary	Spitzbergen, Norway	217
Pollen	Cretaceous	Tertiary	Ural Mts., USSR	217
Pollen	Cretaceous	Tertiary	West Siberia, USSR	217
Dinoflagellate	Cretaceous	Tertiary	Australia	170
Nannoplankton	Cretaceous	Tertiary	Glogow, Poland	171
Foraminifers	Cretaceous	Tertiary	Alabama, USA	172
Nannoflora	Cretaceous	Tertiary	Zinda Pir, W. Pakistan	173
Pollen	Cretaceous	Pleistocene	Yenisei, USSR	168
Nannoflora	Cretaceous	Tertiary	Tang-E-Bijar, Iran	174

Type of Fossil:	"Proper" Age:	Found In:	Location	Reference
Foraminifers	Cretaceous	Jurassic	Scania, Sweden	175
Pollen	Cretaceous	Cambrian	Holy Cross Mts., Poland	176
Nannoplankton	Cretaceous	Tertiary	Ukraine, USSr	177
Pollen	Cretaceous	Tertiary	Utah, USA	149
Spores	Cretaceous	Pleistocene	Newfoundland, Canada	150
Spores	Cretaceous	Tertiary	Newfoundland, Canada	150
Pollen	Cretaceous	Tertiary	West Germany (?)	151
Spores	Cretaceous	Tertiary	Czechoslovakia	152
Foraminifers	cretaceous	Quaternary	Adelie Coast, Antarctica	73
Foraminifers	Cretaceous	Tertiary	Austria	74
Foraminifers	Cretaceous	Tertiary	Sweden	75
Nannoplankton	Cretaceous	Tertiary	Alps. Carpathians, Europe	76
Nannofossils	Cretaceous	Tertiary	west, southwest Pacific	77
Foraminifers	Cretaceous	Pleistocene	Louisiana, USA	78
Pollen	Cretaceous	Tertiary	Venezuela	79
Pollen	Cretaceous	Tertiary	Wyoming, USA	79
Foraminifers	Cretaceous	Tertiary	California, USA	80
Foraminifers	Cretaceous	Pleistocene	Netherlands	81
Foraminifers	Cretaceous	Pleistocene	Utah, USA	81
Foraminifers	Cretaceous	Tertiary	England	75
Pollen	Jurassic	Permian	Sariz, Antalya, Turkey	82
Spores	Jurassic	Tertiary	Isle of Mull, Scotland	83
Spores	Jurassic	Precambrian	Ukraine, USSR	84
Coccoliths	Jurassic	Silurian	North Africa	153
Pollen	Jurassic	Tertiary	Utah, USA	149
Spores	Jurassic	Cretaceous	Newfoundland, Canada	150
Pollen	Jurassic	Tertiary	Newfoundland, Canada	150
Dinoflagellate	Jurassic	Cretaceous	California, USA	154
Dinoflagellate	Jurassic	Tertiary	California, USA	154
Spores	Jurassic	Cretaceous	Perth Basin, Australia	89
Spores	Jurassic	Triassic	Northeast Siberia, USSR	218
Pollen	Jurassic	Pleistocene	Yenisei, USSR	168
Spores	"Mesozoic"	Pleistocene	North Caspia, USSR	166
Spores	Jurassic	Permian	Northeast Siberia, USSR	218
Spores	Jurassic	Cambrian	Northeast Siberia, USSR	218
Spores	Triassic	Cretaceous	British Columbia, Canada	238
Spores	Triassic (and/or older)	Jurassic	Yukon, Canada	178
Spores	Triassic Cretaceous		Alberta, Canada	219
Ostracodes	Triassic	Tertiary	Hungary	85
Spores	Triassic	Jurassic	Andoya, Norway	86
Conodonts	Triassic	Cretaceous	Cameroons	87
Conodonts	Triassic	Jurassic	Japan	87
Spores	Triassic	Tertiary	Kutch, India	155
Pollen	Triassic	Tertiary	Newfoundland, Canada	150
Pollen	Triassic	Tertiary	Utah, USA	149
Spores	Triassic	Cretaceous	Perth Basin, Australia	89
Pollen	"Mesozoic"	Tertiary	Atlantic Ocean	88
Spores	"Mesozoic"	Tertiary	Hungary	83
Brachiopods	Permian	Triassic	Salt Range, Pakistan	142
Conodonts	Permian	Triassic	Salt Range, Pakistan	142
Pollen	Permian	Triassic	Utah, USA	149
Spores	Permian	Cretaceous	Kutch, India	156
Conodonts	Permian	Triassic	Akasaka, Japan	157
Spores	Permian	Tertiary	Kerala, India	155
Spores	Permian	Tertiary	Kutch, India	155
Spores	Permian	Cretaceous	Madhya Pradesh, India	155
Pollen	Permian	Tertiary	Nagaland, India	158
Spores	Permian	Cretaceous	Alberta, Canada	219

Type of Fossil:	"Proper" Age:	Found In:	Location	Reference
Spores	Permian	Cretaceous	British Columbia, Canada	238
Fusulinids	Permian	Triassic	Mine, Japan	180
Pollen	Permian	Triassic	Southeast Siberia, USSr	220
Pollen	Permian	Tertiary	Assam, India	158
Pollen	Permian	Tertiary	Maghalaya, India	159
Spores	Permian	Triassic	Somerset, England	90
Pollen	Pemian	Jurassic	Kutch, India	91
Pollen	Permian	Tertiary	South Australia	91
Pollen	Permian	Cretaceous	Victoria, Australia	91
Ammonoids	Permian	Triassic	KapStosch, Greenland	92
Spores	Permian	Tertiary	Hungary	83
Spores	Permian	Cretaceous	Perth Basin, Australia	89
Spores	Permian	Devonian	Canning Basin, Australia	89
Spores	Carboniferous	Jurassic	Poland	93
Spores	Carboniferous	Jurassic	Sweden	93
Spores	Carboniferous	Jurassic	England	93
Spores	Carboniferous	Jurassic	Scotland	93
Spores	Carboniferous	Jurassic	Latvia, USSR	93
Spores	Carboniferous	Jurassic	Denmark	93
Spores	Carboniferous	Jurassic	Baltic Russia	93
Spores	Carboniferous	Tertiary	Arkansas, USA	94
Spores	Carboniferous	Cretaceous	Perth Basin, Australis	89
Fusulinids	Carboniferous	Pleistocene	Utah, USA	81
Pollen	Carboniferous	Triassic	Devon, England	95
Spores	Carboniferous	Precambrian	Onega, USSR	96
Spores	Carboniferous	Cretaceous	Limburg, Netherlands	91
Pollen	Carboniferous	Cretaceous	Krakow, Poland	97
Pollen	Carboniferous	Tertiary	Hungary	84
Pollen	Carboniferous	Tertiary	Alabama, USA	84
Pollen	Carboniferous	Triassic	Donets Basin, USSR	84
Pollen	Carboniferous	Jurassic	Donets Basin, USSR	84
Spores	Carboniferous	Jurassic	Germany	93
Spores	Carboniferous	Permian	Pakistan	93
Acritarchs	Carboniferous	Tertiary	Newfoundland, Canada	150
Pollen	Carboniferous	Devonian	Missouri-Iowa, USA	160
Pollen	Carboniferous	Cretaceous	Montana, USA	221
Pollen	Carboniferous	Tertiary	Montana, USA	221
Conodonts	Carboniferous	Devonian	Graz/Styria, Austria	222
Pollen	Carboniferous	Triassic	Southeast Siberia, USSR	220
Spores	Carboniferous	Cretaceous	Alberta, Canada	219
Conodonts	Carboniferous (lower)	Carboniferous (medial)	New Mexico, USA	223
Spores	Carboniferous	Cretaceous	British Columbia, Canada	238
Spores	Carboniferous	Permian	Devonshire, England	179
Spores	Carboniferous	Triassic	Devonshire, England	179
Fusulinids	Carboniferous	Triassic	Mine, Japan	180
Spores	Carboniferous	Pleistocene	Ohio, USA	181
Spores	Carboniferous	Permian	Yukon, Canada	182
Brachiopod	Carboniferous	Permian	Yukon, Canada	182
Pollen	Carboniferous	Permian	Caucasia, USSR	183
Trilobite	"Late Paleozoic"	Tertiary	Utah, USA	184
Spores	Carboniferous	Jurassic	West Arctic Islands, Canada	185
Conodonts	Devonian	Carboniferous	Missouri, USA	186
Spores	Devonian	Carboniferous	Canning Basin, Australia	179
Spores	Devonian	Permian	East West Australia	179
Crinoids	Devonian	Carboniferous	Pamir Mts., USSR	187
Conodonts	Devonian	Silurian	Texas, USA	188
Algal Cysts	Devonian	Cretaceous	Northeastern Brazil	189
Pollen	Devonian	Permian	Caucasia, USSR	183

Type of Fossil:	"Proper" Age:	Found In:	Location	Reference
Spores	Devonian	Carboniferous	Yukon, Canada	182
Spores	Devonian	Permian	Yukon, Canada	182
Spores	Devonian	Pleistocene	Ohio, USA	181
Spores	Devonian	Jurassic	West Arctic Islands, Canada	185
Spores	Devonian	Cretaceous	British Columbia, Canada	238
Spores	Devonian (medial)	Devonian (upper)	Belorussia, USSR	224
Conodonts	Devonian	Carboniferous	Kansas, USA	225
Conodonts	Devonian	Carboniferous	New Mexico, USA	223
Conodonts	Devonian	Carboniferous	Graz/Styria, Austria	222
Conodonts	Devonian	Tertiary	Graz/Styria, Austria	222
Conodonts	Devonian	Pleistocene	Graz/Styria, Austria	222
Brachiopods	Devonian (lower)	Devonian (medial)	California, USA	226
Conodonts	Devonian	Carboniferous	Kazakhstan, USSR	227
Conodonts	Devonian	Silurian	Arkansas, USA	228
Chitinozoans	Devonian	Jurassic	East Bhutan	229
Conodonts	Devonian	Carboniferous	Nevada, USA	161
Foraminifers	Devonian	Carboniferous	Nevada, USA	161
Spores	Devonian	Carboniferous	Bolivia	143
Conodonts	Devonian	Carboniferous	Devonshire, England	162
Acritarchs	Devonian (and/or older)	Carboniferous	Newfoundland, Canada	150
Spores	Devonian	Triassic	Devon, England	95
Conodonts	Devonian	Triassic	Kockatea, Australia	89
Spores	Devonian	Cretaceous	Otorowiri, Australia	89
Thelodont, Acanthodian Fish Scales	Devonian	Permian	Canning Basin, Australia	89
Spores	Devonian	Cretaceous	Perth Basin, Australia	89
Pollen	Devonian	Carboniferous	Oklahoma, USA	145
Spores	Devonian	Carboniferous	Belgium	98
Conodonts	Devonian	Permian	Texas, USA	99
Algal Cysts	Devonian	Quaternary	Lake Michigan, USA	100
Conodonts	Devonian	Carboniferous	Texas, USA	101
Conodonts	Devonian	Carboniferous	Oklahoma, USA	102
Acritarchs	Silurian (and/or older)	Devonian	Witney, England	163
Acritarchs	Silurian	Devonian	Belgium	98
Conodonts	Silurian	Carboniferous	Oklahoma, USA	102
Pollen	"Paleozoic"	Precambrian	Krivyi Rog, USSR	84
Pollen	"Paleozoic"	Tertiary	Atlantic Ocean	88
Spores	"Paleozoic"	Tertiary	Bolivia	143
Spores	"Paleozoic"	Pleistocene	North Caspia, USSR	166
Spores	Silurian	Ordovician	Ohio, USA	230
Chitinozoans	Silurian	Jurassic	East Bhutan	229
Plant tissue	Silurian (and/or younger)	Ordovician	Oklahoma, USA	190
Acritarchs	Silurian	Carboniferous	Ballyvergin, Ireland	191
Crinoids	Silurian	Carboniferous	Pamir Mts., USSR	187
Algal Cysts	Silurian	Cretaceous	Northeast Brazil	189
Conodonts	Silurian	Devonian	Texas, USA	188
Spores	Silurian(?)	Devonian	Belorussia, USSR	224
Chitinozoans	Silurian	Cretaceous	Alaska, USA	231
Conodonts	Ordovician	Silurian	Ontario, Canada	192
Conodonts	Ordovician	Devonian	Minnesota, USA	193
Acritarchs	Ordovician	Tertiary	Lough Neagh, Ireland	194
Conodonts	Ordovician	Silurian	Missouri, USA	195
Conodonts	Ordovician	Silurian	Carnic Alps, Italy	196
Conodonts	Ordovician	Silurian	Quibec, Canada	196

Type of Fossil:	"Proper" Age:	Found In:	Location	Reference
Conodonts	Ordovician	Silurian	Central Texas, USA	188
Diacrodians	Ordovician	Precambrian	Saxony, East Germany	197
Conodonts	Ordovician	Silurian	Illinois, USA	198
Conodonts	Ordovician	Silurian	Illinois, Indiana, USA	199
Acritarchs	Ordovician	Carboniferous	Ballyvergin, Ireland	191
Acritarchs	Ordovician	Silurian	Ohio, USA	200
Acritarchs	Ordovician	Silurian	New York, USA	200
Conodonts	Ordovician	Silurian	Central Siberia, USSR	232
Conodonts	Ordovician	Silurian	Southeast Indiana, USA	233
Conodonts	Ordovician (lower)	Ordovician (medial)	West Texas, USA	234
Graptolites Conodonts, Ostracodes	Ordovician	Devonian	Missouri, USA	103
Acritarchs	Ordovician	Silurian	Belgium	98
Acritarchs	Ordovician	Devonian	Netherlands-Belgium	98
Conodonts	Ordovician	Carboniferous	Oklahoma, USA	102
Spores	Ordovician	Carboniferous	Oklahoma, USA	94
Conodonts	Ordovician	Cretaceous	Colorado, USA	104
Acritarchs	"Lower Paleozoic"	Triassic	Devonshire, England	95
Archeocyathids	Cambrian	Permian	Dwyka, South Africa	105
Trilobites	Cambrian (lower)	Cambrian (medial)	Bornholm, Denmark	202
Acritarchs	Cambrian	Silurian	Comeragh, Ireland	191
Acritarchs	Cambrian	Caroniferous	Ballyvergin, Ireland	191
Trilobites	Cambrian	Devonian	Bielsko-Mogilany, Poland	203
Acritarchs	Cambrian	Ordovician	Shropshire, England	204
Algae	Precambrian	Cambrian or Ordovician	Verkhoyansk, USSR	205
Spores	Precambrian	Devonian	Saratov, USSR	235

time. The finding of Paleozoic spores in the Precambrian of Kirvyi Rog, USSR, was rationalized away by Krassilov[84] in a similar way. In another situation, strata was thought to be Precambrian because of its incredible lithologic similarity to proven Precambrian rock until Carboniferous spores were found in the strata (ref. 96). An alternate view proposed was that Carboniferous spores were downwashed into Precambrian and that the stratum was not a solitary Carboniferous outlier. These examples are further support for the controversial work of Burdick[106], who reported gynosperm and angiosperm pollen from the Precambrian formations of the Grand Canyon.

"Reworked" and "downwashed" forms are microfossils, it is claimed, because such small forms are resistant to erosion, transport, etc. However, many entries in Table 1 (92, 89, 103, 142, 148 and 105) involve macrofossils. The fact that most anomalously-occurring fossils are microfossils may be because anomalously-occurring macrofossils are more likely to be considered legitimate stratigraphic extensions rather than "reworking". Another explanation is provided by the Creationist-Diluvialist paradigm. Microfossils, being minute, would be less capable of differential escape and be less subject to Flood-water sorting than macrofossils. The common situation of anomalously-occurring pollen and spores may be evidence that *all* fossil plants were mutually contemporaneous (as indeed demanded by the Creationist-Diluvialist paradigm) but that pollen and spores, being mobile, were transported by wind and water far beyond the restricted ecological zones of these antediluvian plants.

2. Ubiquity and Significance of Anomalously-Occurring Fossils

Bramlotte and Sullivan[107] wrote; "To recognize such redistribution where it has occurred and yet not to invoke this explanation to account for unexpected extensions of the life range of a species presents a serious problem requiring much critical attention."

In a coexistence of Devonian with Permian fossils (ref. 89), Veevers[89] asked: "Are the Permian spores due to infiltration or are the Devonian fossils reworked in Permian deposits?"

Venkatachala[91] wrote: "Palynological fossils of older ages are *commonly* encountered amidst younger assemblages." (italics added).

Muir[108] remarked: "In any kind of ecologic study, false conclusions could be drawn, while *the havoc reworking could play with stratigraphy is immense,*" (italics added)

The occurrence of Carboniferous spores in Jurassic (ref. 93) is so common all over Europe that the author Windle[93] proposed that it must have a unified continent-wide explanation. He suggested that it does not mean that hidden remnants of Carboniferous floras survived into the Jurassic but that continent-wide orogenesis during Jurassic times in Europe caused much Carboniferous strata to be eroded away in Jurassic times.

Concerning "reworked" forms, Stanley[143] said: "These secondary grains usually are present in larger numbers in both marine and non-marine sediments than most workers would like to admit."

Comment: From all the statements cited above it can be seen that anomalous fossils cannot be dismissed as being rare or being only trivial localized occurrences. Stanley's statement hints that many instances of anomalous fossils go unpublished.

3. Lack of Independent Evidence in Many Cases of "Reworking"

Concerning the presence of Permian ammonoids in the Triassic of East Greenland, Teichert and Kummel[92] wrote: "We consider it most probable that some of the Permian faunal elements in the lowest Triassic formations have been brought into that environment as argillaceous boulders, that once coming to rest, dissolved, leaving *well-preserved fossils* that were rapidly buried in the coarse sediment and in a free state were transported very little." (italics added)

In speaking of ostracodes, conodonts, and graptolites of Ordovician age occurring in nodules within Devonian rock, Chauff[103] said: "Phosphate nodules reworked into younger strata may display little evidence of transport."

Concerning solitary "reworked" conodonts, Pokorny[109] wrote: "Allochthonous species can sometimes be recognized by their state of preservation (degree of weathering, wear, different colour or lustre) but sometimes it is possible to distinguish them from autochthonous ones by X-ray examination. Because of their great durability even this may not be successful."

Concerning some conodont speciments that were ascribed to "reworking", Lindstrom[101] said: ". . . these specimens were selected for quality. Indeed the illustrations show perfect, complete specimens."

Muir and Sargeant[113] said:

"Detection of reworking is a difficult problem. In some cases, reworked spores are better preserved than indigenous ones, in other cases worse. They may stain more or less, be more or less crushed, be older or younger, or be the only spores present."

Recently, Brasier[144] wrote: "Yet another disadvantage to the stratigrapher is the ease with which coccoliths are reworked into younger sediments without showing outward signs of wear."

Comment: The statements make it clear that many (if not most) forms which are considered to be reworked do not show any special morphological evidence for having been reworked. "Reworking" is thus solely a rationalization because the fossil has occurred where it *is* not "supposed" to. (see below).

4. Preconceptions of "Proper" Stratigraphic Occurrence Often Sole Justification for Claims of "Reworking"

Hass[110] wrote: "Differences in the physical appearance (color, preservation, luster) of associated speciments are indicators of a mixed fauna; but *the recognition of a mixed fauna is chiefly dependent upon one's knowledge of the true stratigraphic range of each kind of discrete conodont.*" (italics added)

In writing also of conodonts, Lindstrom[111] said: "If they do not fit into the patterns of conodont evolution established for the immediately older and younger beds, they may be in a secondary position. This is, however, a reasoning that one must use very cautiously, for there is the risk that it may lead to a vicious circle . . . Fifthly, one should use every opportunity to check the conodont sequence against index fossils belonging to other groups, as for instance trilobites or ammonoids."

Wilson[112] wrote: "In many cases the contaminants are difficult to recognize and to demonstrate the source of their origin; however, *the incompatible ages of the fossils and sequences of fossil ages are useful criteria in distinguishing mixed palynological deposit.*" (italics added)

Comment: The statements cited make it very clear that "reworking" is a convenient rationalization for "out of place" fossils. It can be and is capriciously invoked. The circular reasoning in assigning fossils to stratigraphic ranges and then turning around and explaining away occurrences of these fossils not fitting the prescribed stratigraphic ranges is obvious. As in so many other areas of the entire evolutionary-uniformitarian paradigm, only information fitting within narrow preconceived notions is accepted.

5. The Reason for the Preponderance of "Rework" over "Downwash" Situations

Goebel[136] said: "The youngest group of conodonts in a mixed fauna establishes the age of the fauna."

Comment: In an anomalous coexistence of fossils of different "ages", it is usually assumed that younger fossils yield the true age of the rock rather than the older ones. This accounts for the relative paucity of "downwash" situations.

6. Convenient Rationalization for Occurrences of Phanerozoic Microbiotas in Precambrian "Basement"

After noting how Phanerozoic contaminants can be seen in Precambrian "basement" along joints and intergranulars of the latter, Zoubek[137] recently wrote: "However, we are not so well aware of one fact typical of the polymetamorphic Precambrian basement of Phanerozoic orogens: during younger orogenesis, usually a metamorphic recrystallization of rocks of the Precambrian basement took place. Consequently, the former cracks and other mechanical discontinuities are often perfectly healed, and the "immigrated" microfossils become, in this way, an integral part of the older newly recrystallized rocks. Thus, isolated finds are to be considered with precaution in polymetamorphic terrains."

Comment: Whenever Phanerozoic microbiotas are found in Precambrian rock, they are dismissed as "contamination". Absence of textural and structural evidence can be rationalized away by claiming that it was erased in a later metamorphism. The attention and warning given by Zoubek hints that such finds of Phanerozoic biotas in Precambrian rock (as the find of Bur-

V. GEOLOGIC-PROCESS RATES AND UNIFORMITARIANISM: SCIENTIFIC AND PHILOSOPHICAL CONSIDERATIONS

1. Introduction: Uniformitarianism—Its Ramifications and Its Fallacious Foundations

In a recent article on uniformitarianism, Bushman[114] said: "Because identifiable cause and effect relationships that require long periods of time to develop according to the operation of natural law can be followed in sequence, we must conclude that the earth is millions of years old." Bushman[114] also cited Hutton (1896): "'Not only are no powers to be employed that are not natural to the globe, no action to be admitted except those of which we know the principle; and no extraordinary events to be alleged in order to explain a common appearance.'"

In an introduction to the reprint of a 19th century geology book, Wright[115] wrote: "Was the world created in six days, perhaps in the year 4004 B.C.? Or must long periods of geologic time be postulated, if we are to explain the thick layers of sediments and the evidences of crumpled rocks in the earth's crust."

Khain[116] wrote: "Having arrived at the conclusion that the present aspect of the Earth's surface is a result of major changes, Lomonosov could not reconcile it with official length of the Earth history, according to Biblical lore (5000 years)... In order to escape the wrath of churchmen, he refers to the numerous contradictions in the views of the Church on that subject, as militating against an acceptance of Church chronology, which is 'neither a dogma of the faith nor a pronouncement of the Councils." By contrast, the church father Clement (cited by Sparks[117]) declared: "Study the sacred Scriptures, which are true and are given by the Holy Spirit. Bear in mind that nothing wrong or falsified is written in them." (1 Clement 45:2-3)

Comment: The attitude of the church father Clement towards Scripture is in striking contrast to the attitude of many believers today. Indeed, the statements of Lomonosov make it clear that Creation, the Flood, and the youth of the earth have never been clearly exposited and defended by the church. Small wonder then that evolution and uniformitarianism had been so quickly and nearly universally accepted.

The other statements illustrate the scientific—as opposed to theological—fallacies of uniformitarianism. The statements of Bushman, Wright, and Khain make it clear that uniformitarian thought considers all processes on earth operating very slowly so that the earth must then be immensely old. The statements of Hutton show the circularity and narrowness of uniformitarian thought. Strong anti-supernatural preconceptions rule over what processes are considered "natural to the globe" and which are "extraordinary" (one should note that the Creation and Flood are "natural to the globe" and that, if anything, it is the present situation of very slow geologic change that is "extraordinary").

Hitchcock (as discussed by Wright), Lomonosov, Hutton, and Bushman all fail to consider that geologic processes have operated at far greater rates in the past than at present. The first statements of Bushman also illustrate the uniformitarian fallacy of believing that sequences on earth (such as fossil horizons forming geologic periods) have time significance.

Clues Pointing to the Rapid Formation of Regionally-Distributed Metamorphic Rocks

Krauskopf[118] said: "On the other hand solid-solid reactions are generally slow, especially reactions between silicates, and quite possibly the reactions necessary for metamorphism would not take place appreciably even in geologic time without the help of fluids. The argument hinges on reaction rates, *about which quantitative information is meager.*" (italics added)

Krauskopf[119] also wrote: "If equilibrium is so much the general rule, a troublesome question presents itself. After a rock has been metamorphosed at high temperature and pressure, it must undergo a gradually decreasing temperature and pressure in order to appear finally in surface outcrops. Why doesn't its composition readjust itself so as to be in equilibrium with the lower temperature-pressure conditions? How are the high-temperature mineral assemblages preserved? Why, to put it baldly, do we ever find metamorphic rocks at all?"

After dismissing the argument that there is so little retrograde metamorphism because orogenic uplift is so much faster than compression as being special pleading, Krauskopf[119] suggests: "For regional metamorphism a better explanation is suggested by the fact that this kind of metamorphism is generally an accompaniment to orogenic movement. Perhaps reactions can occur only during the movement itself, in response to intimate crushing and granulation of the rock; perhaps reaction ceases when movement ceases, preserving the mineral assemblage formed during the orogeny."

Comment: Some uniformitarians have charged that millions of years are necessary for the formation of regional metamorphic rock. The first statement of Krauskopf makes it clear that there is relatively little knowledge about metamorphic reaction rates. Not only is the uniformitarian argument, at best, unproven, but there are major difficulties in understanding the formation of metamorphic rocks in the context of the evolutionary-uniformitarian paradigm.

In uniformitarian thought, mountains are built by compressive tectonic forces over tens of millions of years and the uplift also requires tens of millions of years. Metamorphic rocks should all revert (at the very low P-T conditions of the earth's surface) more or less back to conditions of low P-T. The fact that there is so little such retrograde metamorphism is understood in the light of the Creationist-Diluvialist paradigm. The "intimate crushing and granulation of the rock" is the result of cataclysmic tectonism during the Flood year and the centuries following. Metamorphism is thus not the result of millions of years of deep rock burial and exhumation, but the result of unique cataclysmic tectonic stresses. When these stresses suddenly stopped, there was no more impetus for metamorphic reactions to continue and so metamorphic minerals did not usually revert in the low P-T conditions.

In uniformitarian thought, mountains are built by compressive tectonic forces over tens of millions of years and the uplift also requires tens of millions of years. Metamorphic rocks should all revert (at the very low P-T conditions of the earth's surface) more or less back to conditions of low P-T. The fact that there is so little such retrograde metamorphism is understood in the light of the Creationist-Diluvialist paradigm. The "intimate crushing and granulation of the rock" is the result of cataclysmic tectonism during the Flood year and the centuries following. Metamorphism is thus not the result of millions of years of deep rock burial and exhumation, but the result of unique cataclysmic tectonic stresses. When these stresses suddenly stopped, there was no more impetus for metamorphic reactions to continue and so metamorphic minerals did not usually revert in the low P-T conditions.

3. Factors Contributing to the Rapid Cooling of Superposed Lava Flows

Williams and McBirney[120] said: "Because they have low thermal conductibility and high heat capacity, lavas of even moderate thickness are well insulated and cool slowly. It is often possible to walk on the crust of a moving lava that is red-hot only a few centimeters below the surface. Indeed, lavas have been observed to retain high temperatures as long as 5 or 6 years after their eruption ... Rainfall can *greatly accelerate* heat losses from the surface of lavas. Ault and his co-workers (1961) found that heavy rains, totaling 4.44 inches in a period of 6 days, lowered the temperature-profile of a cooling Hawaiian lava by about 50° C to a depth of about a meter." (italics added)

Comment: Some uniformitarians have claimed that superposed lava flows interbedded with sedimentary rock need immense periods of time to form. The straw man they erected had an implicit presumption that lavas (or plutons, for that matter) need be cooled before they can be covered with a sedimentary layer. The fact that lavas can be touchably cool at the surface and yet furnace-hot at such a shallow depth implies that only the *upper crusts* of lavas and plutons need have cooled during the Flood year itself. Multiple extrusions of lava interbedded with sediment could therefore have taken place during the Flood year. Water-cooling helped.

4. Evidences for Rapid Cooling of Thick Intrusive Igneous Bodies

Warren[121] (cited by Plumstead) discussed the finding of uncooked plant spores near an igneous intrusion in Antarctica. He wrote: "Usable spore and pollen remains have now been obtained from a number of localities and a number of horizons in the Beacon Group, in some places relatively close to thick intrusive sills, and in one instance from a thin sedimentary lens interbedded with lavas or sills."

In describing a situation from the Australian Triassic, Hamilton *et. al.*[122] wrote: "The breccia pipes near Sydney contain numerous inclusions of coal ... The coal both in the breccia pipes and in the peripheral contorted zones is of bituminous rank, which is evidence that it has not been heated above quite modest temperatures ... None of the coal shows evidence of profound thermal alteration."

In noting that contact metamorphism around ultramafic intrusives is much less than expected from theoretical temperatures of such intrusives, Krauskopf[123] said: "Perhaps some ultramafic material is intruded as a 'crystal mush', an aggregate of solid crystals lubricated by compressed water vapor, with the temperature never rising higher than a few hundred degrees."

Comment: The cited examples are evidence that thick igneous bodies were never as hot as widely believed and cooled rapidly (since the baking effect—as amply demonstrated by the everyday example of grilling meat—is a function of the intensity of heat and the duration of the heating). Thick igneous bodies may all have had an early magmatic stage before the Flood, with the minerals at the higher temperature scale of the Bowen's Reaction Series already having been created in a crystalline state. This would greatly reduce the temperatures of intra-Flood and post-Flood intrusions because the higher-temperature minerals would have been pre-crystallized and because the associated heat of phase change would not have had to be expended.

5. Evidence for Cataclysmic Formation of Thick Igneous Complexes

Irvine[127], in describing an ultramafic complex from Alaska, USA, wrote: "Many igneous intrusions show layering formed by gravitational accumulation of crystals that is, both in variety and detail, remarkably similar to the bedding of sedimentary rocks ... Overall, stratification is developed intermittently through an original vertical thickness of 2 miles. Individual layers have been traced for 300 feet, and one continuously layered section is 1,500 feet thick and extends 1,000 feet ... The layering has undoubtedly formed because of magmatic currents *during extremely unstable conditions*." (Italics added)

Comment: This situation provides a most interesting phenomenon: crystals being deposited from a flood of magma the way clastic particles are deposited from a flood of water. The large scale of these turbidite-like sedimentary structures in the igneous body attests to the large-scale effect of cataclysmic magma flow. Such cataclysmic flows or mobilizations of large amounts of magma are perfectly consistent with the cataclysmic origin of igneous bodies during Creation Week and during flood-related tectonomagmatic events. Irvine[127] added that: "Such layering occurs in most compositional types of intrusions but especially in mafic and ultramafic bodies." The Alaskan example cited above is therefore far from an isolated instance.

6. Paramount Significance of the Young Earth Concept

Clebsch[14] wrote: "The displacing of the particular creation-myth itself pales in importance beside the change in human self-awareness that was involved in stretching their planet's age from a mere sixty centuries to many millions of centuries."

Comment: Many believers (even some of those with otherwise Creationist sympathies) tend to regard the earth's age as a very peripheral issue. The statement of Clebsch shows that the very opposite is true. The issue of the age of the earth is as important as the Creation-evolution issue itself!

7. Uniformitarianism Not Capable of Being Empirically Substantiated

Dickinson[124] wrote: "Uniformitarian thinking compels us to recognize, in the record of the rocks, the slow unfolding of diverse sequence of events whose full display is *beyond our immediate experience.*" (italics added)

Comment: Uniformitarians often belittle and ridicule Creationists for believing in "unempirical" events such as the Creation and the Flood. The statement of Dickinson makes it clear that *all* accounts of past earth history are unempirical! The Creation and Flood are no more "unempirical" than a slowly-operating earth over long periods of time. Uniformitarian concepts and their consequences are certainly unempirical, and all the more so when one considers the numerous subsidiary hypotheses and special pleading necessary to maintain the acceptance of, say, the evolutionary-uniformitarian geologic column.

8. Uniformitarianism Rests Entirely on Faith

Hunt[125], a uniformitarian, proposed that a pattern of glacial erratics in southwest Canada indicates mile-high tidal waves caused by a passing meteor near earth about 11,000 years ago (such an idea, apart from the time-scale involved, would not be of direct use to the Creationist-Diluvialist paradigm unless there was post-Flood extraterrestrially-induced catastrophism).

In replying to criticism of his paper, Hunt[126] subsequently wrote: "The commentary of L. E. Jackson Jr., essentially a summary of conventional thought on emplacement of the Erratics Train, opens with the assertion, itself *an acclamation of uniformitarian faith,* that extensive published studies 'easily explain these deposits as products of unassisted earth processes.' He then proceeds to ignore the two principal problems requiring explanation, and which my hypothesis answers." (italics added)

Comment: Uniformitarians frequently claim that their positions are "matters of pure science" while those of the Creationist-Diluvialist are "matters of faith." The reply by Hunt makes it evident that uniformitarian claims rest upon faith, particularly the well-worn premise that presently-operating processes at presently-operating rates under presently-operating conditions account for geologic features of the earth.

9. Religion Not the Cause of Superstitious Ideas About Earth History

Williams[128] wrote: ". . . Voltaire was only half-joking when he supposed that the oyster shells found by travellers in the Alpine passes were the result of the passage of generations of pious pilgrims on their way to Rome."

Comment: At one time, superstitious ideas about fossils were widely held. Historical geology textbooks commonly (but subtly) associate such superstitions with the strong religious convictions of past ages. The statement of Williams notes that Voltaire, the noted infidel, held superstitious views about fossils. Superstition was therefore universal and cannot be blamed on religious belief.

10. Uniformitarian Prejudices against Catastrophic Processes

Pattison et. al.[129] wrote: "With Rhodes, we do not regard the changes in the distribution and evolution of floral and faunal elements as either unduly rapid or abnormal, and consequently do not feel impelled to invoke *catastrophic or otherwise abnormal* causes to explain them." (italics added)

Comment: The equating of catastrophic processes with abnormality reveals the prejudice against catastrophism in contemporary uniformitarian geologic thought.

11. Need for Broader thinking in the Formulation of Geologic Theories

Jodry[130] said: Every few years a new geologic concept is proposed that is so well conceived and so forcefully presented, by authors of such unimpeachable reputation that it is almost universally accepted and applied. But such complete acceptance obscures the fact that *other concepts may apply,* and that *more than one geologic process may produce a given rock.* (italics added)

Comment: Some geologists, upon learning that the author is a Creationist-Diluvialist, said: "Uniformitarianism works, so why invoke anything else?" I answered that it "works" only via special pleading and endless subsidiary hypotheses. Besides (as Jodry's statement cited above indicates) the fact that a geologic theory is widely accepted and seems to "work" does not give one a right to be closed-minded towards alternative viewpoints.

12. Repressive Influence of Uniformitarianism Upon Geologic Thought

Marvin[131] wrote: "The hypothesis of continental drift is the other issue in which the uniformitarian outlook discouraged serious investigation . . . The hypothesis of continental drift appeared to be so antiuniformitarian in spirit that many geologists reacted as they might to a personal insult."

Comment: Irrespective of whether or not one accepts plate tectonics, one can see that much can be learned about the process of geologic thought from it. Uniformitarians love to boast of their position as being intellectually liberating. The statement of Marvin shows that uniformitarianism is actually an intellectual straitjacket. Whether or not a concept in geology is or isn't accepted depends heavily on whether or not it agrees with the sacrosanct dogma of uniformitarianism.

13. Pervasive Intolerance Towards Minority Viewpoints in Uniformitarian Geology

Nitecki et. al.[132] did a study involving a survey of 215 professional American geologists concerning their position towards the "new global tectonics." 87% accepted it (about half considering it "essentially established" and slightly less than half considering it "fairly well established"), and the remaining 12% rejected it as being "inadequately proven". 22% accepted in 1961.

They also noted: "It is significant that those least favorable to the theory were no less and no more familiar with the literature than those who have most recently accepted the theory." (So it can not be claimed that opponents of plate tectonics are less informed than proponents). A significant number of the 12% who reject plate tectonics are younger geologists, so it can not be claimed that skepticism towards plate tectonics comes only from the older geologists trained when plate tectonics was not generally accepted.

Especially significant is the following conclusion: "... those who have recently accepted the theory did so in an atmosphere of general acceptance that does not seem to require that they weight all the evidence themselves."

In a book review, Neumann[133] wrote: "Plate tectonics has been an enormous stimulant to geology and paleogeography. No doubt remains that such processes operate today on a global scale... Although many of its effects remain to be documented, continuing skepticism seems pointless and curious. Thus the final paper of the volume, 'Epilogue: a Paleozoic Pangea' by Boucot and Gray, might be compared to a sophisticated justification of the principles underlying the Flat Earth Society, and seems more a parody than a serious scientific essay."

Comment: From the research of Nitecki *et al.*, it is evident that a significant minority of informed geologists (at least 12%) reject plate tectonics in 1977. Yet plate tectonics is increasingly being presented as proven fact (statements of Neumann). This illustrates the uniformitarian intolerance towards minority viewpoints. If minority viewpoints *within* uniformitarianism can be summarily rejected, how much more so Creationism-Diluvialism and it magnitudes-greater radicalism and vastly smaller minority (than 12%)! The fact that most uniformitarians accept continental drift not from weighing evidence but from "jumping on the Bandwagon" and "following the crowd" has important implications in showing how theories become accepted in geology. Very likely it was a similar shunning of minority viewpoints and "following the crowd" that led to the rejection of Creationism-Diluvialism and swallowing of the claim of Hutton, Lyell, and Darwin over 1½ centuries ago.

While on the subject of plate tectonics, it should be noted that Creationists-Diluvialists beware of "jumping on the bandwagon." Most of the presumed evidences for the "new global tectonics" are squarely evolutionary-uniformitarian and so have no meaning in the Creationist-Diluvialist paradigm. The argument from paleobiogeography, for example, has meaning only if one accepts geologic periods and evolution: acceptance of paleoclimatological arguments also requires acceptance of geologic periods. The vital ocean-floor arguments (magnetic "stripes", ocean-bottom biostratigraphy, K-Ar results from submarine lavas) all require acceptance of geologic periods, geomagnetic reversals, and radiometric dating.

14. Atheistic Character of Uniformitarianism

Concerning the pioneer uniformitarian James Hutton, Marvin[134] wrote: "He was accused of atheism..."

In describing a pioneer uniformitarian who should perhaps qualify as an equal of Hutton and Lyell, Aprodov[135] said "This tribute praises Lomonosov for his purported employment of the dialectic concept to the evolution of the earth. The claim that Lomonosov's interpretations of the earth are materialistic is absolutely true, as are all other scientific interpretations. Whether he is the father of materialism in Russian science is of little importance because all sound science is materialistic... The statements that he first applied historical analyses to explain geologic phenomena and that he established the principle of actualism 70 years before Lyell would be brought to the attention of those unfamiliar with the contributions of the great Russian scientist."

Comment: While Hutton was not an atheist (he was said to be a Deist) his position is atheistic and his critics were correct in pointing this out. Deism and atheism are not far separated: a Supreme Being who does nothing is not greatly different from a Supreme Being who does not exist at all.

It goes without saying that Soviet Communist ideology is openly materialistic and militantly atheistic. The fact that uniformitarianism gets such an enthusiastic endorsement in Soviet Communist ideology (statements of Aprodov) is an excellent indicator of the atheistic foundations of uniformitarianism.

15. Fallacies of the Geologic Column: Extended Stratigraphic Ranges

Glaessner[136] wrote: "Index fossils combine short stratigraphic range with wide geographic distribution. The main problem in the use of index fossils is the finality with which their stratigraphic ranges can be established."

In proposing the resolution of a stratigraphic problem, Berry and Boucot[137] suggested: "... to let the genus *Stricklandia*, previously unknown beneath the Silurian, extend its range down into the Ordovician."

Smit[71] wrote: "It may even be possible that the Dinosaurs of the Pognacium are partly of Tertiary age..."

Comment: Extending the brachiopod *Stricklandia* and allowing dinosaurs to range beyond Cretaceous are two more examples of stratigraphic range extentions; reducing further the credibility of the geologic column.

References

AG—American Association of Petroleum Geologists Bulletin
AJ—American Journal of Science
CA—Bulletin of Canadian Petroleum Geology

EL—Elsevier Scientific Publishing Co., Amsterdam, New York
GA—Geological Society of America Abstracts with Programs
GR—Grana Palynologica
IG—International Geology Review
JP—Journal of Paleontology
JW—John Wiley and Sons, New York, Toronto
MI—Revue Micropaleontologie
MP—Micropaleontology
NA—Nature
PA—Palaeogeography, Palaeoclimatology, Palaeoecology
PS—Pollen et Spores
PY—Geophytology
RE—Creation Research Society Quarterly
RP—Review of Paleobotany and Palynology
RU—Geologische Rudschau
SP—Journal of Sedimentary Petrology
US—Proceedings of the Ussher Society

[1] Woodmorappe J. 1980. An Anthology of Matters Significant to Creationism and Diluviology: Report 1. RE 16(4)209-219.
[2] Brownlow A. H. 1979. Geochemistry. Prentice-Hall, Englewood Cliffs, New Jersey, p. 262-4.
[3] Darlington P. J. 1980. Evolution for Naturalists. J. W., p. 33.
[4] Davies G.F. 1979. Proof in Science. Transactions American Geophysical Union. 60(52)1044.
[5] Frey R. W. 1971. Ichnology: The Study of Fossil and Recent Lebenspurren. Louisiana State University School of Geosciences Miscellaneous Publication 71-1, p. 105-6.
[6] Darlington. 1980. op. cit., p. 24.
[7] ibid, p. 168.
[8] Anderson J.K. and H.G. Coffin. 1977. Fossils in Focus. Zondervan Pub. Co., Grand Rapids, Michigan, pp. 76-81.
[9] Waterhouse J.B. 1979. The Role of Fossil Communities in the Biostratigraphical Record and in Evolution (in Gray J. and A. J. Boucot. 1979. Historical Biogeography, Plate Tectonics, and the Changing Environment. Proceedings of the 37th Annual Biology Colloqium, Oregon State University Press). p. 249-50.
[10] Darlington. 1980. op. cit., p. 76.
[11] ibid, p. 31.
[12] ibid, p. 32.
[13] Heinlein R. cited by W.J.M. Mackenzie. 1979. Biological Ideas in Politics. St. Marten's Press, New York, p. 11.
[14] Clebsch W.A. 1979. Christianity in European History. Oxford University Press, New York, p. 245.
[15] Darlington. 1980. op. cit., p. 243-4.
[16] Littell F.H. 1976. The Macmillan Atlas History of Christianity. MacMillan Pub. Co., New York, London, p. 104.
[17] Hoffmann P. 1979. Hitler's Personal Security. Pergamon Press. London and Cambridge, Mass., p. 264.
[18] Darlington. 1980. op. cit., p. 46.
[19] Clebsch. 1979. op. cit., p. 246.
[20] Darlington. 1980. op. cit., p. 233.
[21] ibid, p. 15, 233.
[22] Shaver R.H. 1974. Structural Evolution of Northern Indiana during Silurian Time (in 1974. Silurian Reef Evaporite Relationship: Michigan Basin Geological Society Field Conference) p. 63.
[23] Woodmorappe. 1980. op. cit., p. 213-14.
[24] Shinn E.L. Recent Intertidal and Nearshore Carbonate Sedimentation around Rock Highs, E. Quatar, Persian Gulf (in Purser B.H. 1973. The Persian Gulf. Springer-Verlag, Berlin, New York). p. 197.
[25] Squires R.L. 1976. Pseudoreefs in the San Andres Formation, Guadaloupe Mountains, New Mexico. GA 8:634-5.
[26] Carozzi A.V. and J.B. Hunt. 1960. Fore-Reef Petrography of the Silurian Richvalley Reef, Indiana. SP 30:209.
[27] McGovney J.E. 1978. Deposition, Porosity, Evolution, and Diagenesis of the Thornton Reef (Silurian) Northeastern Illinois. University of Wisconsin-Madison PhD Thesis. p. 90.
[28] Hinman E.E. 1968. A Biohermal Facies in the Silurian of Eastern Iowa. Iowa Geological Survey Report of Investigations 6:1, 2.
[29] Kahle C.F. 1974. Nature and Significance of Silurian Rocks at Maumee Quarry, Ohio, (in 1974. op. cit.), p. 33.
[30] Twenhofel W.A. 1950. Coral and Other Organic Reefs in Geologic Column. AG 34:182-202.
[31] ibid. p. 183.
[32] Lane N.G. 1970. Crinoids and Reefs (in Yockelson E.L. 1970. Proceedings of the North American Paleontological Convention) p. 1438-40.
[33] Klovan J.E. 1974. Development of Western Canadian Devonian Reefs and Comparison with Holocene Analogues. AG 58:791.
[34] Twenhofel, 1950. op. cit., p. 186.
[35] Klovan. 1974. op.cit., p. 788.
[36] Philcox M.E. 1970. Geometry and Evolution of the Palisades Reef Complex, Silurian of Iowa. SP 40:195.
[37] Ager D.V. 1963. Principles of Paleoecology McGraw Hill, New York, p. 86-7.
[38] Manten A.A. 1965. Review of Ager (ref. 37 above). PA 1:175.
[39] Hekker R.F. 1965. Introduction to Paleoecology. El, p. 51.
[40] Pray L.C. 1976. Guidebook for a Field Trip in Silurian Reefs, Interreef Facies, and Faunal Zones of Northern Indiana and Northeastern Illinois. North-Central Section, Geological Society of America and Western Michigan University, p. 12.
[41] Keith J.W. 1970. Tectonic Control of Devonian Reef Sedimentation, Alberta. AG 54:854.
[42] Anstey R.K. and J.F. Pachut. 1976. Functional Morphology of Animal Colonies by Comparison with Sand Dune Paradigms. GA 8:124-5.
[43] Mountjoy E. 1980. Some Questions About the Development of Upper Devonian Carbonate Buildups (Reefs), Western Canada. CA 28:316.
[44] Anderton R., Bridges P.H., Leeder M.R., and B.W. Sellwood. 1979. A Dynamic Stratigraphy of the British Isles George Allen and Unwin. London, Boston, Sydney, p. 144.
[45] Stanley G.D. 1979. The Early History of Mesozoic Reef Building: Evidence from North America and Europe. GA 11(7)523.
[46] Stanley G.D. 1979. Paleoecology, Structure, and Distribution of Triassic Coral Buildups in Western North America. University of Kansas Paleontological Contribution Article 65, p. 46.
[47] Philcox M.E. 1971. Growth Forms and Role of Colonial Coelenterates in Reefs of the Gower Formation (Silurian) Iowa. JP 45:343-4.
[48] Fursich F. Th. 1980. Preserved life positions of some Jurassic bivalves. Palaeontologische Zeitschrift 54:292.
[49] Neumann A.C. 1981. Waulsortian Mounds and Lithoherms Compared. AG 65:965.
[50] Jenkyns H.C. 1971. Speculations on the Genesis of Crinoidal Limestones in the Tethyan Jurassic. RU 60(2)480-1.
[51] Savage J.M. 1969. Evolution. Holt, Rhinehart and Winston, Inc., New York, Chicago, Montreal, London, Sydney, pp. v-vi.
[52] Valentine J.W. 1977. General Patterns of Metazoan Evolution (in Hallam A. 1977. Patterns of Evolution. El.), p. 35.
[53] Ruse M. 1973. The Philosophy of Biology. Hutchinson and Co., London, p. 96.
[54] ibid., p. 107.
[55] Hallam A. 1977. Preface (in Hallam. 1977. op. cit.). p. v-vi.
[56] Ruse, 1973. op.cit., p. 111.
[57] Laporte L.F. 1980. Human Origins and the East African Evidence. Paleobiology 6:139.
[58] Paul C.R.C. 1977. Evolution of Primitive Echinoderms (in Hallam. 1977. op.cit.), p. 130.
[59] van Heyningen. 1968. Tetanus. Scientific American. April 968, p. 71, 73.
[60] Gretener P.E. 1977. On the Character of Thrust Faults With Particular Reference to the Basal Tongues. CA 25:110.
[61] Burdick C. 1971. Streamlining Stratigraphy (in Lammerts W.E. 1971. Scientific Studies in Special Creation. Presbyterian and Reformed Pub. Co.) pp. 128-135.
[62] Burdick C.L. and H. Slusher. 1969. The Empire Mountains—A Thrust Fault? RE 6(1)49-54.
[63] Gretener P.E. 1972. Thoughts on Overthrust Faulting in a Layered Sequence. CA 20:590-1.
[64] Gretener, 1977. op.cit., p. 111.
[65] Rezvoy D.P. 1971. Problem of "thrust sheets" in the Alay Range. IG 13:1736.
[66] ibid., p. 1741.
[67] Ager D.V. 1974. Storm Deposits in the Jurassic of the Moroccan High Atlas. PA 15:83-93.

⁶⁸Gretener. 1977. op.cit., p. 116.
⁶⁹Woodmorappe. 1980. op.cit., p. 212.
⁷⁰Gartner S. and B. Lodz. 1972. Reworking and Apparent reworking of Neogene fossil plankton. MP 18:117.
⁷¹Smit J. 1979. Microdium, Its Earliest Occurrence and other Considerations. MI 22:46.
⁷²Chene J.D.E., De Klasz I.D.E. and E.E. Archibong. 1978. Biostratigraphic Study of the Borehold UJO-1, SW Nigeria, with special Emphasis on the Cretaceous Microflora. MI 21:125.
⁷³Domack E.W., Fairchild W.W., and J.B. Anderson. 1980. Lower Cretaceous sediment from East Antarctic continental shelf. NA 287:625-6.
⁷⁴Gohabandt K.H.A. 1967. The geologic age of the type locality of Pseudotextularia elegans (Rzehal). MP 13:70.
⁷⁵Haynes J. and A.R.M. El-Naggra. 1964. Reworked Upper Cretaceous and Danian planktonic foraminifera in the type Thanetian. MP 10:354-5.
⁷⁶Bramlette M.N. and E. Marteni. 1964. The great change in the calcarous nannoplankton fossils between the Maestrichtian and Danian. MP 10:294.
⁷⁷Thiede J. 1981. Reworking of Upper Mesozoic and Cenozoic central Pacific deep-sea sediments. NA 289:667-70.
⁷⁸Otvis E.G. and W.D. Brock. 1976. Massive Long-Distance Transport and Redeposition of Upper Cretaceous Planktonic Foraminifers in Quaternary Sediment. SP 46:978.
⁷⁹Funkhouser J.W. 1969. Factors that Affect Sample Reliability (in Tschudy R.H. and R.A. Scott. 1969. Aspects of Palynology). JW, p. 101.
⁸⁰Bramlette M.N. and F.R. Sullivan. 1961. Coccolithophorids and related nannoplankton of the early Tertiary in California. MP 7:133.
⁸¹Jones D.J. 1956. Introduction of Microfossils. Harper Bros, New York, pp. 260-7.
⁸²Akyol E. 197?. Palynologie Du Permien Inferieur De Sariz (Kayseri) Ed De Pamucak Yaylasi (Antalya-Turquie) et Contamination Jurassique Observee, Due Aux Ruisseaux "Pamucak" et "Goynuk". PS 17(1).
⁸³Phillips L. 1974. Reworked Mesozoic Spores in Tertiary Leaf-Beds on Mull, Scotland. RP 17:221.
⁸⁴Krassilov V.A. 1972. (translated 1975). Paleoecology of Terrestrial Plants. JW, p. 56-7.
⁸⁵Van Morkhoven F.P.C.M. 1962. Post-Paleozoic Ostracoda. EL., Vol I, p. 126.
⁸⁶Sturt B.A., Dalland A., and J.L. Mitchell. 1979. The Age of the Sub Mid-Jurassic Tropical Weathering Profile of Andoya, Northern Norway and the Implications for the Late Palaeozoic Paleogeography in the North Atlantic Region. RU 68(2)527.
⁸⁷Muller K.J. and L.C. Mosher. 1971. Post-Triassic Conodonts (in Sweet W.C. and S.M. Bergstrom. 1971. Symposium on Conodont Biostratigraphy. Geological Society of America Memoir 127:467.
⁸⁸Wilson L.R. 1964. Recycling, Stratigraphic Leakage, and Faulty Techniques in Palynology. GR 5:429.
⁸⁹Veevers J.J. 1976. Early Phanerozoic Events on and alongside the Australasian-Antarctic Platform. Journal of the Geological Society of Australia 23:203, 190-1.
⁹⁰Warrington G. 1979. A Derived Late Permian Palynomorph Assemblage from the Keuper Marl (Late Triassic) of West Somerset. US 1979, 4(3)299.
⁹¹Venkatachala B.S. 1969. Palynology of the Mesozoic Sediments of Kutch. W. Indian, Reworked Permian Pollen from the Upper Jurassic Sediment—A Discussion. Palaeobotanist 18(1)45.
⁹²Teichert C. and B. Kummel. 1972. Permian-Triassic Boundary in the Kap Stosch Area, East Greenland. CA 20(4)659.
⁹³Windle T.M.F. 1979. Reworked Carboniferous Spores: An Example from the Lower Jurassic of Northeast Scotland. RP 27:174-5.
⁹⁴Wilson. 1964. op.cit. p. 428
⁹⁵Warrington G. 1971. Palynology of the New Red Sandstone Sequence of the South Devon Coast. US 1971 2(4)307-8.
⁹⁶Danilov M.A. and G.V. Kanev. 1979. Discovery of Paleozoic spores in the Central part of the synclonorium of the Vetreny belt of the Baltic Shield. IG 21:1128-9.
⁹⁷Muir M.D. 1967. Reworking in Jurassic and Cretaceous Spore Assemblages. RP 5:148-50.
⁹⁸Bless J.M. et.al. 1980. Pre-Permian Depositional Environments Around the Brabant Massif in Belgium, the Netherlands, and Germany. Sedimentary Geology 27(1)72, 35, 8-9.
⁹⁹Belinken F.H. 1975. Leonardian and Guadalupian (Permian) Conodont Biostratigraphy in Western and Southwestern United States. JP 49:301-2.
¹⁰⁰Bader R.H. 1981. Stratigraphy, Paleontology, and Palynology of Holocene Deposits in Southwestern Lake Michigan. Unpublished MS Thesis, Northeastern Illinois University, p. 85-90.
¹⁰¹Lindstrom M. 1946. Conodonts EL., p. 72.
¹⁰²Graysom R. 1981. Distribution and Significance of Allochthonous Conodonts in Early Pennsylvanian rocks of eastern Oklahoma. GA 13(6)279.
¹⁰³Chauff K.M. 1978. Recovery of Ordovician conodonts by hydrochloric acid from phosphate nodules reworked into the Sulphur Springs Formation (Devonian) in Missouri, USA. Geological Magazine 115(3)207-8.
¹⁰⁴Dr. Stif Bergstrom, Professor of Geology, Ohio State University, Columbus, Ohio, 1980, personal communication.
¹⁰⁵Rozanov A. Yu. and F. Debrenne. 1974. Age of Archaeocyathid Assemblages. AJ 274:845.
¹⁰⁶Burdick C.L. 1966. Microflora of the Grand Canyon. RE 3(1)38-50. also: 1972. Progress Report on Grand Canyon Palynology. RE 9(1)25-30.
¹⁰⁷Bramlette and Sullivan. 1961. op.cit., p. 133.
¹⁰⁸Muir. 1967. op.cit., p. 153.
¹⁰⁹Pokorny V. 1965. Principles of Zoological Micropaleontology. Vol. II, Pergamon Press, Oxford, New York, p. 67.
¹¹⁰Hass W.J. 1962. Conodonts (in Hass W.H., Hantzchel W., Fisher D.W., Howell B.F., Rhodes F.H.T., Muller K.J., and R.C. Moore. 1962. Miscellanea. Treatise on Invertebrate Paleontology W38.
¹¹¹Lindstrom. 1964. op.cit., p. 73-4.
¹¹²Wilson. 1964. op.cit., p. 427.
¹¹³Muir M.D. and W.A.S. Sarjeant. 1977. Editors Comments (in Muir M.D. and W.A.S. Sarjeant. 1977. Palynology: Part 1., Dowden, Hutchinson, and Co., Stroudsburg, Pennsylvania) p. 223.
¹¹⁴Bushman J.R. 1981. Uniformitarianism According to Hutton. Journal of Geological Education 29:32-3.
¹¹⁵Wright C. 1975. Introduction (to Hitchcock E. 1851, reprinted 1975. The Religion of Geology and Its Connected Sciences. Regina Press, New York) p. 4m.
¹¹⁶Khain V. Ye. 1963. Lomonosov and Modern Geology. IG 5:714.
¹¹⁷Sparks J.M. 1978. The apostolic Fathers. Thomas-Nelson Pub. Co., Nashville, New York, p. 41-3.
¹¹⁸Krauskopf K.B. 1979. Introduction of Geochemistry. McGraw Hill Book Co., New York, Sydney, Tokyo, Toronto, p. 435 (2nd Edition).
¹¹⁹ibid, p. 440-1.
¹²⁰Williams H. and A.R. McBirney. 1979. Volcanology Freeman, Cooper, and Co., San Francisco, p. 121-2.
¹²¹Plumstead E.P. 1964. Palaeobotany of Antarctica (in Adie R.J. 1964. Antarctic Geology. JW) 654.
¹²²Hamilton L.H., Helby R., and G.H. Taylor. 1970. The Occurrence and Significance of Triassic coal in the Volcanic Necks near Sydney. Royal Society of New South Wales Journal and Proceedings 102 (parts 3 and 4) p. 169.
¹²³Krauskopf. 1979. op.cit., p. 342.
¹²⁴Dickinson W. 1971. Uniformitarianism and Plate Tectonics. Science 174:107.
¹²⁵Hunt C.W. 1977. Catastrophic termination of the Last Wisconsin Ice Advance, Observations in Alberta and Idaho. CA 25:456-67.
¹²⁶Hunt C.W. 1978. Reply to L.E. Jackson. CA 26:167.
¹²⁷Irvine T.N. 1964. Sedimentary Structures in Layered Ultrabasic Rocks. AG 48:533.
¹²⁸Williams L.P. 1973. Review of Steno as Geologist. AJ 273:351.
¹²⁹Pattison J., Smith D.B., and G. Warrington. 1973. A Review of Late Permian and Early Triassic Biostratigraphy in the British Isles (in Logan A. and L.N. Hills. 1973. The Permian and Triassic Systems and Their Mutual Boundary. CA Memoir 2) p. 256.
¹³⁰Jodry R.L. 1969. Growth and Dolomitization of Silurian Reefs, St. Clair County, Michigan. AG 53:957.
¹³¹Marvin V.B. 1973. Continental Drift. Smithsonian Inst. Press, Washington, D.C., p. 37-8.
¹³²Nitecki M.H., Lemke J.L., Pullman H.W., and M.E. Johnson. 1978. Acceptance of plate tectonic theory by geologists. Geology 6:661-4.
¹³³Neumann R.B. 1981. Review of Historical Biogeography. JP 55(3)693.
¹³⁴Marvin. 1973. op.cit., p. 36.
¹³⁵Aprodov V.A. 1963. Main Features of Philosophical Materialism in Geological Works of M.V. Lomonosov. IG 5:698.
¹³⁶Glaessner M.F. 1967. Principles of Micropaleontology. Hafner Publ. Co., New York, London, p. 225.

[137]Berry W.B.N. and A.J. Boucot. 1973. Correlation of the African Silurian Rocks. *Geological Society of American Special Paper* 147:60.

[138]Cronin J.E., Boaz N.T., Stringer C.B., and Y. Rak. 1981. Tempo and mode in hominid evolution. *NA* 292, p. 115.

[139]*ibid*, p. 114-115, 119.

[140]*ibid*, p. 116.

[141]Woodmorappe J. 1978. The Cephalopods in the Creation and the Universal Deluge. *RE* 15(2)97-8. especially Figure 3(i).

[142]Kummel B. and C. Teichert. 1970. Stratigraphy and Paleontology of the Permian-Triassic Boundary Beds, Salt Range and Trans-Indus Ranges, West Pakistan (*in* Kummel B. and C. Teichert. 1970. *Stratigraphic Boundary Problems*) U. of Kansas Special Publication No. 4), p. 53.

[143]Stanley E.A. 1969. The Occurrence and Distribution of Pollen and Spores in Marine Sediments (*in* Bronnimann P. and H.H. Renz. 1969. *Proceedings of the First International Conference on Planktonic Microfossils*. Vol. II. E. Brill, Leiden, pp. 641-3.

[144]Brasier M.D. 1980. *Microfossils*. George Allen and Unwin, London, p. 47.

[145]Wilson R.L. 1976. Palynology and evidence for the origin of the Atokak Formation (Pennsylvanian) rocks in the type area. *GA* 8(1)72-3.

[146]Sadowska A. 1975. Uwage o prowadzonych ostatnio badaniach palinologicznych trozecinzedowych osadow ilastych w Sudetach. *Kwartalnik Geologiczny* 19:725-31 (in Polish).

[147]Herngreen G.F.C. 1976. A probable erroneous identification of *Tricolporopollenites distinctus* Groot et Penny from lowermost Cretaceous strata in the eastern Netherlands. *RP* 22:345.

[148]Lytshev G.F. 1978. A New Early Oligocene Beaver of the Genus *Agnatocastor* from Kazakhstan. *Paleontological Journal*. 12:542.

[149]Stanley E.A. 1966. the Problem of Reworked Pollen and Spores in Marine Sediments. *Marine Geology* 4:403.

[150]Barss M.S., Bujak J.P., and G.L. Williams. 1979. Palynological Zonation and Correlation of Sixty-Seven Wells, Eastern Canada. *Geological Survey of Canada Special Paper* 78-2, 118p.

[151]Balme B.F. 1970. Palynologyof Permian and Triassic Strata in the Salt Range and Surghar Range, West Pakistan (*in* Kummel and Teichert. 1970., *op.cit*.) p. 382.

[152]Soukup J., Konzalova M., and Z. Lochmann, 1975. Relikty Cenomanu a toronu V. podlozi miocenu mezy Jarkovem a Novym Sedlem nad Bilanous. *Prague: Ustredni Ustav Geologicky Ventnik* Vol. 50(3)141-52 (in Czech).

[153]Shumenko S.I. and M.E. Kaplan. 1978. Discovery of Coccoliths in Lower Triassic Deposits of Siberia. *Doklady: Earth Sciences* 240-226.

[154]Evitt W.R. and S.T. Pierce. 1975. Early Tertiary Ages from the coastal belt of the Franciscan Complex, Northern California. *Geology* 3:433-6.

[155]Kar P.K. 1980. Permian Miospores in the Miocene Sediments of Kutch, Gujarat. *PY* 10:171.

[156]Saxena R.K. 1979. Reworked Cretaceous spores and pollen grains from the Matanomadh Fromation (Palaeocene) Kutch, India. *Palaeobotanist* 26:167-74.

[157]Matsuda T. 1980. Discovery of Permian-Triassic mixed conodont assemblage from the Akasaka Limestone, Gifu Prefecture. *Geological Society of Japan Journal* 86:41-44 (in Japanese).

[158]Dutta S.K. 1978. A Note on the Significance of the Discovery of Gondwana Palynomorphs in Rocks of Assam, Nagaland, and Meghalaya. *PY* 7:131.

[159]Dutta S.K. 1979. Recycled Permian Palynomorphs in Upper Cretaceous Rocks of Juintia Hills, Meghalaya. *PY* 8:250-1.

[160]Beach T.L. and J.B. Urban. 1974. The investigation of Carboniferous Stratigraphic leaks from Missouri and Iowa. *Geoscience and Man* 9:91-2.

[161]Hose R.K., Wrucke C.T., and A.K. Armstrong. 1979. Mixed Devonian and Mississippian conodont and foraminiferal faunas and their bearing on the Roberts Mountains Thrust, Nevada, *GA* 11:446.

[162]Matthews S.C. and J.M. Thomas. 1974. Lower Carboniferous conodont faunas from North-East Devonshire. *Palaeontology* 17:382.

[163]Richardson J.B., and S.M. Rasul. 1978. Lower Devonian spores and reworked acritarchs from the Witney Borehole, southern England, and their geological implications. *Palynology* 2:231.

[164]Haq B. 1971. Paleogene Calcareous Nannoflora, Part III Oligocene of Syria. *Stockholm Contributions to Geology* 25(3).

[165]Wilson, G.J. 1976. Permian Palynomorphs from the Mangarewa Formation, Productus Creek, Southland, New Zealand, *New Zealand Journal of Geology and Geophysics* 19:137.

[166]Chiguryayeva A.A. and K.V. Voronina, 1961, Data on Upper Pleistocene Vegetation in the North Caspian Region. *Doklady: Earth Science Sections* 131:362.

[167]Alvarez W. and W. Lowrie. 1981. Upper Cretaceous to Eocene pelagic limestones of the Scaglia Rossa are not Miocene Turbidites. *NA* 294:246-7.

[168]Arkhipov S.A. and O.V. Matvayeva. 1961. Spore Pollen Spectra of the Pre-Samarovo Anthropogene Deposits of the Glaciation Zone in the Yenisei Region of the West Siberian Lowlands. *Doklady: Earth Science Sections* 135:1204.

[169]Scott, R.A., Willians W.L., Craig L.C., Barghoorn E., Hickey L.J., and U.D. MacGinitie. 1972. *American Journal of Botany* 59:886-96.

[170]Morgan R. 1977. Elucidation of *Diconodium* and Its Stratigraphic Range in Australia. *Palynology* 1:175.

[171]Odrzywolska-Bienkowa E. and K. Pozaryska. 1981. Micropaleontological Stratigraphy of the Paleogene in the Fore-Sudetic Monocline. *Bulletin de L'Academie Polonaise Des Sciences-Series des sciences del la terre* XXIX(1)26.

[172]Olsson R.K. 1970. Planktonic Foraminifera From Base of Tertiary Millers Ferry, Alabama, *JP* 44:600.

[173]Haq B. 1971a. Paleogene Calcareous Nannoflora Part I. The Paleocene of West-Central Persia and the Upper Paleocene, Eocene of West Pakistan. *Stockholm Contributions to Geology* 25(1)8.

[174]*ibid*., p. 13.

[175]Norling E. 1981. Upper Jurassic and Lower Cretaceous Geology of Sweden. *Geologiska Foreningens i Stockholm Forhandlingar* 103:256.

[176]Pozaryska W., Vidal G., and W. Brochwicz-Lewinski. 1981. New Data on the Lower Cambrian at the southern Margin of the Holy Cross Mts (SE Poland). *Bulletin de L'Academie Polonaise Des Sciences—Series des Sciences de la terre* XXIX(2)170.

[177]Grigorovich A.S. *et al*. 1974. Experiment in investigation of nannoplankton from varicolored flysch in the zone of the Marmorosh Cliffs, Ukrainian Carpathians. *IG* 16:853.

[178]Pocock S. 1972. Palynology of the Jurassic Sediments of Western Canada, Part 2. *Palaeontographica* 137B, p. 136.

[179]Palyford G. 1976. Plant microfossils from the Upper Devonian and Lower Carboniferous of the Canning Basin, Western Australia. Palaeontographica 158B, p. 54, 61.

[180]Matsumoto T. 1978. Japan and Adjoining Areas (*in* Moullade M. and AE.EM. Nairn (1978) *The Phanerozoic Geology of the World II*, Part A, Mesozoic. El.), p. 92.

[181]Winslow M.R. 1962. Plant Spores and Other Microfossils from Upper Devonian and Lower Mississippian Rocks of Ohio. *US Geological Survey Professional Paper* 364, p. 1.

[182]Barss M.S. 1972. A Problem in Pennsylvanian-Permian Palynology of Yukon Territory. *Geoscience and Man* IV, p. 70.

[183]Budzinskiy Yu.A. and K.A. Belov. 1973, Absolute age of rocks in central part of the Caucasian piedmont. *IG* 15:194.

[184]Scott *et. al*. 1972. *op. cit*., p. 892.

[185]Tan J.T. 1980. Late Triassic-Jurassic Dinoflagellate Biostratigraphy, Western Arctic Islands, Canada. *Palaeontology* 4:253.

[186]Canis W.F. 1968. Conodonts and Biostratigraphy of the Lower Mississippian of Missouri. *JP* 42:553-4.

[187]Pashkov B.R. 1979. The Sarezk Series of the eastern part of the Central Pamir (Carboniferous-Upper Devonian). *IG* 21:53.

[188]Miller R.H. 1972. Silurian Conodonts from the Llano Region, Texas. *JP* 46:556-64.

[189]Muller H. 1966. Palynological Investigations of Cretaceous Sediments in Northeastern Brazil (*in* J.E. van Hinte. 1966. *Proceedings of the Second West African Micropaleontological Colloquium*. E.J. Brill, Leidin), p. 131, 134.

[190]Jacobson S.R. 1979. Acritarchs as Paleoenvironmental Indicators in Middle and Upper Ordovician Rocks from Kentucky, Ohio, and New York. *JP* 53:1201.

[191]Smith D.G. 1981. Progress in Irish Lower Palaeozoic Palynology. *RP* 34:141-6.

[192]Rexroad C.B., and L.V. Richard. 1965. Zonal Conodonts from the Silurian Strata of the Niagaran Gorge. *JP* 39:1219.

[193]Ethington R.L. and W.M. Furnish. 1962. Silurian and Devonian Conodonts from Spanish Sahara. *JP* 36:1259.

[194]Wilkinson G.C., Bazley R.A.B., and M.C. Boulter. 1980. The geology and palynology of the Oligocene Lough Neagh Clays, Northern Ireland. *Journal of the Geological Society of London* 137:72.

[195] Thompson T.L. and J.R. Sutterfield. 1975. Stratigraphy and Conodont Biostratigraphy of Strata Continguous to the Ordovician-Silurian Boundary in Eastern Missouri. *Missouri Geological Survey Report of Investigations* 57(part 2), p. 70.

[196] Barnes C.R. and M.L.S. Poplewska. 1973. Lower and Middle Ordovician Conodonts from the Mystic Formation, Quebec, Canada. *JP* 47:769, 789.

[197] Loeblich A.R. and H. Tappan. 1978. Some Middle and Late Ordovician Microphytoplankton from Central North America. *JP* 52:1237.

[198] Liebe R.M. and C.B. Rexroad. 1977. Conodonts from Alexandrian and Early Niagaran rocks in the Joliet, Illinois Area. *JP* 51:844.

[199] Rexroad C.B. and J.B. Droste. 1982. Stratigraphy and Conodont Paleontology of the Sexton Creek Limestone and the Salomie Dolomite (Silurian) in Northwestern Indiana. *Indiana Geological Survey Special Report* 25, p. 4.

[200] Cramer E.H. Carmen M.D., and R.D.D. Cramer. 1972. North American Silurian Palynofacies and Their Spatial Arrangement: Acritarchs. *Palaeontographica* 138B, p. 146.

[201] Mound M.C. 1968. Upper Devonian Conodonts from Southern Alberta. *JP* 42:459.

[202] Berg-Madsen V. 1981. The Middle Cambrian Kalby and Borregard Members of Bornholm, Denmark. *Geblogiska Foreningens i Stockholm Forhandlingar* 103:227.

[203] Konior K. 1980. W. Sprawie "dolnokambryjskiego" wieku nisszej czesci utworow dolnodewonskich z glebolich wiercen obszaru Bielsko-Mogilany. *Kwartalnik Geologiczny* 24(5)501.

[204] Turner R.E. 1980. Ordovician Acritarchs from the Type Section of the Caradoc Series. *Palynology* 4:253.

[205] Konstantinovskiy A.A. 1976. Stratigraphy of Upper Precambrian and lower Paleozoic deposits of the Chersk Range. *IG* 18:1031.

[206] Kugler H.G. and C.M.B Caudri. 1975. Geology and Paleontology of Soldado Rock, Trinidad (West Indies). *Eclogae Geologicae Helvetiae* 68:29.

[207] Chaproniere G.C.H. 1981. Australasian mid-Tertiary larger foraminiferal associations and their bearing on the East Indian Setter Classification. *BMR Journal of Australian Geology and Geophysics* 6:147.

[208] Kecskemeti T. 1981. The Eocene/Oligocene Boundary in Hungary in the Light of the Study of Larger Foraminifera. *PA* 36:254.

[209] Bombita G. and A. Rusu. 1981. New Data on the Eocene/Oligocene Boundary in the Romanian Carpathians. *PA* 36:214-7.

[210] Pflug A. 1963. Review of: *The Palynological Age of Diapiric and Bedded Salt in the Gulf Coastal Province* by Ulrich Jux.

[211] Sher A.V., Virina Ye. I, and V.S. Zazhigan. 1977. The Stratigraphy, Paleomagnetism, and Mammalian Fauna of the Pliocene and Lower Quaternary Deposits around the Lower Reaches of the Kolyma River. *Doklady Earth Science Sections* (English translations; hereafter abbreviated *DE*) 234:123.

[212] Scandone P. et.al. 1981. Mesozoic and Cenozoic Rocks from Malta Escarpment (Central Mediterranean). *AG* 65:1313.

[213] Ehrlich A. and S. Moshkovitz. 1978. Distribution of Calcareous Nannofossils and Siliceous Microfossils in the Plio-Pleistocene Sediments of the Mediterranean Coast and offshore of Israel (a Preliminary Study). *Israel Journal of Earth Sciences* 27:66.

[214] Shumenko S.I. 1973. Calcareous Nannoplankton from Cretaceous-Paleogene Boundary Beds of the Crimea. *DE* 209:72.

[215] Baramov V.I., and L.L. Ankhadeyeva. 1968. A Palynological Description of the Pliocene around Kazan. *DE* 181:106.

[216] Srivastava S.K. 1981. Evolution of Upper Cretaceous Phytogeoprovinces and Their Pollen Flora. *FP* 35:164.

[217] Chilonova A.F. 1981. Senonian (Late Cretaceous) Palynofloral Provinces in Cricumpolar Areas of the Northern Hemisphere. *RP* 35:322.

[218] Malkov B.A. and V.A. Gustemesov. 1976. Jurassic Fossils in Kimberlite of the Olenek Uplift and Age of Kimberlite Volcanism in the Northeastern Part of the Siberian Platform. *DE* 229:67.

[219] Brideaux W.W. 1971. Palynology of the Lower Colorado Group, Central Alberta, Canada. *Palaeontographica* 135:64-5.

[220] Tuchkov I.I. 1973. New Data on the Age of the Fresh-Water Sandstone Conglomerate Formation of the Aldan River Basin. *DE* 209:44.

[221] Ryder R.T. and H.T. Ames. 1970. Palynology and Age of Beaverhead Formation and their Paleotectonic Implications in Lima Region, Montana-Idaho. *AG* 54:1167.

[222] Ebner F. 1980. Conodont Localities in the Surroundings of Graz/Styria. Second European Conodont Symposium-Ecos II. *Austria: Geologische Bundesanstalt Abhandlungen* 35:105-125.

[223] Lane H.R. 1974. Mississippian of Southeastern New Mexico and West Texas—A Wedge-on-Wedge relation. *AG* 58:279.

[224] Kedo G.I., Neleryata N.S. and S.A. Kruchek. 1975. Redeposited Spores in Devonian Sediments of Belorussia. *DE* 225:74-6.

[225] Geobel E.D. 1968. Mississippian Rocks of Western Kansas. *AG* 52:1746-7.

[226] Boucot A.J., Hazzard J.C., and J.G. Johnson. 1969. Reworked Lower Devonian Fossils, Nopah Range, Inyo County; California. *AG* 53:163-7.

[227] Aristov V.A. and A.S. Alekseyev. 1976. Late Tournaisean Conodonts from the *Scaliognathus anchoralis* Zone of East Kazakhstan. *DE* 229:231.

[228] Craig W.W. 1969. Lithic and Conodont Succession of Silurian Strata, Batesville District, Arkansas. *Geological Society of America Bulletin* 80:1627.

[229] Pantic N., Hochuli P.A., and A. Gansser. 1981. Jurassic palynomorphs below the main central thrust of East Bhutan (Himalayas). *Eclogae Geologicae Helvetiae* 74(3)891.

[230] Gray J. and A.J. Boucot. 1972. Palynological Evidence Bearing on the Ordovician-Silurian Paraconformity in Ohio. *Geological Society of America Bulletin* 83:1299.

[231] Carter C. and S. Laufeld. 1975. Ordovician and Silurian Fossils in Well Cores from North Slope of Alaska. *AG* 59:440.

[232] Moskalenko T.A. 1972. Conodonts from Llandoverian Deposits of the Siberian Platform. *DE* 204:236.

[233] Nicoll R.S. and C.B. Rexroad. 1966. Conodont Zones in Salomie Dolomite and related Silurian Strata of Southeast Indiana. *AG* 50:630.

[234] Suhm R.W. and R.L. Ethington. 1975. Stratigraphy and Conodonts of Simpson Group (Middle Ordovician) Beach and Baylor Mountains, West Texas. *AG* 59:1133.

[235] Tikhiy V.N. and M.S. Stanichkova. 1973. Age of the Kazanla Suite and the Beginning of Devonian Sedimentation on the Russian Platform. *DE* 211:94.

[236] Goebel. 1968. *op.cit.*, p. 1742.

[237] Zoubek V. 1981. Introduction to "Micropaleontology of the Precambrian and Its Importance for Correlation." *Precambrian Research* 15(1)5.

[238] Swain, S.N., 1981. Review of Western and Arctic Canadian Biostratigraphy. Earth Science Reviews 16(4): 368.

The Essential Nonexistence of the Evolutionary-Uniformitarian Geologic Column: A Quantitative Assessment

THE ESSENTIAL NONEXISTENCE OF THE EVOLUTIONARY-UNIFORMITARIAN GEOLOGIC COLUMN: A QUANTITATIVE ASSESSMENT

JOHN WOODMORAPPE[*]

Received 2 July, 1980

This article is a systematic and quantitative demonstration of global distributional tendencies of the evolutionary-uniformitarian geologic column.

Maps have been drawn to show the worldwide distributions of all ten geologic periods on all seven continents, and such maps have also been drafted to show complete segments of the geologic column in place.

Calculations have been performed to measure successional tendencies of geologic periods over the earth. For example, it has been found that two-thirds of the earth's land surface has 5 or fewer of the 10 geologic periods in place, and only 15-20% of the earth's land surface has even 3 geologic periods appearing in "correct" consecutive order.

These and similar findings have been briefly related to the Creationist-Diluvialist paradigm.

PLAN OF THIS ARTICLE

Introduction
I. Procedures Utilized in Measuring the Factuality of the Geologic Column.
II. Exposition and Discussion of Results.
III. Probable Diluvial Significance of Global Stratigraphic Trends.

Introduction

A major foundation of the evolutionary-uniformitarian paradigm is the geologic column. This column, presented as fact, purports to demonstrate that the earth and its life have been evolving and that the earth's sedimentary strata contain the resultant biochronologic "onion skins."

A bit of wisdom for geologists is given by Park and MacDiarmid[1], who said: "The final test of all theories and hypotheses in geology is their applicability in the field." Accordingly, Creationists-Diluvialists have long pointed out that the evolutionary-uniformitarian geologic column does not correspond to reality and that fossiliferous strata must be understood as non-evolutionary, mutually-contemporaneous, cataclysmically-formed Noachian Flood deposits.

The works of Price,[2] Nelson,[3] Whitcomb and Morris,[4] and Burdick[5] have called attention to the fact that geologic periods rocks tend to be absent, inconsistent in their stratigraphic successional order from place to place, and all exhibiting some tendency to rest directly upon Precambrian "basement." Clark,[6] on the other hand, noted that there are places where much of the geologic column can be seen in place and in "correct" order.

The purpose of this work is to examine the earth's land surface (although oceans and continental shelves are also considered) in order to determine the degree of correspondence of the geologic column with reality. This study of "the geologic column as it really is" involves the quantitative measure of: 1) the tendency of the earth's land surface to have rocks of many alleged geologic periods in place versus the opposite tendency, and 2) the actual modes of stratigraphic succession of rocks attributed to different geologic periods.

I. Procedures Utilized in Measuring the Factuality of the Geologic Column

Data were gathered in order to present maps showing the distribution on all seven continents of rocks attributed to all ten geologic periods; a separate map for each geologic period (Maps 1-10).

Fidelity to areas on the maps was guaranteed by using Lamberts Azimuthal Equal-Area Projection on the base map used throughout this work. Maps of separate continents were reduced and placed next to each other to eliminate oceanic areas so that the largest possible dimensions of continents capable of being fitted onto a *Quarterly* page could be utilized. The use of separate continents, obtained from several sources,[7,10] reduces the perspective distortion that would result had the continents been bunched together from a single world map. (The bunching together of continents, a space-saving measure, is not at all intended to be an endorsement of the "new global tectonics.") The base map has a scale of one inch to 1530 miles, or one centimeter to 969 kilometers.

Much of the basic data for this work is derived from the works of a Soviet team of geologists headed by Alexander Ronov. They compiled data on all ten geologic periods, showing distribution, thicknesses, gross petrologic compositions, and inferred paleoenvironmental conditions of formation. Volcanic and volcano sedimentary formations were included, but evidently not postorogenic granites. Recently, Ronov *et. al.*[11] pointed out that some of their much earlier works are in need of revision. Accordingly, they have been updated partly by more recent works. Recent data for Antarctica has been added.

The geology of Antarctica is poorly known for east Antarctica because of the glacial cover. It is, however, probably mostly exposed Gondwana Shield. Whatever Phanerozoic rocks there are unaccounted for beneath the ice cap would not change the figures in this work by

---
[*]John Woodmorappe has an M.S. in Geology, and a B.A. in both Geology and Biology.

more than several per cent.

The errors in this work should favor the uniformitarian geologic column. Only a small part of the lower, medial, or upper portions of a geologic period need be present for the area to count as having that geologic period represented there. One thickness category used by the Ronov et. al. team, 0-100 meters, permitted areas to be exaggerated because scattered outliers would give entire areas having them credit for having those geologic periods.

It also appears that the Ronov et. al. team was generous in giving credit as to the representation of geologic periods to areas having rocks metamorphosed beyond biostratigraphic recognition. For instance, Ben-Avraham and Emery[12] wrote that Indonesion geology is poorly known and that the oldest fossiliferous formations are middle and late Paleozoic, resting upon crystalline schists. Yet from maps 1-3 it is evident that the Ronov et. al. team gave blanket credit for Indonesia having complete Lower Paleozoic despite the unknown biostratigraphic age of the schists.

The presumed geologic periods necessarily differ in their areal extent because there is no reason to believe that sedimentation and tectonic rates would be constant. Geologic periods also differed considerably in duration, ranging from only 40 million years for Silurian and Triassic to 100 million years for the Cambrian according to the scale of Braziunas.[13]

Data for making the map for world Cambrian (Map 1) were taken from Ronov et. al.[14,15] except that an addition was made for the Antarctic Cambrian.[16] The work of Ronov et. al.[17,18] on the Ordovician was used to draw that map (Map 2), while another work by Ronov et. al.[19,20] was utilized to show the world's Silurian (Map 3). No Ordovician or Silurian rocks are known from Antarctica, but Maps 2 and 3 have been drawn to show areas where Ordovician and Silurian is suspected, according to Burrett,[21] and Veevers,[22] and Elliot.[23] Recent palynostratigraphic work by Kyle[24] has cast unfavorable light upon another area being Silurian, and so it has not been shown as Silurian.

The map for Devonian (Map 4) used partly an old work by Ronov et. al.[25] which was extensively updated using the works of House,[26] Spasskiy,[27] Brinkmann,[28] Kummel,[29] Brown et. al.,[30] Hermes,[31] Cook and Bally,[32] Miall,[33] and Churkin.[34] Sources for Antarctic Devonian are Boucot,[35] Elliot,[36] and Barrett et. al.[37]

Carboniferous data for making the map (Map 5) were taken from Ronov et. al.[38] and from the more recent works of Churkin,[39] Brown et. al.,[40] Hill,[41] Ross,[42] Rocha-Campos,[43] Ross,[44] Meyerhoff,[45] Kummel,[46] Cook and Bally,[47] and Stocklin.[48] The map for Permian (Map 6) was drafted utilizing the work of Ronov et al.[49] on Permian, as well as more recent works by Rocha-Campos,[43] Meyerhoff,[45] Stocklin,[48] Audley-Charles,[50,51] Meyerhoff and Meyerhoff,[52] Miall,[53] Brown et. al.,[54] Oftedahl,[55] Kummel,[56] Gobbett,[57] Cook and Bally,[58] and Churkin[63]. The PermoCarboniferous of Antarctica was drafted using data from Barrett,[37,59,60] Elliot,[61] and Kemp.[62]

Ronov et. al.s'[64] work on the Triassic was used in drawing Map 7, with minor modifications from the works of Audley-Charles,[51] and Brown et. al.[65] Sources of Antarctic Triassic data used were from the works of Barrett et. al.,[37] Plumstead,[66] and Elliot.[67] Map 8 (Jurassic) was drawn utilizing the work on Jurassic by Ronov et.al.[68] with a minor updating based on Audley-Charles.[51] Antarctic Jurassic data came from Elliot.[69,70]

The map for Cretaceous (Map 9) was drawn from the maps in the work of Ronov et. al.,[71,72] except that additions for Antarctica were made from the works of Elliot,[23,69] Grikurov et. al.,[73] and Drewry.[74] (Drewry presented geophysical evidence for the presence of sedimentary basins under the east Antarctic ice cap, and both Drewry and Grikurov had conjectured the presence of Cretaceous-Tertiary formations there. These areas are shown in Maps 9 and 10 with question marks.)

Ronov et. al.'s[75,8] works on the Tertiary were used to construct the map (Map 10). Sources of Antarctic Tertiary besides the previously-mentioned works of Grikurov et. al.,[73] and Drewry[74] (plus the estimates of Behrendt[79]), were the works of Elliot[69] and Dort.[80]

The intersections of Map 1 (Cambrian), Map 2 (Ordovician), and Map 3 (Silurian) were drafted as Map 11 (Complete Lower Paleozoic). The intersections of Map 4 (Devonian), Map 5 (Carboniferous), and Map 6 (Permian) were drawn as Map 12 (Complete Upper Paleozoic). Map 13 (Complete Paleozoic) is the intersection of Maps 11 and 12, while Map 14 (Complete Mesozoic) is the intersection of Map 7 (Triassic), Map 8 (Jurassic), and Map 9 (Cretaceous). Map 15 (Complete Geologic Column) is the draft of the intersection of Map 10 (Tertiary) with Maps 13 and 14.

The earth's land surface was divided into 967 squares of equal area (each square being 252×252 miles or 406×406 kilometers) for the purpose of performing calculations on the areal tendencies of geologic periods. These square areas were given Cartesian coordinates (but letters instead of numbers) as is shown over a blank base map (Map 16).

A transparency identical to the one superimposed over a blank base map (Map 16) was thus superimposed over Maps 1 through 10, and the preponderant presence or absence of rocks attributed to the ten geologic periods in these 967 square areas was entered into Table 1.

The numbers of geologic periods lithologically represented in each square were written after every row in Table 1. All the raw data in Table 1 was expressed as a percentage of all 967 squares (the entire earth's land surface). Figure 1 was drawn to show what percentage of the earth's land surface has how many geologic periods represented. Figure 2 shows what percentage of the earth's land surface has what particular sequence of consecutive geologic periods represented. Table 2 shows what percentage of a given geologic period (in terms of total subaerial terrestrial areal distribution) rests on what geologic period that is older than that given geologic period. This includes the percentages (of all ten geologic periods) resting directly upon Precambrian "basement."

II. Exposition and Discussion of Results

From maps 11, 12, and 14, it is obvious that the

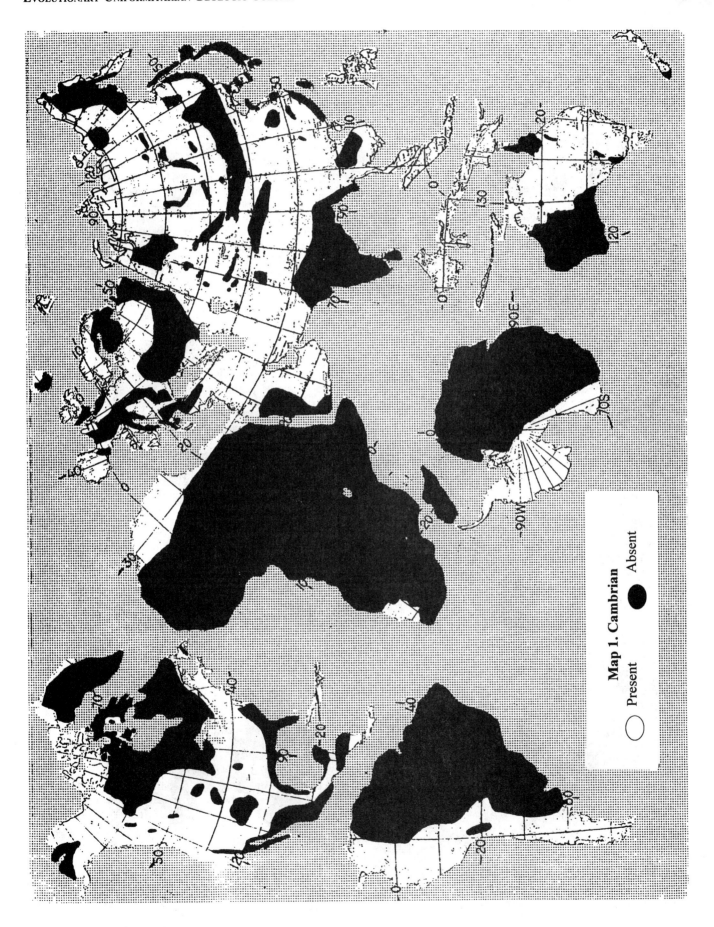

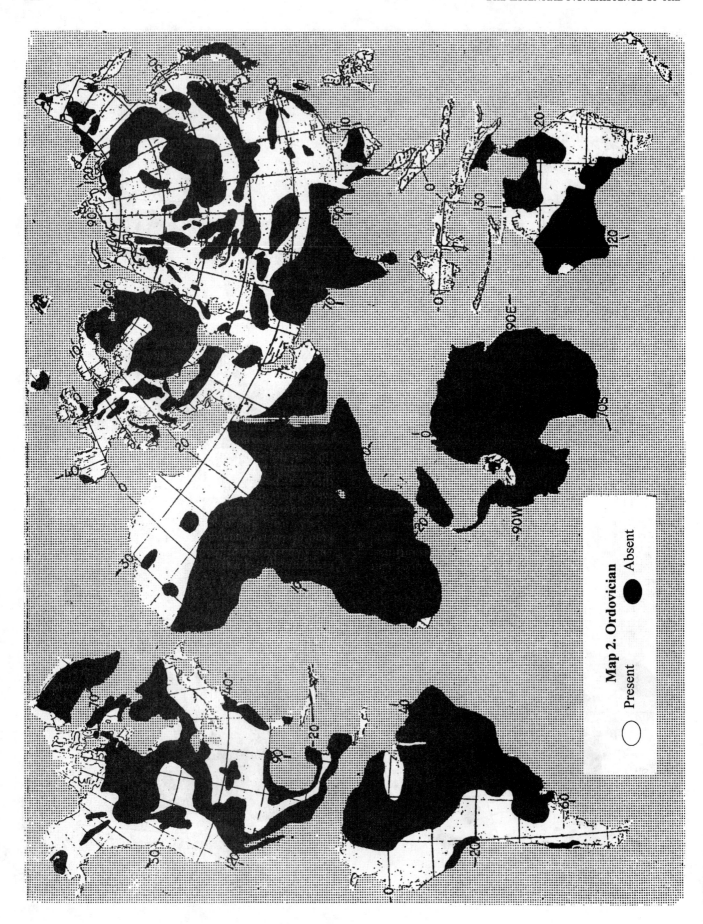

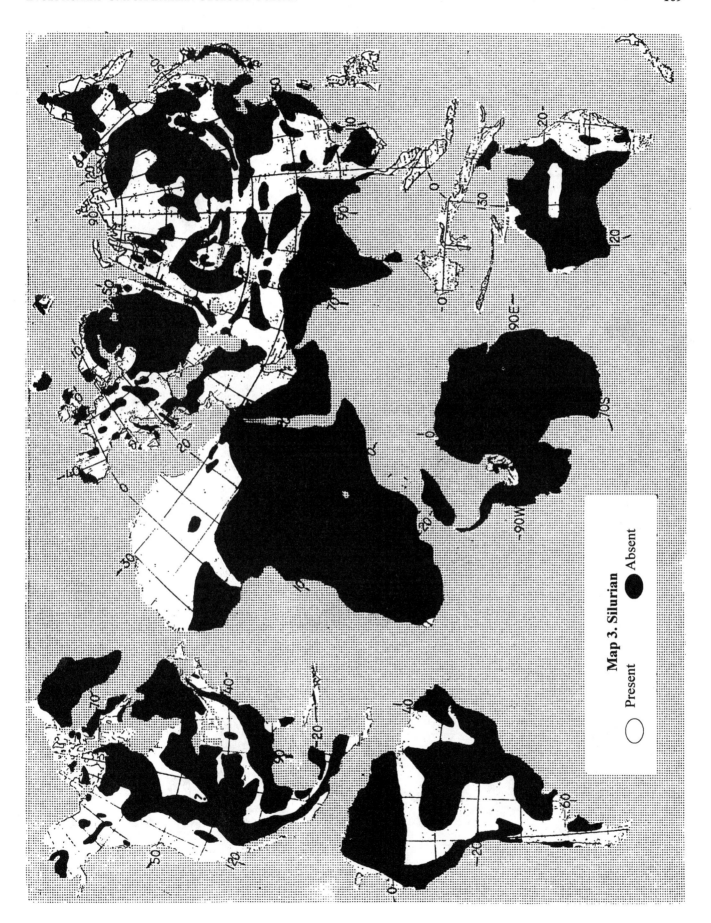

Map 3. Silurian

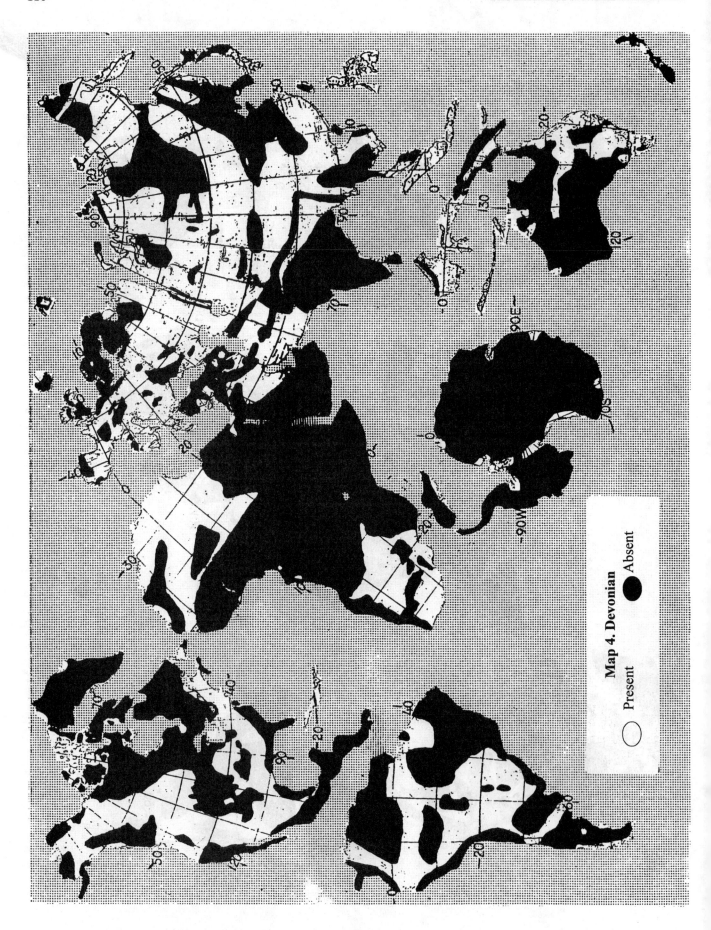

Map 4. Devonian

EVOLUTIONARY-UNIFORMITARIAN GEOLOGIC COLUMN

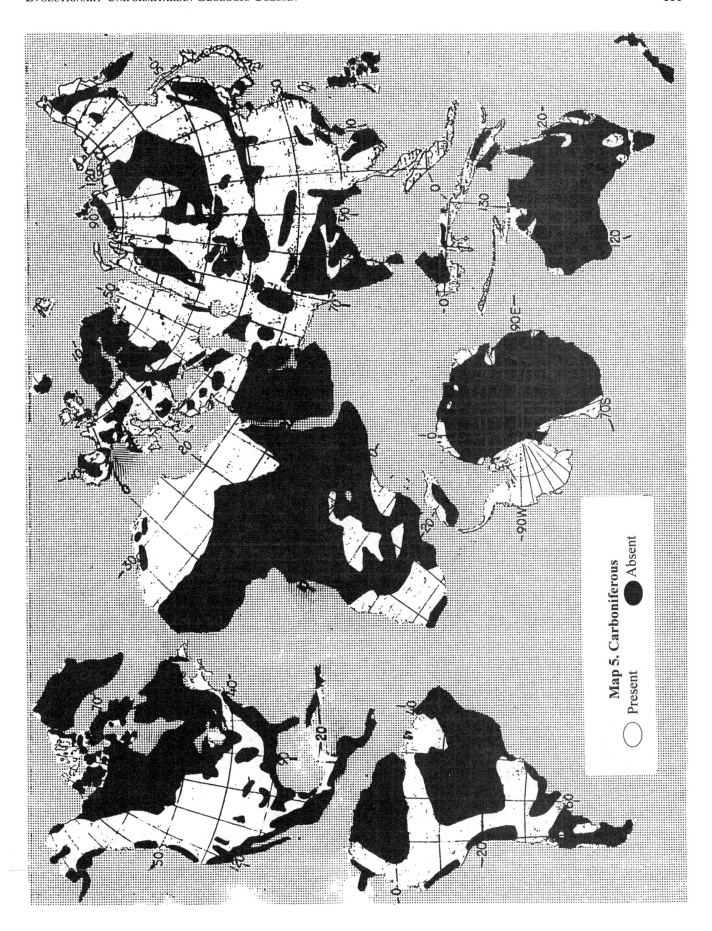

Map 5. Carboniferous
○ Present ● Absent

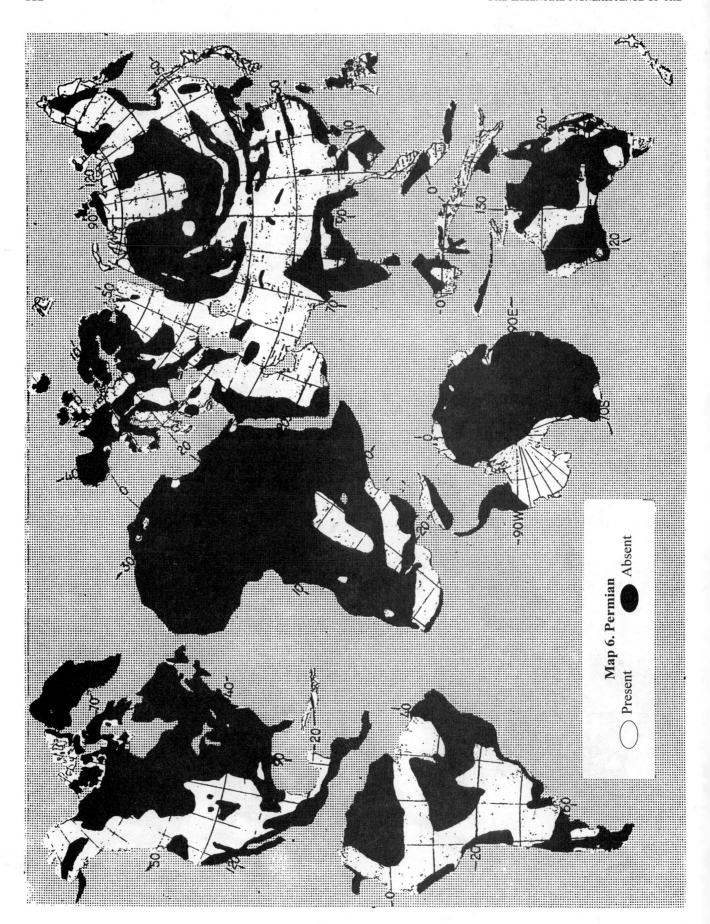

Map 6. Permian
Present ○   Absent ●

EVOLUTIONARY-UNIFORMITARIAN GEOLOGIC COLUMN 113

Map 7. Triassic
○ Present ● Absent

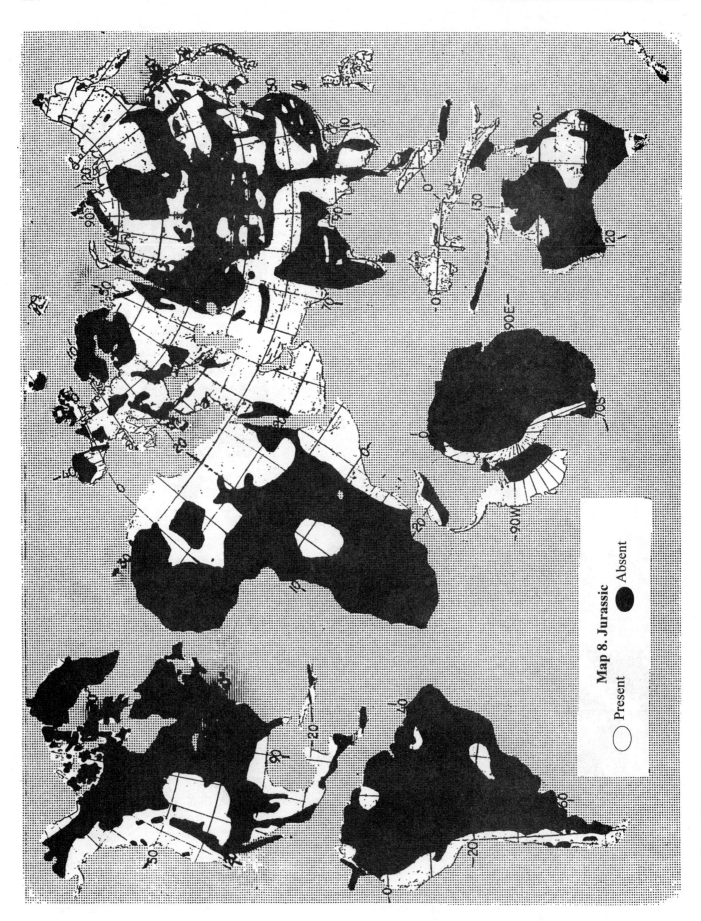

EVOLUTIONARY-UNIFORMITARIAN GEOLOGIC COLUMN

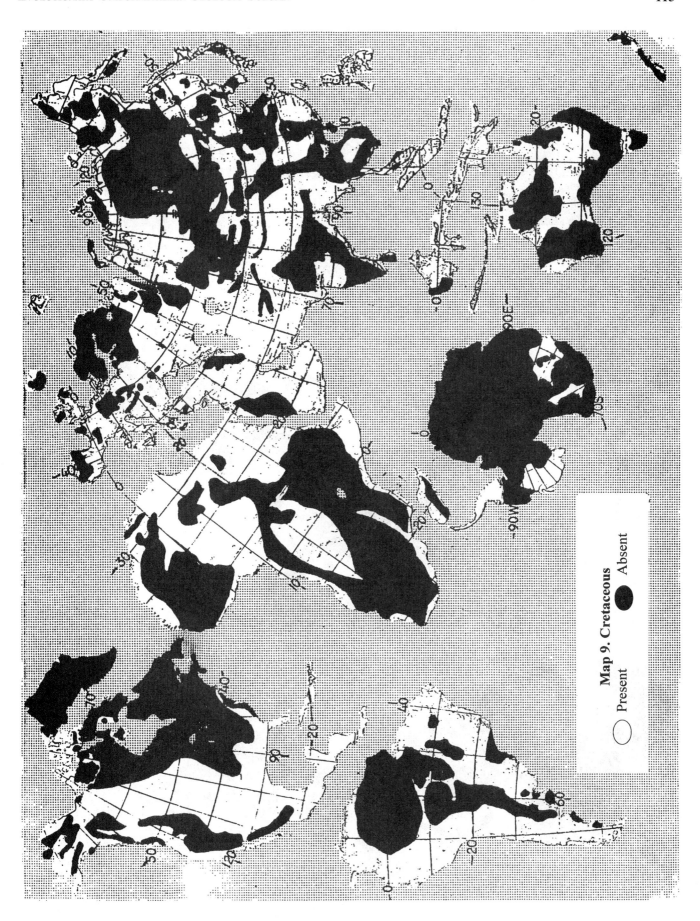

Map 9. Cretaceous
○ Present
● Absent

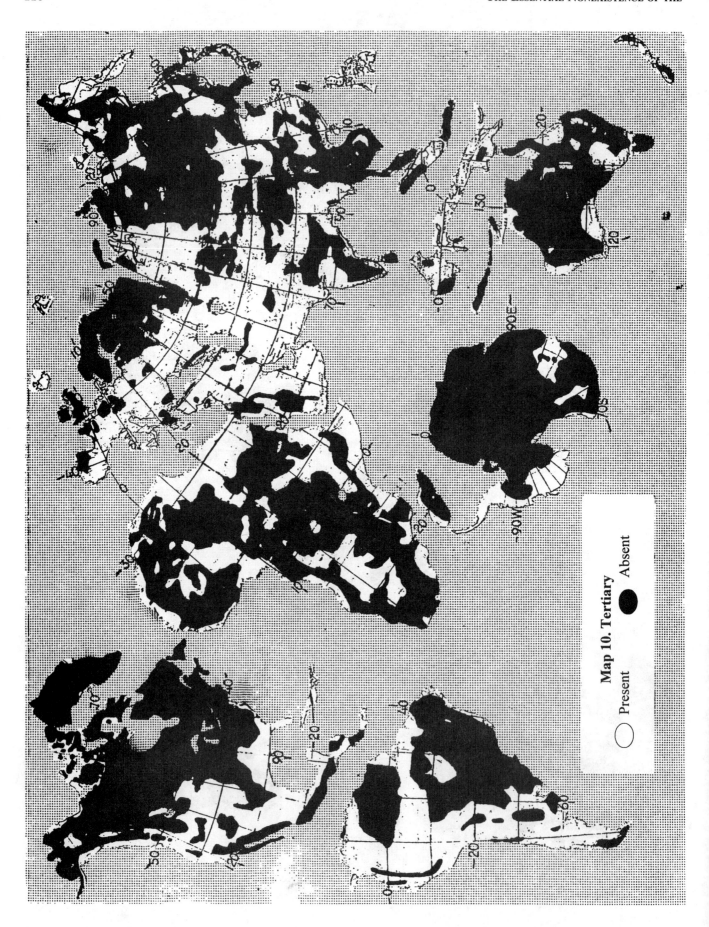

Map 10. Tertiary
○ Present
● Absent

EVOLUTIONARY-UNIFORMITARIAN GEOLOGIC COLUMN

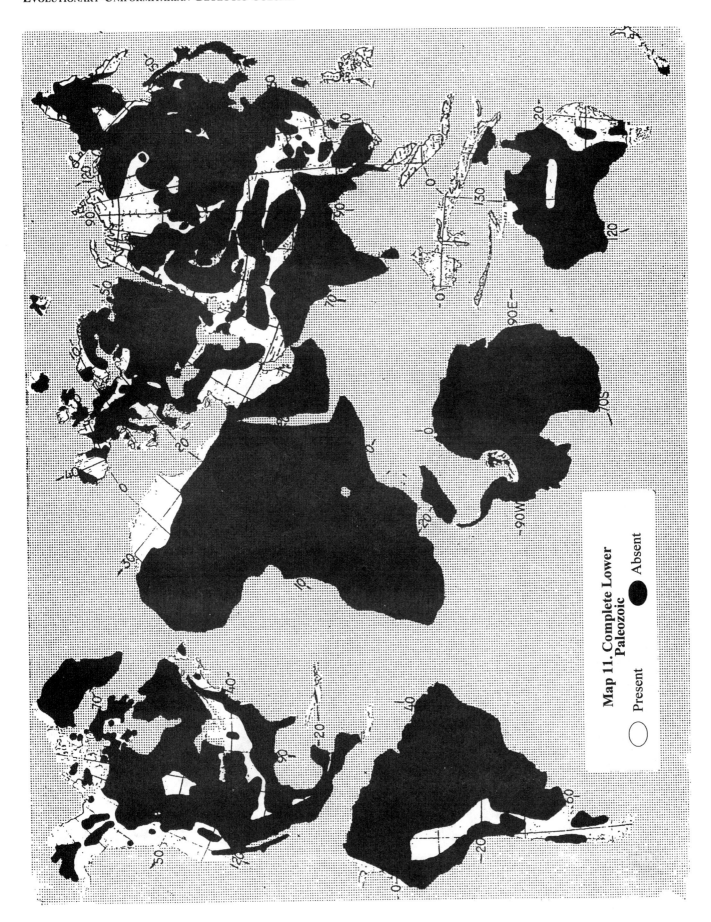

Map 11. Complete Lower Paleozoic
○ Present   ● Absent

Evolutionary-Uniformitarian Geologic Column

Map 13. Complete Paleozoic
○ Present
● Absent

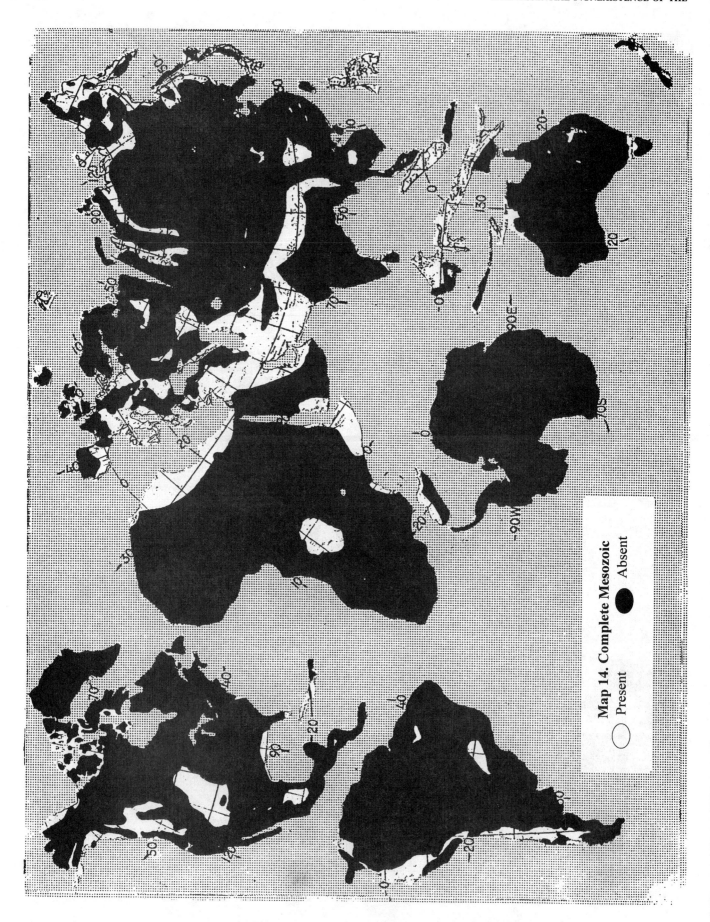

Evolutionary-Uniformitarian Geologic Column

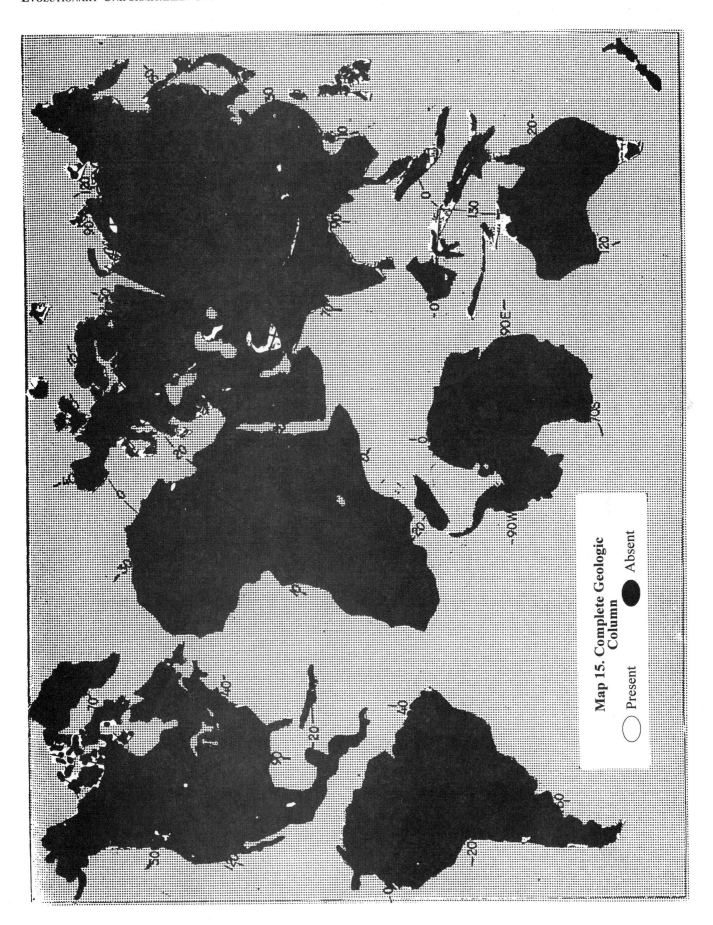

Map 15. Complete Geologic Column
○ Present ● Absent

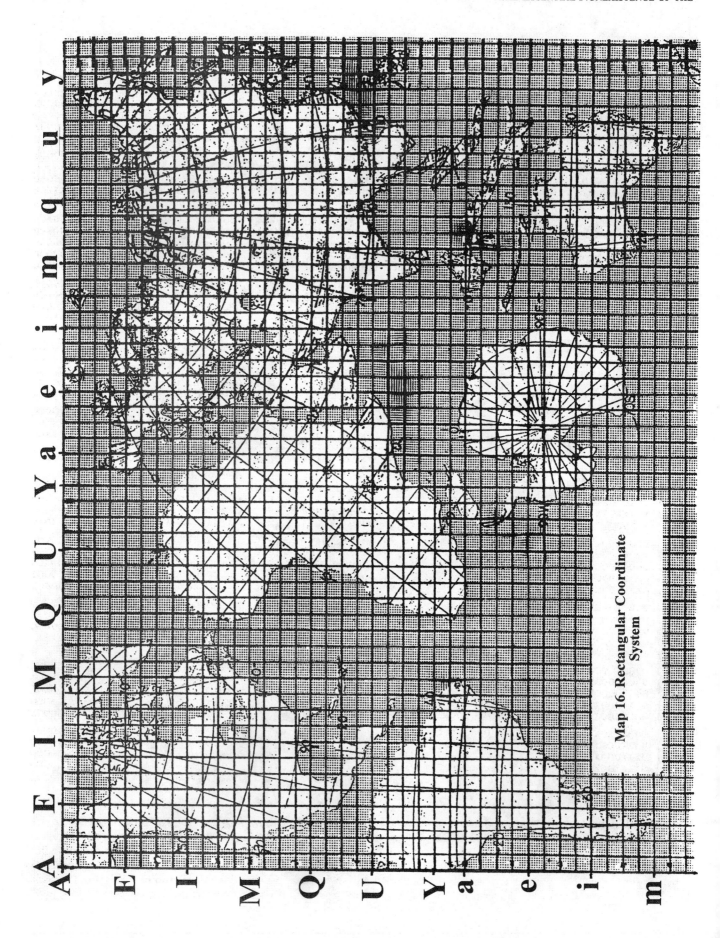

Map 16. Rectangular Coordinate System

123

Coor.	COSDCPTJCT#	Coor.	COSDCPTJCT#	Coor.	COSDCPTJCT#	Coor.	COSDCPTJCT#	Coor.	COSDCPTJCT#
Africa		(a,M)	0X00X0XXXX6	(V,S)	000000XXX03	(T,X)	000X00X00X3	(E,E)	XXXX00X0X06
		(b,M)	XXXX0XOXXX8	(W,S)	000000X0X02	(U,X)	000X00XX003	(F,E)	XXX00000X04
(S,H)	0XXXX000XX6	(R,N)	00000000000	(X,S)	00000000000	(V,X)	0000XXX0003	(G,E)	XXX00000003
(T,H)	XXXX0000X05	(S,N)	00X00000001	(Y,S)	00000000000	(W,X)	00000XXX002	(H,E)	XXXX0000004
(U,H)	XXXX0000XX7	(T,N)	00000000XX2	(Z,S)	000000000X1	(X,X)	0000000X0X5	(I,E)	XXX00000003
(V,H)	XXXXXX0X007	(U,N)	00000000XX2	(a,S)	000000X0001	(Q,Y)	X000X000002	(J,E)	0X000000001
(W,H)	XXXX00XXXX8	(V,N)	0000000XX02	(b,S)	000XX000X03	(R,Y)	000XX000X03	(M,E)	00000000000
(R,I)	0XX0X000X04	(W,N)	000X000X003	(S,T)	000000000X1	(S,Y)	000XX00X0X4	(N,E)	00000000000
(S,I)	0XX0X000005	(X,N)	0XXX00000X4	(T,T)	000000X0XX3	(T,Y)	000X0X0003	(C,F)	XXX0XXXXX08
(T,I)	00X0X000002	(Y,N)	0XX0000X003	(U,T)	000000XXX03	(U,Y)	00000XX0002	(D,F)	X0XXX0X0005
(U,I)	XXXXX000005	(Z,N)	0XX0X00XXX6	(V,T)	000000XXX03	(V,Y)	000X0X0003	(E,F)	XXXX0000X05
(V,I)	XXXXX000005	(a,N)	0XX0X00XXX6	(W,T)	00000X00X02	(W,Y)	00000000000	(F,F)	00X00000X02
(W,I)	XXXXX00XX07	(b,N)	0000X0XXXX5	(X,T)	0000XX00002	(X,Y)	0000000X001	(G,F)	00000000000
(X,I)	XXXXX0XXXX9	(c,N)	XX0XX0XXXX8	(Y,T)	00000000000	(Q,Z)	000XXX00003	(H,F)	00000000000
(Y,I)	XXXXX0XXXX9	(S,O)	00000000XX2	(Z,T)	000000X00X2	(R,Z)	000XXX0003	(I,F)	00000000000
(Q,J)	0XX00000XX4	(T,O)	00000000XX2	(a,T)	000000XX0X3	(S,Z)	000XXXX0004	(K,F)	00X00000001
(R,J)	00X00000001	(U,O)	00000000X01	(b,T)	000000XX0X3	(T,Z)	000XXXX0004	(M,F)	00000000000
(S,J)	0XXXX000004	(V,O)	00000000X01	(R,U)	000000000X4	(U,Z)	000X00X0XX4	(N,F)	00000000000
(T,J)	0XXXX000004	(W,O)	00000000X01	(S,U)	000000X0001	(Q,a)	0000XXX0X03	(C,G)	XXX0XXXX007
(U,J)	0XXXX000004	(X,O)	000X0000XX3	(T,U)	000000X0X02	(R,a)	000XXXX0004	(D,G)	X0XXXXX007
(V,J)	XXXXX00XX07	(Y,O)	0X000000001	(U,U)	0000X0XXXX5	(S,a)	000XXXX0X05	(E,G)	000X0000X02
(W,J)	XXXXX00XX07	(Z,O)	0X00X00XX04	(V,U)	0000XXXXX05	(T,a)	000XXXX0XX6	(F,G)	00X00000X02
(X,J)	XXXXX0XXXX9	(a,O)	0X00X00XXX5	(W,U)	0000XX00002	(W,a)	00000XXXXX5	(G,G)	00000000000
(Y,J)	XXXXX0XXXX9	(b,O)	0000000XXX3	(X,U)	00000000000	(X,a)	0000X0X0X03	(H,G)	00000000000
(P,K)	00XX00000X3	(T,P)	00000000XX2	(Y,U)	00000000X1	(Y,a)	000XX000002	(I,G)	0X000000001
(Q,K)	00XX0000002	(U,P)	00000000X01	(Z,U)	000000XX0X3	**North America**		(K,G)	00000000000
(R,K)	0XXX000003	(V,P)	00000000X01	(a,U)	000000XX002	(C,B)	XXX0XXX006	(C,H)	XXX00XXXX07
(S,K)	0XXX00000X4	(W,P)	00000000X01	(b,U)	000000XX002	(J,B)	XXXX0000004	(D,H)	XX0XXXXXX08
(T,K)	0XXXX0000X5	(X,P)	00000000X01	(c,U)	000000XXXX4	(K,B)	X0X00000002	(E,H)	X00XX0X0X05
(U,K)	0XXXX00XX06	(Y,P)	0000000X001	(d,U)	000000XXXX4	(L,B)	00000000000	(F,H)	X0000000X02
(V,K)	00XXX000X04	(Z,P)	0000X00XX03	(Q,V)	0000X000001	(A,C)	00X0XX00X04	(G,H)	00000000000
(W,K)	0XXXX000004	(a,P)	0000X00XX03	(R,V)	000XX000002	(B,C)	000X0X00X03	(H,H)	00000000000
(X,K)	0XXX0XXXX8	(b,P)	00000000X01	(S,V)	000000X00X2	(C,C)	XXXXXXX0007	(K,H)	00000000000
(Y,K)	XXXXX0XXXX9	(T,Q)	00000000X01	(T,V)	000X0X0XX4	(D,C)	XXX0XXXXXX9	(L,H)	00000000000
(P,L)	000000000X1	(U,Q)	00000000000	(U,V)	000X0X00X3	(H,C)	XXXXXXXX09	(C,I)	XXXXXXXX0X9
(Q,L)	00000000000	(V,Q)	00000000X01	(V,V)	0000XX00002	(I,C)	XXXXX0X0X07	(D,I)	XXX0XXXX007
(R,L)	00000000000	(W,Q)	000000X001	(W,V)	00000000000	(J,C)	XXX00000003	(E,I)	X0XX0000X04
(S,L)	00000000000	(X,Q)	00000000000	(X,V)	00000000000	(K,C)	00000000000	(F,I)	X00X0000X03
(T,L)	00X00000X2	(Y,Q)	00000000X01	(Y,V)	000000X0X02	(L,C)	00000000000	(G,I)	00X00000001
(U,L)	0XX0X00XXX6	(Z,Q)	0000000XX02	(Z,V)	000000X0X02	(M,C)	00000000000	(H,I)	00X00000001
(V,L)	0XX0X000X04	(a,Q)	0000000XX02	(a,V)	000000XXXX4	(N,C)	0X0X00XX004	(K,I)	00000000000
(W,L)	0X)XX000003	(b,Q)	00000000X01	(b,V)	000000XXXX4	(A,D)	XXXX0XXXX08	(L,I)	00000000000
(X,L)	0XXXX000X05	(S,R)	000X0000X02	(c,V)	000000XXXX4	(B,D)	XXXX000XX07	(M,I)	00000000000
(Y,L)	0XXXX0XXXX8	(T,R)	00000000000	(Q,W)	0000X000001	(C,D)	XXXXX000X7	(C,J)	XXXXXXXXX10
(Z,L)	0XX0X0XXXX7	(U,R)	000000X0X02	(R,W)	000XX0000X3	(D,D)	XXX0XXX006	(D,J)	X0XXX0X006
(Q,M)	00000000000	(V,R)	000000X0X02	(S,W)	000X00000X2	(F,D)	XXX00000XX5	(E,J)	XX0XX000X05
(R,M)	00000000000	(W,R)	00000000000	(T,W)	000X00000X2	(G,D)	XX0X0X00X05	(F,J)	XX0X000XX05
(S,M)	00000000000	(X,R)	00000000X01	(U,W)	000XXX00003	(H,D)	XXXXX0X007	(G,J)	XXXX0000X05
(T,M)	00X0000XXX4	(Y,R)	00000000000	(V,W)	000X000001	(I,D)	XXX00000X04	(H,J)	00000000000
(U,M)	00X0000XXX4	(Z,R)	000000000X1	(W,W)	00000000X01	(L,D)	00000000000	(I,J)	0X000000001
(V,M)	000X000XX03	(a,R)	0000000X001	(X,W)	000X000001	(M,D)	00000000000	(J,J)	0XX00000002
(W,M)	00XX000X04	(b,R)	000000X0XX3	(Y,W)	000XX0XXX5	(N,D)	00000000000	(K,J)	00000000000
(X,M)	0XXX00XX05	(S,S)	000000000X2	(Q,X)	X000X000001	(C,E)	X0X0XX0X06	(L,J)	00000000000
(Y,M)	0XXXX0X0X6	(T,S)	000000000X1	(R,X)	000XX000002	(D,E)	XX0XX0X0X06	(M,J)	00000000000
(Z,M)	0XX0X00XXX6	(U,S)	000000XXX03	(S,X)	000X00000X2			(N,J)	X0XX000003

Table 1. This table is a compilation of raw data used for all of the calculations in this article. As shown in Map 16, the earth's land surface was divided into 967 equal areas; each such square having a coordinate which can be found in Map 16 and in this table. (Here in the columns under "Coor.") Maps 1 through 10 were scanned for the same 967 coordinated areas to provide the data which comprise this table. The letters "C, O, S, D,. . ., etc.," indicate the geological periods: Cambrian, Ordovician, Silurian, Devonian, Carboniferous, Permian, Triassic, Jurassic, Cretaceous, and Tertiary. "X"'s indicate the preponderant presence of the represented geological periods in a given square; "O"'s their absence. Under "#" is listed the number of periods present.

Coor.	COSDCPTJCT#	Coor.	COSDCPTJCT#	Coor.	COSDCPTJCT#	Coor.	COSDCPTJCT#	Coor.	COSDCPTJCT#
(B,K)	XXXX000XXX7	(D,R)	00000000XX2	(F,a)	000X0000001	(C,1)	XXXX000XX06	(a,h)	X000XX00003
(C,K)	XXX0X0XXXX8	(E,R)	X0XX0XXXX8	(G,a)	000000000X1	(C,m)	XXX0000XX05	(b,h)	X00X0XXX0X6
(D,K)	X0XXXX00005	(E,S)	00X000X0X03	(H,a)	000000000X1			(c,h)	X00XX0X0Q05
(E,K)	XX0XX00XX06	(F,S)	XX00XX0XXX7	(I,a)	00X0XXX0X05	**Antarctica**		(d,h)	00000000000
(F,K)	XXXXX0XXXX9	(H,S)	0000XX0XXX5	(J,a)	00X00XX0003			(e,h)	00000000000
(G,K)	XX0XX0XXX07	(L,S)	XXXX0XXXXX9	(K,a)	00000000X01	(b,a)	0000XX00002	(f,h)	00000000XX2
(H,K)	00000000000	(E,T)	00X0000XX03	(L,a)	00000000XX2	(c,a)	0000XX00002	(g,h)	00000000XX2
(I,K)	00000000000	(F,T)	0000000XX02	(B,b)	XX0XXXXXXX9	(a,b)	0000XX00002	(h,h)	00000000000
(J,K)	00000000000	(G,T)	0000XXX00X6	(C,b)	XXXXXX00XX8	(b,b)	00000000000	(Y,i)	X000XX0X0X5
(K,K)	0X000000001	(H,T)	XX00XX0XXX7	(D,b)	XXXX0000XX7	(c,b)	00000000000	(Z,i)	X000XX0X0X5
(L,K)	0X0XX000003	(H,U)	00000000XX2	(E,b)	00000000000	(d,b)	0000XX00002	(a,i)	X000XX0X0X5
(M,K)	XXXXX000005	(I,U)	000000X0X02	(F,b)	000X00X0002	(e,b)	0000XX00002	(d,i)	X00XXX00XX6
(B,L)	XXX0XX0XXX8	(I,V)	XXX0000XX05	(G,b)	000000000X1	(f,b)	0000XX00002	(e,i)	00000000000
(C,L)	XXXXXXX0007			(H,b)	000000000X1			(f,i)	00000000XX2
(D,L)	XX0XXXXX007	**South America**		(I,b)	00X00000X02	(a,c)	0000000X001	(g,i)	00000000XX2
(E,L)	XX00XXXXXX8			(J,b)	00000000X01	(b,c)	00000000000	(h,i)	00000000XX2
(F,L)	0000XXXXXX6	(A,V)	XXX000X0XX6	(K,b)	000000X0X02	(c,c)	00000000000	(d,j)	X000X0X003
(G,L)	X0000000X02	(B,V)	XX0XXX00XX7	(A,c)	X00000000X2	(d,c)	00000000000	(e,j)	00000000XX2
(H,L)	00000000000	(C,V)	00000000XX2	(B,c)	X0000X000X3	(e,c)	00000000000	(f,j)	00000000000
(I,L)	00000000000	(D,V)	00000000XX2	(C,c)	0XXX0X0XXX7	(f,c)	00000000000	(g,j)	00000000000
(J,L)	0XX00000002	(E,V)	00000000XX2	(D,c)	0XXXXX00XX7	(g,c)	0000XX00002	(d,k)	X000X0XX004
(K,L)	0XX00000000	(F,V)	000000X0X02	(E,c)	0XXX000002	(w,d)	0000X000XX3	(e,k)	X00X0XX0XX6
(L,L)	XX0X0000003	(A,W)	X000XXX0XX6	(F,c)	00XX0000002	(x,d)	0000X000XX3	(f,k)	00000000000
(M,L)	XXXX0000005	(B,W)	X00XXXX0XX7	(G,c)	000X0X00X03	(a,d)	X00XXX0X005	(g,k)	00000000000
(B,M)	0XX0000XXX5	(C,W)	000000X00X2	(H,c)	000000X0X02	(b,d)	00000000000		
(C,M)	0XXXXXX00X7	(D,W)	000000X0X01	(I,c)	00X00X0X04	(c,d)	00000000000	**Australia**	
(D,M)	000XXXXXXX7	(E,W)	000000XX002	(J,c)	00000000X01	(d,d)	00000000000		
(E,M)	0000XXXXXX6	(F,W)	000000XX002	(K,c)	00000000XX2	(e,d)	00000000000	(q,f)	X0000000X02
(F,M)	0X0XXXXXX7	(G,W)	00000000000	(C,d)	XX00000XXX5	(f,d)	00000000000	(o,g)	XX0XX00005
(G,M)	0XX0X000X04	(H,W)	000000000X1	(D,d)	XXXXXXX0X08	(g,d)	000XX000002	(p,g)	XXX00000003
(H,M)	0X0XX00003	(A,X)	XX0XXXX XX8	(E,d)	XXXX0X000X6	(x,e)	X000X00XXX5	(q,g)	X0X00000X03
(I,M)	0XXX000003	(B,X)	XX0000X0XX5	(F,d)	00XX0000002	(y,e)	X000X00XXX5	(r,g)	X0000000X02
(J,M)	0XXX000003	(C,X)	000000X001	(G,d)	000X00X0X04	(z,e)	XXXXX00X006	(l,i)	0X00XX00XX5
(K,M)	0XXX000003	(D,X)	000000X0X01	(H,d)	000X0XXXX05	(a,e)	0000X00X002	(m,i)	00000000000
(L,M)	0X0X0000002	(E,X)	00000000000	(I,d)	00000XX0X03	(b,e)	00000000000	(n,i)	00000000000
(B,N)	X0X000X0X04	(F,X)	000X0000001	(J,d)	00000000X01	(c,e)	00000000000	(o,i)	00000X00X02
(C,N)	X00XXX00X6	(G,X)	0X0X0000002	(C,e)	XX00000XXX5	(d,e)	00000000000	(p,i)	X0000X00002
(D,N)	000XXXX0X05	(H,X)	000000000X1	(D,e)	XX0XXXX0XX8	(e,e)	00000000000	(q,i)	XX00000002
(E,N)	0000X0X0XX4	(A,Y)	XX0XXXXXXX9	(E,e)	0XXXXXX00X7	(f,e)	00000000000	(r,i)	XX0X0000X04
(F,N)	XX00XX00XX6	(B,Y)	XXXXX0X0XX8	(F,e)	00XX0X00003	(g,e)	00000000000	(s,i)	X00X000XX04
(G,N)	X00XXX00004	(C,Y)	0XXX0000X5	(G,e)	000XXXX0X05	(x,f)	X000X00X0X4	(t,i)	X00X00XXX05
(H,N)	XX0X0000003	(D,Y)	00XXX0000X4	(H,e)	000XX00003	(y,f)	XXXXX0X007	(u,i)	XXXXXXXX0X9
(I,N)	XXXXX000005	(E,Y)	00XXX000008	(I,e)	00000000000	(z,f)	XXXXX00005	(m,j)	00000X00X02
(J,N)	XXXXX000005	(F,Y)	00XXXX000X5	(C,f)	XX00XX0XXX7	(a,f)	X000XX0X004	(n,j)	00000000000
(K,N)	X00X00X0003	(G,Y)	0XXXX000X6	(D,f)	XXX0XX00XX7	(b,f)	00000000000	(o,j)	000000000X1
(C,O)	X00X0X00003	(H,Y)	000XXX00X5	(E,f)	0XXXX000X6	(c,f)	00000000000	(p,j)	X0000X00X03
(D,O)	XX0XXX00XX7	(I,Y)	00XXXX0XX7	(F,f)	000XX00XX5	(d,f)	00000000000	(q,j)	X0000X00X03
(E,O)	0XX0XXX0XX7	(J,Y)	00XXXX0XX7	(G,f)	000XX00X04	(e,f)	00000000000	(r,j)	XX000X00X04
(F,O)	XX00XX00X05	(A,Z)	XX0XXXXXX9	(H,f)	0000XX0X04	(f,f)	00000000000	(s,j)	X000000X03
(G,O)	XXX0X000X05	(B,Z)	XXXXX0XXX8	(C,g)	XXXXXXXXX10	(g,f)	00000000000	(t,j)	XXX0000XX05
(H,O)	XX000000XX4	(C,Z)	XXXX0XX0X7	(D,g)	XXX00XX0XX7	(h,f)	00000000000	(u,j)	XXX000XXX06
(I,O)	XX0XX00X05	(D,Z)	000XXXX0X4	(E,g)	00000XX00X3	(X,g)	X000XX0X0X5	(v,j)	XXXXX00X006
(J,O)	XX0X0000003	(E,Z)	000XXX00X4	(F,g)	000XXXX0X05	(Y,g)	X000XX0X0X5	(m,k)	00000000000
(K,O)	00000000XX2	(F,Z)	000XXX000X5	(G,g)	00000X00X02	(Z,g)	X000XX00003	(n,k)	00000000000
(C,P)	00XX0000002	(G,Z)	00000000000	(C,h)	XXX0X0XXX8	(a,g)	X00XXX000X4	(o,k)	000000000X1
(D,P)	XX0XXX00XX7	(H,Z)	00000000000	(D,h)	XXXXX00XX8	(b,g)	X00XX0X004	(p,k)	00000000XX2
(E,P)	XX0XXXX0XX8	(I,Z)	0XXXXXXX08	(E,h)	00000X00XX3	(c,g)	00000000000	(q,k)	00000X00001
(F,P)	XX0XX00X05	(J,Z)	00XXXX0XX7	(F,h)	00000000XX2	(d,g)	00000000000	(r,k)	00000000001
(G,P)	XX0XX00XX7	(K,Z)	00000000X01	(C,i)	XXX00X0XXX7	(e,g)	00000000000	(s,k)	XX0000X04
(H,P)	XX0X00XXXX7	(A,a)	XX0000XXX05	(D,i)	XX0000XX5	(f,g)	00000000000	(t,k)	XX0X0000X04
(I,P)	0X0X000XXX5	(B,a)	XXX0XX0XX7	(E,i)	XXX0000XX6	(g,g)	00000000000	(u,k)	XXX000X0X05
(J,P)	00X00000XX3	(C,a)	XXX0XX00XX7	(C,j)	XX00XXXXX8	(h,g)	000X0000001	(v,k)	XXXXX000005
(D,Q)	000X000000X2	(D,a)	0000X000XX3	(D,j)	XX00000XX6	(Y,h)	X000XX0XXX6	(n,l)	000000000X1
(E,Q)	XXXXXX0XXX9	(E,a)	000X0000001	(C,k)	XXX000X0X6	(Z,h)	X000XX0XXX6	(r,l)	X000000001
(F,Q)	0000000XXX3			(D,k)	XX000XX0XX6			(s,l)	X00000000X2

Evolutionary-Uniformitarian Geologic Column

Coor.	COSDCPTJCT#
(t,1)	XXX0XX000X6
(u,1)	XXX00000003
(s,m)	0000XX000X3
(t,m)	XXXX0X00005
(u,m)	XXXXX000005
(t,n)	XXXX0X00005

Eurasia

Coor.	COSDCPTJCT#
(f,B)	000000000X1
(j,B)	XX00XXXXXX8
(v,C)	XXX0XXXX007
(w,C)	0X0000XXXX5
(c,D)	XXXX0000004
(d,D)	XXX0XX0XX07
(e,D)	XX0X0000003
(v,D)	X0X0XXXX006
(w,D)	0X0X00XXXX6
(x,D)	0XXXXXXXXX9
(Z,D)	XX00000002
(a,E)	XXXXXXXXXX9
(c,E)	XX0XX00XX06
(f,E)	XX000000002
(g,E)	XXX00000003
(h,E)	XX000000002
(i,E)	X0000000001
(p,E)	XXX00000003
(u,E)	X000X0XX0X5
(v,E)	X00XXXXX006
(w,E)	0X0X0XXXX005
(X,E)	0XXXXXXXX9
(Z,F)	XXXX00XXX8
(a,F)	0XXX0XXXX8
(b,F)	0XX00000XX4
(c,F)	0XXXX0XXX8
(d,F)	XX0XX0XX006
(e,F)	X0X00XXXX7
(f,F)	XXX00000003
(g,F)	XX000000002
(h,F)	00000000000
(i,F)	00000000000
(j,F)	000X0000001
(m,F)	XXXXXX0XX08
(p,F)	XXXX000X005
(q,F)	XXXXXXXXX09
(r,F)	X0X0XXXXXX8
(s,F)	X0XXXXXXXX9
(t,F)	XXXXXXXX09
(u,F)	0X0XX0XX0X6
(v,F)	XX0XXXXXXX9
(w,F)	0XXXXXX08
(x,F)	0XXX0XXXX8
(c,G)	0XXXX0XXX8
(d,G)	000000X00X1
(e,G)	XXXXXXXXX10
(f,G)	XXXXX0XXXX9
(g,G)	XXXX0000004
(h,G)	00000000000
(i,G)	00000000000
(j,G)	X000XX0X004
(k,G)	0X0XXX0XX06
(m,G)	XXXXX00XX8
(n,G)	XXXXXXXXX10
(o,G)	XXXXXX0XXX9
(p,G)	XXXXXXXX09
(q,G)	X0XXXX00005
(r,G)	X0000000001
(s,G)	X00X000X0X4
(t,G)	X00XXXXXX07
(u,G)	X00XXXXX006
(v,G)	0X0XXXXX006
(w,G)	0X0XX0XXXX7
(y,G)	XX0X00XX0X6
(c,H)	0XXXXXXXX9
(d,H)	XXXX0X0XX8
(e,H)	000XXX0XX5
(f,H)	XXXX000XXX7
(g,H)	X0XX000X004
(h,H)	XX0XX00XX06
(i,H)	X0OXXX0XX07
(j,H)	X000XXXXX06
(k,H)	0X0XXX0XX06
(l,H)	XXXXX000XX7
(m,H)	XXXXXXXXXX9
(n,H)	XX00XXX0XX7
(o,H)	XXXXX00XXX8
(p,H)	XXXXXXX0007
(q,H)	XXXXX00006
(r,H)	XXX00X00004
(s,H)	XX00000X0X4
(t,H)	X00000XX0X4
(u,H)	X00X0XXX0X6
(v,H)	X0X00X00003
(c,I)	0X0XXXXXXX8
(d,I)	XXXXX00XXX8
(e,I)	000XXXXXXX7
(f,I)	00X00000XX3
(g,I)	000X000XXX4
(h,I)	X00XX00XX05
(i,I)	X00XXXXX006
(j,I)	000XX000X04
(k,I)	0X0XXXX0004
(l,I)	XX0XX000XX6
(m,I)	00X000XXXX5
(n,I)	0000X00XXX4
(o,I)	0XXXX00XXX8
(p,I)	XXXXXXXX007
(q,I)	XXXXXXX0X08
(r,I)	X0X0XXXX005
(s,I)	XXX0000X004
(t,I)	XXX0000X003
(u,I)	00000000000
(v,I)	X0XX0X00004
(c,J)	0XXXXXXX08
(d,J)	000X0000XX3
(e,J)	X0XXXXXXXX9
(f,J)	00X00000XX3
(g,J)	000XXXXXX7
(h,J)	000XX000XX5
(i,J)	000XXX000X6
(j,J)	000XX000X05
(k,J)	000XX000002
(l,J)	XXXXX000XX7
(m,J)	XXXX00000X7
(n,J)	000X000XXX4
(o,J)	XXXXX000XX7
(p,J)	XXX00X00004
(q,J)	XXX00X000X5
(r,J)	XX000X00003
(s,J)	XX00X000003
(t,J)	X000X000002
(u,J)	X00XX000003
(v,J)	XX0X0XX0005
(w,J)	X0000XXXX05
(x,J)	0XXX0XXXX8
(d,K)	0000XXXXXX6
(e,K)	XX0XXX0XXX8
(f,K)	X0XXX0XXX8
(g,K)	000XX00XX5
(h,K)	000XXXXXX7
(i,K)	000XXXXXX7
(j,K)	000XX0XXX6
(k,K)	XXXXX0000X6
(l,K)	XXXX0000XX6
(m,K)	XXXXX000XX7
(n,K)	X0XXX000XX7
(o,K)	XXXXXX0XX08
(p,K)	X0XX000X004
(q,K)	XXX0X00X005
(r,K)	XXX00X00004
(s,K)	00000000000
(t,K)	X00X0X0003
(u,K)	X00XXX0X005
(v,K)	XXXXX0XX08
(w,K)	000XXXXXXX7
(x,K)	XXX000XX006
(y,K)	0XXXXXX0X07
(d,L)	XX0XXXXXX8
(e,L)	00000XXXXX5
(f,L)	X000XXXXXX7
(g,L)	X0XXXXXXX7
(h,L)	X0XXXXXXXX8
(i,L)	X00XX00XXX7
(j,L)	XX0XXXXXX8
(k,L)	XXXXX000XX7
(l,L)	XX0XX0000X5
(m,L)	XX0XX0000X5
(n,L)	0X0XX0000X4
(o,L)	XXXXX000XX6
(p,L)	X00X0000002
(q,L)	X0000XXXXX6
(r,L)	X0000000001
(s,L)	00000000X01
(t,L)	000XXXXXX06
(u,L)	XX0XX000X05
(v,L)	0XX0X00X0X5
(w,L)	X0000X0XXX5
(x,L)	XXXX0XXXX08
(z,L)	0000X0XXX5
(e,M)	XXX0X0XXX8
(f,M)	X0X00XXXXX7
(g,M)	X00X000XXX5
(i,M)	XXXXX0XXXX9
(k,M)	XXXXX000XX7
(l,M)	XX0XX00XX5
(m,M)	X0XX000XX03
(n,M)	X0XX00000X4
(o,M)	000XX000X3
(p,M)	XXX0000004
(q,M)	X0XXX00000X4
(r,M)	X00XX00003
(s,M)	X00XX000003
(t,M)	X00XX00XXX6
(u,M)	0X00XX00003
(v,M)	00X0XX000X4
(w,M)	00X00X000X3
(x,M)	00X00XX0X04
(z,M)	000XXX0XXX6
(d,N)	XXX000XXXX7
(e,N)	XXXXXX0XXX9
(f,N)	XXX0XXXXX9
(g,N)	X0XXX0XX0X8
(i,N)	X0000XXXXX6
(j,N)	XXXXX0XXX9
(k,N)	XXXXX000XX8
(l,N)	0X0XX0000X4
(m,N)	XXXX0000004
(n,N)	XX0XX0000X5
(o,N)	000XXX0XXX6
(p,N)	0X)XXX))XX6
(q,N)	XX0XX0000X5
(r,N)	0X0XXX00XX5
(s,N)	XX0XX000XX6
(t,N)	0XXXX0000X6
(u,N)	X00X0XX00X3
(v,N)	00X00X0X0X4
(w,N)	XX00X00XX05
(x,N)	XX00X00XX05
(z,N)	0X00XX0X0X5
(d,0)	XX000X00X03
(e,0)	XXXX0XXXX9
(f,0)	XXX00XXXXX8
(g,0)	X0000XX0XX4
(h,0)	XXXX0XXXX8
(i,0)	XXXXX0XXXX9
(j,0)	XXX0XXXXX9
(k,0)	XXXXX0XXXX9
(l,0)	XX0XXX0XX8
(m,0)	XXXXXX0XX08
(n,0)	X0X00X000X4
(o,0)	X0X0000X0X4
(p,0)	XXXXXX00006
(q,0)	X0XXXX00005
(r,0)	00XX0X000X4
(s,0)	000000000X01
(t,0)	X00X0XXXX5
(u,0)	XX0X0XXXX7
(v,0)	XX00X0XX0X6
(x,0)	XX00000002
(z,0)	000XXXXXX07
(d,P)	0X0000000X2
(e,P)	XXXX0XXXXX9
(f,P)	XXXXX0XXXX9
(g,P)	XX0X0XXXX8
(h,P)	XXXXXXXXXX10
(i,P)	X00XXXXX07
(j,P)	X00XXXXXXX5
(k,P)	X0000XX0X5
(l,P)	00XXXXXXX8
(m,P)	X0XXX0XXX8
(n,P)	000X0X00XX4
(o,P)	0X0XXX0XX7
(p,P)	0XXX00XX0X5
(q,P)	0XXXX0XX07
(r,P)	XXXXX0XX006
(s,P)	XXXXXX0XX08
(t,P)	XX00XXXXX8
(u,P)	XX00XX000X5
(v,P)	XX00X0XX0X6
(d,Q)	0X00000X0X3
(e,Q)	0X000X0XXX5
(f,Q)	XXXX0XXXXX9
(g,Q)	XXXXX0XXXX9
(h,Q)	XXXXXXXXXX10
(i,Q)	XXXXXXXXXX10
(j,Q)	X0X00XXXXX7
(k,Q)	X0X0UXXXXX7
(l,Q)	XXX0XXXXX9
(m,Q)	XXX0XXXXXX9
(n,Q)	X00X000XXX5
(o,Q)	X00XX0XX06
(p,Q)	X0XXXX00X06
(q,Q)	XXXXXX00X07
(r,Q)	XXXXXXX0007
(s,Q)	XXXXXXX0007
(t,Q)	XX0XXX00005
(u,Q)	X000XX000X4
(v,Q)	0000XX00002
(w,Q)	0XXXXX000X6
(d,R)	0000000X001
(e,R)	X0000X0XXX5
(f,R)	XXX00XXXXX8
(h,R)	XX0XXXXXX9
(i,R)	X00XXXXXXX8
(j,R)	X000XXXXX7
(k,R)	X000XXXXX7
(l,R)	X00000000X2
(m,R)	XXX00XX00X6
(n,R)	XXXX0XXXXX9
(o,R)	X00XXXXXXX8
(p,R)	X00XXXXX06
(q,R)	XXX00XXXXX8
(r,R)	XXXXXXX0X08
(s,R)	XXXXXXX0X9
(t,R)	XXXXXXX0X08
(u,R)	XXX0X0XX0X7
(v,R)	0X0XXXX0X06
(w,R)	X00XXX00XX6
(d,S)	000000XXXX4
(e,S)	X0000X0XXX5
(f,S)	X0000X0XXX5
(g,S)	XX000XXXXX7
(h,S)	XX00XXXXXX8
(j,S)	X000XXXXXX7
(k,S)	0000XX0XXX5
(l,S)	00000000000
(m,S)	00000000000
(n,S)	X00000X00X3
(o,S)	XXXXXX00X8
(p,S)	XXXXXX0XX9
(q,S)	XXXXXXX0X8
(r,S)	XX0XXXXXX8
(s,S)	XX0XXXX007
(t,S)	X0XXXX0XX8
(u,S)	X0XXXX0006
(v,S)	XXXXXXX0X08
(w,S)	000XX000X03
(x,S)	0000X000X02
(d,T)	000000XXXX4
(e,T)	X000000XXX4
(f,T)	X0000X0XXX5
(g,T)	XX000X0XXX6

Coor.	COSDCPTJCT#	Coor.	COSDCPTJCT#	Coor.	COSDCPTJCT#	Coor.	COSDCPTJCT#	Coor.	COSDCPTJCT#
(k,T)	0000000XXX3	(o,U)	00000000XX2	(m,W)	0000X00XXX4	(k,b)	XXXXXXXX008		
(l,T)	00000000XX2	(q,U)	0000000XXX3	(s,W)	X000XXXX005	(n,b)	XXXX0XXXX9		
(m,T)	00000XX0002	(r,U)	XX00XXXX006	(t,W)	XX0XX0X0005	(u,b)	XXXX0X0X07		
(n,T)	0000XX00002	(s,U)	0XXXX0X0005	(u,W)	XX00000X003	(r,c)	XXX0XXXXX9	(m,h)	00000000000
(o,T)	000000000X1	(t,U)	XX0XX0X0005	(y,W)	XXXX00XXXX8	(s,d)	XXX000XXX6	(n,h)	00000X00X02
(p,T)	000X0000XX3	(u,U)	XXX0000X0X5	(l,X)	00000000XX2	(t,d)	XXX0XXXXX08	(o,h)	XX0X0X0XX05
(q,T)	X000000XXX4	(x,U)	XXXXX00X0XX7	(m,Y)	0000X000001	(s,e)	X00X0000XX4	(p,h)	XX0X0X00004
(r,T)	XX00XXXXX07	(l,V)	00000000000	(s,Y)	XXXXXX0X08	(t,e)	X000000XXX4	(q,h)	X0000000001
(s,T)	XXXXXXX0007	(m,V)	0000000X001	(m,Z)	XXXX00XXXX8			(r,h)	XXXX0000004
(t,T)	XXXXXXXX008	(n,V)	00000000XX2	(r,Z)	XXXXX0XXX08			(s,h)	XX0X000XX05
(u,T)	XXXX00X00X6	(r,V)	X000XXX0004	(s,Z)	XXXXX0XXX08	**New Zealand**		(t,h)	000X0X0XX04
(v,T)	0X0X00XXXX6	(s,V)	0XXX0XXXX07	(l,a)	XXX000XXXX7			(u,h)	X0XXXX000X6
(l,U)	00000000XX2	(t,V)	000X00X0X03	(m,a)	XXXX00XXX07	(y,p)	XXX00XXX0X7		
(m,U)	0000XX00002	(y,V)	XXX00X0XXX7	(s,a)	XXXXXXXXX10	(z,o)	XXX00X0X005		
(n,U)	00000000000	(l,W)	00000000000	(t,a)	XXXXXXXXX10				

earth's land surface is hard-pressed to produce even 3 of the 10 geologic periods in "correct" consecutive order.

The quantitative data is particularly revealing. From Figure 1, it is evident that nearly 13% of the earth's land surface has 5 geologic periods represented (irrespective of their order or identity) while slightly less than 1% has all 10 periods simultaneously in place. From the cumulative frequency curve in Figure 1, it can be seen that 42% of earth's land surface has 3 or less geologic periods present at all; 66% has 5 or less of the 10 present; and only 14% has 8 or more geologic periods represented at all.

Individual geologic periods' coverage of the earth's land surface range from a high of just over 51% for Cretaceous (Fig. 2) to a low of only 33% for Triassic. Sequences of consecutive geologic periods cover far less area. Sequences of 3 consecutive geologic periods and their per cent terrestrial areal coverage are: complete Lower Paleozoic 21%, complete Upper Paleozoic 17%, complete Mesozoic 16%. For 6 consecutive geologic periods, one finds only 5.7% of the earth's land surface covered by complete Paleozoic, and only 4.0% covered by complete Upper Paleozoic/complete Mesozoic.

The overall failure of geologic periods to be numerically abundant in most places on earth and their even greater failure to occur in consecutive sequences is significant enough, but it can be seen from Table 2 that where geologic periods' rocks *do* exist they often fail to rest "properly." A significant percentage of every geologic period's rocks does not overlie rocks of the next older geologic period. In fact, only a *bare majority* of Cretaceous, Jurassic, and Devonian "properly" overlie the next older geologic period's rocks (Jurassic, Triassic, and Silurian, respectively). Some percentage of *every* geologic period rests directly upon Precambrian "basement," especially high percentages of Ordovician (23.2%) and Devonian (18.6%) doing so.

There apparently are regions on earth where all ten geologic periods can be found superposed (Map 15). Frank and Fuchs[81] presented stratigraphic correlation charts for Himalayan Geology in West Nepal. A complete geologic column was shown, but Cambrian and Ordovician were shown as uncertain. In the lower Himalayas, the Triassic was shown as unknown, and some slates were questionably placed in a range from Precambrian to Ordovician. In the lower Himalayas/Kashmir region, a complete column is shown except for the Triassic, but there are major gaps in the Ordovician and the Silurian.

Another example of a major region apparently possessing the entire geologic column in place is a part of the Bolivian Andes, described by Lohmann.[82] The area is called the Northern Antiplano; a region approximately 68°-70°W and 16°-18°S. But even here, the Cambrian and Ordovician are uncertain for lack of nearby outcrops. The Cordilleran area nearby (66°-68°W and 16°-20°S) also has an apparently complete column, but Jurassic strata are said to be scarce throughout the area.

It has already been demonstrated that the apparent completeness of the geologic column in Indonesia stems from ascribing metamorphosed "basement" to the entire Lower Paleozoic. Similarly, it has been pointed out by Ray and Achayya[91] that in eastern Burma, western Thailand, and Malaya the Permian and Mesozoic rest upon a folded and metamorphosed "basement" of inferred early Paleozoic. The alleged completeness of the geologic column shown for Cuba (Map 15) based upon data of the Ronov *et. al.* team can be questioned for the same reason. The work of Hatten[92] demonstrates how poorly known the pre-Mesozoic geology of Cuba is: Mesozoic rocks overlying metamorphosed equivalents of presumed Paleozoic strata.

All ten geologic periods are undoubtedly represented in the Swiety Krzys (Holy Cross) Mountains of south-central Poland (20-22E and 50-52N) as is evident from the local map of the mountain region enclosed in the work of Ksiazkiewicz, Samsonowich, and Ruhle.[93] In fact, the presence of all or almost all of the geologic column (in terms of sedimentary lithologies that are unambiguously biostratigraphically dated) can be shown to occur over much of Poland.[94,95] But only a very small percentage of the earth's land surface has most or all of the geologic column in place (Map 15, Figure 1 and 2).

Continental shelf data were not included in this article, but Ronov *et. al.*[11] noted that submerged shelves cover only 14% of the total continental area. Examples are now cited which show that they tend to have few and non-consecutive geologic periods represented. The report on the North Sea by Kent and Walmsley[83] reported only post-Devonian sedimentary strata, with usually only 4 or 5 of the 6 geologic periods represented per borehole.

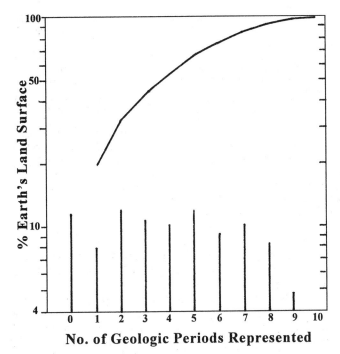

Figure 1. This semilogarithmic plot forms a histogram showing the tendency, by numerical abundance, for geological periods to be represented over the earth's land surface. The cumulative frequency curve above shows the per cent of the earth's land surface which has that many geological periods represented, or fewer. For example, 13% of the earth's surface has 5 of the 10 geological periods represented; while 66% has 5 or fewer represented. No more than about 1%, too little to show up on the graph, has all 10 periods.

The eastern North American continental shelf[84] from Florida to Nova Scotia has Jurassic, Cretaceous, and Tertiary rock, with a much smaller presence of Triassic. Link[85] pointed out that continental shelves tend to be dominantly Mesozoic and Cenozoic.

Continental data of this article would be completely overwhelmed by oceanic data if it were included. Blatt et. al.[86] said: "Almost all of the sediment preserved in modern ocean basins is younger than Triassic." The inclusion of oceanic data would therefore greatly increase the percentages of few and recent geologic periods. The percentages of many present geologic periods and many consecutive geologic periods, already minor, would become vanishingly small.

The more the earth's surface fails to display the vaunted evolutionary-uniformitarian geologic column in terms of actual presence and "correct" stratigraphic layering of geologic periods rocks, the more the geologic column passes into the realm of fantasy. Concerning geologic time, Douglas[87] wrote: "Time can only be established through recognition of events, as time itself is not measurable without an event to mark its passing. Furthermore, geologic events have a reality only through their manifestation in the rock record."

Recently, Gingerich[96] wrote: "The study of organic evolution is both a geological and a biological subject. Evolution means change, change implies time, and the great sweep of life history is recorded in sedimentary rocks and measured in geological time." The

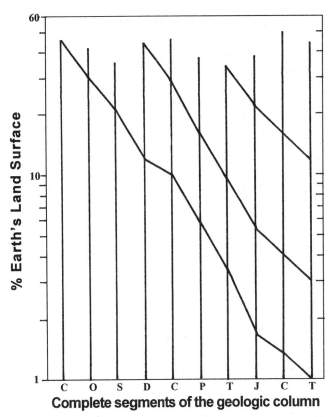

Figure 2. This semilogarithmic plot shows the tendency of the earth's land surface to have sequences of consecutive geological periods in place. The vertical bars form a histogram showing the per cent of the terrestrial areal coverage (on land) of individual geological periods. The curves show the decline of area covered with each younger geological period added. For example, 47% of the earth's land surface has Cambrian alone, 31% has both Cambrian and Ordovician, and only 21% has Cambrian, Ordovician and Silurian simultaneously represented.

Table 2. This table shows gross stratigraphic successional relationships: the per cent of each geological period overlying directly every older geological period. For example, 65.7% of Carboniferous lies upon Devonian, 10.3% upon Silurian, 5.82% upon Ordovician, 7.16% upon Cambrian, and 11.0% directly upon Precambrian "basement".

	C	J	T	P	C	D	S	O	C	P^C
Tertiary	72.0	8.78	3.93	4.62	3.23	1.85	0.231	0.231	0.693	4.39
Cretaceous		51.6	12.1	8.87	4.84	3.23	2.82	6.05	0.806	9.68
Jurassic			54.8	18.9	11.4	5.05	3.19	1.86	0.798	4.00
Triassic				58.4	18.4	5.63	3.75	0.938	0.938	11.9
Permian					76.3	10.5	6.08	1.10	2.76	3.30
Carboniferous						65.7	10.3	5.82	7.16	11.0
Devonian							52.1	20.5	8.83	18.6
Silurian								78.4	10.5	11.1
Ordovician									76.8	23.2
Cambrian										100

significance of the geologic column to evolution is therefore obvious. Yet not only does this column basically not exist, but even where geologic periods' rocks *do* occur, their biostratigraphic basis *itself* is arbitrary. One need only consult the volume edited by Harland[97] to note that most fossil taxons overlap from a few to several of the ten geologic periods.

Of course, the absence of geologic periods is claimed to be a result of non-deposition during those periods in the regions of their absence, or to subsequent erosion. This is self-serving because there is no deterministic reason why the earth's land surface should (or should not) become *everywhere* depositional sometime within any span of several tens of millions of years comprising each geologic period. The claim of non-deposition and erosion during geologic periods begs the question, because it does not face the question whether or not these geologic periods ever existed in the first place.

Since only a small percentage of the earth's surface obeys even a significant portion of the geologic column, it becomes an overall exercise of gargantuan special pleading and imagination for the evolutionary-uniformitarian paradigm to maintain that there ever *were* geologic periods. The claim of their having taken place to form a continuum of rock/life/time of ten biochronologic "onion skins" over the earth is therefore a fantastic and imaginative contrivance.

III. Probable Diluvial Significance of Global Stratigraphic Trends

This section briefly considers how the findings of this work and related findings affect the Creationist-Diluvialist paradigm. Since it has already been demonstrated that the geologic column is not supported by what actually is found on earth, the principle of Occam's Razor favors the Creationist-Diluvialist paradigm because of its intrinsic abrogation of all concepts of evolution, geologic periods, and geologic time.

The fact that most of the earth's land surface has few of the ten geologic periods represented (Fig. 1) means that Diluviology needs to explain the stratigraphic separation of only a few fossil groupings over most terrestrial areas. Furthermore, the fact that most represented geologic periods tend not to be consecutive further implies that only a few groupings of fossil types exist over most areas. This says nothing of the long range of most fossil taxons.

The fact that Devonian, Jurassic, and Cretaceous appear to have special liberties not to rest on the next older geologic periods (Table 2) may have Diluvial significance. The Devonian has the first large-scale diversity of land fossils (although all geologic periods consist overwhelmingly of marine fossils), and its liberty may be due to the ecological independence of its terrestrial fossils from the almost wholly marine Lower Paleozoic as well as the poverty of the Lower Paleozoic in the Gondwana continents (Maps 1, 2, and 3).

The successional liberties of Jurassic and Cretaceous (Table 2) may support the position that they are post Flood and that their fossil populations are therefore truly successional and not part of the mutually-contemporaneous Flood-buried continuum of the older geologic periods.

The works of Ronov et. al.[11,88] and Schwab[89] consider volumes, compositions, etc., of rocks from the viewpoint of geologic periods. No geologic period has a monopoly on any type of lithology. However, submarine volcanogenic rocks decrease going from global Paleozoic to global Mesozoic. Terrestrial volcanogenic deposits increase drastically going from Paleozoic to Mesozoic and Cenozoic.

This may reflect the global tendency for Paleozoic rocks to have been formed during the Flood so that Volcanism was primarily submarine, while Mesozoic and Cenozoic have been formed during late Flood and post Flood conditions so that volcanism was subaerial rather than subaqueous in nature.

Those strata which are at or near the earth's surface at any given localities are the likeliest candidates for having formed under late and post Flood conditions. For instance, where Jurassic has a thick mantle of Cretaceous-Tertiary rock overlying it, the Jurassic in that case probably was a mid-Flood deposit. Jurassic with nothing overlying it, on the other hand, probably was late-Flood to post-Flood in origin. The fact that world outcrop areas of geologic periods decline exponentially going stratigraphically downward[90] probably reflects the ever-decreasing probability of older geologic periods' rocks to be primarily late-Flood and post-Flood in origin.

References

AG-American Association of petroleum Geologists Bulletin
EL-Elsevier Scientific Publishing Co., Amsterdam, London, New York
GE-Geology
GN-Proceedings of the International Gondwana Symposium
IG-International Geology Review
PA-Palaeogeography Palaeoclimatology Palaeoecology
SE-Sedimentary Geology
SO-Sovietskaia Geologiia
TE-Tectonophysics

[1] Park C. F. and R. A. MacDiarmid. 1975. *Ore Deposits*. Freeman and Co. San Francisco, p. 2 (3rd Edition)
[2] Price G. M., 1923. *The New Geology.*, Pacific Press, California. Pp. 288, 296, 610-619.
[3] Nelson B. C., 1931, 1968, *The Deluge Story in Stone*. Bethany Fellowship Pub. Co., Minnesota, p. 146, 150-151.
[4] Whitcomb J. C. and H. M. Morris, 1961. *The Genesis Flood*. Baker Book House, Michigan, 17th printing, p. 135-6.
[5] Burdick C., 1971. Streamlining Stratigraphy (in Lammerts W. E., 1971. *Scientific Studies in Special Creation*. Presbyterian and Reformed, New Jersey, p. 127-8.
[6] Clark H. W., 1968. *Fossils, Flood, and Fire*. Outdoor Pictures, California, Pp. 53-5.
[7] Freeman O. W., and J. W. Morris, 1958. *World Geography* McGraw Hill, New York, Toronto, London, plates following p. 8.
[8] _____ 1978. *CBS News Almanac*. Hammond Almanac Co., New Jersey, p. 519, 523.
[9] _____ 1972. *Encyclopedia Britannica*. Index and Atlas, p. 7, 43, 97.
[10] _____ 1972. *The Earth and Man*. Rand McNally and Co., New York, Chicago, San Francisco, p. 157.
[11] Ronov A. B., Khain V. E., Balukhovskiy A. N., and K. N. Seslavinskiy. 1980. Quantitative Analysis of Phanerozoic Sedimentation. SE 25:312.
[12] Ben-Avraham Z. and K. O. Emery. 1973. Structural Framework of Sunda Shelf. AG 57:2327.
[13] Braziunas T. F. 1975. A geological duration chart. GE 3:342-3.
[14] Ronov A. B., Seslavinskiy K. B., Khain V. Ye. 1974. Kembriiyskie Litologicheskiye Formatsii Mira. SO 1974 (12) 29, 23, 17 (in Russian).
[15] Ronov A. B., Seslavinskiy K. B., and V. Ye. Khain. 1977. Cambrian

Lithologic Associations of the World. *IG* 19:379, 385, 391 (translation of preceeding).

[16] Clarkson P. D., Hughes C. P. and M. R. A. Thompson. 1979. Geological Significance of a Middle Cambrian Fauna from Antarctica. *Nature* 279:791.

[17] Ronov A. B., Khain B. E., Seslavinskiy K. B. 1976. Ordovikskie Litologicheskie Formatsii Mira. *SO* 1976 (1) 11 (in Russian).

[18] Ronov A. B., Khain B. E., and K. B. Seslavinskiy. 1976. Ordovikskie Lithologic Associations of the World. *IG* 18:1399 (translation of preceeding).

[19] Khain B. E., Ronov A. B., Seslavinskiy K. B. 1977. Siluriyskie Litologicheskie Formatsii Mire. *SO* 1977 (5) 29 (in Russian).

[20] Khain B. E., Ronov A. B., and K. B. Seslavinskiy. 1978. Silurian Lithologic Associations of the World. *IG* 20:256 (translation of preceeding).

[21] Burrett C., 1973. Ordovician Biogeography and Continental Drift. *PA* 13:161-201.

[22] Veevers J. J., 1976. Early Phanerozoic Events on and Alongside the Australasian-Antarctic Platform. *Journal of the Geological Society of Austrlia* 23(2)184.

[23] Elliot D. H., 1975. Tectonics of Antarctica: a review. *American Journal of Science* 275-A p. 54 (table).

[24] Kyle R. A., 1977. Devonian Palynomorphs From the Basal Beacon Supergroup of South Victoria Land, Antarctica. *New Zealand Journal of Geology and Geophysics* 20 (6)1082, 1096-8.

[25] Ronov A. B., B. E. Khain. 1954. Devonskiye Litologicheskiye Formatsii Mira. *SO* 1954 sb. 41 c52, 56, 60 (in Russian).

[26] House M. R., 1971. Devonian faunal distributions (*in* Middlemiss F. A., Rawson P. F., and G. Newall. 1971. Faunal Provinces in Space and Time. *Geological Journal Special Issue No. 4*), p. 84.

[27] Spasskiy H. Ye., Pubatolov, B. H., Kravtsov A. G., 1973. Paleobiograficheskiye Raonirovaniye Ranne i sredno Devonskih Morey Zimnol shara. *Stratigrafiya nizhnevo i srednovo Devona*. Tom II, 232, 234, 236 (in Russian).

[28] Brinkmann R., 1976. *Geology of Turkey* EL, p. 23.

[29] Kummel B. 1970. *History of the Earth*. Freeman and Co. San Francisco (2nd Edition). p. 206.

[30] Brown D. A., Campbell K. S. W., and K. A. W. Crook. 1968. *The Geological Evolution of Australia and New Zealand*. Pergamon Press, Oxford, Sydney, Toronto, New York, etc., pp. 120, 132, 140-2.

[31] Hermes J. J., 1968. The Papuan Geosyncline and the Concept of Geosynclines. *Geologie en Mijnbouw* 47, p. 82-3.

[32] Cook T. D., and A. W., Bally. 1975. Stratigraphic Atlas of North and Central America. Shell Oil Company/Princeton University Press, New Jersey., p. 68, 69, 70.

[33] Miall A. D., 1973. Regional Geology of Northern Yukon. *Bulletin of Canadian Petroleum Geology* 21 (1) 103, 105.

[34] Churkin M. 1973. Paleozoic and Precambrian Rocks of Alaska and Their Role in its Structural Evolution. *United States Geological Survey Professional Paper* 740, p. 22, 26.

[35] Boucot A. J., Doumani G. A., and G. F. Webers. 1967. Devonian of Antarctica (*in* Oswald D. H., 1967. *International Symposium on the Devonian System* Vol. I. Alberta Society of Petroleum Geologists, Calgary, Alberta. p. 640.

[36] Elliot. 1975. *op. cit.*, p. 67.

[37] Barrett P. J., Grindley G. W., and P. N. Webb. 1972. The Beacon Supergroup of East Antarctica. (*in* Adie R. J., 1972. *Antarctic Geology and Geophysics* Universitetsforlaget, Oslo). p. 320.

[38] Ronov A. B., Khain B. E., 1955. Kamyenougolniye Litologicheskie Formatsii Mira. *SO* 1955. sb. 48, p. 92-93 (map inserts) (in Russian).

[39] Churkin. 1973. *op. cit.*, p. 28, 40.

[40] Brown *et. al.* 1968. *op. cit.*, p. 156, 165.

[41] Hill D., 1973. Lower Carboniferous Corals (*in* Hallam A. 1973. *Atlas of Palaeobiogeography*) p. 135.

[42] Ross C. A. 1973. Carboniferous Foraminiferida (*in* Hallam. 1973. *op. cit.*,) p. 127, 128, 131.

[43] Rocha-Campos A. C., 1973. Upper Paleozoic and Lower Mesozoic Paleogeography, and Paleoclimatological and Tectonic Events in South America (*in* Logan A. and L. V. Hills. 1973. *The Permian and Triassic Systems and Their Mutual Boundary*. Canadian Society of Petroleum Geologists, Calgary, Alberta) p. 400.

[44] Ross C. A., 1967. Development of Fusulinid (Foraminiferida) Faunal Realms. *Journal of Paleontology* 41 (6) 1342-3.

[45] Meyerhoff A. A., 1970. Continental Drift: Implications of Paleomagnetic Studies, Meteorology, Physical Oceanography, and Climatology. *Journal of Geology* 78(1) 26-7.

[46] Kummel. 1970. *op. cit.* p. 207.

[47] Cook and Bally. 1975. *op. cit.*, p. 91, 106.

[48] Stocklin J. 1968. Structural History and Tectonics of Iran: A Review. *AG* 52: 1234.

[49] Ronov A. B., Khain B. E., 1956. Permskiye Litologicheskiye Formatsii Mira. *SO* 1956. Sb. 54, p. 31-36. (in Russian).

[50] Audley-Charles M. G., 1965. Permian Paleogeography of the Northern Australia-Timor Region. *PA* 1:300-1.

[51] Audley-Charles M. G., 1978. The Indonesian and Phillipine Archipelagos (*in* Moullade M. and A. E. M. Nairn. 1978. *The Phanerozoic Geology of The World*. Vol. II. EL.), p. 177-192.

[52] Meyerhoff A. A., and H. A. Meyerhoff. 1974. Tests of Plate Tectonics. (*in* Kahle C. F., 1974. Plate Tectonics-Assessments and Reassessments. *AG Memoir 23*) p. 82.

[53] Miall. 1973. *op. cit.*, p. 109.

[54] Brown *et. al.* 1968. *op. cit.*, p. 181.

[55] Oftedahl C., 1976. Northern End of European Continental Permian. The Oslo Region (*in* Falke H. 1976. The Continental Permian in Central, West, and South Europe. Reidel Pub. Co., Holland). p. 3.

[56] Kummel. 1970. *op. cit.*, p. 196.

[57] Gobbett D. J., 1973. Permian Fusulinacea (*in* Hallam. 1973. *op. cit.*, p. 153, 155-6.

[58] Cook and Bally. 1975. *op. cit.*, p. 132-3.

[59] Barrett *et. al.* 1972. *op. cit.*, p. 327.

[60] Barrett P. J., 1970. Stratigraphy and Paleogeography of the Beacon Supergroup In the Transantarctic Mountains, Antarctica (*in* _____ 1970. 2nd *GN*) p. 256.

[61] Elliot. 1975. *op. cit.*, p. 72-3, 86.

[62] Kemp E. M., Balme B. E., Helby R. J., Kyle R. A., Palyford G., and P. L. Price. 1977. Carboniferous and Permian Palynostratigraphy in Australia and Antarctica: a review. *BMR Journal of Australian Geology and Geophysics* 2:202.

[63] Churkin. 1973. *op. cit.*, p. 32, 40.

[64] Ronov A. B., Khain B. E., 1961. Triasovive Litologicheskiye Formatsii Mira. *SO* 1961 (1) 30, 34, 36, 39, 43. (in Russian).

[65] Brown *et. al.* 1968. *op. cit.*, p. 224, 228.

[66] Plumstead E., 1964. Palaeobotany of Antarctica (*in* Adie R. J., 1964. *Antarctic Geology* North-Holland Pub. Co. New York) p. 638.

[67] Elliot D. H., 1975. B. Gondwana Basins of Antarctica (*in* Campbell K. S. W. ed., 1975. *Gondwana Geology* 3rd *GN*) p. 513.

[68] Ronov A. B., Khain B. E., 1962. Yoorskiye Litologicheskiye Formatsii Mira. *SO* 1962 (1) 20, 28. (in Russian).

[69] Elliot. 1975. *op. cit.*, p. 70, 78-9, 86.

[70] *ibid.*, p. 72-3.

[71] Khain B. E., Ronov A. B., Balukhovskiy A. H., 1975. Myelovive Litologicheskiye Formatsii Mira. *SO* 1975 (11) 18, 30 (in Russian).

[72] Khain B. E., Ronov A. B., Balukhovskiy A. H., 1976. Cretaceous Lithologic Associations of the World. *IG* 18:1276, 1288 (translation of preceeding).

[73] Grikurov G. E., Ravich M. G., and D. S. Soloviev. 1972. Tectonics of Antarctica. (*in* Adie. 1972. *op. cit.*,) p. 458.

[74] Drewry D. J. 1976. Sedimentary Basins of the East Antarctic Craton from Geophysical evidence. *TE* 36:302-312.

[75] Ronov A. B., Khain B. E., Balukhovskiy A. H., 1978. Paleogenoviye Litologicheskiye Formatsii Kontinentov. *SO* 1978 (3) 18, 26, 36 (in Russian).

[76] Ronov A. B., Khain B. E., and A. H. Balukhovskiy. 1979. Paleogene Lithologic Associations of the Continents. *IG* 21:422, 430, 440 (translation of preceeding).

[77] Khain B. E., Ronov A. B., Balukhovskiy A. H. 1979. Neogenoviye Litologicheskiye Formatsii Kontinentov. *SO* 1979 (10) 8, 29 (in Russian).

[78] Khain B. E., Ronov A. B., and A. H. Balukhovskiy. 1980. Neogene Lithologic Associations of the Continents. *IG* 22: (in press) (translation of preceeding).

[79] Behrendt J. C., 1979. Speculations on Petroleum Potential of Antarctica. *AG* 63:418.

[80] Dort W. 1972. Late Cenozoic Volcanism in Antarctica. (*in* Adie. 1972. *op. cit.*,) p. 650.

[81] Frank W., and G. R. Fuchs. 1970. Geological Investigations in West Nepal and Their Significance For the Geology of the Himalayas. *Geologische Rundschau* 59(2) 566.

[82] Lochmann H., 1970. Outline of Tectonic History of Bolivian Andes. *AG* 54(5) 735-9.

[83] Kent P. E., and P. J. Walmsley. 1970. North Sea Progress. *AG* 54 (1) 169-171.

[84] Poag C. W., 1978. Stratigraphy of the Atlantic Continental Shelf

and Slope of the United States. *Annual Review of Earth and Planetary Sciences* 6:251-280.

[85] Link W. K., 1970. Petroleum and Continental Drift. *AG* 54(1) 182.

[86] Blatt H., Middleton G., and R. Murray. 1980. *Origin of Sedimentary Rocks.* Prentice-Hall, Englewood Cliffs, New Jersey., 2nd Edition, p. 18.

[87] Douglas R. J. W., 1980. On the Age of rocks and Precambrian time scales. *GE* 8:168.

[88] Ronov A. B., Khain V. Ye., Balukhovskiy A. N., and K. B. Seslavinskiy. 1977. Changes in distribution, volumes, and rates of deposition of sedimentary and volcanogenic deposits during the Phanerozoic (within the present continents). *IG* 19(11) 1297-1300.

[89] Schwab E. L., 1978. Secular trends in the deposition of sedimentary rock assemblages-Archean throughout Phanerozoic time. *GE* 6:532.

[90] Blatt H., and R. L. Jones. 1975. Proportions of Exposed Igneous, Metamorphic, and Sedimentary Rocks. *Geological Society of America Bulletin* 86:1085-8.

[91] Ray K. K., and S. K. Achayya. 1976. Concealed Mesozoic-Cenozoic Alpine Himalayan Geosyncline and Its Petroleum Possibilities. *AG* 60(5)800.

[92] Hatten C. W., 1967. Principal Features of Cuban Geology. *AG* 51:780-803.

[93] Ksiazkiewicz M., Samsonowich, and E. Ruhle. 1968. *An Outline of Geology of Poland.* The Scientific Publications Foreign Cooperation Center of the Central Institute for Scientific, Technical, and Economic Information, Warsaw, (enclosed map).

[94] _____ 1970. *Geology of Poland.* Wydawnictwo Geologiczne, Warsaw, Vol. 1, pt. 1, p. 160, 235, 316, 325-7, 380, 410-4, 458-60, 528, 534, enclosed map.

[95] _____ 1976. *Geology of Poland.* Wydawnictwo Geologiczne, Warsaw, Vol. 1, pt. 2, p. 36, 202, 242, 334, 517, 570.

[96] Gingerich P. D., 1980. Evolutionary Patterns In Early Cenozoic Mammals. *Annual Review of Earth and Planetary Sciences.* 8:407.

[97] Harland W. B., (ed.). 1967. The Fossil Record: a symposium with documentation. *Geological Society of London/Palaeontological Association,* 827 p.

An Anthology of Matters Significant to Creationism and Diluviology: Report 1

AN ANTHOLOGY OF MATTERS SIGNIFICANT TO CREATIONISM AND DILUVIOLOGY: REPORT I

John Woodmorappe*

Received 9 August, 1979.

This report focuses on points of interest to Creationists and Diluvialists, gleaned primarily from the geological sciences.

The main topics mentioned include: biological evolution; the need of apologetics and consequences of a lack of apologetics; the lingering influence of Charles Lyell; examples and types of mixing of fossils in the geological time scale; the subjectivity of fossil species and genera; the artificiality of the geological column; evidences that many alleged ancient fossil reefs are in fact not reefs; critiques of uniformitarian claims about sedimentary environments; miscellaneous evidence against the validity of the geological column; and the way in which errors, similar to those in geology, can arise in archeology.

Outline

I. SOME BRIEF NOTES ON BIOLOGICAL EVOLUTION.
II. PHILOSOPHICAL ISSUES RELATED TO UNIFORMITARIAN VS. DILUVIAL GEOLOGY.
III. OVERLAP OF FOSSILS IN THE GEOLOGIC RECORD.
IV. SOME GEOLOGICAL RAMIFICATIONS OF THE CREATION AND THE FLOOD.
V. CRITIQUE OF ALLEGED ANCIENT SEDIMENTARY ENVIRONMENTS: EMPHASIS ON REEFS AND DELTAS.
VI. SOME EVIDENCES AGAINST THE VALIDITY OF THE GEOLOGICAL TIME SCALE.
VII. ARCHEOLOGY AND PREHISTORY.

The scientific papers written by one individual are only a small fraction of what can be written about scientific tenets in an all encompassing paradigm such as the evolutionary-uniformitarian or the Creationist-Diluvialist paradigm. The active study of the professional scientific literature brings to light many such tenets.

It is useful for the tenets to be available to other workers even if they have not been studied to the point of writing separate works on them. This report is intended to be a collection of miscellaneous findings, primarily in the geological sciences, which are of especial significance in the negation of the evolutionary-uniformitarian paradigm and the establishment of the Creationist-Diluvialist paradigm.

The advantage of such a report is that it presents tenets in a more organized and more convenient form than scattering them over several CRS *Quarterlies* in the "Letters to the Editor" Column. Some of the tenets presented in this report go along with the findings of other Creationists and Diluvialists, others serve as sequels to topics covered in the 3 main works of the author published in the March 1978, September 1978, and September 1979 numbers of the *Quarterly*, while still others are brand new lines of evidence. All the findings can be used by other scholars to incorporate in their works on given topics, while others it is planned will be incorporated by the author in his future works.

I. SOME BRIEF NOTES ON BIOLOGICAL EVOLUTION

1. Hostility Towards Teleology

Woodfield[1] wrote: "Modern science is on the whole hostile to teleological explanations. That they are obscurantist and unemperical has been the dominant view among scientists since the Renaissance... The most common criticisms of teleology nowadays are either that they are animistic, i.e. they assume that the thing being explained has a mind, or that they tacitly invoke a supernatural being who directs the course of events."

Comment: Animism is a pagan view. If Woodfield's statement is correct, then a strongly atheistic value system dominates the sciences of origin. This is evident by the fact that postulating a supernatural becomes "obscurantist and unempirical" while the most tenuous and unobserved materialistic-mechanistic views are never recognized as being obscurantist and unempirical. It is high time that scientists recognize that a (however unempirical) teleological explanation can be as scientific as a (however unempirical) materialistic scenario for the unobservable past.

2. The Cambrian Explosion: An Enigma for Evolution

Towe[2] wrote: "One of the most striking and enigmatic aspects of paleontology has been the sudden appearance of advanced and diversified metazoan organisms in the early Cambrian. This subject has been the object of considerable research and speculation and numerous hypotheses have been proposed to explain the phenomenon."

Comment: This is yet another statement indicating the magnitude of difficulty that the Cambrian explosion holds for evolution.

3. Similar Environment An Insufficient Explanation for "Convergent Evolution"

In commenting on the claim that evolution to fit a similar environment actually explains "convergence", Riedl[3] said: "Parallel environmental factors may account for most of these but not very likely for all of them." He then proposed an evolutionary system that involves some kind of feedback mechanism.

Comment: Recognition of the difficulty for evolution caused by "convergence" has been discussed and documented by the author in his previous article[4] on Cephalopods. It is more reasonable to accept that the Creator used similar morphologies on otherwise very different forms of life rather than contend that similar forms arose twice independently.

4. Morphology of Carnivores Before the Fall and Curse

Salvadori and Florio[5] write: "Equipped with powerful jaws and long molar teeth typical of carnivores, the panda eats enormous quantities of fibrous bamboo

---
*John Woodmorappe, B.A., has studied both Biology and Geology.

shoots, although occasionally it hunts small mammals to complete its diet."

Comment: The panda helps explain what presently-carnivorous animals may have been partly like before the Fall and the Curse. It is commonly supposed that enormous morphological changes had been wrought on animals to make them carnivorous. The example of the panda suggests that morphological changes were not as great as behavioral changes. Cats, dogs, bears, etc. may have had long molar teeth as they have today since the Creation, but the sharp teeth may have been used on bamboo shoots instead of on the flesh of prey.

II. PHILOSOPHICAL ISSUES RELATED TO UNIFORMITARIAN VS. DILUVIOLOGICAL GEOLOGY

1. The Need for Effective Apologetics

In contrasting identifiers (those college students who practice the religion of their upbringing) with apostates (those college students who have totally rejected the religion of their upbringing), Caplovitz and Sherrow[6] note: "In every religion, the label 'intellectual' is substantially more popular with the apostates."

Comment: Herein lies the fruit of years of neglect by Christians of apologetics. The philosophy of atheistic evolutionary humanism has been allowed to so totally saturate the academic scene that it is considered unintellectual to be a believer. Solid presentation of apologetics and the defense of the Gospel and of Scriptures is long overdue. Happily, Creationism on campus has had a major impact, but much more is needed.

2. Fallacies in Claims of Repression of Scientific Inquiry

Lewis[7] writes: "As a result of intensive research since the beginning of the present century, the traditional image of the Middle Ages as an epoch of sterile subservience to the authority of the Church and of Aristotle in scientific matters has been destroyed. Instead, it is now recognized as a period during which scholars became capable of wide ranging and subtle, albeit habitually inconclusive, speculation upon topics commonly supposed to be the distinctive concern of early modern science."

Comment: Unbelievers frequently attack religious belief for allegedly having repressed scientific inquiry and having hindered the development of science. This statement of Lewis from a scientific journal (*Nature*), if accurate, indicates that there was substantial science in the Middle Ages and that the Church did not repress scientific inquiry.

3. Atheistic Consequences of Uniformitarian Geology

Writing in a geologic journal, Campbell[8] said: "Time was when hope of heaven and fear of hell exercised a major influence on man's thinking and conduct. Today, with the declining influence of revealed religion (a decline for which many hold science in general and geology in particular responsible) man seems to possess fewer or even no guides for his thoughts and actions. Thus it is philosophy that he must turn for help, for it is philosophy that seeks to define the underlying principles of the universe."

Comment: Uniformitarianism and the denial of God and his works go together. This was discussed in detail by the author in his work[9] on cyclic sedimentation. The drift towards meaninglessness and to man-centered humanistic philosophy follow as a consequence.

4. Charles Lyell on Catastrophism

Long ago, the pioneer uniformitarian geologist Lyell[10] said "Never was there a dogma more calculated to foster indolence, and to blunt the keen edge of curiosity than this assumption of the discordance between the ancient and existing causes of change."

Comment: From his own experience with dozens of geology professors, the author can definitely conclude that never has a dogma fostered more uninquisitiveness, narrow-mindedness, and even some hostility than the dogma of uniformitarianism. When asked of catastrophism on a more than local scale but not mentioning the Flood or the geological time scale, one professor said: "Let's stick with modern facies analysis... They work. Why drag in catastrophism?" Another said "When are you going to start thinking like a scientist?" (He has since moderated, and respects the author and his works). Another said and still says curtly: "Worldwide Flood.. Hah! ... are you bringing that _____ up again?" Still another said: "Diluvialism is like the plogiston theory... gone forever... no finds could ever bring it back." (This is fallacious. While the author knows of no chemist who upholds the plogiston theory and who would write a paper supporting it, we all know dozens of professional geologists who are Diluvialists and many who have written scientific papers supporting that position.)

5. Persistence of Lyellian Influence on Modern Geology

In a review of a recent book B.J.S.[11] said: "... when we realize that so many of our accepted and frequently unquestioned beliefs descend directly from observations made by Lyell."

Comment: Mentioning Lyell is not only of historical interest. Psychology may have outgrown Freud, but in many instances geology has not yet outgrown Lyell.

6. The Interpretative Nature of Geological Sciences

Von Herzen[70] said: "Where many physical variables are relevant over a broad range of time and space scales as for most earth science hyptheses, formal 'proof' becomes difficult or impossible. The validity of an hypotheses then becomes a subjective judgement, either individually or by many persons, and is frequently dependent on the way the original hypothesis is framed."

Comment: The oft-repeated uniformitarian claim that their's is the only valid and scientific way of looking at earth history is therefore presumptuous, especially when it is realized that uniformitarianism rests upon rationalistic premises.

7. Present Uniformitarian Attitudes Towards Catastrophic Processes

Kumar and Sanders[74] wrote: "Moreover, as the geologic philosophy of what might be called *uniformitarianist catastrophism* continues to gain adherents and respectability, the study of storm deposits in the geologic record is losing much of its former 'instant stigma'." Uniformitarianist catastrophism describes the kind of thought expressed by Lyell (1830); it refers to the effects of catastrophic storms within the context of a uniformitarian view of geologic history." (italics theirs)

Comment: It is evident that there long has been (and to some extent still is) resistance towards the acceptance of *any* catastrophic processs. Where tolerated, catastrophic processes are admitted only as numerous, local in nature, fitting within the geologic time scale, and having occurred only within the context of ancient analogs of modern sedimentary environments.

8. A Case of Reversed Cause-Effect

Dietrich[12] wrote: "The quest for knowledge about geological history began when man first wondered about his environment. Some historians think the biblical story about the Great Flood may even have been framed in answer to questions dealing with man's finding fossil seashells on high and dry land."

Comment: This is like saying that heavy perspiration causes hot weather. He proposed that fossils caused the Flood (belief in the Flood.) Actually, it was the Flood which caused fossils. Ancient post-Diluvian man may have been aware of the Flood as the cause of fossils to a greater extent than has been supposed.

9. A Principle True for Both Uniformitarianism and Diluvialism.

Glikson[13] said: "In dealing with rocks formed when the world was less than half of its present age, a strict adherence to the doctrine of uniformitarianism is considered unjustified."

Comment: He is, of course, saying that conditions on the earth for the first 2.2 billion years may have been substantially different from the last 2.2 billion years. This principal amazingly also applies in the Creationist-Diluvialist paradigm. The earth was half its present age at about the time of Moses. Uniformitarianism has applied only since then, again only for the last half of time of the earth's existence.

III. OVERLAP OF FOSSILS IN THE GEOLOGIC RECORD

1. Increased Stratigraphic Ranges for Vertebrate Fossils

Wills[14] wrote: "... Period names ... are losing much of their old significance through the repeated discovery of fossils which overlap the old boundaries ... 'Carboniferous' amphibia now range right through the Permian and well up into the Trias. Many 'Triassic' reptiles range up into the Jurassic; and some also downwards even to low in the Permian. Recently a tortoise carapace not hitherto known below the Eocene (or one remarkable similar) turned up in Triassic rocks."

Comment: It is obvious that many fossil groups are found to have longer stratigraphic ranges than had been earlier supposed. In his cephalopod[15] paper, the author documented some long-ranging cephalopod taxons.

2. Mammalian Footprints in Carboniferous Rock?

Sarjeant[71] wrote: "Barkas considered it 'not improbable' that the former tracks were those of a 'small, broad, four-legged mammal' ... they figure as footprints of uncertain systematic character ... the tracks described by Barkas have received no subsequent attention, nor have any others been reported from Northumberland."

Comment: The possibility of mammalian tracks in Carboniferous rocks is radical because "primitive" mammals are not supposed to have appeared until Late Triassic-100 million years later. "Advanced" mammals are not supposed to have appeared until early Tertiary-200 million years later.

3. Horse Hoofprints in Devonian Rock?

Sarjeant[72] also wrote: "His illustration shows circular impressions with a raised central region, indeed rather like horse hoofmarks but quite *unlike* the footprints of any animal likely to have been in existence in the Devonian." (italics his)

Comment: Horses are not supposed to have appeared until the Tertiary, 300 million years after the Devonian. The effect of preconceived notions about stratigraphic ranges is obvious in the above statements by Sarjeant. While isolated footprint evidence is tenuous, it is admitted that the prints from the Old Red Sandstone do look like horse hoofprints.

4. A Possible Drastic Increase in the Stratigraphic Range of Frogs

Cameron and Estes[16] discussed the riddle of fossil "tadpole nests." They believe that they are poorly preserved. But the main argument they advance for these structures not being tadpole nests is the fact that these structures are found in Silurian rock while the earliest frogs are Jurassic. They also consider no other vertebrate in existence in Silurian times to be capable of having made such structure.

Comment: Although somewhat tenuous, this may be an evidence for increasing the stratigraphic range of frogs from Jurassic-Recent to Silurian-Recent. The profound gap between the first appearance of body fossils and first appearance of trace fossils would be an evidence against geologic time; the gap favoring a short or nonexistent time difference between Silurian and Jurassic rocks.

5. Progressive Increase in Stratigraphic Ranges is a Rule

Raup[17] testified that it is much more probable as a result of continued fossil collecting to increase the stratigraphic range of a known taxon rather than discover a new short-range taxon.

Comment: It is therefore clear that the instances cited above are far from isolated. The fact that extension of stratigraphic ranges with further collecting is the major trend indicates that the entire fossil record is becoming more random with time; that overlap of all fossils is continually increasing.

6. Stratigraphic Range Extension Of Conodonts Ascribed To "Time Warps"

Sandberg[18] reported finding typically Middle Devonian conodonts in Early Devonian rock, and Ordovician conodonts in Late Devonian rock. These were called "nearly perfect heterochronous heteromorphs" and were not considered to be simple extensions of stratigraphic ranges nor reworkings but "time warps." When conodonts are found in rocks "younger" than they are "supposed" to be, this is supposed to be evolutionary atavism (genetic throwbacks to earlier forms). By contrast, when conodonts appear in much older rock than is "proper" for them, then this is supposed to be a case of them being precursors of evolutionary change ("anticipation" of forms that will appear in the distant future).

Comment: This "time warp" concept illustrates the absurd lengths to which rationalizations for "out of place" fossils will go.

7. "Reworking" An Alleged Cause for Mixed Paleofloral Remains

Vanguestaine[19] said: 'Siegenian and Emsian acritarchs in the Synclinorium of Dinant originate from redistributed Silurian and Ordovician sediments... On the southern flank Silurian and Ordovician are mixed in some localities."

Comment: "Reworking" is a common rationalization for fossils found where they are not "supposed" to be. In his article on cyclic sedimentation[20], as well as in the one on cephalopods,[21] the author gave several other examples of alleged reworking.

8. Expected State of Preservation Reversed in Alleged Reworking

Mildenhall and Wilson[22] wrote: "Abundant well-preserved Cretaceous pollen and spores occur in all samples studied... Rare Pliocene pollen and spores occur..."

Comment: If there is found a difference in state of preservation of fossils considered to have been reworked versus those formed penecontemporaneously with sedimentation, then it is expected that reworked fossils will be in a worse state of preservation than the others because the former were transported in the reworking. This case is interesting because the allegedly reworked fossils are in a better state of preservation than the others which give the "true" age of sediment! "Reworking" rationalizations for mixed fossils are plastic.

9. Subjectivity of Fossil Species and Genera

North[23] said: "Except in standard sections, more fossils are identified wrongly than rightly."

Comment: The fact that geologists routinely "misidentify" index fossils should be an excellent indicator of the subjectivity of fossil species and genera.

10. Disparity Between Morphologic and Evolutionary Genera

In discussing genera within the Bivalve Family Lucinidae, Bretsky[24] wrote: "... the major phenetic clusters showed little similarity to the genera recognized on phylogenetic grounds. Some of the major clusters represented only part of one genus, others combined two or more genera, and others were an assortment of species from several genera."

Comment: Such artificiality of phylogenetic genera shows that evolution is fantasy.

Furthermore, all claims of stratigraphic ranges rest upon the taxonomy of the fossils involved. The statements by North and by Bretsky illustrate the subjectivity of fossil species and genera; a point made by the author in his work on cephalopods.[25] North was speaking about fossil cephalopods (ammonoids used in biostratigraphy) whereas Bretsky's statement shows that it applies also to Bivalves.

11. Circularity in the Geologic Time Scale

North[26] wrote: "The paleontological time-scale rests squarely on the law of superposition, independent of any theory or assumption. From this unassailable foundation; the paleontologist became for more than a century the arbiter of all stratigraphic organization. But for geologists, the law of superposition presupposes the existence of decipherable geological sections, and every geological section must have a top and a base. The paleontological succession was pieced together from hundreds of such sections, the tops and bases of which had been established by geologists on the ground. The paleontologists' wheel of authority turned full circle when he puts this process into reverse and used his fossils to determine tops and bottoms for himself. In the course of time he came to rule upon stratigraphic order, and gaps within it, on a world-wide basis."

Comment: A certain degree of circularity is evident here. The Law of Superposition can only apply if there are objective sections which can be superposed, and sections exist only if there are tops and bottoms delineated for them. But tops and bottoms can be set only if there is an order of fossils. The order of fossils is determined by the superposition of sections, and the superposition of sections is determined by their tops and bottoms which are deduced by the order of fossils.

12. Circularity in Delineation of Biostratigraphic Ranges

Bond and Bromley[73] said: "On the basis of these field relations, the Gokwe Formation is significantly younger than the Stormberg Basalts of the Karroo System, but older than the Kalahari Beds. Thus it is younger than the early part of the Jurassic and older than the Mid-Tertiary. This is a wide bracket which by paleontological means may be narrowed considerably. The presence of remains of dinosaurs rules out any age later than the end of the Mesozoic."

Comment: Circular reasoning is evident here. Dinosaurs are known to be Mesozoic because they are only found in Mesozoic rocks, and rocks are Mesozoic if they contain dinosaurian remains.

13. Patchwork in Formation of the Geologic Column

Krassilov[27] said: "The field geologist is seldom so lucky as to find out the sequential relations of all rocks in his area without reference to adjacent areas or more distant regions." He also said: "The regional stratigrapher is supposed to define existing divisions of rocks and then to relate them to the international scale which stands for standard reference. Thus, regional classification can be conceived as natural and the international scale as artificial at least outside the stratotype area."[28]

Comment: This illustrates how the geologic column is conceptual. Regional stratigraphy is not left to itself but is put together into a worldwide "onion skin" system of geologic ages. But the geologic column is admittedly artificial as applied to any one single area on earth. The geologic column is not as much read from the rock as it is read *into* rock.

IV. SOME GEOLOGICAL RAMIFICATIONS OF THE CREATION AND THE FLOOD

1. The Creation of the Earth and Differentiation of the Crust

Johnston[29] said: "Pillow lavas... are common in many parts of the Canadian Shield."

Comment: Since pillow lavas form underwater, this implies that there was a great deal of water during formation of the Shield areas. The Shield areas, being composed primarily of igneous and metamorphic rocks, and (with few exceptions) unfossiliferous, probably were formed during Creation Week. The formation of pillow lavas in Shield regions suggest the process of Creation whereby God differentiated the waters from the land (Genesis 1:9-10). Since most of the Shield is composed of granitic rocks, and granitic magma contains as much as 20% dissolved water, it is not difficult to envision the differentiation being first of all the driving off of waters from the formless crust so that it could crystallize. Much water would still be available during associated volcanism, enabling pillow lavas to form. Finally, the waters drained off the Shield areas to form permanent, dry land.

2. Impact Craters or Unusual Volcanism During Creation Week?

Nininger[30] commented: "The majority of impact craters are old—too old for the material that made them to be preserved in recognizable form; but coesite, stishovite, impactite, shatter-cones and shock metamorphoses are more durable, and they are not common products of vulcanism."

Comment: The fact that meteoritic material is not preserved at or near alleged meteoritic craters in the Precambrian may be an evidence against such a mode of origin, especially in the context of a proven young earth. Because the process of formation of the earth and its crust took God only the first few days of Creation Week, some action may have involved an explosive transfer of material from one part of earth into another via "autobombardment." The process of formation of impact craters may have been volcanism so intense that large chunks were injected into suborbital, ballistic trajectories. The craters thus formed indirectly from volcanism; the petrology and minerology reflecting the impact whereas the absence of meteoric material indicates that it was terrestrial, not extra-terrestrial, material that formed the craters.

3. Primacy of Submarine Volcanism During the Flood

Moore[31] said: "Pillow lavas, produced as fluid lava cools underwater, is the most abundant volcanic rock on earth . . ."

Comment: The significance of submarine volcanism during the Creation Week has already been discussed. Noteworthy is the fact that submarine volcanism was also predominant during the Flood; water being everywhere and lava usually coming into contact with it during eruptions.

4. Tectonics and Metamorphism During the Flood

Sawkins *et. al.*[32] wrote: "Geologists who study metamorphic rocks have long realized that enormous amounts of thermal energy are required to convert sediments and volcanics of regional extent from their original condition into high-grade metamorphic assemblages. Clearly, the requisite heat must be supplied from below, but by what physical means was it delivered to the site of metamorphism? The conduction of heat through rock is so slow that the thermal conduction of heat from deeper regions is simply inadequte to account for the geometry of many metamorphic belts.

Comment: The authors are proposing that heat from plutonism helped cause metamorphic belts; having noted that geothermal heating is insufficient. Metamorphic rocks find an easy explanation in a Diluviological context. Mountains being built in months instead of tens of millions of years would have trapped tremendous amounts of heat from tectonomagnetism. The heat would have metamorphosed the rocks; having been incapable of dissipating quickly because so much was generated in such a short time.

5. More Evidence For Catacylsmic Deposition

Milici *et. al.*[33] write: "Figure 24 shows a tree that was buried to a depth of 4.6m (15 ft.) Because the tree is in growth position and shows no root regeneration, it probably was buried very quickly, certainly before it could decay."

Comment: This is yet another example of a tree trunk being buried upright over many feet of sediment, demonstrating very rapid burial. Similar examples and other lines of evidence for cataclysmic burial have been discussed by the author in his works as well as by other Diluvialists. The finding of evidences for a young earth is crucial in showing that one, not many, cataclysms formed these deposits.

6. Disturbance of Marine Ecology During the Flood

McKerrow[34] wrote: "It can be concluded from the above discussion that Mesozoic, Tertiary, and modern bottom-dwelling communities are more intimately linked with sediment type than those in the Paleozoic (except for some early Palaeozoic communities which were dominated by the deposit-feeding trilobites.")

Comment: This find is very amenable to a Diluvian interpretation. The poor connection between fossils and lithology may be an evidence that they were not *in situ* seas but Flood deposits. The Paleozoic thus represents early Flood deposition with catastrophic mixing of marine fossils with sediment. By the time the rocks labeled Mesozoic were being deposited, the Flood was receding, sedimentation was slower, and more organisms could find habitat in regions of energy appropriate for them.

V. CRITIQUE OF "ANCIENT SEDIMENTARY ENVIRONMENTS": EMPHASIS ON REEFS AND DELTAS

1. Evidence Against Evaporitic Origins of Salts

Meynen[35] wrote: "Paradoxical as it appears, it is at times difficult to judge the aridity and humidity of a climate from the fossil plants. This is because the buried plants are mainly those growing near or in the basins. For example, the Kazanian flora . . . contains only mosses, ferns and sphenopsids . . . i.e. obviously hydrophyllic forms. However, the flora-bearing beds alternate in this section with rocks indicative of an arid climate (gypsums, dolomites).

Comment: This interbedding of evaporites (so-called) with plants may be an evidence against these salts having arisen from evaporation of drying-out seas over immense periods of time.

2. Uniformitarian Presuppositions and "Ancient Reefs"

Braithwaite[36] said: "The spur to recognize fossil reefs came initially from their investigation in modern seas and, at the time their study first began, the simple presence of corals was felt to be evidence of 'a reef.' "

Comment: This illustrates how the "present is the key to the past" dogma of uniformitarianism causes a natural tendency to attribute ancient rocks to present processes. In particular, it illustrates how fossiliferous limestone can be attributed to ancient reefs by reading reefs into the rock because reefs exist today. The whole issue of alleged "ancient reefs" has been discussed by Austin[37] (former pseudonym Nevins).

3. Plasticity and Accommodation of Alleged Reef Characteristics

Rigby[38] said: "Any model for recognition of reefs in the geologic record must allow for considerable variation in relief, size, shape, biological composition, and facies relations."

Comment: It is evident from the statement that all of the criteria for recognition of reefs are quite plastic. It thus becomes facile to take characteristics of limestones and attribute a reef environment to the rocks. The great variation in criteria that have been used may indicate that reefs have been read into, not out of, the rocks.

4. Difficulty of Proving "Ancient Reefs."

Philcox[39] wrote: Wave resistance is used in many definitions as a criterion for 'reef.' It is therefore important to clarify what wave resistance means, whether it can be recognized in ancient reefs, and what effect the use of this criterion has on our thinking... Diagnosis of the wave-resistance capacity of ancient buildups is difficult."

Comment: Most definitions of a reef include wave-resistance caused by the binding action of organisms that grow there. This indicates that it is very difficult to prove that ancient rock was one a wave-resistant sediment, and—by implication—a reef.

5. Most "Ancient Reefs" Admitted Not To Be Reefs

Wilson[40] said: "Organic framework construction is known to be important in the middle Paleozoic and in some situations from Jurassic to Holocene times. Most organic buildups in the geologic record are in no sense organic frame-built reefs."

Comment: Even by uniformitarian standards, much of what has been attributed to reefs isn't reef at all.

6. Silurian "Reefs" of US Midwest Are Not Reefs

Despite advocating a reef origin, Lowenstam[41] wrote· "Regarding the reef builders, unless the *Stromatactis*-like forms are algal, the sole reef builders were all coelenterates. Since bottom-cementing habits had not as yet been acquired in Niagaran times by Foraminifera, pelyecypods, gastropods, or barnacles, there was a noticeable lack of these accessory elements which today contribute to the strengthening of the framework constructed by the reef builders... Today the adnate tubes of polychaete worms form an important strengthening agent of the framework erected by the reef-builders. Though this habit among polychaetes was already developed in Niagaran times, as shown by *Spirorbis* in the inter-reef habitats (Lowenstam 1948), the niche range had not then been extended to reefs...

Among the reef dwellers, we miss altogether elements such as the calcareous algae, Foraminifera, and echinoids which today make up a major portion of, and occasionally the bulk of, the reef sands. The niche of *Halimeda* of today was, in Niagaran times, occupied by the similarly jointed crinoids, that of the pelecypods largely by brachiopods. Today reef crinoids are few and eluetherozoic in habit, as are all reef echinoderms; in the Niagaran the Pelmatozoa were exclusively sessile forms. This is further due to the fact that as of today we have no evidence of asteroid or holothuroid representation on the Niagaran reefs...

It is perhaps worth stressing that the enormous biomass of the Niagaran reefs was, as that of modern reefs, basically dependent upon a plant foundation. Yet so far, we have no definite records of this vital element of the Niagaran reefs."

Comment: The Niagaran (Middle Silurian) "Reefs" of the Great Lakes region are probably just as famous as the Permian Capitan Reef of Texas, discussed by Austin.[37] Upon close examination, the statements of Lowenstam render the reef interpretation untenable. It is evident that very many forms of life expected to be present on reefs are absent. The claim that these forms had not evolved to fit the reef niches by Silurian time may be viewed as a rationalization. The absence of so many reef organisms from the Silurian "reefs" is thus decisive evidence against them being reefs. Also noteworthy is the fact that Braithwaite[36] noted the problematic nature of *Stromatactic* and, based on his studies, concluded that *Stromatactis* was mud-based and could not have served as a framebuilder or former of a wave-resistant structure.

7. "Ancient Reefs" are Actually Flood Deposits

Wilson[42] wrote: "By analogy with present shelf depths, whose Holocene sediments much resemble ancient limestones, we can safely assume that in the geologic past, shelves and platforms hundreds of miles wide were covered with water only a few tens of meters deep. Again, by analogy with actualistic models, it is hardly conceivable that tidal currents and wave action in such widespread and shallow seas could have been very effective. Yet many deposits across the North American craton contain uniform sequences of rock types for hundreds of miles and such deposits commonly include beds of clean quartz sandstone and pelletoid or oolitic lime grainstones! These are clearly the result of wave and/or current activity and yet are widely distributed over thousands of square miles. On such flatbottomed, shallow seas not much wave energy can be generated and tidal effects are severely restricted (Keulegan and Krumbein, 1950). It is thus probable that such deposits were formed by shoreward progradation of shorelines and offshore oolitic bars."

Comment: Instead of being evidence for progradation and lateral migration of the environment, the presence of evidence for high energy deposition over wide areas may be evidence against the usual uniformitarian interpretation of ancient shallow seas. The Flood can easily account for the widespread presence of high-energy conditions of deposition.

8. Subjectivity in "Sedimentary Environment" Designations

Friedman and Sanders[43] said: "Although the test of a geologist's skill is the depth of perceptiveness and logic used in interpreting rocks in terms of their formative processes and environments of deposition, *even the most astute geologist commonly is unable to arrive at a unique interpretation of a given rock*. (italics theirs)

Comment: Such difficulty of "sedimentary environment" diagnosis is often attributed to complicating factors. An alternate view would be that difficulty is caused by the fundamentally erroneous presuppositions and methodology of uniformitarian geology which attributes all ancient rock to environments now in operation (as seas, rivers, deltas, etc.), when in actuality these are Flood deposits.

9. Sedimentary Structures From a Wide Variety of Hydrologic Conditions

Suttner[44] wrote: "Also, no longer are primary structures routinely directly associated with depositional environments in interpretation of facies motifs. Instead, there is broader recognition of the fact that sedimentary structures are products of processes and that processes overlap specific environments."

Comment: A wide variety of hydrologic conditions can produce given sedimentary structures. The uniformitarian claim that ancient sedimentary rock can be objectively and compellingly assigned to analogs of environments now in operation grows weaker as does the claim that only processes now in operation can uniquely explain sedimentary structures in ancient sedimentary rock.

10. Tenuous Basis For Many "Sedimentary Environment" Claims

Harbaugh[45] said: "All too commonly, geologists who study stratigraphic sequences attempt to interpret the depositional environments of the strata in terms of only one or two variables—such as water depth or wave energy."

Comment: Many "sedimentary environment" designations are weak even from a uniformitarian viewpoint. Variables such as water depth and wave energy easily dovetail with a Diluvian interpretation.

11. Contradictory "Sedimentary Environment" Designations

Adams and Patton[46] wrote: "Various interpretations— offshore bar, beach, eolian, and fluvial,—have been made for the depositional environment of the Lyons Formation. Although attempts have been made to apply a single environmental origin to the entire lateral exposure, it now seems probable that the formation was deposited in several closely related but distinct environments, and that these account for the observed lateral and vertical changes within the formation."

Comment: The fact that the whole gamut of sedimentary environments has been attributed to the Lyons Formation (a well-sorted, mature sandstone) may be an indicator of the fallaciousness of attributing ancient rock to sedimentary processes now in operation. The whole issue of "sedimentary environments" was discussed by the author in his work on cyclic sedimentation.[47] Contradictory designations may indicate that these rocks were not formed in analogs of present environments, but rather that Flood deposition simulated an assortment of environmental characteristics.

12. Differences Between Ancient and Modern Sedimentation

Lundegard et. al.[48] said: "The Brallier depositional sequence differs significantly from existing submarine-canyon-fan models in that it lacks large-scale radial dispersal patterns as well as canyon and channeled inner-fan facies. Rather than radial progradation, characteristic of a large, stable submarine fan, uniform progradation from multiple sources ... In spite of the paucity of modern analogs for such a depositional system, the Brallier Formation and other ancient examples attest to the significance of turbidite sedimentation in deltaic settings."

Comment: This is a clear example of ancient sediments not quite resembling modern ones. Multiple-source progradation may be an evidence that the above-discussed sediment is not an ancient delta but a Flood deposit; torrents of water flowing from various regions and covering a large area. The prominence of turbidites is especially suggestive of large-scale Flood deposition.

13. Widespread, High-Energy Sedimentary Veneers Unlike Today's

Newell[49] wrote: " ... most present configurations (topography, chemistry, circulation, climate,) are strikingly unlike those that must have prevailed when the Paleozoic and Mesozoic limestone seas spread over immense and incredibly flat areas of the world ... Closely comparable epeiric seas probably do not exist today ... Supratidal and intertidal mud flats are today much more restricted in area than they must have been at times of widespread Paleozoic and Mesozoic limestone seas."

Comment: The widespread sedimentary rock veneer on top of the cratonic basement may be an indicator that these rocks resulted not from shallow seas of which there are few presently, but from Flood deposition.

14. Disturbance of Modern Sedimentation

Swift et. al.[50] said "Thus, a palimpsest sediment is one which exhibits petrographic attributes of an earlier depositional environment and in addition, petrographic attributes of a later environment ... Palimpsest sediments, like relict sediments on modern shelves, are more common in the rock record than is commonly realized."

Comment: The fact that an ancient sedimentary rock shows the attributes of more than one environment may be an indicator of the fallaciousness of attributing ancient sediments to analogs of modern environments. Rather, the mixing of environmental attributes indicates that the properties of flowing Floodwater simultaneously resembled processes taking place in several different modern environments.

15. Mixing of Trace Fossil Assemblages and Sediment Type

Chamberlain[51] said: " ... *Chondrites* and *Zoophycus* are more extensive in deep-water deposits than previously thought." In speaking of sedimentary rock facies, Seilacher[52] wrote: "*Glossifungites* facies may occur in deeper and *Zoophycus* facies in shallower positions due to local channeling or restrictions, but *Nereites* facies seems always to be restricted to the deepest zone."

Comment: It is commonly taught that trace fossil assemblages show an ichnofacies zonation due to organisms having inhabited different regions of a sedimentary environment. Thus, the zonation from shallow, agitated waters to deep, tranquil waters is reflected by the zonation: *Skolithos*, *Glossifungites*, *Cruziana*, *Zoophycus*, and *Nereites*. The statements of Chamberlain and Seilacher indicate that the ichnofacies zonation is sometimes mixed. Such zonation therefore is not exact and need not imply the onetime presence of analogs of modern environments. The overlap and even mixing of ichnofacies may have resulted from organisms of different subenvironments having become mixed during Flood deposition.

VI. SOME EVIDENCES AGAINST THE VALIDITY OF THE GEOLOGIC TIME SCALE

1. The Vaunted Supremacy of the Geologic Column

After considering Archbishop Ussher's famous chronology as being a search for the limits of time, McLaughlin[53] commented: "Today we continue Archbishop Ussher's search, but we now use a far more accurate source of information than his Bible. From the beginning of its history, the earth has maintained a detailed geological journal . . . "

Comment: the items already presented in this report and especially those about to be presented in this section demonstrate the fallacy of McLaughlin's statement.

2. Significance of Paraconformities

After describing some trace fossils from Ordovician rocks near Sinat, Iraq, Seilacher[75] commented: "The Silurian is missing and the boundary to the overlying Devonian was for a long time arbitrary due to the sandy and unfossiliferous nature of the passage beds. A sudden shift in the trace fossil record from a *Nereites* to a *Skolithos* community has here proved to be a better boundary criterion than lithology and seemingly conformable relationship."

Besides listing several paraconformities, Newell[54] said: "The Devonian rests paraconformably on Cambrian rocks over much of Montana (Sloss and Laird 1947). Many geologists would term this a disconformity, but over large areas its recognition and evaluation depends solely on fossils. Every experienced biostratigrapher can cite other examples of such paraconformities."

Comment: Many Diluvialists (for example, Whitcomb and Morris[55]) have taken note of paraconformities. In the latter example cited above, the Ordovician and Silurian are "missing" and there is little or no evidence for the alleged 150 million years of nondeposition and erosion. The absence of geologic ages with no evidence of sedimentary discontinuity is thus evidence against the validity of the Phanerozoic geologic column. Noteworthy from the latter cited statement is the fact that such paraconformities are common.

3. Gross Disparity in Extrapolation of Sedimentation Rates

Wilson[56] wrote: "Considerable discrepancy exists when such rates obtained from deposition on modern tidal flats and reefs are applied to thicknesses of ancient neritic strata. For example, the Great Bahama Bank should have 35000-50000 m of post-Cretaceous sediment instead of 4500± (Goodell and Garman, 1969, p. 528). Since these rates do not jibe with the rates of deposition of even thickest known ancient carbonate deposits of comparable environment, we assume that the carbonate producing system operates intermittently and is very sensitive."

Comment: The fact that present sedimentation rate must be reduced by a factor of 10 in order to make it compatible with the amount of time supposed to have elapsed since the Cretaceous may instead by interpreted as evidence against the validity of geologic time and geologic ages. Even the amount that there is need not imply millions of years of formation, because sedimentation rate was magnitudes greater during the Flood than at present.

4. Near Absence of Major Placers in Pre-Tertiary Sedimentary Rock

". . . most placer deposits of economic value are in rocks of Tertiary age or younger. . . "[57]

Comment: The paucity of placers in the geologic record is explained by claiming that they were in regions or erosion and therefore were not preserved. Alternatively, their near-absence may indicate that the fossil record is not the result of normal sedimentation, but of cataclysmic Flood sedimentation where not many igneous and metamorphic parent rocks are eroded to yield placers. Their abundance in Tertiary reflects post-Flood sedimentation, whereas Precambrian placers date back to the Creation.

5. Paucity of Ecological Relationships Among Ancient Life

Ansich and Gurrola[58] wrote: "Much of present day ecological study is concerned with biological interactions and relationships among organisms. Such interactions, e.g., commensalism, predation, parasitism, and competition for space, are rarely preserved and, at best, can only be inferred from the data available to the paleontologist."

Comment: These lacks are attributed to the non-preservation as fossils of most organisms that lived. Another reason may be that fossils do not represent ancient *in situ* seas but Flood deposits, where ancient seas were stirred up and their contents deposited on land. Lack of solid ecological evidences from most fossil groups may thus be evidence against geologic periods and geologic time.

6. Unnatural Basis For Geologic Periods

Rodgers[59] said: "Detailed analysis of the 'fine structure' of the Taconic Orogeny combats the dogma that orogenies are sharp, discrete events punctuating the geologic record (separating periods and abruptly terminating geosynclinal sedimentation) and suggests instead that they reflect 'random walk' processes within the earth."

Comment: The fact that orogenic processes fail to show well-defined effects within geologic periods may be yet another evidence of the fallaciousness of uniformitarian historical geology. The fact that an orogeny may affect rocks of a geologic period and yet the same orogeny not effect rocks of the same period in the same general region may be an indicator of the non-existence of geologic periods.

7. Paleontological Difficulty Caused by Acceptance of Geologic Time

Hewitt and Hurst[60] described the cephalopod *Aegoceras* and the fact that its size increases going stratigraphically upward. It was considered that it is either an evolutionary growth rate increase or a temperature-induced increase in growth rate. If accepted as evolutionary, then the time required would necessitate the section having been deposited as a rate of only 5mm./1000 years, which is less than pelagic oozes and this is inferred to be an offshore shelf region where a sedimentation rate of 1mm/1yr. should be expected, unless 99% of time was occupied by unseen disconformities. For this reason an ecological, not evolutionary, explanation was favored for the stratigraphic size increase.

Comment: The impossibly low sedimentation rate that would have to be accepted if the *Aegoceras* progression were evolutionary is a *reductio ad absurdum* of claims of evolution in stratigraphic section.

8. Evidence Against Geologic Time From Fossil Diversity

Raup[61] said: "There are about 250,000 different species of fossil plants and animals known... In spite of this large quantity of information, it is but a tiny fraction of the diversity that actually lived in the past. There are well over a million species living today and known rates of evolutionary turnover make it possible to predict how many species *ought* to be in our fossil record. That number is at least 100 times the number we have found." (italics his)

Comment: This very small (relative to today) amount of fossil species is much more easily explained by the fossils representing mutually-contemporaneous life that was buried during the Noachian Deluge than by successive evolving populations over hundreds of millions of years. The paucity of fossil species indicates one population giving rise to all fossils, not countless populations.

9. Persistence of Drainage Patterns

Friedman and Sanders[62] wrote: "Two contrasting kinds of results have come from modern paleocurrent studies. On the one hand, slopes such as that which presently inclines southward toward the Gulf of Mexico from Minnesota, have maintained this orientation for at least several hundred million years. On the other hand, paleoslopes such as those inferred from the Cenozoic strata of the French Maritime Alps, have completely reversed inclination since the Pliocene Epoch. Presumably, such slope reversals have resulted from rapid subsidence to great depths of much of the Mediterranean Basin."

Comment: Owing to the ease of topographic changes on the earth, the lack of change of paleoslope over alleged several hundred million years in the continental interior of the US may be evidence favoring a very short period of deposition for these deposits rather than several hundred million years. Such constant paleocurrent direction is thus more in accord with a brief but global Flood than it is with geologic ages taking place on a constantly changing earth.

10. "Extra" Sections Within and Between Geologic Periods

Mintz[76] said: "The problem has always been complicated by the fact that the strata which are missing in the type area but found elsewhere are not part of the original definition of any of the units, and hence great arguments have arisen as to the time period to which they ought to be assigned. To this day geologists are still involved in the process of fixing the period boundaries."

Comment: Stratigraphic sections keep turning up which could theoretically be attributed to additional geologic periods between or within existing ones. However, Mintz[77] noted that geologic periods are arbitrary divisions of (supposed) geologic time, so the modern practice is to integrate these "extra" sections into existing geologic periods instead of erecting new ones. The appearance of "extra" sections indicates that stratigraphic order of fossils is more random than previously known, and that geologic periods are even less credible than before.

11. Significance of Alleged Overthrusts and "Wrong"-Order Strata

In a book on fossils, Spocyznska[78] wrote: "Certain fossils are found only in particular strata, and these latter are composed only of certain specific types of rock. These characteristic fossils are called 'indicator fossils' and, as this term implies, they act as pointers to a particular period of geologic time. It will be seen from this that if fossils A and B are fund only in beds of, say, Carboniferous age, should they turn up in combination with Permian rocks which date from a much later period, one must draw the conclusion that movements of the earth's crust have thrown these earlier beds up. The intrusion of Carboniferous fossils A and B among C and D which are found only in Permian strata is therefore a definite indicator of earth movements at some time or other."

Comment: This illustrates how overthrusts are assumed if fossil strata appears in "wrong" order. The statement about fossil mixing being a "definite indicator" of overthrust hints at lithologic and structural evidences being inconclusive; overthrusts accepted only because of "wrong" order. This point is made by many Creationists (for example, Read[79]).

12. Volcanic Evidence Against Geologic Time

In describing a situation from New South Wales, Australia, Dulhunty and McDougall[80] wrote: "In some places... the younger flows overlap the Garrawilla Lavas, making differentiation difficult... Petrographically, the lavas of the two age groups appear... to be remarkably similar, consisting of alkali olivine basalts, but the Mesozoic rocks tend to be somewhat altered."

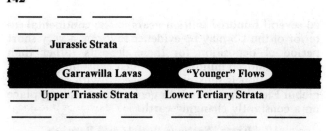

Figure 1. This diagram illustrates an excellent line of evidence against geologic time. Lower Jurassic lava flows (left) are nearly identical to nearby Tertiary flows (right). This suggests that both are really one contemporaneous extrusion of magma. The 120 million years of time which is supposed to separate Lower Jurassic from Tertiary is thus shown to be totally fictitious. (See Section VI, no. 12, for documentation.)

Comment: Two groups of lava flows, one Jurassic and the other Tertiary (see Figure 1) are nearly identical in composition. This argues for one lava flow only, drawing Jurassic and Tertiary into contemporaneity.

13. Possible Evidence Against Geomagnetic Reversals

The eminent British geophysicist Tarling[81] wrote: "It is now generally accepted that most of Europe and North America were contiguous from Devonian to Cretaceous-Tertiary times. On this basis, the paleomagnetic data of the two continents should also be consistent when the continents have been placed in their previous relationship. In fact there are serious, consistent differences between the paleomagnetic data on most acceptable reconstructions while reconstructions based on the paleomagnetic data alone do not result in viable continental reconstructions. Such observations suggest either that the paleomagnetic data still contain consistent errors, the geomagnetic model is wrong, or that the actual continental relationships were radically different to all extant models."

Comment: These grave difficulties in using paleomagnetic data may be evidence against the uniformitarian view that the earth's magnetic field has been constantly reversing itself and been in existence for millions of years over geologic ages. This would be in accordance with the work of Barnes[82], who proposed a short-lived non-reversing geomagnet and consequent young earth.

14. Skepticism Among Uniformitarian Geologists Towards Radiometric Dating

In a published discussion following their paper, Sabine and Watson[63] said: "Mr. Webster Smith ... regarded the atomic dating method (except in respect to carbon) as still very tentative especially where the older rocks were concerned and where discordant and even absurd results were quite common. There were records of granites which atomically were older than other granites that they intruded ... argon was all too prone to be either deficient, wholly absent, or even too high; in such cases the author 'adjusted' his figures."

Comment: There is some skepticism among uniformitarians towards radiometric dating; a point made and documented towards the end of the author's work[64] on radiometric geochronology.

15. Disregard For Radiometric Results Inconsistent With Others

Cahen and Snelling,[65] after discarding an anomalously old K-Ar result, said: "... the only reason for ... rejecting the age ... was that it was out of keeping with the three other apparent ages of biotites of the same Series ..."

Comment: This illustrates how discrepant results in radiometric dating are conveniently disregarded. In his work,[64] the author extensively documented the fact that dates are disregarded not only when they are in conflict with some favored value(s), but also when they conflict with accepted values for biostratigraphic positions.

16. Selective Publication of Dating Results

Bath[66] said: "Unpublished work by the author on Silurian shales from Pembrokeshire and the Welsh Borderlands has shown that such rocks can define isochrons giving ages significantly younger than the time of deposition adduced from faunal evidence."

Comment: Besides indicating that Rb-Sr isochrons from shales are often anomalous, this also illustrates that discrepant results are often (or usually) not published; two points documented previously.[64]

17. Failure of Radiometric Dating in Precambrian Stratigraphic Studies

After describing one group of geologists who uncritically accepted radiometric dating, Salop[67] said: "... the other group, in the face of many discrepancies between radiometric and geological data, tended to reject this technique in stratigraphic studies, or accepted it with great caution."

Comment: The fact that Pre Cambrian Dating results are often very contradictory has been discussed towards the end of the author's work.[64] This statement by Salop illustrates the consequences of such erratic dates.

18. The K-Ar Isochron Method Has Its Own Problems

Shafiqullah and Damon[68] said: "The $^{40}Ar/^{36}Ar$ vs. $^{40}K/^{36}Ar$ isochrons are valid only when all samples of the system under consideration have the same nonradiogenic argon composition. If this condition does not hold, invalid ages and intercepts are obtained. Models 2-9 yield isochron ages that are too high, too low, or in the future, sometimes by orders of magnitude."

Comment: It is often claimed that the K-Ar isochron method is superior to the conventional K-Ar method because the former measures, not assumes, the initial $^{40}Ar/^{36}Ar$ ratio, and can overcome and "excess argon" problem. This statement indicates that isochrons can form that are discrepant and absurd, and—like conventional results—are subject to open-system rationalizations whenever discrepant and unwanted results are obtained. The Bourinot Group (volcanics; Cambrian, ref. 372) in Table 1 of the author's work[64] is an example of an isochron that is "too young," and this is attributed to thermally-induced argon loss with homogenization of the remainder during the alleged heating event on the rock.

VII. ARCHEOLOGY AND PREHISTORY

1. Fallacy in Dating of Prehistoric Man and His Alleged Cultural-Technical Evolution

Brown[69] said: "In archeology it is now realized, despite long resistance, that dating and classification by means of technical typology, for example by stone tools, is no longer possible in many cases. The Acheulian stone industries of Africa are possibly as old as 1.4 Ma, but in Atlantic and Mecditerranean Europe they are rare until mid-way through the Brunhes Epoch."

Comment: This finding is very much in line with the work of Creationists who long insisted that technical typology is not an evolutionary stage in man but a progressive restoration of technology among Noah's immediate descendents in the first few centuries after the Flood. The difficulties in using technical typology for dating reflects the fallacious evolutionary premises that hold to human evolution.

References

AG —American Association of Petroleum Geologists Bulletin
CA —Cambridge University Press
CR —Creation Research Society Quarterly
EC —Palaeogeography, Palaeoclimatology, Palaeoecology
EL —Elsevier Scientific Publishing Company, Amsterdam, New York
ES —Earth Science Reviews
GA —Geological Society of America Abstracts with Programs
GL —Journal of Geological Society of London
JG —Journal of Geology
JP —Journal of Paleontology
LE —Lethaia
SP —Journal of Sedimentary Petrology
UN —Transactions of the American Geophysical Union

[1] Woodford A. 1976 *Teleology*, CA, p. 3
[2] Towe K. M. 1970. Oxygen-Collagen Priorty and the Early Metazoan Fossil Record. *Proceedings of the National Academy of Science* 65(4)781
[3] Riedl R. 1977. A Systems-Analytical Approach to Macro evolutionary Phenomena. *Quarterly Review of Biology* 52(4)353
[4] Woodmorappe J. 1978. The Cephalopods in the Creation and the Universal Fossil Deluge. *CR* 15(s) 98-100
[5] Salvadori F. B. and P. L. Florio. 1978 (English Edition) *Rare and Beautiful Animals*. Newsweek Books, New York, p. 62
[6] Caplovitz D. and F. Sherrow. 1977. *The Religious Drop-Outs*. Sage Library of Social Research 44:57 (Sage Press, Beverly Hills, London)
[7] Lewis C. J. T. 1975. Review of *A Source Book in medieval Science* by E. Grant. *Nature* 254:465
[8] Campbell I. 1972. Search for the Phiosophers Stone. *GA*(7)467
[9] Woodmorappe J. 1978. A Diluvian Interpretation of Ancient Cyclic Sedimentation. *CR* 14(4)191-2
[10] Lyell C. 18? Evolution (in Thatcher O.j. 1902. *The Ideas that have Influenced Civilization in The Original Documents*. IX:231
[11] B. J. S. 1973. *Review of: Charles Lyell: The Years to 1841: The Revolution in Geology* by L. G. Wilson. *Economic Geology* 68(4)581
[12] Dietrich R. V. 1970. *Geology and Virginia*. University Press of Virginia, Charlottesville, p. 159.
[13] Glikson A. Y. 1979. Early Precambrian Tonalite-Trondhjemite Sialic Nuclei. *ES* 15(1)2
[14] Wills C. j. 19970. The Triassic succession in the central Midlands. *GL* 1265:287
[15] Woodmorappe, 1978. (cephalopods) *op. ct.*, p. 101
[16] Cameron B. and K. Estes. 1971. Fossil and Recent "Tadpole Nests" A Discussion. *SP* 41:171-8
[17] D. Raup. 1979., Chicago: Field Museum of Natural History. Curator. Personal Communicatation.
[18] Sandberg C. A. 1977. Conodont Time Warps-Dilemma for Conodont Dating and their possible Solutions. *GA* 9:648-9
[19] Vanguestaine M. 1979. Remainiements D'Acritarches Dans Le Siegenien Et L'Emsien (Devonien Inferieur) Du Synclinorium De Dinant (Belgique). *Annales de la Societe Geologique de Belique* Tome 101:243
[20] Woodmorappe J. 1978. (cyclic sedimentation) op. cit., p. 192.
[21] Woodmorappe JK. 1978 (cephalopods) *op. cit.*, p. 102-103-104
[22] Mildenhall D. C. and G. J. Wilson. 1978. redeposited Lower and Upper Cretaceous palynomorphs from the Mangere Formation, Mangere Island, Chatham Islands, New Zealand. *New Zealand Journal of Geology and Geophysics* 21(5)661-2
[23] North F. K. 1964. The Geological Time-Scale. *Royal Society of Canada Special Publication* 8:7
[24] Bretsky S. S. 1969. Disparity Between Phenetic and Phylogenetic Classifications at Supraspecific Levels. *GA* 1(7)19
[25] Woodmorappe. 1978. (Cephalopods) op. cit., p.100
[26] North. 1964. *op. cit.*, p.5
[27] Krassilov V. 1978. Organic Evolution and natural stratigraphic classification. *LE* 11(2)101
[28] *ibid*, p. 100
[29] Johnston W. G. Q. 1969. Pillow Lava and Pahoehoe: A Discussion. *JG* 77:730
[30] Nininger H. H. 1979 (Book Review) *ES* 14(4)363
[31] Moore J. G. 1975. Mechanism for Formation of Pillow Lava. *American Scientist* 63:269
[32] Sawkins F. J., Chase C. G., Darby D. G., and G. Rapp Jr. 1978. *The Evolving Earth*. Macmillan and Co., New York, London, p. 245-6
[33] Milici R. C., Briggs G., Knox L. M., Sitterly P. D., and A. T. Stattler. 1979. The Mississippian and Pennsylvanian (Carboniferous) Systems in the United States-Tennessee. *United States Geological Survey Professional Paper* 1110-G32-4
[34] McKerrow W. S. 1978. *The Ecology of Fossils* MIT Press, CA, p. 287
[35] Meynen S. V. 1969. The Continental Drift Hypotheses in the Light of Carboniferous and Permian Paleoflora. *Geotectonics* 4:290
[36] Braithwaite C.J. R. 1973. Reefs: Just a Problem of Semantics. *AG* 57(6)1106
[37] Nevins S. (formerly-used pseudonym for Dr. Steven Austin). 1972. Is The Capitan Limestone a Fossil Reef? *CR* 8(4)231-248
[38] Rigby j. K. 1969. Reefs and Reef Environments. *AG* 53:738
[39] Philcox M. E. 1970. Reefs and Wave Action. *AG* 54:864
[40] Wilson J. L. 1970. Facies in Organic Buildups in Epeiric Seas and Along Shelf Margins *AG* 54:875
[41] Lowenstam H. 1957. Niagaran Reefs in the Great Lakes Area. *Geological Society of America Memoir* 67:247
[42] Wilson J. L. 1975. *Carbonate Facies in Geologic History* Springer-Verlag, Berlin, Heidelberg, New York, p. 47
[43] Friedman G. M. and J. E. Sanders. 1978. *Principles of Sedimentology* John Wiley and Sons, New York, Toronto, Brisbane, p. 196
[44] Suttner. L. J. 1979. Recent Trends and Advances in Sedimentology. *Journal of Geological Education* 27:17
[45] Harbaugh J. W. 1970. Computer Simulation of Marine Sedimentation. *AG* 54:555
[46] Adams J. and J. Patton. 1979. Sabkha-Dune Deposition in The Lyons Formation (Permian) Northern Front Range, Colorado. *Mountain Geologist* 16(2)47
[47] Woodmorappe. 1978. (cyclic sedimentation) *op. cit.*, p. 197-200
[48] Lundegard P. D., Samuels N. D., and W. A. Pryor. 1979. Upper Devonian Turbidites of Central and Southern Appalachian Basin-A Prodeltaic Clastic Ramp. *AG* 63:488-9
[49] Newell N. D. 1967. Paraconformities (in Teichert C. and E. L. Yochelson. 1967. Essays in Paleontology and Stratigraphy. *University of Kansas Special Publication* 2) p. 357
[50] Swift D. J. P., Stanley D. J., and J. R. Curray. 1971. Relict Sediments on Continental Shelves: A Rconsideration. *JG* 79:343
[51] Chamberlain C. K. 1975. Trace Fossils in DS DP Cores of the Pacific *JP* 49(6(1074-96
[52] Seilacher A. 1967. Bathymetry of Trace Fossils. *Marine Geology* 5:418
[53] McLoughlin J. C. 1979. *Archosauria: A New Look at the Old Dinosaur* Viking Press, New York. p. 1
[54] Newell. 1967. op. cit., p. 355
[55] Whitcomb J. C. J. and H. M. Morris. 1961 *The Genesis Flood*. Baker Book House, Michigan, USA, pp. 207-211
[56] Wilson. 1975. op. cit., p. 16
[57] _____ 1974. *The Earth's Physical Resources* Block 3: Mineral Deposits. Open University Press, Milton Keynes, England, p. 48
[58] Ansich W. I. and r. A. Gurrola. 1979. Two Boring Organisms In Lower Mississippian Community of Southern Indiana. *JP* 53:335
[59] Rogers J. 1971. The Taconic Orogeny. *Geological Society of America Bulletin* 82:1141
[60] Hewitt R. A. and J. M. Hurst. 1977. Size Changes in Jurassic liparoceratid ammonites and their stratigraphical and ecological significance. *LE* 10(4)300
[61] Raup D.M. 1979. Conflicts Between Darwin and Paleontology. *Field Museum of Natural History Bulletin* January 1979, p. 22
[62] Friedman and Sanders. 1978. op. cit., p. 415

63. Sabine P. A. and J. Watson. 1965. (Introduction to) Isotopic age-Determinations of Rocks from the British Isles, 1955-64. *GL* 12:525
64. Woodmorappe J. 1979. Radiometric Geochronology Reappraised. *CR* 16(2)102-129, 147
65. Cahen L. and N. J. Snelling. 1966. *Geochronology of Equatorial Africa*. North Holland Publishing Co, Amsterdam, p. 141
66. Bath A. H. 1974. new isotopic age data on rocks from the Long Mynd, Shropshire. *GL* 130:570
67. Salop L. J. 1977. *Precambrian of the Northern Hemisphere*. EL, p. 3
68. Shafiqullah M. and P. E. Damon. 1974. Evaluation of K-Ar Isochron methods. *Geochimica et Cosmochimica Acta* 38:1355-6
69. Bowen D. A. 1978. *Quarterly Geology*. Pergamon Press, Oxford, New York, Toronto, p. 193
70. Von Herzen R. P. 1979. Reply to Tatsch. *UN* 60(34)626
71. Sarjeant W. A. J. 1974. History and Bibliography of theStudy of Fossil Vertebrate Footprints In the British Isles. *EC* 16:328, 330
72. *ibid.*, p. 325
73. Bond G. and K. Bromley. 1970. Sediments with the Remains of Dinosaurs near Gokwe, Rhodesia. *EC* 8:320
74. Kumar N. and J. F. Sanders. 1976. Characteristics of Shoreface Storm Deposits: Modern and Ancient Examples. *SP* 46(1)146
75. Seilacher. 1967 *op. cit.*, p.416
76. Mintz L. W. 1972. *Historical Geology*. Charles Merill Pub. Co., Ohio.
77. *ibid.*, p. 32-3
78. Spoczynska J. O. I. 1971. *Fossils: A Study in Evolution*. Rowman and Littlefield, Totowa, New Jersey, p. 13-14
79. Read J. G. 1979. *Fosils, Strata and Evolution*. Scientific-Technical Presentations, California, 64 p.
80. Dulhunty J. A. and I. McDougall. 1966. Potassium-Argon Dating of Basalts in the Coonabarabran-Guddenah District, New South Wales. *Australian Journal of Science* 28(10)393-4
81. Tarling D. H. 1979. The Reliability of Phanerozoic Paleomagnetic data from Europe. *UN* 60(32)569
82. Barnes T. G. 1973. *Origin and Destiny of the Earth's Magnetic Field* ICR Technical Monograph No. 4, 64 P.

Radiometric Geochronology Reappraised

RADIOMETRIC GEOCHRONOLOGY REAPPRAISED

JOHN WOODMORAPPE*

Received 11 December 1978

The use of radiometric dating in Geology involves a very selective acceptance of data. Discrepant dates, attributed to open systems, may instead be evidence against the validity of radiometric dating.

A systematic and critical review of dating applications is presented; emphasis being placed on the geologic column. Over 300 serious discrepancies are tabulated. It is, however, demonstrated that most discrepant results are not published. Discrepant dates capriciously relate to petrography and regional geology.

Neither internal consistencies, mineral-pair concordances, nor agreements between different dating methods necessarily validate radiometric dating.

The large spread of values for igneous and metamorphic rocks (especially of the Precambrian) may indicate artificial imposition of time-values upon these rocks.

Plan of this Article

Introduction
I. Phanerozoic Geochronology: Selection Amidst Contradictions.
  A. Introduction to Phanerozoic Geochronology
  B. The Selective Publication of Dating Results
  C. Direct Radiometric Dating of Sedimentary Rocks
  D. Open-System Rationalizations for Discrepant Igneous Dates As Deduced from Petrography and Regional Geology
  E. Radiometric Dating of Biostratigraphically-Bracketed Igneous Rocks
    1. Effusives (Tuffs and Bentonites) and Extrusives (Lava Flows)
    2. Intrusives (Hypabyssal and Plutonic Rocks)
II. Highlights of Precambrian and Non-Biostratigraphic geochronology.
  A. Consistency and Concordances Among Radiometric Dates
  B. Radiometric Violations of Superpositional and Cross-Cutting Relationships
  C. Age Values for Igneous and Metamorphic Terranes

Introduction

Radiometric dating is supposed to confirm and quantify the evolutionary-uniformitarian claims of geologic ages with biological evolution. All speculations of a legendary Flood, a local Flood, or a tranquil Flood placed aside, both evolutionist-uniformitarians and Creationist-Diluvialists will agree that there is no evidence of a truly global Flood if uniformitarian time scales are accepted in any way. Once divested of all the time claims imposed upon it, the fossiliferous rock testifies to the Noachian Deluge, and all life (fossil and extant) is then mutually contemporaneous as is demanded by a literal six (24 hr.) day Creation. Therefore acceptance of radiometric dating abrogates not merely a young earth, but also the six-day Creation and the Universal Deluge.

Creationist-Diluvialists have uncovered dozens of powerful evidences for an Earth age of only several thousand years, and these scientific evidences are summarized by Morris.[212] Reasons for questioning the validity of radiometric dating are presented in the major works by Whitcomb and Morris,[213] Cook,[214] Slusher,[215] and Wilkerson.[216] These works center on the many unproven assumptions behind radiometric dating. This work, by contrast, seeks critically to evaluate the claims of radiometric dating via a geological approach; the author believing that dating is best understood in its geologic context.

The intended theme of this work is best portrayed by the following statements by Goldich:[217] "Fifteen years ago, radiometric age determinations on minerals and rocks were so startling that 'absolute age' became a password. Intensive research with successive improvements in the K-Ar, Rb-Sr, and U-Pb methods, however, revealed that geologic processes influence isotopic systems and that the age measurements are analytical values that commonly require geological interpretation." Thus open systems caused by geologic events ("geological interpretation") is claimed as a cause of spurious dates, but a less self-congratulatory view would be that only a *select* number of dates are accepted as indicating the true age of the rock; others explained away as having become open systems.

Owing to the breadth of the topic of radiometric dating, this work will not consider any extraterrestrial dating, nor marine dating, and little Pleistocene dating. Attention will be focused upon the dating of biostratigraphically-defined materials in the Phanerozoic. The dating of the fossil record is thus emphasized over the Pre-Cambrian because it is the Phanerozoic which is a pivotal point in both the evolutionist-uniformitarian and Creationist-Diluvialist paradigms, and because that is where radiometric dating can be compared with fossil dating.

I. Phanerozoic Geochronology: Selection Amidst Contradictions

A. Introduction to Phanerozoic Geochronology

The values given for geologic periods in terms of the time come from dated material that is biostratigraphically defined. It will become obvious that actually grossly contradictory results are obtained from materials of the same geologic periods and that some values are accepted as true, while others are explained away.

It can not be said that discrepancies are primarily caused by the poorer analytical equipment available in the late 1950's/early 1960's as contrasted with that of the late 1970's. Recently, Waterhouse[315] said: "Improved laboratory techniques and improved constants have not reduced the scatter in recent years. Instead the uncertainty grows as more and more data is accumulated . . . ."

Table 1 is a compilation of over 300 different sets of dates that are in gross conflict with one another and with expected values for their indicated paleontological positions. How unwanted and discrepant dates are ra-

---
*John Woodmorappe, B.A., has studied both Biology and Geology.

Table 1. The phanerozoic time scale: major inconsistencies. This is a compilation of over 350 different radiometric dates that are very anomalous with respect to accepted values for their biostratigraphic positions. At left are the values for geological periods (and subdivisions of them) in millions of years, taken from radiodates that are considered to be correct. The numerical values for the subdivisions of the time scale are those of Braziunas.[211] To the right of them are given the discrepant results. How these dates are explained away is extensively discussed in the text.

The asterisk (*) denotes a date from a biostratigraphically-bracketed igneous body; non-asterisk items are direct dates on sedimentary rock. Dates for sedimentary rocks are obtained from separated authigenic minerals: g-glauconite; il-illite. Directly dating fossils (ca-calcite) is not considered to be reliable (only 2 examples are given) and is hardly ever done. Dates from igneous rocks are obtained from entire samples (wr- whole-rock; wri- whole rock isochron) or from mineral separates (mi-mineral isochron; b-biotite; s-sanidine; l-lepidolite; m-muscovite; p-phlogopite; h-hornblende; arf-arfvedsonite; z-zircon; mo-monzanite; ch-chevkinite; hu-hutchettolite; gl-undevitrified volcanic glass (1 example only given); pl-plagioclase feldspar). Information not given is denoted: ng.

Age Expected	Age Obtained	Method & Material	Common or Formational Name/Locality	Reference
			TERTIARY	
1	81	K-Ar g	sandstone/Yenisei, USSR	1
5	1-10.6	K-Ar s, b	*Bailey Ash/California, USA	2
9.5	13-31	K-Ar s	*tuff/Nevada, USA	322
10	95	K-Ar wr	*basalt/Nigeria	3
10	153±10	K-Ar h	*Nogales Fm. (tuff)/Arizona, USA	4
11	6.4-7.6 / 7.2	K-Ar ng / Rb-Sr mi	*Mount Capanne granodiorite/Elba Island, Italy	5
12	30-36	K-Ar g	ng/ng	6
23	3.4	K-Ar wr	*Suta Volcanics/Solomon Islands	391
<25	30-40	Rb-Sr wri	*volcanics/Saudi Arabia	7
27	31-43	K-Ar wr	*Velolnyk suite (volcanics)/Kamchatka, USSR	392
30	21.6	K-Ar s	*tuff/Oregon, USA	8
38	21-2	K-Ar g	sediments/New Zealand	9
42	18-36	K-Ar g	sediment/California, USA	10
42	31	K-Ar g	sediment/Hordhorn, W. Germany	11
47	24-153	K-Ar b	*tuffs/Wyoming, USA	12
52	39	K-Ar g	Winona sand/Gulf Coast, USA	13
55	200-280	Pb^{207}/Pb^{206}	ores/Wyoming, USA	393
60	38	K-Ar g	ng/Gulf Coast, USA	13
60	35-110 / 42-113 / 114-430	Pb^{206}/U^{238} / Pb^{207}/U^{235} / Pb^{207}/Pb^{206}	Front Range ores/Colorado, USA	14
<65	49-271	K-Ar wr	*Mafic Dykes/Mull, Scotland	323
65	46	K-Ar b	*Ruby Star Granodiorite/Arizona, USA	15
65	46	K-Ar g	sediment/California, USA	16
65	260±54	Rb-Sr wri	*Loch Uisg Granophyre/Scotland	324
>65	42-48	K-Ar wri	*andesite/British Columbia, Canada	17
<70	500	Rb-Sr wri	*basalts/Patagonia, Argentina	394
<70	494±20 / 756±80 / 794±50	Pb^{206}/U^{238} z / Pb^{207}/U^{235} z / Pb^{208}/Th^{232} z	*granites/Utah, USA	18
<70	2750 / 300	Rb-Sr wri / Pb^{207}-Pb^{206}	*volcanics/Inner Hebrides, Scotland	18
<70	90-1330	Rb-Sr wri	*basalts/western USA	20
<70	290, 400	Rb-Sr wri	*andesite/Peru	20

CRETACEOUS

70	103	K-Ar ng	*Brazeau FM. (bentonite)/Alberta, Canada	21
>70	9-31	Rb-Sr b	*Mandi granite/Manikaran, India	395
72	12-21 20	K-Ar ng Rb-Sr wri	*Pamir-Shugnan granites/India	22
74	104±8	K-Ar g	Tinton Sands/New Jersey, USA	23
>74	47	K-Ar h	*Ginger Ridge granodiorite/Jamaica	24
<75	490-770	Pb^{207}/Pb^{206}	ore/Nigeria	325
<75	120, 481	K-Ar h	*diorite/Colorado, USA	60
75	142	K-Ar p	*kimberlite/Orange Free State	25
77	105±3	K-Ar g	Woodbury Fm./New Jersey, USA	23
>82	45, 55	K-Ar wr	*Mifune Group (tuffs)/Kyusha, Japan	326
~90	~250	Rb-Sr wri-il	Pierre Shale/South Dakota, USA	26
96	44, 76-8	K-Ar g	sediment/Brezina, Czechoslovakia	27
96	66-116	K-Ar g	sandstone/Georgia, USSR	28
99	42 59-82	K-Ar il K-Ar g	*Viking Fm. (bentonite)/Alberta, Canada	29
100	142±10	K-Ar g	ng/ng	6
100	68-73	U^{238}/Pb^{206}	*Mbeya Carbonatite/Tanganyika	30
100	170-8	K-Ar il	clays/Texas, USA	31
100	22-1343	K-Ar b, h, wr	*lamprophyre dikes/New Zealand	396
101	75-115	K-Ar g	Clearwater Fm./Alberta, Canada	32
101	70	K-Ar g	sediment/north Caucasus, USSR	27
>102	72-143	K-Ar b	*Eagle Granodiorite/British Columbia, Canada	33
105	70-80	K-Ar b	Santa Lucia Plutons/California, USA	327
>105	61	K-Ar b	*Tres Guanos Quartz Monzonite/Cuba	34
>112	58-143	K-Ar b, h	*Peninsular Ranges Batholith/California, USA-Mexico	35
115	31	K-Ar g	sediment/Salzgitter, W. Germany	36
118	88	K-Ar g	sediment/Esciagnelles, France	37
~120	60-68, 129-150	K-Ar wr	*trachybasaltic dikes/Cordoba, Argentina	328
<120	241	K-Ar wr	*diabase dyke/Franklin, Canada	38
>122	71-149	K-Ar wr	*Isfjorden diabases/Advantdalen, Norway	329
<130	235-382	Pb^{207}/Pb^{206} z	*granites/Japan	39
<130	500	Rb-Sr wri	*volcanics/James Ross Island, Antarctica	40
~130	70 70	K-Ar wr Rb-Sr wri	*volcanics/Chile	330

JURASSIC

±130	74±5	Rb-Sr wri	*Serie Tobifera volcanics/Chile	331
>130	74	K-Ar wr	*Rock Hill Basalt/Nevada, USA	41
>130	95, 120	K-Ar wr	*La Teta Lava/Colombia	332
<140	163, 186	K-Ar ng	*Coast Range Batholith/Alaska, USA	42
140	275±20	K-Ar h	*Pearse Peak Diorite/Oregon, USA	43
~140	435-780	Rb-Sr il	Shale/Great Basin, Australia	333

140	105-171	K-Ar b, h	*Nelson Batholith/British Columbia, Canada	44,
	104-150	Rb-Sr wri		397
140	95-160	K-Ar b, h	*Wallowa Batholith/Oregon, USA	45
140	50	K-Ar wr	*basalt/Caucasus Mts., USSR	398
>140	3-23	Rb-Sr m, b	*Mt. Bukulia Granite/Yugoslavia	334
145	106, 117	K-Ar g	sediment/Milne Land, Greenland	32
145	77-180	K-Ar ca	Malm Limestones/Bavaria, W. Germany	46
148	86-118	K-Ar g	Fernie Fm./Alberta, Canada	47
>150	114-78	K-Ar wr	*Apoteri volcanics/Guyana-Brazil	48
152	26	K-Ar g	sediment/Braunschweig, W. Germany	49
>153	67-240	K-Ar b, h, m	*granites/West Malaysia	50
>153	110	Rb-Sr mi	*pegmatite/New Zealand	399
~155	61±2	K-Ar h	*Novatak Glacier pluton/Alaska, USA	335
~155	1020±320	Rb-Sr wri	*volcanics/Dronning Maud Land, Antarctica	51
157	92-145	K-Ar wr	*volcanics/Zuidwal, Netherlands	336
160	140-165	K-Ar b	*Carmel Fm. (bentonite)/Utah, USA	52
	85-163	Rb-Sr b		
160	70-109	K-Ar wr	*Rajmahal Traps (lavas)/India	400
~160	78	K-Ar b	*Chocolate Fm. (volcanic breccia)/Peru	337
165	106±6	K-Ar b	*granodiorite/British Columbia, Canada	53
165	21	K-Ar g	sediment/Coston Del Vette, Italy	54
165	63, 138-178	K-Ar b	*Topley intrusions/British Columbia, Canada	55
165	176, 228	Rb-Sr wri-il.	Bedford Canyon Fm. (shale)/California, USA	425
165	109-165	K-Ar wr	*Volcanics/North Sea, England	426
<170	7-250	K-Ar wr	*Basalts/Victoria Land, Antarctica	427
<170	223	K-Ar b	*granodiorite/Yukon, Canada	56
170	78-90	K-Ar b, h	*Chigmit Mountains Batholith/Alaska, USA	57
~175	265	Rb-Sr wri	*rhyodacite/California, USA	338
178	134-76	K-Ar wr	*Kirkpatrick Basalt/Victoria Land, Antarctica	58
<180	296-302	K-Ar wr	*dolerite/Midlands, England	59
<180	2661	Pb^{206}/U^{238} z	*Independence dikes/California, USA	401
	2774	Pb^{207}/U^{235} z		
	2860	Pb^{207}/Pb^{206} z		
~180	70, 120-250	K-Ar b	*granite porphyry, pegmatite/Caucasus Mts., USSR	61

TRIASSIC

~180	120-25	K-Ar wr	*Serra Geral Fm. (lavas, diabase)/Brazil	339
180	105-175	K-Ar b	*granite/Billiton, Indonesia	62
180	170-265	K-Ar b, h	*Guichon Creek Batholith/British Columbia, Canada	63
<185	244±15	K-Ar wr	*mafic dike/Idaho, USA	64
185	291	K-Ar ng	micaceous sandstone/Nilgiri, India	65
185	137-219	K-Ar wr	*Karroo volcanics/Lesotho	66
~185	186-1230	K-Ar pl, wr	*diabase dikes/Liberia	402
<190	286	K-Ar h	*Tulameen ultramafic complex/British Columbia, Canada	67
~190	79, 140	K-Ar wr	*Watchung Basalt/New Jersey, USA	68

~190	151-201	K-Ar wr	*lavas/Connecticut, USA	69
~190	178-217, 328±24	K-Ar wr	*North Mountain Basalt/Nova Scotia, Canada	70
~190	80	K-Ar b	*Mt. St. Elias pluton/Alaska, USA	340
~190	139-217	K-Ar b	*Hotailuh Batholith/British Columbia, Canada	71
~190	89-199 104, 144	K-Ar b, h Rb-Sr wri	*Klotassin Batholith/Yukon, Canada	72
<195	152-235	K-Ar wr	*Foum Zguid dolerite, lavas/Morocco	73
195	126	K-Ar gl	*tuff/Alaska, USA	74
<200	246±10	K-Ar p	*kimberlite/Siberia, USSR	428
<	170, 259	Pb^{206}/U^{238} z	*diorite/California, USA	429
<200	212-278	K-Ar wr	*gabbroids/Tien Shan, USSR	430
~200	102-110	K-Ar wr	*diabase dikes/Arctic Canada	431
~200	100-220	K-ar ng	intrusives/New Zealand	432
200	118±10	K-Ar m	*pegmatite/Neyriz, Iran	433
~200	61-181	K-Ar wr	*volcanics/Othris, Greece	434
~200	98-107	K-Ar b	*granites, quartz monzonites/British Columbia, Canada	75
200	43-188	K-Ar ca	bone/Nordwurttemberg, W. Germany	46
200	20-218 295±20	Pb^{207}/U^{235}	Pitchblende, Chinle Fm./Arizona, USA	76
<204	270±45 360	Rb-Sr wri	*Ferrar Dolerite/Victoria Land, Antarctica	77
<205	363±15	Rb-Sr wri	*Inas Granite/Malaysia	78
~210	114-214	K-Ar ng	shale/Nilgiri, India	65
~210	175-237	Rb-Sr wri	*Predazzo granite/Dolomites, Italy	79
~220	131±5	K-Ar wr	*rhyolite/Peru	341
>225	188-270	K-Ar wr	*Korvunchana Series (basalts, tuffs, ejecta)/Tunguska, USSR	80
230	231-280	K-Ar wr	*Semeitau lavas/Kazakhstan, USSR	81

PERMIAN

230	241-75	K-Ar h	*Yakuno gabbros/Japan	342
>230	71-118	Pb^{207}/Pb^{206}	ore/Lenterios, Portugal	82
>230	97 111	Pb^{207}/U^{235} Pb^{206}/U^{238}	ore/Wittichen, W. Germany	82
>230	146-57	Pb^{207}/Pb^{206}	ore/La Crouzille, France	82
>230	87	K-Ar b	*granite/Kocaeli, Turkey	83
240	165	K-Ar g	limestone/Sokolka, USSR	84
240	173-243	Rb-Sr wri-il	Estrada Nova Fm. (shale)/Brazil	343
~240	175	K-Ar g	sediment/Kirovsk, USSR	85
240	310±70	Pb^{207}/Pb^{206}	ore/Thickley, Scotland	86
<250	346	K-Ar ng	*granite/Peru	87
~250	182	K-Ar wr	*trachyandesite lava/Caucasus Mts., USSR	85
~250	337±61	Pb^{207}/Pb^{206}	Jachymov Pitchblende/Bohemia, Czechoslovakia	88
250	165-263	K-Ar wr	*rhyolite, tuff/Zechstein, E. Germany	344

250	136-245	Rb-Sr b	*volcanics/New Zealand	89
~250	96±6	K-Ar b	*lavas/Oman, Iran	435
~250	1380	Rb-Sr wri	*Croydon Volcanics/Queensland, Australia	436
>250	130-182	K-Ar wr	*volcanics/Auk Field, England	437
>255	209	K-Ar b	*Mitaki Granite/Japan	90
<260	330-380	K-Ar p	*Stockdale kimberlite pipes/Kansas, USA	91
<260	294-351	Rb-Sr wri	*Nychum volcanics/Queensland, Australia	345
~260	165	K-Ar g	sediment/Vestspitsbergen, Norway	92
260	259-315 216	K-Ar b Th^{232}/Pb^{208}	*Oslo Series (subvolcanics)/Norway	93
260	219-280	Rb-Sr wri	*Barhalde Granite/Schwarzewald, W. Germany	94
~265	363	Rb-Sr il	Stearns Shale/Kansas, USA	95
265	193±7 230±11	K-Ar wr K-Ar b	*Filipowice Tuff/Krakow, Poland	96
265	252-374 164-595	K-Ar il Rb-Sr il	Eskridge, Stearns Shale/Kansas, USA	97
265	229-74	K-Ar wr	*Lizzie Creek Volcanics/Queensland, Australia	98
~265	245±220	Rb-Sr mi	*Donnersberg Rhyolite/Saar, W. Germany	99
<270	2700	U-Pb z	*Diorite dike/California, USA	100
<270	368±18	K-Ar b	*peridotite/Pennsylvania, USA	101
<270	1180±60	Rb-Sr wri	*Rose Dome Granite/Kansas, USA	102
<270	831	K-Ar h	*Tortilla Quartz Diorite/Arizona, USA	346
~270	280-300, 390	Rb-Sr wri	*Pine Mountain Granite/British Honduras	103
270	155	K-Ar g	sediment/ng	149

CARBONIFEROUS

>270	150-280	K-Ar b	*granites/northern Italy	104
>270	120-280	K-Ar b	*granites/Zabaikal, Mongolia	105
>270	157, 194 162, 220	Pb^{206}/U^{238} z Pb^{207}/U^{235} z	*granite/Switzerland	403
280	255, 859	Rb-Sr wri-il	Madera Fm. (shale)/New Mexico, USA	404
280	330-441	Rb-Sr il	underclays/Illinois, USA	405
280	190	K-Ar wr	*basalt, gabbro/Franklin, Canada	406
280	385	Pb^{207}/Pb^{206} z	*Monti Orfano granite/Italy	407
<285	305-475	Pb^{207}/Pb^{206}	ores/Limburg, Netherlands	106
290	220±25	Rb-Sr wri	*Wamsutta Fm. (rhyolite)/Massachusetts, USA	107
290	70-140 85	Pb^{207}/Pb^{206} K-Ar wr	*Mrzyglod diabase/Krakow, Poland	108
<300	390	U-Pb z	*porphyritic intrusions/British Columbia, Canada	109
300	139-315	K-Ar wr	*Whin Sill (diabase)/Northumberland, England	110
300	233	K-Ar il	shale/Missouri, USA	111
300	114-385	K-Ar il	shale/Iowa, USA	111
~300	85	K-Ar wr	*tuffs/Andscollo, Argentina	112
300	318-456 338-511	K-Ar il Rb-Sr il	underclays/Pennsylvania-Ohio, USA	113

300	230-55	K-Ar g	Veraya Tier sediments/Kuibyshev, USSR	114
>300	220-354	K-Ar ng	*Old Crow Batholith/Yukon, Canada	115
>300	20-350	K-Ar m, b	*granites/Caucasus Mts., USSR	116
~310	185-315	K-Ar wr	*Toadstone Lava/Derbyshire, England	117
310	162-262	K-Ar wr	*volcanics/Silesia, Poland	408
310	145	K-Ar b	*granite/Caucasus Mts., USSR	347
310	282-367	K-Ar wr	*basalt/Quebec, Canada	118
310	238	K-Ar wr	*felsite lava/Primorye, USSR	85
320	266	K-Ar wr	*porphyritic lava/Kara Mazar, USSR	85
>320	245-330	K-Ar arf., s	*Han Bogdo Batholith/Mongolia	119
330	267-312	K-Ar wr	*Waterswallows Sill/Derbyshire, England	120
>330	265±7	Pb^{206}/U^{238} z	*Sicker volcanics/British Columbia, Canada	121
	263±7	Pb^{207}/U^{235} z		
	244±20	Pb^{207}/Pb^{206} z		
>330	225	K-Ar b	*Vallorcine Granite/Aiguilles Rouges, Switzerland	122
334	250-78	K-Ar g	shale/Texas, USA	123
<335	420	Rb-Sr wri	*granites/Montagne Noire, France	124
~340	126±30	K-Ar wr	*pumice tuff/Nottinghamshire, England	117
<340	440±60	Pb^{207}/Pb^{206}	*Vosges granites/Col de Grosse Pierre, France	125
<340	780	Pb^{207}/U^{235} z	*granite/Saxony, E. Germany	126
	2500	Pb^{207}/Pb^{206} z		
<340	347-500	K-Ar b	*granodiorites/Balkhash, USSR	409
340	247	K-Ar w	*dacitic lava/Aral, USSR	410
340	240-330	K-Ar ng	*granites/east Ural Mts., USSR	127
>342	268-373	Rb-Sr wri	*Nictaux granites/Nova Scotia, Canada	128
343	288, 307	K-Ar b	*Kuttung lavas/New South Wales, Australia	348
~345	154	K-Ar il	Rocky Mountain Fm./Alberta, Canada	47
>345	165	K-Ar ng	arkosic sandstone/Alexander I Land, Antarctica	129
<350	240-1630	K-Ar wr	*diabases/Georgia, USA	356
<350	400-453	K-Ar wr	*kimberlites/Siberia, USSR	411
350	222-284	K-Ar s	*Exshaw Fm. (bentonite)/Alberta, Canada	130
350	315±7	Rb-Sr wri	*tuffs/Aljustrel, Portugal	349

DEVONIAN

~350	240-330	K-Ar b	*granites/Aral, USSR	131
>350	242±10	K-Ar b	*granite/Langkawi Island, Malaysia	50
	218±11	Rb-Sr wri		
>350	275±11	K-Ar b	*granite/Alaska, USA	132
360	290-330	K-Ar b	*granite/Queensland, Australia	133
	286	Rb-Sr wri		
~360	248-375	K-Ar b	*biotite granite/Maine, USA	134
~360	249-93	K-Ar b	*binary granite/Maine, USA	134
~360	247±8	K-Ar b	*garnetiferous granite/Maine, USA	134
>360	239-306	K-Ar m	*Andover Granite/Massachusetts, USA	134
>360	221-241	K-Ar m, b	*Fitchburg Pluton/Massachusetts, USA	134
>360	260±8	K-Ar b	*Peabody Granite/Massachusetts, USA	134

>360	230±8	K-Ar b	*Esmond Granite/Rhode Island, USA	135
~370	173±16	Rb-Sr wri-il	Ponta Grossa Fm. (shale)/Brazil	350
370	308	K-Ar wr	*Lovozero Suite (basalt)/Kola, USSR	136
>370	307	K-Ar b	*granite/Bungonia, Australia	137
372	300-333	K-Ar wr	*Hoy Lavas (basalts)/Scotland	138
<375	440	Rb-Sr wri	*Webhannet, Lyman plutons/Maine, USA	139
375	285	K-Ar g	sediment/Saskatchewan, Canada	85
~375	246-306 241-335	K-Ar b Rb-Sr b	*granites/New Hampshire, USA	140
375	350-450	Pb^{206}/U^{238}	Chattanooga Shale/Tennessee, USA	141
<377	439±22	K-Ar b	*peridotite/New York, USA	101
380	310 305	Rb-Sr wri Rb-Sr g-mi	*Tioga Bentonite/Virginia, USA	142
380	266	K-Ar wr	*porphyritic lava/Kara Mazar, USSR	85
380	278-83	K-Ar wr	*nepheline syenite, tuff/USSR	85
~380	500-900	Pb^{207}/Pb^{206}	galena/Kazakhstan, USSR	143
390	475±80	Rb-Sr wri	*Irizar Granite/Victoria Land, Antarctica	351
390	303±18	Rb-Sr g	Carlisle Center Fm./New York, USA	144
390	295±9 605±83	K-Ar m Rb-Sr wri	*Kinsman Quartz Monzonite/New Hampshire, USA	134 140
395	238	K-Ar wr	*porphyritic lava, tuff/USSR	145
400	364-510	K-Ar b	*Shap adamellite/Westmorland, England	146,7
400	212, 360	K-Ar b	*Gocup Granite/New South Wales, Australia	352
400	321	Rb-Sr g	sediment/ng	148
400	240	K-Ar g	sediment/ng	149
400	401, 484	K-Ar h	*granite/Newfoundland, Canada	353

SILURIAN

400	290±15	K-Ar wr	*basalts/Aral, USSR	412
~400	114-190 315, 354 339±12	K-Ar b K-Ar h Rb-Sr wri	*Hikami granite/Japan	150
400	173±4	K-Ar wr	*rhyolite tuff/Florida, USA	438
400	264-380	K-Ar ng	*Mount Peyton Batholith/Newfoundland, Canada	439
>400	339±5	K-Ar m	*granite/Peru	440
>400	189±4	K-Ar b	*migmatite/Columbia	441
>400	250-400	Pb^{207}/Pb^{206}	galena/Kazakhstan, USSR	151
>400	330 238±30	Pb^{206}/U^{238} z Pb^{208}/Th^{232} z	*Cape Granite/Republic of South Africa	152
~405	247	K-Ar g	Binnewater Sandstone/New York, USA	153
~410	230, 550	Pb^{207}/Pb^{206}	galena/Perthshire, England	154
>410	245-88	K-Ar wr	*Vent Rhyolite/Wales	155
<420	1243	Rb-Sr wri	*Ornakam-Moldhesten granite/Norway	354
<420	411-493	Rb-Sr wri	*Murrumbidgee Batholith/New South Wales, Australia	413, 414
420	344-39	Rb-Sr il	Marblehead illite/Wisconsin, USA	405
420	326	K-Ar il	Bertie Fm./New York, USA	156

420	332	K-Ar il	Camillus Fm./New York, USA	157
420	345	Rb-Sr wri	*Newbury volcanics/Massachusetts, USA	355
420	369-519	K-Ar il	Colonus Shale/Skane, Sweden	158
>420	25-180, 410	K-Ar ng	*migmatites/Bulgaria	357
<430	635-94	K-Ar p	*kimberlite/east Siberia, USSR	159
430	363	K-Ar il	Rochester Shale/New York, USA	157
430	349	K-Ar b	*granite/Terskei Alatau, USSR	145
~430	314±9	K-Ar wr	*diabase/Wales	358
430	289	K-Ar il	State Circle Shale/Canberra, Australia	160
430	111-235	Rb-Sr wri-il	Trombetas Fm. (shale)/Brazil	359
>430	310-630	K-Ar wr	*Natkusiak Fm. (basalt lavas)/Franklin, Canada	161, 2
435	357	K-Ar il	Williamson Shale/New York, USA	163
435	367	K-Ar il	Sodus Fm./New York, USA	163
435	113-26 275±11 300-70	Rb-Sr b Rb-Sr m Rb-Sr wri	*granites/Ax-les-Thermes, France	164
<440	409, 563 432, 556 535-47 444, 555	Pb^{206}/U^{238} z Pb^{207}/U^{235} z Pb^{207}/U^{235} z Pb^{208}/Th^{232} z	*Dale City Quartz Monzonite/Maryland, USA	165, 6
<440	400-1300	K-Ar h, b	*Furuland Granite/Sulitjelma, Norway	360
440	267-84	K-Ar il	Chimney Hill Limestone/USA	167

ORDOVICIAN

~440	510±10	Pb^{207}/Pb^{206} z	*Deadman's Bay Granite/Newfoundland, Canada	168
>440	306±10	Rb-Sr wri	*Oporto Granite/Portugal	415
>440	168	K-Ar b	*granite/Caucasus Mts., USSR	416
447	445, 337	K-Ar b	*Bail Hill Volcanics/Dumfriesshire, Scotland	169
447	294, 416	K-Ar wr	*Alcaparrosa Fm. (lavas)/San Juan, Argentina	361
<450	540±50	Rb-Sr wri	*Oughterard Granite/Ireland	170
<450	564±24	Rb-Sr wri	*Ben Vuirich Granite/Scotland	171
<450	580±20	K-Ar h	*Rosetown Pluton/New York, USA	172
<450	542-62 1310-1490	K-Ar b Rb-Sr wri	*granite/Idaho, USA	173
<450	425-540	K-Ar b	*diorites/Kazakhstan, USSR	409
450	350-540	K-Ar il	Sylvan Shale/USA	232
450	272±13	K-Ar wr	*Vent Rhyolite/Wales	155
~450	308, 420	Rb-Sr wri-il	Anse du Veryach Series (shale)/France	362
450	402±25 566±75	K-Ar b Rb-Sr b	Utica shales/Quebec, Canada	174
~450	329±10	K-Ar b	*Waits River Fm./New Hampshire, USA	134
~450	273-340	K-Ar b	*Albee Fm./New Hampshire, USA	134
>450	372±6	K-Ar m	*Main Donegal Granite/Ireland	175
>450	315 300	K-Ar b K-Ar h	*Ellicott City granodiorite/Maryland, USA	176
~460	360	K-Ar ng	*granite porphyry/Caucasus Mts., USSR	363
~460	375-471	Rb-Sr g	sediment/ng	148

470	379	K-Ar s	*Mystery Cave bentonite/Minnesota, USA	364
470	350, 423	Rb-Sr il	claystones/Kentucky, USA	405
470	362, 371	K-Ar h, wr	*gabbro/Cockermouth, England	417
>470	300-420	Rb-Sr wri	*granites/Newfoundland, Canada	418
475	493-584	K-Ar b, h	*granite/eastern Canada	177
475	390 ± 19	K-Ar m	*Walloomsac Fm./Connecticut, USA	135
~475	376, 450	K-Ar wr	*diorite, gabbro/Quebec, Canada	365
475	362	K-Ar g	sediment/Georgia, USSR	145
475	44	Rb-Sr b	*bentonite/Tennessee, USA	366
475	300, 400-900	Rb-Sr il-mi	Lowville Limestone/New York, USA	113, 178
475	297	K-Ar g	*bentonite/Ostergotland, Sweden	179
<480	590	Rb-Sr wri	*Cooma Granite/New South Wales, Australia	180
~480	247-487	K-Ar ng	*Tangriseau microgranite/Wales	367
480	305 ± 10	K-Ar wr	*Warboys diorite/Huntington, England	181
<480	458-548	K-Ar h	*Ophiolite (mafic-ultramafic complex)/Newfoundland, Canada	368
495	355 ± 20	Rb-Sr g	sediment/Falkoping, Sweden	144
495	495, 362	Rb-Sr wri-il	shales/New Zealand	419
~500	407-11	K-Ar wr	*olivine diabase dike/Ontario, Canada	182
500	380 ± 35	Rb-Sr g	sediment/Stenbrottet, Sweden	183

CAMBRIAN

~500	345	K-Ar wr	*trachyandesites/Normandy, France	369
>500	392-584	Rb-Sr g	sediment/ng	148
~510	412 ± 60	K-Ar g	Murray Shale/Tennessee, USA	184
510	411-50 413-33	K-Ar g Rb-Sr g	Franconia Fm./Wisconsin-Minnesota, USA	185
>510	396-569	Rb-Sr wri	*Tioueiine granite/Ahaggar, Algeria	186
~520	225-400 330-430 720-920	Pb^{206}/U^{238} Pb^{207}/U^{235} Pb^{207}/Pb^{206}	kolm (alum shale)/Gullhogen, Sweden	187
~520	373-500	Rb-Sr wri-il	Erguy Kerity Fm. (shales)/France	362
~530	391-4	K-Ar b	*Windyhills Granite/Aberdeen, Scotland	188
540	340	K-Ar g	sediment/ng	149
<550	635-694	K-Ar -	*kimberlite/Siberia, USSR	428
550	404	K-Ar il	sediment/Bohemia, Czechoslovakia	442
~550	?-460	K-Ar wr	*felsic volcanics/Georgia, USA	443
~550	282-1097	K-Ar il	Conasauga Shale/Virginia, USA	189
~550	300-413	K-Ar g	sediment/Alberta, Canada	371
550	393	K-Ar il	Riley Fm./Texas, USA	190
~550	830-1160	Rb-Sr wri	*norites/Scotland	191
>550	284 ± 5	K-Ar wr	*Sledgers Fm. (lava)/Victoria Land, Antarctica	420
<560	764	Pb^{207}/U^{238} ch	*granite/North Baikal, USSR	192
565	346-390	Rb-Sr g	sediment (Mt. Whytte Fm.)/USA	183
~565	760 790	K-Ar g Rb-Sr g	sediment/Northern Territory, Australia	193

565	411-891 394	K-Ar wr K-Ar wri	*Bourinot group (volcanics)/Nova Scotia, Canada	372
>565	400±40	Pb^{208}-Th^{232} z	*Carlton rhyolite/Oklahoma, USA	373
>565	460	K-Ar b	*granite/Oklahoma, USA	194
568	447	K-Ar il	sediment/Shropshire, England	11
570	406	K-Ar g	sediment/Wyoming, USA	16
570	1145±98	Rb-Sr wri	*andesite/Suldal, Norway	374
570	645, 708	K-Ar g	Gros Ventre Shale/Wyoming, USA	195
570	399	K-Ar g	sediment/Narke, Sweden	179
570	395-511	K-Ar wr	*Antrim Plateau Volcanics/Victoria River, Australia	196
570	400-542	Rb-Sr g	Flathead Sandstone/Montana, USA	197
~570	490±15	ng ng	*Palmer Granite/Adelaide, Australia	375
570	385	K-Ar g	sediment/Byelorussia, USSR	198
575	395-413	K-Ar g	Cavell Fm./Alberta, Canada	47
575	413	K-Ar g	sediment/Gotland, Sweden	54
~580	894±58	Rb-Sr wri	*granite/Massachusetts, USA	199
580	457	K-Ar il	sediment/England	11
580	439	K-Ar g	Chilhowee Group/Tennessee, USA	190
580	334	K-Ar wr	*Lighthouse Cove Fm. (basalts)/Newfoundland, Canada	200
590	393-442	K-Ar il	Rome Fm./Virginia, USA	201
590	436	K-Ar g	sediment/Oland, Sweden	195
~600	486±22	K-Ar b	*granite/Pechora, USSR	202
600	479±20	K-Ar b	*Fairville Granite/New Brunswick, Canada	203
600	452-529	Rb-Sr wri-il	shales/Shropshire, England	421

UNDER CAMBRIAN

>600	350-420	K-Ar g	sediment/Kazakhstan, USSR	371
>600	320-380	K-Ar b	*granite/Danmarkshaven, Greenland	376
>600	500±17	Rb-Sr wri	*Hoppin Hill Granite/Massachusetts, USA	204
>600	100-550	K-Ar wr	*Akitkan Fm. (volcanics)/Baikal, USSR	377
>600	489±10	K-Ar b	*Athis granite/Normandy, France	205
>600	350-440	K-Ar b	*Carnsore Granodiorite/Ireland	378
>600	480±10	K-Ar m	*pegmatite/Wadi Hawashia, Egypt	206
>600	373-8	K-Ar b	*granite/Kviteseid, Norway	379
>600	480 320	Pb^{208}-Th^{232} mo Pb^{206}-U^{238} mo	*granite/Gebel Dara, Egypt	206
>600	450±120 370±25	Pb^{208}-Th^{232} z Pb^{206}-U^{238} z	*riebeckite granite/Gebel Gharib, Egypt	206
>600	420 410 370	Pb^{208}-Th^{232} hu Pb^{207}-U^{235} hu Pb^{206}-U^{238} hu	*granite/east Siberia, USSR	126
>600	34	K-Ar m	*granite porphyry/New Mexico, USA	380
>600	470±18	K-Ar il	sediment/Algeria	207
>600	485-570	K-Ar h	*diorites/Worcestershire, England	381
>600	460	Rb-Sr wri	*Wooltana Volcanics (basalt)/Australia	208

>600	340-412	K-Ar b	*granites/Norway	382
>600	482 ± 30	Rb-Sr wri	*Bull Arm Volcanics/Newfoundland, Canada	209
>600	401 ± 16	K-Ar wr	*mafic dykes/South West Africa	383
>600	370 ± 38	Rb-Sr wri	*Coldbrook group (volcanics)/New Brunswick, Canada	210
>600	461-540	K-Ar b	*Mancellian granites/Chausey Islands, France	384
>600	300 ± 5	Rb-Sr wri	*volcanics/Morocco	385
>600	440	K-Ar h	*charnockite/India	386
>600	510	Rb-Sr wri	*granite/Montagne Noire, France	422
>600	350-400	Rb-Sr wri	*granites/Alaska, USA	444
>600	368-1200	K-Ar m, b	*Forsayth Batholith/Queensland Australia	445

tionalized away is the topic for subsequent sections in this work.

Many other dates could have been listed, but Table 1 is limited to dates which approach 20% discrepancy: being either 20% "too young" or "too old" for their biostratigraphical positions. Many are over 30% discrepant. A 20% discrepancy means that an indicated date is off by at least one geologic period in the lower Mesozoic and off by two geologic periods in the early Paleozoic. From Table 1, it is evident that a 330 million year date is obtained for Carboniferous rock, but that the same value is often obtained for rocks as old as Cambrian. Viewed another way, Devonian rocks give "true" values near 375 million years, but also values of 220 ("properly" Triassic) and 450 ("properly" Ordovician).

The arbitrariness of the practice of selecting some values as being true and disregarding others which conflict with them was recently admitted by Waterhouse,[315] who commented: "It is, of course, all too facile to 'correct' various values by explanations of leakage, or initially high concentrates of strontium or argon. These explanations may be correct, but they must first be related to a time line or 'cline of values' itself subject to similar adjustments and corrections on a nonstatistical, nonexperimental basis."

Table 1 does not include the many anomalous dates from those minerals that have grown in disrepute with respect to radiometric dating. K-feldspars usually give K-Ar ages that are "too young," and this is attributed to argon loss associated with exsolution and perthitic growth. Only sanidine is considered reliable. Because of their K-feldspars, whole-rock dating of acidic intrusive igneous rocks is avoided, and mica or amphibole separates are used instead. Minerals such as beryl, cordierite, and zeolite often give erratic K-Ar ages attributable to isotope fractionation. With few exceptions, Table 1 is confined to datings on material that is considered to be reliable.

Very many igneous bodies which have been dated yield a wide spread of discordant dates, but they could not be entered into Table 1 because they have little or no biostratigraphic definition of their relative ages and thus escape violating any biostratigraphy. Others were entered into Table 1 as "greater than," "less than," or "approximately equal to" some biostratigraphic limit: dates being so anomalous that they violated even their liberal biostratigraphic limits. The heavy attention given to K-Ar dating in Table 1 reflects the overwhelmingly preponderant use of this method over all others combined; and it also serves to balance the heavy emphasis placed on U-Th-Pb dating by other Creationist scholars.

The uniformitarians may contend that there are many more dates in agreement with accepted values than there are anomalies such as all of Table 1. Even if this were true, it would not appear to be an overwhelming majority, and a significant minority of discrepant dates would probably be sufficient to discredit all of radiometric dating. Since most igneous bodies have wide biostratigraphic limits,[219] it is difficult to tell that a date is not anomalous because it could take on many different values and not be anomalous.

As a matter of fact, the number of determinations actually used to define "correct" values for the geologic column are fewer than the anomalies comprising Table 1, except for the Cenozoic and Cretaceous. Armstrong[316] compiled a listing of dates that are considered to be reliable time-points for the Phanerozoic; the list presumably up to date as of 1976-1977. There are 260 pre-Cenozoic dates compiled, but 98 of the 260 are Cretaceous, and many determinations have only partial biostratigraphic brackets.

Some estimates would make dates in agreement with accepted values a minority among all dates obtained in the Phanerozoic. That direct sediment dates (on glauconite) agree within plus or minus 10% of accepted values approximately half the time is estimated by Obradovich and Peterman.[218] Afanass'yev[219] considers dates from biostratigraphically bracketed intrusives to be less reliable than glauconite dates, while Armstrong and Besancon[220] point out that most dates on basaltic lavas are discordant. Combining these estimates, it may be that somewhat less than half of all dates agree within 10% of accepted values for their respective biostratigraphic positions.

B. The Selective Publication of Dating Results

An objective comparison between the number of fitting vs. the number of anomalous dates in the Phanerozoic is hindered (if not prevented) by the fact that anomalous dates frequently (or usually) are not reported in scientific journals.

Many researchers directly or indirectly imply that they are not reporting discrepant dates that they have obtained. An implication of there being unreported K-Ar ages that are "too young" and attributed to argon loss is made by Mitchell and Reen,[221] who write: "Alternatively, if the *reported* ages can be argued to be free of argon loss . . ." (emphasis added here and elsewhere unless otherwise indicated). A contrast between proper/published and anomalous/unpublished dates is evident in the following statements by Twiss and DeFord:[222] ". . . some of the ages do not correspond with stratigraphic position. The *reported* ages in millions of years are . . ." Armstrong[223] writes: "The dates *reported* are all consistent with observed geologic relationships." Polevaya *et.al*[224] write: "The *published* results attest to the possibility of obtaining rather reliable ages for glauconites by the argon method."

In a recollection of how anomalous dates re-occurred, Curtis *et. al.*[225] said: "As a result I suggest that it would be important to report all anomalous results." This is not done. Most sobering of all is the following recent statement by Mauger:[12] "In general, dates in the 'correct ball park' are assumed to be correct and are published, *but those in disagreement with other data are seldom published nor are discrepancies fully explained.*"

There is a tendency to leave unpublished the results which conflict with those of other investigators or which disagree with accepted values. Thus, a certain reluctance to provide a non-fitting date seems to be the case in this report by Forman:[226] "The remarkably congruent date obtained for the Tiburon Peninsula eclogite with that . . . for the Cazadero tectonic blocks is very pleasing (147 m.y. *versus* 135 to 150 m.y.). Thus it is a little untidy to report 106 m.y. for the age of the amphibolite on Catalina Island."

C. Direct Radiometric Dating of Sedimentary Rocks

Sedimentary rock samples are not indiscriminately dated because the constituent detrital particles would ostensibly give the age of the source region(s), and not the time since sedimentation. Alternatively, some value between that of provenance and of sedimentary rock formation might be obtained. Dating of fossil material and limestone samples has been abandoned because in nearly every case (as ref. 46; Jurassic and Triassic, Table 1) the K-Ar ages are far too low, and this is attributed to the low argon-retentive properties of calcite.

Sedimentary rocks are dated by the K-Ar and Rb-Sr methods utilizing the authigenic minerals glauconite and illite. In dating glauconites, Rubinstein[227] commented: ". . . we often get anomalously high figures." This is supposedly caused by minute amounts of allogenic contaminant incorporated within the glauconite. Usually, however, glauconite K-Ar and Rb-Sr values are "too young" for their biostratigraphic position. Many such anomalously young dates are entered into Table 1.

Rationalizations for these discrepant dates have centered upon claims that glauconite is vulnerable to becoming an open system via spontaneous recrystallization, leaching, or slight heating. Although some correlation of anomalous dates with permeability of host rock and depth of burial were found by Morton and Long,[228] they otherwise stated: "Several factors were studied which could correlate with open-system behavior: grain morphology, extent of weathering, percent expandable layers, recrystallization caused by former deep burial, and permeability of the host rock . . . . None of these factors, singly or in combination, was infallibly useful to predict which of the Llano glauconites will have behaved as an open or closed chemical system . . . ." Owens and Sohl[229] found similar crystallinity and compositions between glauconites yielding expected, and anomalous, dates.

Depth of burial is supposed to produce diametrically opposed effects on glauconite dates. Evernden *et. al.*[230] considered deeply-buried glauconites to be less reliable than shallow-buried ones, on the grounds that the former are more likely to be exposed to heating. Holmes[231], by total contrast, contended that shallow-buried glauconites would tend to be less reliable because deeply-buried ones are better shielded from weathering.

Most any discrepant date can be excused on the basis of some presumed system-opening situation. The wide variety of factors which presumably cause open systems and plasticity in attributing anomalous dates to these factors suggest that these are just rationalizations. An alternative view would be that dates are discrepant not because of "open systems" and geologic causes but because radiometric dating is invalid.

K-Ar and Rb-Sr dates on the clay mineral illite produce far greater discrepancies than dates on glauconite. Table 1 includes dates on illite so anomalous that they are several times the "true age" of the rock, or a small fraction of the "true age." For instance, the Conasauga Shale (Cambrian, ref. 189, Table 1) gives us spread of ages ranging from Upper Carboniferous to twice the time since the Cambrian.

"Correct"-age illites are assumed to be authigenic. Those fractions which yield ages "too old" are explained away as being detrital in origin. Finally, those that give ages "too young" for their biostratigraphic positions are claimed to either be the result of argon loss from the authigenic illite, or from the illite being diagenetic and not authigenic in origin. Agreement with accepted values for the biostratigraphic positions may be coincidental. One such K-Ar dating on a shale was described by Hower *et. al.*:[232] "The whole-rock age is quite good for the Upper Ordovician. However, an examination of the K-Ar ages and mineralogy of the various particle sizes shows 'good' age must be at least partly fortuitous."

Rb-Sr mineral dates, like the K-Ar dates, are subject to the same interpretations when the dates turn out discrepant (for example Permian, ref. 95, Table 1). The Lowville Limestone (Ordovician, ref. 113, 178, Table 1) is one of several examples in Table 1 of situations where Rb-Sr illite isochrons are constructed but not accepted as the age of the rock because of their very discrepant values. The ages that were "too old" were supposed to indicate that the Sr isotopes had homogenized in the source region and not in the shale at the time of its deposition. At the same time, "too young" Rb-Sr isochrons were considered to be indicators of diagenetic Sr isotope homogenizations long after deposition of the shale. Still other Rb-Sr isochrons were not even considered to have any geological meaning.

U-Th-Pb methods have been used in dating uranium-bearing shales. They are now not considered to remain closed systems throughout alleged geologic time. The classic Swedish kolm (Cambrian, ref. 187, Table 1) is an outstanding example of erratic dates being attributed to its inability to remain a closed system.

D. Open-System Rationalizations for Discrepant Igneous Dates As Deduced from Petrography and Regional Geology

Radiometric dating finds its greatest utility in the dating of igneous rocks. The section after this one will consider radiometric dating of biostratigraphically-bracketed igneous bodies. This section will review some principles by which discrepant dates are attributed to some processes that would cause open systems. Microscopic studies of rock thin-sections helps determine if the rock has suffered weathering, deuteric alteration, or low grade metamorphism. The author is not denying that a heating, leaching, or weathering event on a rock could cause an open system and thereby make discrepant dates, but wishes to point out that a very plastic and capricious situation exists between the state of the rock and whether the dates from it are "good," or if they are anomalous.

Very many K-Ar dates from altered rocks are "good." For granites, Miller and Mohr[233] said: "But again there is microscopical evidence for appreciable alteration, at least of the feldspar, which must be reconciled with an apparently complete retention of radiogenic argon." Zartman[234] reported that: "... much greater discordances occur in fresh rather than obviously weathered rock." Armstrong[235] stated that: "In two cases altered and fresh samples of the same granite body were analyzed and in both excellent agreement was obtained." No correlation between degree of alteration of quartz monzonite and its ages was found by Bassett et. al.[236] Bell[237] warned: "In interpreting the isotopic ages, it should be remembered that isotopic diffusion does not necessarily imply recrystallization. The danger in confidently correlating certain petrofabric and isotopic events is apparent."

Basalts and other mafic rocks are believed to be especially vulnerable to becoming open systems because of the ease of their alteration. But in dating basalts, Dasch et. al.[238] wrote: "... by analyzing samples of strongly altered (59) and relatively fresh (C20) rock from the same dike; within analytical uncertainly the two samples have identical K-Ar dates." Evans et. al.[239] reported: "Some of the present highly altered rocks gave discrepantly low ages, others, equally altered, did not appear to have suffered extensive argon loss. In addition, low ages were obtained from some apparently very fresh rocks. In fact, no direct quantitative correlation between the presence or amount of the fine-grained secondary minerals of reputed poor retentivity and the apparent age discrepancy observed could be established."

The fact that very many anomalous results by K-Ar dating come from unaltered material is frankly admitted by Durant et. al.:[240] "Although it is obviously better to work with fresh material, the fact that a rock is petrographically fresh is no guarantee that 40Ar loss or excess 40Ar phenomena are absent."

Rb-Sr dating is generally believed to be more resistant to heating than is K-Ar, but more vulnerable to leaching caused by weathering or hydrothermal fluids. Zartman[241] deliberately dated some weathered granite by Rb-Sr, and found: "That these minerals give such a consistent age pattern, indicating a closed system even upon exposure to rather severe weathering, is rather remarkable."

Many discrepant Rb-Sr dates have no alteration on the rock sample dated. Zartman et. al.[242] reported: "An anomalously low Rb-Sr age ... was found for the biotite. Although no effects of metamorphism or alteration are visible in the syenite, some such process undoubtedly disturbed this ... radiometric system." In speaking of all Rb-Sr dating, Faure and Powell[243] generalize: "First of all, there may be no mineralogical or textural evidence to warn the geochronologist that an igneous rock or any of its minerals he is analyzing for an age determination has been altered." In fact, Hamilton[244] wrote: "It is *quite common* for a metamorphic event detected by Rb-Sr dating to leave no obvious imprint on the hand specimen or in thin section."

The implications of all these findings are enormous. Any discrepant date can be explained away, and a heating or weathering event can be invented whenever necessary for this purpose. No evidence need be found because its absence can be attributed to it being strong enough to make the unwanted date discrepant, but too weak to show up in thin section. An illogical situation arises because at one time it is claimed that radiometric dates have withstood obvious alteration of the rock, while at other times they supposedly were so sensitive that they were made discrepant by an event too weak to alter the rock itself. A skeptical view of radiometric dating looks at all these lacks of correlation of alteration and discrepancy of dates as evidence that they are just rationalizatio;ns, and that discrepant dates are not primarily caused by alteration but by the fundamental invalidity of radiometric dating.

Rationalizations for discrepant dates are also formed on the basis of the regional geologic context of the rock sample being dated. The reasoning involved is best described by Evernden and Richards:[245] "Thus, if one believes that the derived ages in particular instances are in gross disagreement with established facts of field geology, he must conjure up geological processes that could cause anomalous or altered argon contents of the minerals."

It is interesting to note cases where the regional geology of the dated sample would indicate the likelihood of a date being discrepant, but it isn't; or else some dates are supposedly disturbed while others strangely are not. In his dating, Zartman[246] wrote: "It is difficult to postulate a mechanism responsible for the low Rb-Sr and K-Ar ages in the pegmatite biotite while the adjacent granitic biotite has not suffered a similar effect." Pankhurst[247] commented: "The writer thinks it unlikely that this biotite alone can have escaped the event so consistently recorded by the remainder, especially since it comes close to a contact with a younger intrusion. It seems that this result is anomalous ...."

On the basis of regional geology, a strong expectation of dates having been made discrepant via a later orogenic event near them was supposed by Eastin and Faure,[248] who found: "A remarkable feature of these volcanic rocks is that they appear *not* to have lost radiogenic Ar^{40} during the Ross Orogeny . . ." (emphasis theirs). Similarly, Haller[249] reported: "The biotite of a polygenic migmatite was found to yield a K-Ar age of 490 ± 12 m.y. The surprising thing about this result is that the isotopic age is not about 400 m.y. since the Caledonian effects appear to be so strong."

Burchart[250] stated: "Numerous cases have been reported in which Sr isotopic ratios have been found extremely sensitive to geologic events . . . which may have been indistinguishable by classic geologic or petrologic methods. The case of the Tatra Mountains serves as an example of just the opposite situation . . . a major event according to petrological record does not always leave its mark on isotopic dates." It is obvious that once again there is a very plastic relationship between anomalous dates, and their regional geology. That discrepant dates very frequently are not related to regional geology and can be explained away without resorting to it is evident in the following statement by Kratts:[251] ". . . an understanding of the true causes of discrepant vaues in concrete cases reveals geologic and geochemical events not discernible by the usual geologic methods."

Any discrepant date can be explained away by claiming that some event has opened the system, while at the same time claiming that the alleged event is not recorded in regional geology if the discrepancy cannot be conveniently fitted in to the geologic history of the region as envisioned by uniformitarians.

Not surprisingly, dates must agree with their biostratigraphic position in order to be considered valid indicators of age. In their datings, Wanless *et. al.*[252] commented: "No stratigraphic evidence is available to confirm or deny this age." Elsewhere, Williams *et. al.*[253] wrote: "The internal consistency demonstrated above is not a sufficient test of the accuracy of the age determinations; they must also be consistent within any age constraints placed on intrusion by fossils in the country rocks."

The investigators dating a Triassic basalt (ref. 70, Table 1) said: "The Mississippian age for sample NS-45 cannot be correct because it is grossly inconsistent with the stratigraphic position of the lavas. No clues as to apparent preferential loss of potassium or gain of excess Ar^{40} from this sample are in evidence from thin section examination." In dating the Ferrar Dolerite (ref. 77, Table 1), the authors wrote: "Rb-Sr analyses of an initial group of hypersthene tholeiites were well aligned on the isochron diagram. They appeared to define an isochron of 270 ± 45 million years. This result is incorrect, since it contradicts a firm stratigraphic control of the age . . . ."

Recently, Hayatsu[423] wrote: "In conventional interpretations of K-Ar age data, it is common to discard ages which are substantially too high or too low compared with the rest of the group or with other available data, such as the geological time scale. The discrepancies between the rejected and the accepted are arbitrarily attributed to excess or loss of argon." In invoking "excess argon" because of biostratigraphically and paleomagnetically discrepant K-Ar results from Pleistocene lavas, Armstrong[424] said: "This is an inherent uncertainty in dating young volcanic rocks; anomalies may be detected only by stratigraphic consistency tests, independent dating techniques, and comparison with the known time scale of geomagnetic reversals during the last 5 m.y. (Cox 1969)."

E. Radiometric Dating of Biostratigraphically-Bracketed Igneous Rocks

1. Effusives (Tuffs and Bentonites) and Extrusives (Lava Flows)

Since extrusives and effusives are depositionally accumulated, the Law of Superposition applies to them. The biostratigraphic bracket is imposed by the fact that they must be younger than what they overlie but older than whatever overlies them.

Volcanogenic ash very frequently yields "too old" K-Ar dates. It is claimed that this is from "excess radiogenic Ar" whereby the parent magma is contaminated with argon heated out of the wall rock (vent) that it is coming through, and that the minerals which crystallize do so too rapidly to degas this contaminant argon. Scarborough[254] writes: "Age information is interpreted carefully because of a distinct tendency for certain ash layers to contain a variable amount of excess argon." Eocene tuffs (ref. 12, Table 1) are one of several such examples.

Other anomalously old K-Ar dates are explained away by claiming contamination of the tuffs by detrital minerals. Christiansen *et. al.*[255] said: "Preliminary K-Ar data from alkali feldspars of the second and third ash-flow sequences yield dates apparently too great, suggesting contamination by Precambrian feldspars." These alleged contaminants cannot readily be distinguished, according to Curtis *et. al.*,[256] and they suggest that: ". . . the thing to do is get a sequence of dates and *throw out those that are vastly anomalous.*"

K-Ar dates from undevitrified volcanic glass (one example in Table 1-ref. 74) are not considered to be reliable. Other igneous materials, when "too young", are claimed to have undergone argon loss (for example, the Filipowice Tuff, Ref. 96). The ejecta of the Korvunchana Series (ref. 80) turned out "too young", and this was ascribed to supposed potassium additions. Unaltered mineral separates from bentonites which are "too young" likewise are attributed to open systems. The Carmel Formation bentonite (ref. 52) gave anomalously young K-Ar and Rb-Sr ages, and this was attributed to hydrothermal effects, even though there is no evidence for it. The Tioga Bentonite (ref. 142), though weathered, amazingly yielded a good Rb-Sr isochron for 15 of 22 samples, although the isochron defined an anomalously young date.

Rationalizations for discrepant dates from lava flows are similar to those for tuffs. "Too old" K-Ar dates from lavas (for instance, basalt, ref. 118) are attributed to "excess argon." K-Ar dates from lavas are commonly discrepantly young, and these results are explained away by claiming some thermally-induced argon loss (as from deuteric alteration or low grade metamorphism) although (as the previous section demonstrated) no evidence for such an alleged event need be found.

Some of the many discrepantly young K-Ar dates from lavas included in Table 1 are: Karroo volcanics (ref. 66), Toadstone Lavas (ref. 117), Hoy Lavas (ref. 138), Antrim Plateau volcanics (ref. 196), etc. That contradictory dates from lavas occur as a rule is stated by Armstrong and Besancon:[220] "Detailed dating studies using pre-Tertiary whole rock basalt and dolerite specimens have been made . . . and limited optimism for the method appears justified. In spite of collection of unweathered samples and precautions taken to discard samples with evident alteration, it is *usual* to obtain a spectrum of discordant dates and to select the concentration of highest values as the correct 'age'." Even this is accommodating because, as Stewart *et. al.*[257] pointed out, a maximum in an age-value spread need not be accepted as any specific event.

Rb-Sr whole-rock isochron dating of lavas has produced its share of surprises. Some lavas have given Rb-Sr dates very much greater than the maximum permissable biostratigraphic age. Examples of this in Table 1 include the Ferrar Dolerite (ref. 77), refs. 19, 20, 40, and 191. It is claimed that these isochrons are not defining the time since the Sr isotope homogenization when the lava crystallized, but are defining some Sr isotope homogenization in the mantle. In supporting this, Pankhurst[258] remarked: "This is nothing short of a revolution in the fundamental principles of Sr-isotope geology."

Rb-Sr isochron dates which are "too young" are explained away by claiming that the isochrons are not the Sr isotope homogenization when the rock formed, but a re-homogenziation caused by some later thermal event on that rock. Some examples of this from Table 1 include the Wooltana volcanics (ref. 208) and the Coldbrook Group (ref. 210).

2. Intrusives (Hypabyssal and Plutonic Rocks)

The law of Cross-Cutting Relationships serves to set biostratigraphic brackets for intrusive rocks. An intrusive rock must be younger than any country rock that it intrudes, but older than any rock unconformably overlying it or containing clasts eroded off that intrusive.

Many discrepant results from intrusives are rationalized away immediately by accepting the dates but reinterpreting the biostratigraphic bracket. For instance, the Rose Dome Granite (ref. 102) was believed to contain Pennsylvanian (Upper Carboniferous)-aged inclusions, but this was dropped when the granite yielded a preCambrian Rb-Sr isochron age. A granite (ref. 173) intrudes a quartzite which was believed to be part of an Ordovician quartzite lithostratigraphic unit. This quartzite is now considered to be preCambrian and not part of the Ordovician quartzite because of PreCambrian dates from the granite.

A granite (ref. 199) gave an Rb-Sr date much greater than the biostratigraphic age of the rock intruded. This discrepancy was explained by claiming that the granite is "thrust in" tectonically, not intrusive. But against this, Gates *et. al.*[259] noted that it is: ". . . not marked by a major fault or an unconformity. Rather . . . seem to be interlayered and concordant throughout the mapped area."

The most common manipulations performed to resolve discrepancies involve the claim of composite plutons. A discrepant date can be explained away by claiming the sample dated is from an earlier or a later phase of the biostratigraphically-bracketed pluton. For instance, the anomalously old date from the Pine Mountain Granite (ref. 103) was explained away by regarding the Rb-Sr date as giving the age of an earlier phase of the Granite. Pertaining to an Ordovician granite (ref. 177), the authors wrote: "1) The Granitic bodies may be composite with an earlier, partly updated phase of latest PreCambrian or Early Cambrian age and a later phase in Middle or Late Ordovician, and 2) The granitic bodies were indeed emplaced during Middle Ordovician time but biotite and hornblende during (Devonian) alteration and/or slight recrystallization have absorbed freed radiogenic argon. *These explanations are not altogether satisfying.*"

The upper biostratigraphic limit on granites imposed by clast-bearing sedimentary strata was disregarded in the case of the granites entered in Table 1 under refs. 150, 203, and (Devonian) ref. 50, when radiometric dates conflicted with it. It was simply claimed that the clasts resulted from erosion of some hidden early or late granites, and not from the main granitic body being dated.

An amazing series of rationalizations were proposed for discrepant U-Th-Pb zircon dates from the Dale City Quartz Monzonite (ref. 165, 166). Seiders[165] accepted the intrusion of this body into a slate, but denied the Ordovician age of the slate claiming that the fossils from it were inconclusive. Higgins[166] accepted the biostratigraphic age-designation of the slate, but claimed that the Monzonite is not intrusive into it, but unconformable. Alternatively, he accepted the intrusive nature of the Monzonite into the slate but suggested that the anomalously old zircon U-Th-Pb dates are caused by inherited (contaminant) lead.

Dates from plutons that turn out discrepant are subject to the same open-system rationalizations as are those from lavas. An examination of Table 1 notes many plutons (for example, refs. 44, 104, 105, and many others) yielding a spread of mutually-contradictory K-Ar dates, some of which spread over several geologic ages.

K-Ar dates that are "too old" for their biostratigraphic positions are attributed to "excess argon" contamination or to an allegedly protracted two-stage magmatic history where the early phase (xenoliths, xenocrysts, or phenocrysts) is supposedly hundreds of millions of years before the later phase (groundmass).

Kimberlites and peridotites often show anomalously old K-Ar dates. Pertaining to the Stockdale kimberlite pipes, the authors wrote (ref. 91): "These ages are indicative of a xenocrystic origin of the micas." Hence the constituent minerals were assumed to have originated deep within the earth hundreds of millions of years before the intrusion. Elsewhere (ref. 101), the constituent biotites were flaky, suggesting an earlier origin, but some did not give anomalously old K-Ar dates. It was claimed that they had been degassed of their previously-accumulated radiogenic argon.

Of especial significance are some Siberian kimberlites (ref. 159), because not only the xenocrysts but also the groundmass gave "too old" K-Ar dates. The ease of explaining away any discrepant date was portrayed when it was claimed that the groundmass got anomalously old by K-Ar when the xenocrysts allegedly shed some of their argon into the magma just as it was crystallizing to form the groundmass.

Anomalously old K-Ar dates are not confined to ultramafic and mafic rocks. Hornblende from the Pearse Peak Diorite (ref. 43) was much "too old" by K-Ar, and this was attributed to its alleged occlusion of "excess argon." Much the same situation occurred in the case of the Rosetown Pluton (ref. 172); among others in Table 1.

K-Ar dates from intrusives often turn out far "too young" for their biostratigraphic positions. This is explained away by claiming that the K-Ar dates from 30 million to 350 million years from certain granites (ref. 116), the 30 million-year dates are supposed to be completely degassed by the heating of the Alpine orogeny, the 350 million-year dates are considered to have survived all the heating events, and the many dates in between are supposed to have no geologic meaning; being supposedly "hybrid ages" of partial argon loss during the Alpine heating event of 30 million years ago.

Thus, discrepantly young K-Ar dates can be claimed to have been completely "reset" if the anomalously young K-Ar date can be matched up with some inferred process (as orogeny) of regional geology, or it can be claimed to be geologically meaningless and giving a value in between the "true" age of the rock and the "true" age of a thermal event if the date does not correspond to either. The Vallorcine Granite (ref. 122) is one example of a "geologically meaningless" K-Ar date. Refs. 34, 127, 104, and 120 are just a few of the many hypabyssal and plutonic rocks of varying compositions that are entered into Table 1, and which give "too young" K-Ar dates. The Klotassin Batholith (ref. 72) is especially significant because the alleged heating event on it which made its K-Ar dates "too young" was supposedly intense enough to have also "rejuvenated" its Rb-Sr isochron date.

Another line of rationalizations for discrepantly young K-Ar dates for intrusives is the claim that they were deeply buried tens or hundreds of millions of years after their crystallization, making them too hot to hold their accumulating radiogenic argon for that amount of time. Some Devonian granites (refs. 134, 135) provide the classic example. It is supposed that they give K-Ar dates accepted for Permian because they were deeply buried from Devonian to Permian time. Zartman et. al.[260] admitted the difficulty of belief in such a prolonged post-crystallization burial of the granites, but accepted it because they found tectonism and regional metamorphism inferred for Permian time to be too localized to account for such a wide area yielding "too young" K-Ar dates from granites. Elsewhere, the batholith under ref. 35 is a similar example of alleged protracted burial.

A variation of the K-Ar method known as the Ar^{40}/Ar^{39} Spectral method has been used on plutons in recent years. It supposedly can distinguish between "excess argon," "rejuvenated," and "true" K-Ar dates because a true K-Ar date will shed its argon uniformly at different temperature fractions, giving rise to a flat spectrum. The disturbed rock or one with "excess argon" will give off very different amounts of argon in different temperature-fractions and give rise to a "stepped" and "saddle" spectrum, respectively.

That this method can invariably distinguish between "excess" and "disturbed" argon contents was questioned by one study.[261] Dallmeyer[262] found that "too young" K-Ar dates yielded undisturbed spectra, so he claimed that the Ar^{40}/Ar^{39} technique is incapable of distinguishing between a true date and a date of very prolonged burial. The claim that a flat spectrum indicates a true date comes to beg the question, because any flat spectrum from a "too young" conventional date can be claimed to be from protracted burial, and thus the whole method ends up avoiding the question that it is purported to answer.

The Ar^{40}/Ar^{39} method of K-Ar dating failed dramatically by giving flat spectra for samples that would virtually certainly be considered to have been disturbed, and for samples containing "excess argon." Biotites were taken from a gneiss in a contact metamorphic zone within a few hundred meters of an igneous intrusion. Conventional K-Ar dates were dramatically younger nearer the intrusion than further. In summarizing the results of the Ar^{40}/Ar^{39} method applied on the biotites, Ashkinadze et. al.[263] wrote: "The spectrum shows no features that would indicate any natural disturbance in the K-Ar system. The plateau levels could be erroneously taken as representing the true age of the specimens if the conclusions...were followed." In another experiment, Ashkinadze et. al.[264] performed the Ar^{40}/Ar^{39} method on biotites that had given absurdly old K-Ar dates. It gave a flat spectrum, failing to indicate any "excess argon."

The Rb-Sr whole-rock method is frequently used on intrusives. There is some subjectivity in isochron construction that enables samples to be chosen so that they define an isochron age in agreement with accepted values for the biostratigraphic position of the intrusives. Two quite different isochrons obtained from the Barhalde Granite (ref. 94) attest to this. An outright case of fudging the Rb-Sr isochron is evident in the following description of the Kinsman pluton (ref. 140) by authors Lyons and Livingston: "The Kinsman Quartz Monzonite for all six isochron points also yields an unsatisfactory isochron of 605 ± 83 m.y. The isochron shown ... however, has been drawn by eliminating sample MK 37-73... the resulting isochron of 411 ± 19... embraces what we consider to be an accurate determination of the age of emplacement of the Kinsman."

Any discrepant Rb-Sr isochron can be explained away by claiming that some points on it don't "belong" on that isochron because they allegedly came from different crustal sources and had different initial Sr^{87}/Sr^{86} ratios. In the case of the Ben Vuirich Granite (ref. 171) the anomalously old Rb-Sr isochron was dismissed as "a spurious result" and attributed to source Sr isotopes not homogenizing.

Other anomalous Rb-Sr isochrons are rationalized away by claiming that they become open systems after the rock formed. In the case of the Cooma Granite (ref. 180), an alleged open system was the cause for the "too old" Rb-Sr isochron. By contrast, "incipient weathering" was the cause of the "too young" Rb-Sr isochron from the Hoppin Hill Granite (ref. 204). Either weathering or deuteric alteration was the alleged cause for the anomalously young isochrons from the Tioueiine granite (ref. 186), and hardly any alteration was visible in petrographic thin sections.

U-Th-Pb methods on intrusives are performed either on separated accessory minerals or on ore bodies that formed during late-magmatic or post-magmatic processes around intrusives. U-Th-Pb dates from pitchblendes and galenas are usually discordant and in conflict with biostratigraphic positions. Some European ores (ref. 82) gave "too young" U-Th-Pb dates because they allegedly were subject to later re-mineralizations. Others (for instance, refs. 106, 143, and 151) are much "too old" for the biostratigraphic positions of their host rocks, and are explained away by claiming that they became contaminated with ancient, "remobilized" lead during their mineralizations.

Zircons are the most common accessory mineral used for U-Th-Pb dating of intrusives. Like ore deposits, discrepant dates from them are attributed to various supposed open-system and contaminating conditions. The Deadman's Bay Granite (ref. 168) is one example of this, as is the Cape Granite (ref. 152). Zartman[265] pointed out that: "... the morphology of the zircon crystal does not always reflect the presence of inherited old radiogenic lead." Thus any anomalous zircon date can be explained away just because it is discrepant.

Some anomalously old U-Th-Pb dates from zircons are supposed to indicate that the zircons themselves are "xenocrystic," or inherited from the country rock through which the magma intruded. Zircons from plutons entered under refs. 18 and 100 are examples of this explanation.

In considering all the dating methods, it should not be supposed that dates which fit accepted values for Phanerozoic systems are the only ones that are consistent. That discrepant dates also are consistent is pointed out by Polyakov et. al.:[266] "Still more important is the fact that 'rejuvenated' dates are nonrandom and recur on a regional or even a global scale." The significance of dates that are internally consistent from different samples of the same outcrop or the same igneous body; and dates which are in agreement by different dating methods will all be discussed in the next section (considering both Phanerozoic and Precambrian consistencies).

II. Highlights of Precambrian and Non-Biostratigraphic Geochronology

A. Consistency and Concordances Among Radiometric Dates

It is often claimed that reliable K-Ar dates can be objectively distinguished from apparent dates because the former will show internal consistence from widely separated samples, while the latter will scatter because an altering event does not affect all regions of the sampling area equally.

In practice, the analytic data is subordinate to geologic "fit" of the dates obtained. Marvin et. al.[267] write: "Many geologists commonly evaluate age determinations only in the light of geologic evidence and do not adequately consider the importance of the analytical data. Admittedly, an analytically valid age may occasionally prove to be geologically spurious because it conflicts with incontrovertible geologic field relations."

There are many instances of dates with good internal consistency being rejected as not giving the correct age of a rock because they conflict with accepted values. In a Precambrian situation, K-Ar dates were much younger than the (presumed correct) Rb-Sr dates, and about the K-Ar dates McKee and Noble[268] commented: "Continuous partial argon loss may have occurred. If this is the case, the consistency of these apparent ages is fortuitous."

Cases of modern lavas giving anomalously old K-Ar dates are so well known to Creationists that they are not repeated here. In one case, authors McDougall et. al.[269] warned against accepting consistent K-Ar dates as necessarily valid, because they found that: "With few exceptions, anomalously old but often internally consistent K-Ar dates were found for the lavas."

The fact that anomalously young K-AR dates can be internally consistent was pointed out by Wetherill.[176] A different argument from consistency involves the supposition that samples of widely differing K^{40} concentrations which yield consistent K-Ar dates are valid because a suite of samples would not have gained or lost Ar^{40} in just the right amounts to be identically proportional to their respective K^{40} concentrations. But a series of anomalously old but correlative K^{40}/Ar^{40} situations was described by Kaneoka and Aoki,[319] who wrote: A Greenland dolerite has also shown an anomalous old age in spite of good correlation in the $^{40}Ar/^{36}Ar$ vs. $^{39}Ar/^{36}Ar$ isochron diagram. These evidences suggest that the excess Ar^{40} is sometimes located in K or K-similar sites." Consistency of K-Ar dates is therefore no proof for their validity.

In Rb-Sr whole-rock dating, it is commonly believed that if the points plotted to define an isochron show good collinearity, then the rock remained a closed system and the date will be valid. But against this view, Goldich[217] wrote: "Linearity of points defining a whole-rock Rb-Sr isochron is not a sufficient criterion of the isochron age being the time of igneous emplacement or crystallization." Elsewhere, Matsuda[270] made a very similar statement.

A profound regional internal consistency for "too young" Rb-Sr isochrons was attributed to orogenic reheating. Fairbain et. al.[271] wrote: "If this is the actual explanation, it is remarkable that it has operated to about the same degree on three volcanic series hundreds of miles apart."

It is widely held that when K-Ar dates on biotite and hornblende are in agreement, the date is valid. Numerous authors approximate the energy of activation for thermally-induced argon loss to be reached at minimum temperatures of 250 °C for biotite and 450 °C for hornblende. Had a pluton cooled so slowly that it yielded spurious K-Ar dates, the hornblende date would be older than the biotite date because the former is cool enough to retain its accumulating radiogenic argon sooner than the latter. Alternatively, had there been a later heating event on the rock, K-Ar biotite/hornblende would be discordant because the biotite alone would be degassed, or (if hot enough to degas both) the hornblende would still be older because again it would be cool enough to retain its argon sooner that the biotite.

That is the theory. In fact, there are many cases of consistent K-Ar dates from these mineral separates that are "too young." The Chigmit Mountains Batholith (ref. 57) is one example. Elsewhere, McDougall and Leggo[272] wrote: ". . . hornblende and biotite K-Ar ages agree in a few cases where the measured ages are too young . . . . Hence, arguments based on consistency of results must be used with caution." Webb and McDougall[273] studied 51 discordant biotite/hornblende pairs and found no statistically significant preference for one to be older than the other. They concluded: "The concordance of biotite and hornblende K-Ar ages could lead to the erroneous conclusion that the date is the time of emplacement or strong metamorphism."

In K-Ar dating of the Peninsular Ranges Batholith (ref. 35), concordant but anomalously young biotite/hornblende pairs were found. The claim made was that the Batholith was deeply emplaced for tens of millions of years after intrusion, making it too hot for either mineral to retain argon. Then it is supposed that it got uplifted so rapidly that it cooled from above 450 °C to below 250 °C so quickly that no difference is shown between biotite and hornblende despite the fact that they are "too young". A similar train of thought was utilized by Wetherill et. al.[274] for a similar find.

The most powerful argument claimed for the validity of radiometric dates is the agreement of results from different dating methods. K-Ar/Rb-Sr whole-rock and U-Th-Pb/Rb-Sr whole-rock agreements appear to be more common than K-Ar/U-Th-Pb and non-concordia U-Th-Pb concordances. It is thus significant that the Rb-Sr isochron can be fudged.

Just as the Rb-Sr isochron can be fudged to get it to fit biostratigraphic evidence (as in the previously-discussed Barhalde Granite (ref. 94) and Kinsman Quartz Monzonite (ref. 140)) so it can be fudged to get it to agree with K-Ar or U-Th-Pb. An indication of some subjectivity in Rb-Sr isochrons was shown when several granites yielded isochrons ranging from 418 to 479 m.y., and when the data were pooled together, a 413±7 m.y. isochron resulted.[275]

A U-Pb zircon date of 2560 m.y. was in disagreement with an Rb-Sr isochron of 1840 m.y. until the latter was subject to the following initial-ratio fudge: "An age of 2510 is obtained by calculating an average rock total and by using a more normal r_i."[278]

The numbers of cases of concordances are no doubt exaggerated by the selective publication of dating results. In a discrepancy, the results of the method most considered correct will be published, and the results of another method ignored. In one granite (ref. 133) K-Ar and Rb-Sr isochron dates are in agreement, but "too young" for the biostratigraphic position. Just as the agreement was minimized by noting the change in Rb-Sr value by a change in initial Sr^{87}/Sr^{86} ratio (changeable from allegedly poor radiogenic Sr^{87} enrichment), so other agreements may be caused by adjusting the initial ratio.

Certain concordances are dissolved after further studies. In one situation described by Higgins,[276] U-Pb and Rb-Sr isochron dates agreed at 425 m.y. for three igneous bodies. One of the igneous bodies was reinterpreted as being much later and its 425 m.y. Rb-Sr mineral isochron dissolved and considered a meaningless result.

Allsopp et. al.[277] wrote: ". . . measurements on one pegmatite sample . . . indicated an apparent Rb-Sr age of 3.4 b.y.; this age was then interpreted as support for the 3.4 b.y. lead-model age . . . . Although analytically correct, the high apparent age . . . is considered anomalous in the light of new measurements . . . ."

Geologic interpretations are frequently changed in order to avoid situations where there would otherwise be a gross conflict between biostratigraphy and results in agreement by two or more radiometric dating methods. Clasts from the Hikami Granite (ref. 150) and an unnamed Devonian granite (ref. 50) were reinterpreted as being derived from a supposedly hidden later granite in order to avoid an anomalously young situation for the agreeing K-Ar/Rb-Sr whole-rock isochron dates. The biostratigraphic age of conglomerate containing clasts of the Pamir-Shugnan Granites (ref. 22) was changed for the very same reason. When K-Ar/Rb-Sr dates on glauconite (ref. 193) were much "too old" for the biostratigraphy but in agreement with each other, the glauconite was claimed to be reworked. Yet it was considered enigmatic because the nearest appropriate source area was much too distant for a soft glauconite pellet to be transported.

At least 2 of the 4 U-Th-Pb dating methods would be simultaneously in agreement with each other and "too old" (for example, ref. 18) if the claim of the "xenocrystic" origin was not accepted.

Concordant results by two or more methods, as the results from only one method, can be rejected as giving the age of the rock and instead considered to have other geologic meaning. Anomalously young but concordant K-Ar/U-Th-Pb dates from the Mrzyglod diabase (ref. 108) are supposed to indicate that the alleged heating event which had completely "rejuvenated" the K-Ar date also was related to a uranium mineralization. Discrepantly young but concordant K-Ar/Rb-Sr isochron from the Klotassin Batholith (ref. 72) are supposed to reflect a later heating event having "rejuvenated" both dates.

Still other agreements between different dating methods are considered to have no meaning whatsoever. In commenting on a K-Ar/Rb-Sr agreement on biotite, Wasserburg and Lanphere[279] said that it ". . . is a case of accidental concordance. That is, the time calculated does not have any meaning in terms of an event." In fact, anomalous agreements of K-Ar/Rb-Sr on biotite are so common that equal daughter-product loss is invoked, as by Webb and McDougall:[280] ". . . the frequent correspondence between Rb-Sr and K-Ar ages on this mineral from all environments suggests that Sr diffuses at similar rates and at the same temperatures as does Ar." K-Ar/Rb-Sr agreements on glauconite that are "too young" are also common (for example ref. 185, the Franconia Formation), and Hurley[281] proposed a similar explanation.

Miller and Kulp[282] described a case of what they considered to be a series of fortuitous agreements between Pb^{206}/U^{238} and Pb^{207}/U^{235} caused by a mixing-in of contaminant lead. Stuckless et. al.[283] found a Rb-Sr isochron in agreement with K-Ar but considered the possibility that the isochron: ". . . may be fortuitous." A granite yielded concordant K-Ar/Rb-Sr isochron values which fell into a time span considered to be a period of tectonomagmatic quiescence. For that reason, Armstrong[284] suggested that the mutually-corroborating results are neither the age of the granite nor a "rejuvenating" event but an in-between value: ". . . by coincidence, nearly concordant but meaningless K-Ar and Rb-Sr dates."

A fortuitious U-Th-Pb/Rb-Sr isochron agreement is supported by Dietrich et. al.,[285] who write: "the 1320 ± 100 m.y. date defined by a whole rock isochron is not known to correlate with any well-recognized sedimentary or tectonic event (It is similar to the discordant 'age' of about 1300 m.y. reported from two probably detrital zircon samples . . . any agreement with these 1300 m.y. 'ages' would appear to be little more than a coincidence.).''

The fact that there are many anomalous dates which agree by two methods and are explained away as mutually rejuvenated or totally fortuitous suggests that concordant dates need not be accepted as an unassailable proof for the validity of radiometric dating. The fact that agreements considered to be fortuitous occur on a routine basis in some cases (Rb-Sr/K-Ar agreements on glauconite and igneous biotite) further encourages belief that all concordances may have a geochemical explanation that has nothing to do with the true ages of the rocks dated.

Of course, it is important to note that comparisons between results of different dating methods are commonly (if not usually) discordant, and are promptly dismissed as open systems. York and Farquhar[387] said: Where the results of comparisons of this sort disagree, it is clear that some sort of transfer of materials into or out of the rock or mineral has taken place. It has also become apparent from the number of published discordant ages that disturbances of this nature are far more common than was formerly realized."

In speaking of the U-Th-Pb methods, Davidson[388] wrote: "Ideally, the isotopic ratios . . . should all give results in good agreement . . . . In practice, very few uranium and thorium minerals have been found to exhibit this concordant pattern of ages, and the much more common discordances between the three or four values have been facilely explained away, as each investigator thought best fitted to local circumstances . . . ."

Hurley and Rand[389] did a comparison of K-Ar and Rb-Sr results from the Precambrian continental crust, using published and some unpublished sources from nearly every continental region on earth. The best-fit line on the Rb-Sr vs. K-Ar graph was K-Ar=.75Rb-Sr, indicating a systematic tendency of K-Ar/Rb-Sr discordance. One fourth of the points were indicative of such discordancy that they plotted at or near the line K-Ar= .5Rb-Sr.

In writing of results of datings in Phanerozoic orogenic belts, Brown and Miller[390] said: "In general, strong discordances can be expected among ages deduced by different methods." Since most dates obtained by the simultaneous application of more than one method come from orogenic belts, it may follow that the majority of comparative datings from the Phanerozoic, as in the Precambrian, are discordant.

B. Radiometric Violations of Superpositional and Cross-Cutting Relationships

Radiometric dates routinely violate common-sense relationships of field geology. It is almost self-evident that in a depositional situation the topmost beds must be at least slightly younger than those below them, and in an intrusive relationship it is the intruding body which must be younger than the body it cuts.

In writing about tuffs (obvious depositional accumulation), Curtis et. al.[286] write: "These beds may appear to be pure . . . yet gave different ages from top to bottom; *the younger age being on the bottom.*" The rationalization invoked for this absurd situation was that the first-deposited were uncontaminated while the upper beds were subject to influx of contaminating detrital minerals.

A tuff yielding a K-Ar date of 40-41 m.y. was found intruded by a dike and sill yielding K-Ar dates of 49-50 m.y.[287] A diorite whose biotite yielded a 157 m.y. K-Ar date is intruded by a quartz diorite yielding a 204 m.y. K-Ar biotite date.[87] Violations of cross-cutting relationships are not exceptional. Hopson[288] states: "This curious relationship, in which the pegmatites give mineral ages older than those from the host rocks, is now known to be common . . . ." These gross anomalies are explained away by claiming that the pegmatite gives older K-r ages than the intruded country rock because the country rock is composed of fine-grained minerals that are more vulnerable to thermally-induced argon loss. This is accommodating, because many coarse-grained pegmatites properly give younger K-Ar and Rb-Sr mineral dates than the fine-grained country rock. Furthermore, Leach et. al.[289] found coarse-grained schists giving ages near 72 m.y. whereas fine-grained varieties gave dates near 123 m.y., and they appealed to ". . . some other process . . . " to explain this.

A different set of violations involves Rb-Sr whole-rock isochrons. The Stony Creek Granite of 610±50 m.y. Rb-Sr isochron age cuts the Monson Gneiss of 444±15 m.y. Rb-Sr isochron age.[290] Remobilization of Sr isotopes during metamorphism was the supposed cause of this.

Violations are especially prominent in Precambrian rocks. A pegmatite yielding Rb-Sr isochrons of 1.7 to 2.7 b.y. cuts the country rock which yields 2.0-2.2 b.y. Rb-Sr isochrons.[291] In the Baltic Shield, metamorphic rocks of the Kola series give Rb-Sr dates of 2.4-2.7 b.y. and maximum K-Ar dates of 2.7-2.8 b.y. They are intruded by basic and ultrabasic rocks comprising the Nonchegorsk Massif, and yield U-Pb dates of 2.9 b.y. and K-Ar dates of 3-3.5 b.y. Elsewhere, granites cutting the Kola Series have given K-Ar dates from 2.8 to 3.6 b.y. and U-Pb dates of 2.8 b.y.[292] In the Guyana Shield, the basement rock yields a 1.8. b.y. Rb-Sr isochron date, while a gabbroic-doleritic mass intruding it gives K-Ar dates up to 3.06 b.y. "Excess argon" in the basic rocks is the rationalization invoked.[293] Space limitations prevent the listing of countless other violations.

Rank absurdities arise from radiometric dating which are explained away in terms of metamorphic processes. For instance, Varadarajan[317] found that the Amritpur Granite of India yields K-Ar dates of 1.3-1.9 b.y. and intrudes metavolcanics of 228 m.y. He rejected others' suggestions that the contact is not intrusive, and proposed that the ages are correct but that the Granite was partially remelted after 228 m.y. ago, and thereby emplaced intrusively into the much younger metavolcanics. In another case,[318] the Kochatev granite massif of 450 m.y. was unconformably overlain by the Zerendinsk Series of 1.2-1.5 b.y. The rationalization invoked for this absurdity was that the granite had been remelted while the overlying formation was unaffected.

Probably the greatest violations of all are radiometric dates many times greater than the accepted 4.5 b.y. age of the earth. They may be a type of *reductio ad absurdum* of accepting billions-of-years ages for rocks. A plagioclase crystal gives an unexplicable age of 4.9 b.y. while everything else around gives Rb-Sr ages of 1.3-1.5 b.y.[294] Mark et.al.[295] found a probably Tertiary basalt yielding an isochron of 10 b.y. The Pharump diabase[296] from the Precambrian of California yielded an Rb-Sr isochron of no less than 34 b.y., which is not only over 7 times the age of the earth but also greater than some uniformitarian estimates of the age of the universe. This super-anomaly was explained away by claiming some strange metamorphic effect on the Sr.

K-Ar ages much greater than inferred earth age are also common. Gerling et. al.[297] called attention to some chlorites yielding K-Ar dates of 7 to 15 b.y. It had been noted that some minerals which yield such dates (as beryl, cordierite, etc.) can be claimed to have trapped excess argon in their channel structures or to have fractioned the Ar isotopes, but none of this can apply to the simple mica-like structure of chlorite. They also pointed out that for the anomalies to be accounted for by excess argon, unreasonably high partial pressures of Ar during crystallization would have to be required. They concluded by suggesting some unknown nuclear process which no longer operates to have generated the Ar. Elsewhere, Galimov[298] suggested that ^{40}K captures an external electron via K-capture. The possibility of K decaying to Ar some other way should be interesting to those Creationist physicists studying the question of how invariant decay rates are.

C. Age Values for Igneous and Metamorphic Terranes

It is most interesting to note many cases where radiometric dating imposes time partitions which appear to be unsupported or contradicted by lithostratigraphic evidence. In dating layers of lava flows, Williams and Curtis[299] wrote: "According to potassium-argon determinations, the exposed volcanic rocks range in age from about 2.5 m.y. to 1.4 m.y. We admit, however, that some age-determinations are difficult to reconcile with observations made in the field. The absence of signs of deep erosion and soil-horizons suggests that accumulation of the Rampart beds was essentially a continuous process, without long periods of volcanic quiet."

Precambrian terranes composed of igneous and metamorphic rock contain unconformities which are supposed to be erosional surfaces generated during orogenic cycles each of which is hundreds of millions of years in duration. Several cycles are supposed to have occurred during the Aphebian (dates as 1.7 to 2.5 b.y.) and Archean (over 2.5 b.y.). Stockwell[300] notes many cases were Aphebian rocks grade directly into Archean, and attributes this to the destruction of the alleged unconformities during gneissification. An alternate view would be that the unconformities never existed, that the hundreds of millions of years never elapsed, and that radiometric dating has artificially segmented continuous strata into fallacious time partitions.

At the same time, Archean rock is unconformably overlain by igneous and metamorphic rock sometimes billions of years younger. Yet Hurley et. al.[301] observe that commonly the radiometric ages are the same on both sides of the unconformity; that Archean basement tends to be as radiometrically young as the overlying material. This is attributed to rejuvenation of Archean as the overlying rocks are formed, but an alternative view is that rocks supposedly billions of years apart in time are uniformly young because radiometric dating is invalid and so are the artificially-erected time designations, and various contrivances are necessary to harmonize lithostratigraphy with radiometric dating and uniformitarian concepts.

In generalizing on all Precambrian geochronology, Peterman et. al.[302] wrote: "... correlations based on lithologic and successional similarities and 'layer cake' stratigraphy generally result in an oversimplification and serious errors." This is supposedly caused by the geologic complexity of Precambrian terranes, but another view is that radiometric dates routinely violate not only lithostratigraphy, but result in a completely artificial imposition of time partitions which have no basis in reality and naturally result in confusion.

Dott and Dalziel[303] wrote: "Lithic correlation of the Baraboo metasedimentary sequence with Animikie rocks in northern Michigan and northern Wisconsin has seemed compelling, for each succession has pure quartzite overlain by a carbonate and iron-bearing interval, which is in turn succeeded by thick slates." The Baraboo rocks have yielded a "meaningless" date near 750 m.y. and a spread of K-Ar and Rb-Sr dates from 1.1 to 1.6 b.y. By contrast, Animikie rocks have given U-Pb and Rb-Sr dates from 1.9 to 2.1 b.y., with an even older correlative of 2.1-2.4 b.y. dates. Are rocks of such similar composition and lithostratigraphy really separated by hundreds of millions of years of time, or is radiometric dating a delusion?

Unnatural time-imposition by radiometric dating may be evident even in rocks that approach 4.0 b.y.— the oldest dates accepted for terrestrial material, In comparing greenstone-granite formations from Greenland that have yielded dates near 3.8 b.y. with other greenstone-granites that give only 2.7-3.3 b.y. dates, Moorbath et. al.[304] commented: "The major volcanic and sedimentary features are essentially indistinguishable from those of younger greenstone belts in North America and southern Africa . . . ."

An extremely plastic acceptance of radiometric dates is admitted for the Precambrian. Sabine and Watson[305] cite: "... the rather subjective task of separating 'true' from 'apparent' ages." Wasserburg and Lanphere[306] appealed to complex Precambrian histories to explain discrepant dates, and commented: "... geochronologic studies commonly lead to a confusing array of data, in which many of the determinations... are not the age of primary metamorphism or intrusions." Barton[307] said: "As in the case with radiometric ages determined from almost any rock unit, it is impossible to establish unequivocally that the ages reported here reflect the time of original crystallization or emplacement of the bodies from which they are derived."

Just as dates in the Phanerozoic must fit accepted values for their biostratigraphic positions in order to be accepted as valid, so also radiometric dates are selectively accepted depending upon the geologic relationships of Precambrian rock. The subordinate nature of radiometric dates even in the Precambrian is evident in the following statement by Pulvertaft:[308] "... poor exposures and low relief have forced geologists to rely too much on isotopic age determinations."

A very prominent feature of Precambrian geochronology is the fact that radiometric dates from the igneous and metamorphic complexes spread over hundreds of millions (or even billions) of years. Odom and Fullagar[309] found the Cranberry Gneiss yielding Rb-Sr isochrons from 670 m.y. to 1.3 b.y. and commented: "... the so-called Cranberry Gneiss might represent a time span equivalent to that of the entire Phanerozoic."

This is not exceptional. Sabine and Watson[310] point out that: "... one outcome of isotopic age-studies had been the demonstration that some metamorphic complexes had been built up by repeated activity over periods of several hundred million years." A skeptical view of radiometric dating would question the incredible lengths of time indicated for these rocks and suggest that there is actually a very self-contradictory spectrum of dates and that this spectrum is not caused by repeated petrogenetical activity but that it actually illustrates the meaningless and invalidity of radiometric dating.

Very self-contradictory age-spreads in the Precambrian are so common that it is claimed there was repeated tectonomagmatic activity going on for billions of years in given regions. Pulvertaft[311] writes: "Reactivation is recognized by most geologists as a common feature of Pre-Cambrian basement areas. What is not always realized is the scale on which it may have taken place."

The Sao Francisco Craton of Brazil shows a K-Ar spread of 1.1 to 1.9 b.y.[312] Elsewhere, Peterman and Hedge[313] found both K-Ar and Rb-Sr mineral dates in a huge mutually-contradictory spread of 80 m.y. to 1.8 b.y. Gerling et. al.[292] reported K-Ar dates from the Baltic Shield spreading from a high of 3.5 b.y. to a low near 1.2 b.y. In the Saamo-Karelion Zone alone, a spread of 1.7 to 2.8 b.y. was encountered.[314]

When stripped of all the claims of reactivation, it is obvious that rocks yield ages which spread over significant fractions of the entire earth's alleged 4.5 b.y. history. Such absurdly contradictory results may be further evidence against the validity of radiometric dating.

Creationists and Diluvialists are not alone in their disbelief of radiometric dating. The fact that radiometric dating lacks credibility even in some uniformitarian circles is evident in the following statement by Houtermans:[320] "Sometimes the dates given by radioactive methods are accepted enthusiastically by the classical geologists, sometimes if these dates do not fit their previously formed hypotheses they come to the conclusion to deny the usefulness of radioactive methods altogether." Similarly, Brown and Miller[321] commented: "Much still remains to be learned of the interpretation of isotopic ages and the realization that the isotopic age is not necessarily the geologic age of a rock has led to an over-skeptical attitude by some field geologists." Whether the skepticism has been excessive or whether it has been insufficient is, of course, a matter of opinion.

Every paradigm has explanations for data that won't fit it. The paradigm crumbles when the explanations are not accepted and considered to be only excusing rationalizations that cover-up the basic failure of the paradigm. Likewise, all of the various open-system explanations for discrepant results may be accepted at face value, or they may be seen as excusing rationalizations that cover up the invalidity of radiometric dating and all the eons of time that it purportedly demonstrates and measures.

References

AA — Geophysical Journal of the Royal Astronomical Society
AG — American Association of Petroleum Geologists Bulletin
AJ — American Journal of Science
AN — Antarctic Journal of the United States
BV — Bulletin volcanogique
CE — Canadian Journal of Earth Sciences
CG — Geological Survey of Canada Paper
CH — Chemical Geology
CO — Proceedings of the International Geological Congress
CT — Contributions to the Geologic Time Scale (American Association of Petroleum Geologists Studies in Geology No. 6)
DE — Doklady: Earth Science Sections (translation from Russian)
EG — Economic Geology
EP — Earth and Planetary Science Letters
GA — Geological Society of America Abstracts with Programs
GB — Geological Society of America Bulletin
GC — Geochimica et Cosmochimica Acta
GE — Geology
GG — BMR Journal of Australian Geology and Geophysics
GH — Eclogae Geologicae Helvetiae
GI — Geochemistry International
GJ — Geochemical Journal (of Japan)
GL — Quarterly Journal of the Geological Society of London
GM — Geological Society of America Memoir
GN — Norges Geologiske Undersokelse Bulletin
GP — Geological Society of America Special Paper
GR — Journal of Geophysical Research
GU — Journal of the Geological Society of Australia
IG — International Geology Review
IN — Journal of the Geological Society of India
IS — Isochron/West: A Bulletin for Isotope Geochronology
JG — Journal of Geology
JR — Journal of Research of the United States Geological Survey
KG — Kwartalnik Geologiczny
LG — Liverpool Manchester Geological Journal
MI — Geologie en Mijnbouw
MP — Contributions to Minerology and Petrology
NA — Nature
NP — Norsk Polarinstitutt Skrifter

NY —New York Academy of Sciences Transactions
OG —Overseas Geology and Mineral Resources
PE —Journal of Petrology
RE —Precambrian Research
RU —Geologische Rundschau
SC —Science
SG —Scottish Journal of Geology (Transactions of the Edinburgh Geological Society prior to 1964)
SP —Journal of Sedimentary Petrology
SR —Colloques Internationaux du Centre National de la Recherche Scientifique
SV —Springer Verlag Publishing Co. Berlin, Heidelberg, New York
TI —Norsk Geologisk Tiddskrift
TS —The Phanerozoic Time Scale (Geological Society of London Special Publication 120s)
UG —United States Geological Survey Professional Paper
UN —Transactions American Geophysical Union
YG —Proceedings of the Yorkshire Geological Society
ZG —New Zealand Journal of Geology and Geophysics

[1] Firsov L. V. and S. S. Sukhorukova. 1968. "Quaternary" Glauconite of Cretaceous Age from the Lower Yenisei. DE 183:101-2.
[2] Bandy, O. L. and J. C. Ingle. 1970. Neogene Planktonic Events and Radiometric Scale, California. GP 124:143
[3] Fisher, D. E. 1971. Excess Rare Gases in a Subaerial Basalt from Nigeria. NA Physical Science 232:60.
[4] Marvin, R. F., Stern, T. W., Creasey, S. C., and H. H. Mehnert. 1973. Radiometric Ages of Igneous Rocks From Pima, Santa Cruz, and Cochise Counties, Southeastern Arizona. UG Bulletin 1379:18.
[5] ———. 1964. Items: Abstracts of Published radiometric and stratigraphical data with comments. TS: 343.
[6] Lipson, J. 1958. Potassium-Argon Dating of Sedimentary Rocks. GB 69:146.
[7] Dickinson, D. R., Dodson M. H., and I. G. Gass. 1969. Correlation of Initial Sr87/Sr86 with Rb/Sr in Some Late Tertiary Volcanic Rocks of South Arabia. EP 6:84-9.
[8] ———. 1964. Op. cit., p. 327-8.
[9] Ibid., p. 352.
[10] Evernden, J. F., Curtis, G. H., Obradovich, J., and R. Kistler. 1961. On the Evaluation of Glauconite and Illite for dating sedimentary rocks by the potassium-argon dating method. GC 23:10.
[11] Ibid., p. 99.
[12] Mauger, R. L. 1977. K-Ar Ages of Biotites from Tuffs in Eocene Rocks of the Green River, Washakie, and Uinta Basins, Utah, Wyoming, and Colorado. (University of Wyoming) Contributions to Geology 15(1):37.
[13] Evernden et. al. 1961. Op. cit., p. 90.
[14] Banks, P. O., and L. T. Silver. 1964. Re-examination of Isotopic Relationships in Colorado Front Range Uranium Ores. GB 75:469-71.
[15] Marvin et. al. 1973. Op. cit., p. 12.
[16] Evernden et. al. 1961. Op cit., p. 91.
[17] Ross, J. V. 1974. A Tertiary Thermal Event in South Central British Columbia. CE 11:1116-7.
[18] Stern, T. W., Newell, M. F., Kistler, R. W., and D. R. Shawe. 1965. Zircon, Uranium-Lead, and Thorium-Lead Ages and Mineral Potassium-Argon Ages of LaSal Mountains, Utah. GR 70:1503-4.
[19] Pankhurst, R. J. 1977. Strontium Isotope Evidence for mantle events in the continental lithosphere. GL 134:264.
[20] Ibid., p. 258.
[21] Folinsbee, R. E., Baadsgaard, H., and J. Lipson. 1960. Potassium-Argon Time Scale. 21st CO (III):10.
[22] Shanin, L. L. 1976. Some patterns of "rejuvenation" in radiometric ages of rocks in the Southwestern Pamir. IG 18:841-2.
[23] Owens, J. P. and N. F. Sohl. 1973. Glauconite from New Jersey-Maryland Coastal Plain: Their K-Ar Ages and Applications in stratigraphic studies. GB 84:2827, 2814-5.
[24] Lewis, J. F., Harper, C. T., Kemp, A. W., and J. J. Stipp. 1973. Potassium-Argon Retention Ages of Some Cretaceous Rocks from Jamaica. GB 84:331-9.
[25] Lovering, J. F. and J. R. Richards. 1964. Potassium-Argon Age Study of Possible Lower-Crust and Upper-Mantle Inclusions in Deep-Seated Intrusions. GR 69:4898.
[26] Dasch, E. J. 1969. Strontium isotopes in weathering profiles, deep-sea sediments, and sedimentary rocks. GC 33:1548.
[27] Polevaya, N. I., Murina, G. A., and G. A. Kazakov. 1961. Utilization of Glauconite in Absolute Dating. NY 91:304.
[28] ———. 1964. Op. cit. p. 383.
[29] Ibid., p. 395.
[30] Snelling, N. J. 1965. Age Determinations on Three African Carbonatites. NA 205:491.
[31] ———. 1964. Op. cit., p. 390.
[32] Ibid., p. 399.
[33] Roddick, J. C. and E. Farrar. 1972. Potassium-Argon Ages of the Eagle Granodiorite, Southern British Columbia. CE 9:596-9.
[34] Meyerhoff, A. A., Khudoley, K. M., and C. W. Hatten. 1969. Geological Significance of Radiometric Dates from Cuba. AG 53:2498.
[35] Krummenacher D., Gastil, R. G., Bushee, J., and J. Doupont. 1975. K-Ar Apparent Ages, Peninsular Ranges Batholith, Southern California and Baja California. GB 86:760.
[36] Evernden et. al. 1961. Op. cit., p. 96.
[37] Hurley, P. M., Cormier, R. F., However, J., Fairbairn, H. W., and W. H. Pinson. 1960. Reliability of Glauconite for Age Measurement by K-Ar and Rb-Sr Methods. AG 44:1807.
[38] Leach, G. B., Lowdon, J. H., Stockwell, C. H., and R. K. Wanless. 1963. Age Determinations and Geological Studies (including isotopic ages—report 4) CG 63-17, p. 56-7.
[39] Banks, P. O. and N. Shimizer. 1969. Isotopic Measurement on zircons from Japanese granitic rocks. GJ 3:25, 27-8.
[40] Pankhurst, R. J. 1977. Op. cit., p. 262-3.
[41] Armstrong, R. L., Speed, R. C., Graustein, W. C., and A. Y. Young. 1976. K-Ar Dates from Arizona, Nevada, Montana, Utah, and Wyoming. IS 16:2.
[42] Lanphere, M. A., MacKevett, E. M. Jr., and W. Stern, 1964. Potassium-Argon and Lead-Alpha Ages of Plutonic Rocks, Bolson Mountain Area, Alaska. SC 145:705-6.
[43] Koch, J. C. 1966. Late Mesozoic Stratigraphy and Tectonic History, Port Orford-Gold Beach Area, Southeastern Oregon Coast. AG 50:52-3.
[44] Wanless, R. H., Stevens, R. D., Lachance, G. R., and C. M. Edmonds. 1968. Age Determinations and Geological Studies: K-Ar Isotopic Ages, Report 8. CG 67-2, part A, p. 49-50.
[45] Armstrong, R. L., Taubeneck, W. H., and P. O. Hales. 1977. Rb-Sr and K-Ar geochronometry of Mesozoic granitic rocks and their Sr isotopic composition, Oregon, Washington, and Idaho. GB 88:399, 403.
[46] Lippolt, H. J. and W. Gentner. 1963. K-Ar Dating of Some Limestones and Fluorites (Examples of K-Ar Ages with low Ar concentrations) (in Radioactive Dating, International Atomic Agency, Vienna) 240-2.
[47] Folinsbee et. al. 1960. Op. cit. p. 11.
[48] Berrange, J. P. and R. Dearnley. 1975. The Apoteri Volcanic Formation-Tholeiitic flows in the Northern Savannas Graben of Guyana and Brazil. RU 64(3):894-5.
[49] Evernden et. al. 1961. Op. cit. p. 93.
[50] Gobbett, D. J. and C. S. Hutchinson (ed.) 1973. Geology of the Malay Peninsula. Wiley-Interscience. New York, London, Toronto. p. 235-7.
[51] Pankhurst. 1977. Op. cit., p. 259.
[52] Marvin, R. F., Wright, J. C. and F. G. Walthall. 1965. K-Ar and Rb-Sr Ages of Biotite from the Middle Jurassic Part of the Carmel Formation, Utah. UG 425 B104-7.
[53] Wanless et. al. 1968. Op. cit. p. 26.
[54] Evernden et. al. 1961. Op. cit. p. 95.
[55] ———. 1964. Op. cit. p. 278-9.
[56] Lowdon, J. A. 1960. Age Determinations by the Geological Survey of Canada: Report 1, Isotopic Ages. GC 60-17, p. 7.
[57] Reid, B. L. and M. A. Lanphere. 1969. Age and Chemistry of Mesozoic and Tertiary Plutonic Rocks in South Central Alaska. GB 80:138.
[58] Fleck, R. J., Sutter, J. F., and D. H. Elliot. 1977. Interpretation of discordant Ar^{40}/Ar^{39} age-spectra of Mesozoic tholeiites from Antarctica. GC 41:17.
[59] Francis, E. H., Smart J. G. O., and N. J. Snelling. 1968. Potassium-Argon Age Determination of an East Midlands Alkaline Dolerite. YG 36:293, 5.
[60] Armstrong, R. L. 1969. K-Ar Dating of Laccolithic Centers of the Colorado Plateau and Vicinity. GB 80:2082-3.
[61] Nalivkin, D. V. 1962 (1973 translation). Geology of the USSR. Oliver and Boyd, London. p. 666.
[62] ———. 1964. Op. cit. p. 320.
[63] White, W. H., Erickson, G. P. Northcote, K. E., Dirom, G. E., and J. E. Harakal. 1967. Isotopic Dating of the Guichon Batholith, British Columbia. CE 4:678.

[64]Armstrong, R. L. and J. Besancon. 1970. A Triassic Time Scale Dilemma: K-Ar Dating of Upper Triassic Mafic Igneous Rocks, Eastern USA and Canada, and Post-Upper Triassic Plutons, Western Idaho, USA. GH 63:21, 3.
[65]Petrushevskiy, B. A. 1975. Problems in geology of the Himalaya. IG 17:714.
[66]Fitch, F. J. and J. A. Miller. 1971. Potassium-argon Radioages of Karroo Volcanic Rocks from Lesotho. BV XXXV (1):69, 74-5.
[67]Leech, et. al. 1963. Op. cit. p. 39.
[68]Kulp, J. L. 1963. Potassium-Argon Dating of volcanic rocks. BV XXVI:254.
[69]Armstrong and Besancon. 1970. Op. cit. p. 20, 22, 26.
[70]Carmichael, C. M. and H. C. Palmer. 1968. Paleomagnetism of the Late Triassic North Mountain Basalt of Nova Scotia. GR 73:2813.
[71]Wanless, R. K., Stevens, R. D., Lachance, G. R., and R. N. Delabio. 1972. Age Determinations and Geological Studies: K-Ar Isotopic Ages, Report 10. CG 71-2:18-21.
[72]LeCouteur, P. C. and D. J. Tempelman-Kluit. 1976. Rb-Sr Ages and a profile of initial Sr^{87}/Sr^{86} ratios for plutonic rocks across the Yukon Crystalline Terrane. CE 13:319-20, 23.
[73]Schermerhorn, C. J. G., Priem, H. N. A., Boelrijk, N. A. I. M., Hebeda, E. H., Verdurmen, E. A. Th., and R. H. Veischure. 1978. Age and Origin of the Messajana Dolerite Fault-Dike System (Portugal and Spain) In the Light of the Opening of the North Atlantic Ocean. JG 86:305, 7.
[74]Brew, D. A. and L. J. P. Muffler. 1965. Upper Triassic Undevitrified Volcanic Glass From Hound Island, Keku Strait, Southeastern Alaska. UG 525-C38.
[75]Wanless et. al. 1968. Op. cit. p. 23-6.
[76]_____. 1964. Op. cit. p. 361.
[77]Compston, W., McDougall, I., and K. S. Heier. 1968. Geochemical Comparison of the Mesozoic Basaltic Rocks Of Antarctica, South Africa, South America, and Tasmania. GC 32:131.
[78]Bignell, J. D., and N. S. Snelling. 1977. Geochronology of Malayan Granites. OG 47:42-3.
[79]_____. 1971. Items: Abstracts of published radiometric and stratigraphical data with comments. TS 1971 supplement, p. 75.
[80]Naumov, V. A., and A. M. Mukhina. 1977. Absolute age of volcanic formations of the Central Siberial Platform. IG 19:951, 954-6.
[81]_____. 1971. Op. cit. p. 53.
[82]Davidson, C. F. 1960. (a). Rejuvenation of Pitchblende in Hercynian Ore Deposits. EG 55:384.
[83]Brinkmann, R. 1976. Geology of Turkey Elsevier, Amsterdam.
[84]_____. 1964. Op. cit. p. 286.
[85]Davidson, C. F. 1960. (b). Some Aspects of Radiogeology. LG 2:236.
[86]_____. 1967. Symposium: The Sub-Carboniferous Basement in Northern England. YG 36:341.
[87]Stewart, J. W., Evernden, J. F., and N. J. Snelling. 1974. Age Determinations from Andean Peru: A Reconaissance Survey. GB 85:1111.
[88]Davidson. 1960. (b). Op. cit. p. 326.
[89]Aronson, J. L. 1968. Regional Geochronology of New Zealand. GC 32:667, 692.
[90]Hayase, I., and S. Nohda. 1969. Geochronology on the "oldest rock" of Japan. GJ 3:50.
[91]Brookins, D. G. and V. J. McDermott. 1967. Age and Temperature of Intrusion of Kimberlite Pipes in Riley County, Kansas. GA 1976, p. 366.
[92]Hurley et. al. 1960. Op. cit. p. 1804.
[93]Neumann, H. 1960. Apparent Ages of Norwegian Minerals and Rocks. TI 40:174-181.
[94]Brooks, C., Wendt, I., and W. Harre. 1968. A Two-Error Regression Treatment and Its Application to Rb-Sr and Initial Sr^{87}/Sr^{86} Ratios of younger Variscan Granitic Rocks from the Schwarzewald Massif, Southwest Germany. GR 73:6071-3.
[95]Chaudhuri, S. and D. G. Brookins. 1967. Rb-Sr Studies of Paleozoic Sediments. UN 48:242.
[96]Borucki, J. and A. Oberc. 1964. Wiek bezwzgledny tufu filipowskiego na podstawie datowan metoda potasowo-argonowa. KG 8:788 (in Polish).
[97]Brookins, D. G. and S. Chaudhuri. 1973. Comparison of Potassium-Argon and Rubidium-Strontium Age Determinations for Eskridge and Stearns Shales (Early Permian) Eastern Kansas. AG 57:523-4.
[98]Webb, A. W. and I. McDougall. 1968. The Geochronology of the Igneous Rocks of Eastern Queensland. GU 15:328.
[99]Lippolt, H. J. Von. 1976. Der Vertreuensbereich permischer Glimmer-Modell-Alter ans dem Saar Nahe Gebett. Neues Jahrbuch fur Geologie und Palaeontologie Monatschefte. Heft8, Jahr 1976, p. 471-5.
[100]Chen, J. H. 1976. U-Pb Isotopic Ages of the Southern Sierra Nevada Batholith and Independence Dike Swarm, California. GA 8(6):810.
[101]Zartman, R. E., Brock, M. R., Heyl, A. V., and H. H. Thomas. 1967. K-Ar and Rb-Sr Ages of some Alkalic Intrusive Rocks from Central and Eastern United States. AJ 265:863.
[102]Bickford, M. E. and D. G. Mose. 1969. Age of the Rose Dome Granite, Woodson County, Kansas. GA-1969, part 2, p. 2.
[103]Bateson, J. H. 1972. New Interpretation of Geology of Maya Mountains, British Honduras. AG 56:962.
[104]Ferrara, G. and F. Innocenti. 1974. Radiometric Age evidences of a Triassic Thermal Event in the Southern Alps. RU 63:573-5.
[105]Rutkowski, E. and J. Borucki. 1965. Pierwsze datowania bezwzgledne (K-Ar) granitoidow Mongolii zachodniej. KG 9:664-5 (in Polish).
[106]Priem, H. N. A., Boelrijk, N. A. I. M., and A. J. H. Boerboom. 1962. Lead Isotope Studies of the Lead-Zinc Deposits in Southern Limburg, the Netherlands. MI 41(10):432, 4.
[107]Bottino, M. L., Fullager, P. D., Fairbairn, H. W., Pinson, W. H. Jr., and P. M. Hurley. 1968. Rubidium-strontium study of the Blue Hills Igneous complex and the Wamsutta Formation Rhyolite, Massachusetts. GA-1968, p. 423.
[108]Borucki, J. and J. Lis. 1966. Sklad izotopowy i wiek bezwzgledny olowiu i galeny obszaru slasko-Krakowskiego. KG 10:926-7 (in Polish).
[109]Monger, J. W. H. 1977. Upper Paleozoic Rocks of the Western Canadian Cordillera and their bearing on Cordilleran Evolution. CE 14:1853.
[110]Fitch, F. J. and J. A. Miller. 1967. The age of the Whin Sill. LG 5:242-4.
[111]Bailey, S. W., Hurley, P. M., Fairbairn, H. W., and W. H. Pinson, Jr. 1962. K-Ar Dating of Sedimentary Illite Polytypes. GB 73:1168.
[112]Valencio, D. A. 1977. The paleomagnetism and K-Ar age of Upper Carboniferous Rocks from Andscollo Province of Neuquen, Argentina. UN 58:744.
[113]Hofmann, A. W., Mahoney, J. W., and B. J. Giletti. 1974. K-Ar and Rb-Sr Data on Detrital and Postdepositional History of Pennsylvanian Clay from Ohio and Pennsylvania. GB 85:640-2.
[114]_____. 1964. Op. cit. p. 283.
[115]Bell, J. S. 1973. Late Paleozoic Orogeny in the Northern Yukon. Proceedings of the Symposium on the Geology of the Canadian Arctic p. 30-1.
[116]Faul, H. 1962. Age and Extent of the Hercynian Complex. RU 52:775.
[117]Sabine, P. A. and J. V. Watson. 1970. Isotopic age-determinations of rocks and minerals from the British Isles, 1967-8. GL 126:392.
[118]Wanless, R. K., Stevens, R. D., Lachance, G. R., and R. N. Delabio. 1974. Age Determinations and Geological Studies: K-Ar Isotopic Ages, Report 11. CG 73-2, p. 73-4.
[119]Kovalenko, V. I., Kuz'min, M. I., Pavlenko, A. S., and A. S. Perfil'yev. 1973. South Gobi Belt of Rare Metal-Bearing Alkali Rocks in the Mongolian Peoples Republic and Its Structural Position. DE 210:77-9.
[120]Stevenson, I. P., Harrison, R. K., and N. J. Snelling. 1970. Potassium Argon Age Determination of the Waterswallows Sill, Buxton, Derbyshire. YG 37:445-6.
[121]Miller, J. E. and R. K. Wanless. 1974. A Paleozoic zircon age of the West Coast Crystalline Complex of Vancouver Island, British Columbia. CE 11:1720-1.
[122]Faul. 1962. Op. cit. p. 771, 4.
[123]Hurley et. al. 1960. Op. cit. p. 1804.
[124]Hamet, J. and C. J. Allegre. 1976. Hercynian orogeny in the Montagne Noire (France). GB 87:1433-4.
[125]Faul, H. and E. Jager. 1963. Ages of Some Granitic Rocks in the Vosges, the Schwarzewald, and the Massif Central. GR 68:3293, 7.
[126]Vinogradov, A. P. and A. I. Tugarinov. 1961 translation. Some Supplementary Determinations of Absolute Age (Towards a Universal Geochronological Scale) DE 134(1-6):919.
[127]Borodina, N. S. Fershtater, G. B., and G. I. Samarkin. 1971. Variscan Tonalite-Granodiorite Association of the Southern Urals. DE 200:32.
[128]Fairbairn, H. W., Hurley, P. M., and W. H. Pinson. 1964. Preliminary Age Study and Initial Sr^{87}/Sr^{86} of Nova Scotia Granitic Rocks by the Rb-Sr Method. GB 75:253.
[129]Grikurov, E. G., Krylov, A. Ya., and Yu. I. Silin. 1967. Absolute age of Some Rocks From the Scotia Arc and Alexander I Land (Western Antarctica) DE 172:21.

¹³⁰Silver, L. T. 1965. Compilation of Phanerozoic Data for Western North America. (in _____ 1965. *Geochronology of North America*. National Academy of Sciences; National Research Council, Washington, D. C.) p. 231.

¹³¹Knyazev, V. S. and O. A. Schnip. 1971. Magmatic Rocks in the base of the Turanian plate. *IG* 13:352.

¹³²Lanphere, M. A., Loney, R. A., and D. A. Brew. 1965. Potassium-Argon Ages of Some Plutonic Rocks, Tenakee Area, Chichagof Island, Southeastern Alaska. *UG* 525:B110.

¹³³Webb and McDougall. 1968. *Op. cit.* p. 317-8.

¹³⁴Zartman, R. E., Hurley, P. M., Krueger, W. H., and B. J. Giletti. 1970. A Permian Disturbance of K-Ar Radiometric Ages in New England: Its Occurrence and Cause: *GB* 81:3361-3.

¹³⁵*Ibid.*, p. 3360.

¹³⁶Tikhonenkova, R. P. 1972. New Data on The Composition and Age of the Lovozero Suite of the Kola Region. *DE* 203:76.

¹³⁷Evernden, J. F. and J. R. Richards. 1962. Potassium-Argon Ages in Eastern Australia. *GU* 9(1):12, 41.

¹³⁸Halliday, A. N., McAlpine, A., and J. G. Mitchell. 1977. The Age of the Hoy Lavas, Orkney. *SG* 13:44, 51.

¹³⁹Gaudette, H. E., Fairbairn, H. W., Kovach, A., and A. M. Hussey. 1975. Preliminary Rb-Sr Whole-Rock Age Determinations of Granitic Rocks in Southwestern Maine. *GA* 7(1):62-3.

¹⁴⁰Lyons, J. B. and D. E. Livingston. 1977. Rb-Sr age of the New Hampshire Plutonic Series. *GB* 88:1808-11.

¹⁴¹_____. 1964. *Op. cit.* p. 269-70.

¹⁴²Fullager, P. D. and M. L. Bottino. 1969. Rubidium-Strontium Age Study of Middle Devonian Tioga Bentonite. *Southeastern Geology* 10:250.

¹⁴³Ordynets, G. A. and G. P. Poluarshinov. 1973. The Isotopic Composition of Lead in Ore Deposits of The Central Ishim Region. *GI* 10(3):521.

¹⁴⁴Hurley *et. al.* 1960. *Op. cit.* p. 1806.

¹⁴⁵Davidson. 1960(b). *Op. cit.* p. 337.

¹⁴⁶Mayne, K. I., Lambert, R. J., and D. York. 1959. The Geological Time Scale. *NA* 183:212.

¹⁴⁷_____. 1964. *Op. cit.* p. 273-4.

¹⁴⁸Cormier, R. F. 1956. Rubidium-Strontium Ages of Glauconite. *GB* 67:1812.

¹⁴⁹Thompson, G. R. and J. Hower. 1973. An explanation for low radiometric ages from glauconite. *GC* 37:1474.

¹⁵⁰Shibata, K. 1974. Rb-Sr geochronology of the Hikami granite, Kitakama Mountains, Japan. *GJ* 8:

¹⁵¹Ordynets and Poluarshinov. 1973. *Op. cit.* p. 520-1.

¹⁵²Allsopp, H. L. and P. Kolbe. 1965. Isotopic age determinations on the Cape Granite and intruded Malmesbury sediments, Cape Peninsula, South Africa. *GC* 29:1115-6.

¹⁵³_____. 1964. *Op. cit.* p. 361.

¹⁵⁴Darnley, A. G. 1964. Uranium-thorium-lead age-determinations with respect to the Phanerozoic Time Scale. *TS*:77.

¹⁵⁵Fitch, F. J., Miller, J. A., and M. Y. Meneisy. 1963. Geochronological Investigations of Rocks from North Wales. *NA* 199:449-50.

¹⁵⁶_____. 1964. *Op. cit.* p. 363.

¹⁵⁷*Ibid.*, p. 364.

¹⁵⁸Klingspor, I. 1976. Radiometric age-determination of basalts, dolerites, and related syenite in Skane, southern Sweden. *Geologiska Foreningens Forhandlingar* 98(3):207.

¹⁵⁹Davidson, C. F. 1964. On Diamantiferous Diatremes. *EG* 59:1371.

¹⁶⁰Compston, W. and R. T. Pidgeon. 1962. Rubidium Strontium Dating of Shales by the Total-Rock Method. *GR* 67:3501.

¹⁶¹Palmer, H. C. and A. Hayatsu. 1975. Paleomagnetism and K-Ar Dating of Some Franklin Lavas and Diabases, Victoria Island. *CE* 12:1444.

¹⁶²Thorsteinsson, R. and E. T. Tozer. 1962. Banks, Victoria, and Stefansson Islands, Arctic Archipelago. *Geological Survey of Canada Memoir* 330:22-3.

¹⁶³_____. 1964. *Op. cit.* p. 365.

¹⁶⁴Jager, E. and H. J. Zwart. 1968. Rb-Sr Age Determinations of Some Gneisses and Granites of the Aston-Hospitalet Massif (Pyrenees). *MI* 47:349, 53, 55.

¹⁶⁵Seiders, V. M. 1976. Age, origin, regional relations, and nomenclature of the Glenarm Series, central Appalachian Piedmont: A reinterpretation: Discussion and Reply. *GB* 87:1519-20.

¹⁶⁶Higgins, M. W. 1976. Reply to V. M. Seiders. *GB* 87:1524, 6.

¹⁶⁷Evernden *et. al.* 1961. *Op. cit.* p. 98.

¹⁶⁸Berger, A. R. and R. S. Naylor. 1975. Isotopic Dates on Zircons From the Deadman's Bay Pluton, Northeastern Newfoundland, and Their Geological Implications. *Geological Association of Canada (Minerological Association of Canada Annual Meeting* 1974) 3rd circular, p. 9.

¹⁶⁹_____. 1971. *Op. cit.*, p. 61.

¹⁷⁰Brown, P. E., Miller, J. A., and R. L. Grasty. 1968. Isotopic Ages of Late Caledonian Granitic Intrusions in the British Isles. *YG* 15:263, 6.

¹⁷¹Pankhurst, R. J. and R. T. Pidgeon. 1976. Inherited Isotope Systems and the Source Region Pre-History of Early Caledonian Granites in the Dalradian Series of Scotland. *EP* 31:55, 6.

¹⁷²Dallmeyer, R. D. 1975. Ar⁴⁰/Ar³⁹ Release spectra of Biotite and Hornblende from the Cortlandt and Rosetown Plutons, New York, and Their Regional Implications. *JG* 83:633.

¹⁷³Armstrong, R. L. 1975. PreCambrian (1500 m.y. old) Rocks of Central Idaho—The Salmon River Arch and Its Role in Cordilleran Sedimentation and Tectonics. *AJ* 275-A, p. 441-3, 7, 453.

¹⁷⁴_____. 1964. *Op. cit.* p. 378.

¹⁷⁵Brown, Miller, and Grasty. 1968. *Op. cit.*, p. 263, 4.

¹⁷⁶Wetherill, G. W. 1966. K-Ar Dating of PreCambrian Rocks (in Schaeffer, O. A., and J. Zahringer. 1966. *Potassium Argon Dating*. *SV*) p. 111.

¹⁷⁷Wanless *et. al.* 1968. *Op. cit.* p. 128.

¹⁷⁸Dasch, E. J. 1969. Strontium Isotopes in weathering profiles, deep-sea sediments, and sedimentary rock. *GC* 33:1545.

¹⁷⁹Evernden *et. al.* 1961. *Op. cit.* p. 94.

¹⁸⁰Pidgeon, R. T. and W. Compston. 1965. The Age and Origin of the Cooma Granite and its Associated Metamorphic Zones, New South Wales. *PE* 6(2):193, 6, 207.

¹⁸¹LeBas, M. J. 1972. Caledonian Igneous Rocks Beneath Central and Eastern England. *YG* 39:76-7.

¹⁸²Brown, C. E. 1975. Problematical age of an Alkaline Analcite-Bearing Olivine Diabase Dike in St. Lawrence County, New York. *GA* 7:31-2.

¹⁸³Herzog, L. F., Pinson, W. H., and R. F. Cormier. 1958. Sediment Age Determination by Rb/Sr Analysis of Glauconite. *AG* 42:726.

¹⁸⁴Holmes, A. 1959. A Revised Geological Time Scale. *SG (earlier form)* 17:195.

¹⁸⁵Hurley *et. al.* 1960. *Op. cit.* p. 1794.

¹⁸⁶Boissonas, J., Borsi and G. Ferrara, Fabre J., Fabries J., and M. Gravelle. 1969. On The Early Cambrian Age of Two late orogenic granites from west-central Ahaggar (Algerian Sahara). *CE* 6:30, 34.

¹⁸⁷Cobb, J. C. 1961. Dating of Black Shales. *NY* 91:312.

¹⁸⁸Bell, K. 1968. Age Relations and Provenance of the Dalradian Series of Scotland. *GB* 79:1180-1, 5.

¹⁸⁹Sedivy, R. A. and B. Broekstra. 1978. K/Ar Age Analysis of Size-Fractioned Samples of Conasauga Shale. *GA* 10:197.

¹⁹⁰Hurley *et. al.* 1960. *Op. cit.* p. 1805.

¹⁹¹Brooks, C., James, D. E., and S. R. Hart. 1976. Ancient Lithosphere: Its Role in Young Continental Volcanism. *SC* 193:1088.

¹⁹²Mirkina, S. L., Zhidkov, A. Ya, and M. N. Golubchina. 1973. Radiologic Age of Alkalic Rocks and Granitoids of the North Baikal Region. *DE* 211:117-8.

¹⁹³Plumb, K. A., Shergold, J. H., and M. Z. Stefanski. 1976. Significance of Middle Cambrian Trilobites from Elcho Island, Northern Territory, Australia. *GG* 1(1):53.

¹⁹⁴_____. 1964. *Op. cit.* p. 377-8.

¹⁹⁵Evernden *et. al.* 1961. *Op. cit.* p. 92.

¹⁹⁶Bultitude, R. J. 1976. Flood Basalts of Probable Early Cambrian Age in northern Australia (in Johnson, R. W. 1976. *Volcanism in Australasia*, Elsevier Co., New York, Amsterdam) p. 7.

¹⁹⁷Chaudhuri, S. and D. G. Brookins. 1969. The Isotopic Age of the Flathead Sandstone, (Middle Cambrian) Montana. *SP* 39:367.

¹⁹⁸Kazakov, G. A. and N. I. Polevaya. 1958. Some Preliminary Data on Elaboration of the Post-Pre-Cambrian Scale of Absolute Geochronology Based on Glauconites. *Geochemistry* 4:379.

¹⁹⁹Brookins, D. G. and S. A. Norton. 1975. Rb/Sr Whole-Rock Ages Along the PreCambrian-Cambrian Contact, East Side of the Berkshire Massif, Massachusetts. *GA* 7:30-1.

²⁰⁰Strong, D. F. and H. Williams. 1972. Early Paleozoic Flood Basalts of Northwestern Newfoundland: Their Petrology and Tectonic Significance. *Geological Association of Canada Proceedings* 24(2):43-4.

²⁰¹Dietrich, R. V., Fullager, P. D., and M. L. Bottino. 1969. K/Ar and Rb/Sr Dating of Tectonic Events in the Appalachians of Southwestern Virginia. *GB* 80:308-9.

²⁰²Kushnareva, T. I. and N. B. Rasskazova. 1978. The Ordovician of the Pechora syneclise. *IG* 20(6):704-5.

²⁰³Wanless, R. K., Stevens, R. D., Lachance, G. R., and R. N. Delabio.

1970. Age Determinations and Geological Studies: K-Ar Isotopic Ages, Report 9. *CG* 69-2A, p. 69.
[204] Fairbairn, H. W., Moorbath, S., Ramo, A. O., Pinson, W. H. Jr., and P. M. Hurley. 1967. Rb-Sr Age of Granitic Rocks of Southeastern Massachusetts and the Age of the Lower Cambrian at Hoppin Hill. *EP* 2:322.
[205] *Ibid.*, p. 325.
[206] Schurmann, H. M. E. 1964. Rejuvenation of Precambrian Rocks Under Epirogenetical Conditions During Old Paleozoic Times in Africa. *MI* 43:196-8.
[207] Bertrand, J. M. L. and R. Caby. 1978. Geodynamic Evolution of the Pan African Orogenic Belt: A New Interpretation of the Hoggar Shield (Algerian Sahara). *RU* 67(2):363, 9.
[208] Compston, W. and P. A. Arriens. 1968. The Precambrian Geochronology of Australia. *CE* 5:566-7.
[209] McCartney, W. D., Poole, W. H., Wanless, R. K., Williams, H., and W. D. Loveridge. 1966. Rb/Sr Age and Geological Setting of the Holyrood Granite, Southeastern Newfoundland. *CE* 3:509.
[210] Cormier, R. F. 1969. Radiometric dating of the Coldbrook Group of southern New Brunswick, Canada. *CE* 6:397.
[211] Braziunas, T. F. 1975. A geological duration chart. *GE* 3:342-3.
[212] Morris, H. M. 1975. The Young Earth. *Creation Research Society Quarterly* 12(1):19-22.
[213] Whitcomb, J. C. Jr., and H. M. Morris. 1961. *The Genesis Flood.* Presbyterian and Reformed Pub. Co., Pennsylvania, p. 331-368.
[214] Cook, M. A. 1966. *Prehistory and Earth Models.* Max Parrish, London, p. 23-66.
[215] Slusher, H. S. 1973. Critique of Radiometric Dating. *ICR Monograph* 2:47 p.
[216] Wilkerson, G. 1976. Review of Uranium-Thorium-Lead Radiometric Dating Methods (*in* D. A. Wagner, ed., 1976. *Student Essays on Science and Creation* 1:50-91).
[217] Goldich, S. S. 1972. Fallacious Isochrons and Wrong Numbers. *GA* 4(4):322.
[218] Obradovich, J. D., and C. E. Peterman. 1968. Geochronology of the Belt Series, Montana. *CE* 5:741.
[219] Afanass'yev, G. D. 1970. Certain Key Data for the Phanerozoic Time-Scale. *GH* 63(1):1.
[220] Armstrong and Besancon. 1970. *Op. cit.* p. 19.
[221] Mitchell, J. G. and K. P. Reen. 1973. Potassium-argon ages from the Tertiary Ring Complexes of the Ardnamurchan Peninsula, Western Scotland. *Geological Magazine* 110:337.
[222] Twiss, P. G. and R. K. DeFord. 1967. Potassium-Argon Dates from Vieja Group, Rim Rock County, trans-Pecos, Texas. *GA*-1967, p. 380.
[223] Armstrong, R. L. 1963. K-Ar Dates from West Greenland. *GB* 74:1189.
[224] Polevaya *et. al.* 1961. *Op. cit.* p. 298.
[225] Curtis, G. H., Savage, D. E., and J. F. Everndern. 1961. Critical Points in the Cenozoic. *NY* 91:349.
[226] Forman, J. A. 1970. Age of the Catalina Island Pluton, California. *GP* 124:41.
[227] Rubinshtein, M. M. 1961. Some Critical Points of the Post-Cryptozoic Geological Time Scale. *NY* 91:364.
[228] Morton, J. P. and L. E. Long. 1978. Rb-Sr Ages of Paleozoic Glauconites From the Llano Region of Central Texas. *GA* 10(1):22-3.
[229] Owens and Sohl. 1973. *Op. cit.* p. 2824.
[230] Evernden *et. al.* 1961. *Op. cit.* p. 83.
[231] Holmes. 1959. *Op. cit.* p. 191.
[232] Hower, J., Hurley, P. M., Pinson, W. H., and H. W. Fairbairn. 1961. Effect of Minerology on K/Ar Age As a Function of Particle Size in a Shale. *GA*-1961. p. 202.
[233] Miller, J. A. and P. A. Mohr. 1964. Potassium-Argon Measurements on The Granites and Some Associated Rocks from Southwest England. *LG* 4(1):116.
[234] Zartman, R. E. 1964. A Geochronological Study of the Lone Grove Pluton from the Llano Uplift, Texas. *PE* 5(3):399.
[235] Armstrong, R. L. 1970. Geochronology of Tertiary igneous rocks, eastern Basin and Range Province, eastern Nevada and vicinity, USA *GC* 34:222.
[236] Bassett, W. A., Kerr, P. F., Schaeffer, D. A., and R. W. Stoenner. 1963. Potassium-Argon Dating of the Late Tertiary Volcanic Rocks and Mineralization of Marysvale, Utah. *GB* 74:218.
[237] Bell. 1968. *Op. cit.* p. 1183.
[238] Dasch, E. J., Armstrong, R. L., and S. E. Clabaugh. 1969. Age of the Rim Rock Dike Swarm, Trans-Pecos, Texas. *GB* 80:1822.
[239] Evans, A. L., Fitch, F. J., and J. A. Miller. 1973. Potassium-Argon age determinations on some British Tertiary igneous rocks. *GL* 129:425.
[240] Durant, G. P., Dobson, M. R., Kokelaar, B. P., MacIntyre, R. M., and W. J. Rea. 1976. Preliminary report on the nature and age of the Blackstones Bank Igneous Centre, western Scotland. *GL* 132:325.
[241] Zartman. 1964. *Op. cit.* p. 390.
[242] Zartman *et. al.* 1967. *Op. cit.* p. 861.
[243] Faure, G. and J. L. Powell. 1972. *Strontium Isotope Geology.* SV., p. 92.
[244] Hamilton, E. I. 1965. *Applied Geochronology.* Academic Press, London and New York, p. 105.
[245] Evernden and Richards. 1962. *Op. cit.* p. 3.
[246] Zartman. 1964. *Op. cit.* p. 387.
[247] Pankhurst, R. J. 1970. The geochronology of the basic igneous complexes. *SG* 6:95.
[248] Eastin, R., and G. Faure. 1971. The Age of the Littlewood Volcanics of Coats Land, Antarctica. *JG* 79:244.
[249] Haller, J. 1971. *Geology of the East Greenland Caledonides.* John Wiley and Sons, New York.
[250] Burchart, J. 1970. The Crystalline Core of the Tatra Mountains: A Case of Polymetamorphism and Polytectonism. *GH* 63(1):55.
[251] Kratts, K. O. 1975. Study of the Precambrian in USSR (progress and problems). *IG* 17(9):1104.
[252] Wanless *et. al.* 1970. *Op. cit.* p. 24.
[253] Williams, I. S., Compston, W., Chapell, B. W., and T. Shirahase. 1975. Rubidium-Strontium Age Determinations on Micas From a Geologically Controlled Composite Batholith. *GU* 22(4):502.
[254] Scarborough, R. B. 1974. Geochronology of Pliocene Vitric Ash Falls in Southern Arizona. *GA* 6(3):249.
[255] Christianson, R. L., Obradovich, J. D., and H. R. Blank. 1968. Late Cenozoic Volcanic Stratigraphy of the Yellowstone Park Region—A Preliminary Report. *GA*-1968, p. 592.
[256] Curtis *et. al.* 1961. *Op. cit.* p. 348-9.
[257] Stewart *et. al.* 1974. *Op. cit.* p. 1108.
[258] Pankhurst. 1977. *op. cit.* p. 259.
[259] Gates, R. M., Martin, C. W., and R. W. Schnabel. 1973. The Cambrian-Precambrian Contact in Northwestern Connecticut and West Central Massachusetts. *GA* 5(2):165.
[260] Zartman *et. al.* 1970. *op. cit.* p. 3369.
[261] _____. 1972. Advances in Geochronometry: Ar^{40}/Ar^{39} technique of potassium-argon dating. *UG* 800-A119-20.
[262] Dallmeyer, R. D. 1977. $^{40}Ar/^{39}Ar$ Age Spectra of Minerals from the Fleur De Lys Terrane in Northwest Newfoundland: Their Bearing On Chronology of Metamorphism Within the Appalachian Cycle. *JG* 85:342.
[263] Ashkinadze, G. Sh., Gorokhovskiy, B. M. I., and Yu. A. Shokolyukov. 1977. Evaluation of Spectral $^{40}Ar/^{39}Ar$ Method of Dating Biotites with Partial Argon Loss. *GI* 13:187.
[264] Ashkinadze, G. Sh., Gorokhovskiy, B. M. I., and Yu. A. Shokolyukov. 1977. $^{40}Ar/^{39}Ar$ Dating of Biotite Containing Excess Argon. *GI* 14:175.
[265] Zartman, R. E. 1978. Reply to V. M. Seiders. *GB* 89:1117.
[266] Polyakov, G. V., Firsov, L. V., Teleshev, A. Ye., and G. S. Fedoseyev. 1972. Potassium-argon age of Early Paleozoic Granitoids of the Eastern Sayans as Determined on Rock and Biotite Samples. *DE* 202:225.
[267] Marvin *et. al.* 1973. *Op. cit.* p. 1.
[268] McKee, E. H. and D. C. Noble. 1976. Age of the Cardenas Lavas, Grand Canyon, Arizona. *GB* 87:1190.
[269] McDougall, I., Polach, H. A., and J. J. Stipp. 1969. Excess radiogenic argon in young subaerial basalts from the Auckland volcanic field, New Zealand. *GC* 33:1485, 1417.
[270] Matsuda, J. 1974. A virtual isochron Rb-Sr for an open system. *GJ* 8:1535.
[271] Fairbairn, H. W., Bottino, M. L., Pinson, W. H., and P. M. Hurley. 1966. Whole Rock Age and Initial $^{87}Sr/^{86}Sr$ ratio of volcanics underlying Fossiliferous Lower Cambrian in the Atlantic Provinces of Canada. *CE* 3:517.
[272] McDougall, I. and P. J. Leggo. 1965. Isotopic Age Determinations on Granitic Rocks From Tasmania. *GU* 12:324-5.
[273] Webb and McDougall. 1968. *Op. cit.* p. 335-6.
[274] Wetherill, G. W., Tilton, G. R., Davis, G. L., Hart, S. R., and C. A. Hopson. 1966. Age Measurements in the Maryland Piedmont. *GR* 71:2148.
[275] Long, L. E. 1978. Rb-Sr Isotope Systems in Caledonian Granites, County Donegal, Ireland. *GA* 10:113-4.
[276] Higgins, M. W. 1973. Superposition of Folding in the Northeastern

276. Maryland Piedmont and Its Bearing on the history and Tectonics of the Central Appalachians. *AJ* 273-A, p. 186.
277. Allsopp, H. L., Ulrych, T. J., and L. O. Nicolaysen. 1968. Dating of some significant events in the history of the Swaziland System by the Rb-Sr isochron method. *CE* 5:610.
278. Peterman, Z. E., Goldich, S. S., Hedge, C. E., and D. N. Yardley. 1972. Geochronology of the Rainy Lake Region, Minnesota-Ontario. *GM* 135:206.
279. Wasserburg, G. J. and M. A. Lanphere. 1965. Age Determination in the Precambrian of Arizona and Nevada. *GB* 76:745.
280. Webb and McDougall. 1968. *Op. cit.* p. 316.
281. Hurley, P. M. 1966. K-Ar Dating of Sediments (*in* Schaeffer and Zahringer. 1966. *Op. cit.*) p. 137.
282. Miller, D. S. and J. L. Kulp. 1958. Isotopic Study of Some Colorado Plateau Ores. *EG* 53:937.
283. Stuckless, J. S., and R. L. Erickson. 1976. Strontium Isotope Geochemistry of the Volcanic Rocks and Associated Megacrysts and Inclusions from Ross Island and vicinity, Antarctica. *MP* 58:120.
284. Armstrong, R. L. 1976. The Geochronometry of Idaho. *IS* 15:11.
285. Dietrich et. al. 1969. *Op. cit.* p. 312.
286. Curtis et. al. 1961. *Op. cit.* p. 344.
287. Yates, R. G., and J. C. Engels. 1968. Potassium-Argon Age of Some Igneous Rocks in Northern Stevens County, Washington. *UG* 600-D246.
288. Hopson, C. A. 1964. The Crystalline Rocks of Howard and Montgomery Counties, Maryland. *Maryland Geological Survey.* 1964, p. 201.
289. Leech et. al. 1963. *Op. cit.* p. 36.
290. Hills, F. A. and E. J. Dasch. 1968. Rb-Sr Evidence for Metamorphic Remobilization of the Stony Creek Granite, Southeastern Connecticut. *GA*-1968, p. 137.
291. Page, R. W. 1976. Reinterpretation of Isotopic Ages from the Hills Creek Mobile Zone, Northwestern Australia. *GG* 1(1):80.
292. Gerling, E., Kratz, K., and S. Lobach-Zhuchenko. 1968. Precambrian Geochronology of the Baltic Shield. 23rd *CO* 4:267.
293. Hebeda, E. H., Boelrijk, N. A. I. M., Priem, H. N. A., Verdurmen, E. A. Th, Verschure, R. H., and M. R. Wilson. 1973. Excess Radiogenic Argon in a Precambrian gabbroic-doleritic mass in Western Surinam. *Fortschritte Der Mineralogie* 50(3):82.
294. Wasserburg and Lanphere. 1965. *Op. cit.* p. 749.
295. Mark, R. K., Lee-Hu C., Bowman, R. E., and E. H. McKee. 1974. Recently (-10^9y) Depleted Radiogenic (87/86 Sr=0.706) Mantle Source of Ocean Ridge-Like Thoileitte, Northern Great Basin. *GA* 6:456.
296. Faure and Powell. 1972. *Op. cit.* p. 101-2.
297. Gerling, E. K., Morozova, I. M., and V. D. Sprintsson. 1968. On the Nature of the Excess Argon in Some Minerals. *GI* 9(6):1090.
298. Williams, H., and G. H. Curtis. 1977. The Sutter Buttes of California: A Study of Plio-Pleistocene Volcanism. *University of California Publications in Geology* 116:41.
299. Stockwell, C. H. 1968. Geochronology of stratified rocks of the Canadian Shield. *CE* 5:696.
300. Hurley, P. M., Fisher N. H., Pinson, W. H., and W. H. Fairbairn. 1961. Geochronology of Proterozoic Granites in Northern Territory, Australia: Part I: K-Ar and Rb-Sr Determinations. *GB* 72:661.
301. Peterman et. al. 1972. *Op. cit.* p. 195.
302. Dott, R. H. and I. W. D. Dalziel. 1971. Age and Correlation of the Precambrian Baraboo Quartzite of Wisconsin. *JG* 80:553.
303. Moorbath, S., O'Nions, R. K., and R. J. Pankhurst. 1975. The Evolution of early Precambrian Crustal rocks at Iscia, west Greenland-Geochemical and Isotopic Evidence. *EP* 27:238.
304. Sabine, P. A. and J. Watson. 1965. (Introduction to) Isotopic age-determinations of rocks from the British Isles, 1955-64. *GL* 121:531.
305. Wasserburg and Lanphere. 1965. *Op. cit.* p. 736.
306. Barton, J. M. Jr. 1977. Rb-Sr Ages and Tectonic setting of some granitic intrusions, coastal Labrador. *CE* 14:1641.
307. Pulvertaft, T. C. R. 1968. The Precambrian Stratigraphy of Western Greenland. 23rd *CO* 4:105.
308. Odom, A. L. and P. D. Fullagar. 1971. Rb-Sr Whole Rock Ages of the Blue Ridge Basement Complex. *GA* 3:336.
309. Sabine and Watson. 1965. *Op. cit.* p. 532.
310. Pulvertaft. 1968. *Op. cit.* p. 104.
311. Tugarinov and E. V. Bibikova. 1971. Geochronology of Brazil. *GI* 8:495.
312. Peterman, Z. E. and C. E. Hedge. 1968. Chronology of Precambrian events in the Front Range, Colorado. *CE* 5:749.
313. Gerling et. al. 1968. *Op. cit.* p. 271.
314. Waterhouse, J. B. 1978. Chronostratigraphy for the World Permian. *CT*, p. 316.
315. Armstrong, R. L. 1978. Pre-Cenozoic Phanerozoic Time Scale-Computer File of Critical Dates and Consequences of New and In-Progress Decay-Constant Revisions. *CT*, p. 74-87.
316. Varadarajan, S. 1978. Potassium-Arton Ages of the Amritpur Granite, District Nainital, Kumaun Himalaya and its stratigraphic position. *IN* 19:380-2.
317. Tugarinov and Bibikova. 1971. *Op. cit.* p. 501.
318. Kaneoka, I. and K. Aoki. 1978. Ar^{40}/Ar^{39} analyses of Phlogopite Nodules and Phlogopite-Bearing Peridotites In South African Kimberlites. *EP* 40:127.
319. Houtermans, F. G. 1966. The Physical Principles of Geochronology. *SR* No. 151, p. 242.
320. Brown, P. E. and J. A. Miller. 1969. Interrpretation of isotopic ages in orogenic belts (in Kent *et. al.* 1969. Time and Place in Orogeny. *GL* Special Publicatin No. 3), p. 137.
321. Marvin, R. F. and J. C. Cole. 1978. Radiometric Ages Compilation A, US Geological Survey. *IS* 22:9-10.
322. Mussett, A. E., Brown, G. C., Eckford, M., and S. R. Charlton. 1972. The British Tertiary Igneous Province: K-Ar Ages of Some Dykes and Lavas, from Mull, Scotland. *AA* 30:406-12.
323. Pankhurst, R. J., Walsh, J. N., Beckinsale, R. D., and R. R. Skelhorn. 1978. Isotopic and Other Geochemical Evidence For the Origin of the Loch UISG Granophyre, Isle of Mull, Scotland. *EP* 38:355.
324. Jacobson, R. R. E., Snelling, N. J., and J. F. Truswell. 1963. Age Determinations In the Geology of Nigeria, With Special Reference To The Older and Younger Granites. *OG* 9:178-9.
325. Shibata, K., Matsumoto, T., Yanagi, T., and R. Hamamota. 1978. Isotopic ages and Stratigraphic Control of Mesozoic Igneous Rocks in Japan. *CT*, p. 160.
326. Mattinson, J. M. 1978. Age, Origin, and Thermal History of Some Plutonic Rocks From the Salinian Block of California. *MP* 67:241-2.
327. Linares, E. and D. A. Valencio. 1975. Paleomagnetism and K-Ar Ages of Some Trachybasaltic Dykes From Rio De Los Molinos, Province of Cordoba, Argentina. *GR* 80:3315.
328. Majar, H. 1972. Geology of the Adventdalen Map Area. *NP* 138:33-4.
329. Halpern, M. 1972. Rb-Sr and K-Ar dating of rocks from southern Chile and West Antarctica. *AN* VII:150.
330. Halpern, M. and G. M. Carlin. 1971. Radiometric chronology of crystalline rocks from southern Chile. *AN* VI:192.
331. MacDonald, W. D. and N. D. Opdyke. 1972. Tectonic Rotations Suggested by Paleomagnetic Results from Northern Colombia, South America. *GR* 77:5723.
332. Faure, G., Kaplan, S., and G. Kulbicki. 1966. Interpretation des mesures d'ages fournies par l'analyse de mineraux argileux d'une se'rie sedimentaire d'Australie. *SR* No. 151, p. 572.
333. Deleon, G., Cewenjak, Z., Martinovic, G., and R. Filipovii. 1966. Age of Mt. Bukulja Granite (Yugoslavia). *SR* No. 151, p. 455, 6.
334. Hudson, T., Plfaker, C., and M. A. Lanphere. 1977. Intrusive Rocks of the Yakutat-St. Elias Area, South Central Alaska. *JR* 6:165.
335. Jeans, C. V., Merriman, R. J., and J. G. Mitchell. 1977. Origin of Middle Jurassic and Lower Cretaceous Fuller's Earths in England. *Clay Minerals* 12:11-12, 41.
336. Noble, D. C. 1977. A Summary of Radiometric Age Determinations From Peru. *IS* 19:10.
337. Bateman, P. C. and L. D. Clark. 1974. Stratigraphic and Structural Setting of the Sierra Nevada Batholith, California. *Pacific Geology* 8:82.
338. McDougall, I. and N. R. Ruegg. 1966. Potassium-argon dates on the Serra Geral Formation of South America. *GC* 30:191-2.
339. Hudson et. al. 1977. *Op. cit.* p. 161, 164.
340. Noble, D. C., Silberman, M. L., Megard, F., and H. R. Bowman. 1978. Comendite (Peralkaline Rhyolite) and Basalt In the Mitu Group (Peru): Evidence For Permian-Triassic Lithospheric Extension in the Central Andes. *JR* 6:454.
341. Shibata, K., Igi, S., Uchiumi, S. 1977. K-Ar Ages of hornblendes from gabbroic rocks in Southwest Japan. *GJ* 11:57-60.
342. Cordani, U. G., Kawashita, K., and A. T. Filho. 1978. Applicability of the Rubidium-Strontium Method to Shales and Related Rocks. *CT*, p. 105.
343. Waterhouse. 1978. *Op. cit.* p. 313.
344. *Ibid.*, p. 311.
345. Banks, N. G., Cornwall, H. R., Silberman, M. L., Creasey, S. C. and R. F. Marvin. 1972. Chronology of Intrusion and Ore Deposition at Ray, Arizona: Part I, K-Ar Ages. *EC* 67:867, 875.

³⁴⁷Afanassiev, G. D. 1966. Interpretation of K-Ar Dating of Rocks From the North Caucasus Folded Area. *SR* No. 151, p. 426.
³⁴⁸Irving, E. 1964. *Paleomagnetism and Its Application to Geological and Geophysical Problems*. John Wiley and Sons, New York, p. 330.
³⁴⁹Priem, H. N. A., Boelrijk, N. A. I. M., Hebeda, E. H., Schermerhorn, L. J. G., Verdurmen, E. A. Th., and R. H. Verschure. 1978. Sr-Isotopic Homogenization Through Whole-Rock Systems Under Low-Greenschist Facies Metamorphism In Carboniferous Pyroclasts at Aljustrel (Southern Portugal) *CH* 21:307.
³⁵⁰Cordani *et. al.* 1978. *Op. cit.* p. 106.
³⁵¹Stuckless, J. S. and R. L. Ericksen. 1975. Rubidium-strontium ages of basement rocks recovered from DVDP hole 6, southern Victoria Land. *AN* X:303, 4.
³⁵²Richards, J. R., Barkas, J. P., and T. G. Vallance. 1977. A Lower Devonian point in the geological timescale. *GJ* 11:47-9.
³⁵³Wanless, R. K., Stevens, R. D., Lachance, G. R., and R. N. D. Delabio, 1974. Age Determinations and Geological Studies: K-Ar Isotopic Ages, Report 12. *CG* 74-2, p. 62-3.
³⁵⁴Andresen, A. and K. S. Heier. 1975. A Rb-Sr Whole rock isochron date on an igneous rock-body from the Stavanger Area, South Norway. *RU* 64:261, 3.
³⁵⁵Spooner, C. M. and H. W. Fairbairn. 1970. Relation of Radiometric Age of Granitic Rocks near Calais, Maine to the Time of the Acadian Orogeny. *GB* 81:3667.
³⁵⁶Dooley, R. E. 1977. K-Ar Relationships in Dolerite Dykes of Georgia. *GA* 9:134.
³⁵⁷Zagorcev, I. S. 1974. On the Precambrian tectonics of Bulgaria. *RE* 1:141.
³⁵⁸Thomas, J. E., Dodson, M. H., and D. C. Rex and G. Ferrara. 1966. Caledonian Magmatism in North Wales. *NA* 209:867-8.
³⁵⁹Cordani *et. al.* 1978. *Op. cit.* p. 101.
³⁶⁰Wilson, M. R. 1971. The Timing of Orogenic Activity in the Bodo-Sulitjelma Tract. *GN* 269:185-8.
³⁶¹Vilas, J. F. and D. A. Valencio. 1978. Paleomagnetism and K-Ar age of the Upper Ordovician Alcaparrosa Formation, Argentina. *AA* 55:143-5.
³⁶²Cordani *et. al.* 1978. *Op. cit.* p. 112.
³⁶³Afanassiev. 1966. *Op. cit.* p. 424.
³⁶⁴Folinsbee, R. E., Baadsgaard, H., and G. L. Cumming. 1963. Dating of Volcanic Ash Beds (Bentonites) by the K-Ar Method. *National Academy of Sciences Publication* 1075, p. 71, 2.
³⁶⁵Sequin, M. K. 1977. Paleomagnetism of Middle Ordovician Volcanic Rocks From Quebec. *Physics of the Earth and Planetary Interiors* 15:364-6.
³⁶⁶Adams, J. A. S., Edwards, G., Henle, W., and K. Osmond. 1958. Absolute Dating of Bentonites by Strontium-Rubidium Isotopes. *GB* 69:1527.
³⁶⁷Fitch, F. J., Miller, J. A., and J. G. Mitchell. 1969. A new approach to radio-isotopic dating in orogenic belts (in Kent *et. al.* 1969. *Op. cit.*), p. 182-3.
³⁶⁸Ross, R. J., Naeser, C. W., and R. S. Lambert. 1978. Ordovician Geochronology. *CT* p. 349.
³⁶⁹Hagstrum, J. T. 1978. Paleomagnetism of Radiometrically Dated Late Precambrian to Cambrian Rocks from the Armorican Massif, France. *UN* 59:1038.
³⁷⁰Kovach, A., Fairbairn, H. W., and P. M. Hurley. 1976. Reconaissance Geochronology of Basement Rocks From the Amazonas and Maranhao Basins in Brazil. *RE* 3:472, 5.
³⁷¹Salop, L. J. 1977. *Precambrian Of The Northern Hemisphere*. Elsevier Scientific Publishing Co., Amsterdam, Oxford, New York, p. 4.
³⁷²Hayatsu, A., and C. M. Carmichael. 1970. K-Ar Isochron Method and Initial Argon Ratios. *EP* 8:71, 3.
³⁷³Tilton, G. R., Wetherill, G. W., and G. L. Davis. 1962. Mineral Ages from the Wichita and Arbuckle Mountains, Oklahoma, and the St. Francis Mountains, Missouri. *GR* 67:4011-12.
³⁷⁴Sigmond, E. M., and A. Andresen. 1976. A Rb-Sr isochron age of meta-andesites from Skorpehei, Suldal, south Norway. *TI* 56:315, 318-9.
³⁷⁵White, A. J. R. 1966. Genesis of Migmatites From the Palmer Region of South Australia. *CH* 1:173.
³⁷⁶Salop. 1977. *Op. cit.* p. 76 and enclosed maps.
³⁷⁷*Ibid.*, p. 5 and enclosed maps.
³⁷⁸Leutwein, F., Sonet, J., and M. D. Max. 1972. The Age of the Carnsore Granodiorite. *Geological Survey of Ireland Bulletin* 1(3): 303-7.
³⁷⁹Broch, O. A. 1964. Age determination of Norwegian minerals up to March 1964. *GN* 228:97.
³⁸⁰Hoggatt, W. C., Silberman, M. L., and V. R. Todd. 1977. K-Ar Ages of Intrusive Rocks of the Central Peloncillo Mountains, Hidalgo County, New Mexico. *IS* 19:4.
³⁸¹Lambert, R. J. and D. C. Rex. 1966. Isotopic Ages of Minerals from the Precambrian Complex of the Malverns. *NA* 209:605, 6.
³⁸²Gee, D. G. 1969. Isotopic age-determinations (in Flood *et. al.* 1969. The Geology of Nordhaustlandet, northern and central parts) *NP* 146:133.
³⁸³Piper, J. D. A. 1975. The Paleomagnetism of Precambrian Igneous and Sedimentary Rocks of the Orange River Belt in Africa and South West Africa. *AA* 40:314-5, 339.
³⁸⁴Adams, C. J. D. 1976. Geochronology of the Channel Islands and adjacent French Mainland. *GL* 132:241.
³⁸⁵Charlot, R. 1976. The Precambrian of the Anti-Atlas (Morocco): A Geochronological Synthesis. *RE* 3:293-7.
³⁸⁶Grasty, R. L. and C. Leelanandam. 1965. Isotopic ages of the basic charnockite and khondalite from Kondapalli, Andhra Pradesh, India. *Minerological Magazine* 35:530-1.
³⁸⁷York, D. and R. M. Farquhar. 1972. *The Earth's Age and Geochronology*. Permagon Press, Oxford, New York, Toronto.
³⁸⁸Davidson, C. F. 1960. Some Aspects of Radiogeology. *LG* 2:314.
³⁸⁹Hurley, P. M. and J. R. Rand. 1969. Pre-Drift Continental Nuclei. *SC* 164 (3885) 1241.
³⁹⁰Brown and Miller. 1969. *Op. cit.* p. 152 (reference 321).
³⁹¹Chivas, A. R. and I. McDougall. 1978. Geochronology of the Koloula Prophyry Copper Prospect, Guadalcanal, Solomon Islands. *EG* 73:682.
³⁹²Pozdeyev, A. I. 1973. Late Paleogene terrestrial volcanism in Koryak Highland and its metallogenic characteristics. *IG* 15:825-6.
³⁹³Ludwig, K. R. 1978. Uranium-Daughter Migration and U/Pb Isotope Apparent Ages of Uranium Ores, Shirley Basin, Wyoming. *EG* 73:31.
³⁹⁴Hawkesworth, C. J., Norry, M. J., Roddick, S. C., and P. E. Baker. 1979. $^{143}Nd/^{144}Nd$, $^{87}Sr/^{86}Sr$, and Incompatible Element Variations in Calc Alkaline andesites and Plateau Lavas From South America. *EP* 42:46, 48.
³⁹⁵Bhanot, V. B., Bhandari, A. K., Singh, V. P., and A. K. Kansal. 1979. Geochronological and Geological Studies On a Granite of Higher Himalaya Northeast of Manikaran, Himachal Pradesh. *IN* 20:92.
³⁹⁶Wellman, P., and A. Cooper. 1971. Potassium-Argon Age of Some New Zealand Lamprophyre Dikes near the Alpine Fault. *ZG* 14:343-4.
³⁹⁷Duncan, I.S., and R. R. Parrish. 1979. Geochronology and Sr Isotope Geochemistry of the Nelson Batholith: A Post-Tectonic Instrusive Complex in Southeast British Columbia. *GA* 11(3):76.
³⁹⁸Afanas'yev, G. D. and F. N. Yefimov. 1976. Petromagnetic Characteristics of gabbroid dike rocks in Northern Caucasus. *IG* 18:174.
³⁹⁹Aronson, J. L. 1965. Reconaissance Rubidium-Strontium Geochronology of New Zealand Plutonic and Metamorphic Rocks. *ZG* 8:416-7.
⁴⁰⁰Agrawal, J. K., and I. A. Sc. Rama. 1976. Chronology of Mesozoic volcanics of India. *Proceedings of the Indian Academy of Sciences* 84A(4):160-1, 9.
⁴⁰¹Chen, J. H. and J. G. Moore. 1979. Late Jurassic Independence dike swarm in eastern California. *GE* 7:129-131.
⁴⁰²Dalrymple, G. B., Gromme, C. S., and R. W. White. 1975. Potassium-Argon Age and Paleomagnetism of Diabase Dikes in Liberia: Initiation of Central Atlantic Rifting. *GB* 86:399, 401.
⁴⁰³Gulson, B. L., and H. Rutishauser. 1976. Granitization and U-Pb Studies of zircons in the Lauterbrunnen Crystalline Complex. *GJ* 10:13, 20.
⁴⁰⁴Mukhopadhyay, B. and D. G. Brookins. 1976. Rb-Sr Whole-Rock Geochronology of the Madera Formation Near Albuquerque, New Mexico. *SP* 46:680, 3.
⁴⁰⁵Chaudhuri, S. and D. G. Brookins. 1979. The Rb-Sr Systematics In Acid-Leached Clay Minerals. *CH* 24:232-6.
⁴⁰⁶Trettin, H. P. and H. R. Balkwill. 1979. Contributions to the tectonic history of the Innuitian Province, Arctic Canada. *CE* 16:765.
⁴⁰⁷Doe, B. R. 1970. *Lead Isotopes* SV, p. 14.
⁴⁰⁸Lis, J. and H. Sylwestrzak. 1977. Geochronologia a geochemiczne zroznicowanie mlodopaleozoicznych wulkanitow dolnego Slaska. *KG* 21:368 (in Polish).

[409]Sobolev, R. N., Borshchevskiy, Yu.A., and V. M. Shulga. 1971. The K-Ar age of Kazakhstan granitoids, as influenced by hybridization. *IG* 13:1746-8.

[410]Knyazev and Schnip. 1971. *Op. cit.* p. 353.

[411]Brakhfogel, F. F. and V. V. Kovalskiy. 1979. Age of the kimberlite bodies of the Siberian Platform. *IG* 21:310-2.

[412]Knyazev and Schnip. 1971. *Op. cit.* p. 351.

[413]Roddick, J. C. and W. Compston. 1977. Strontium Isotopic Equilibration: A Solution to a Paradox. *EP* 34:241.

[414]Roddick, J. C. and W. Compston. 1976. Radiometric Evidence For the Age of Emplacement and Cooling of the Murrumbidgee Batholith. *GU* 23:223.

[415]Priem, H. N. A., Boelrijk, N. A. I. M., M. Verschure, R. H., and E. H. Hebeda. 1967. Isotopic Age Determinations on Granitic Rocks in Northern Portugal. *MI* 46:370-2.

[416]Somin, M. L. 1970. Interpretation of absolute age determinations of crystalline schists of the Main Caucasian range by the argon method. *IG* 12:1227-8.

[417]Rundle, C. C. 1979. Ordovician intrusions in the English Lake District. *GL* 136:29-30, 33.

[418]Bell, K. and J. Blenkinsop, Berger, A. K., and N. R. Jayasinghe. 1979. The Newport granite: its age, geological setting, and implications from the geology of the Northeastern Newfoundland. *CE* 16:265.

[419]Cooper, R. A. 1974. Age of Greenland and Waiuta Groups. South Island New Zealand. *ZG* 17:955, 961.

[420]Dow, J. A. S., and V. E. Neall. 1974. Geology of the Lower Rennick Glacier, Northern Victoria Land, Antarctica. *ZG* 17(3):687, 691-2.

[421]Bath, A. H. 1974. New isotopic age data on rocks from the Long Mynd, Shropshire. *GL* 130:567, 570-1.

[422]Burg, J. P. and P. J. Matte. 1978. A Cross Section through the French Massif Central and the Scope of its Variscan Geodynamic Evolution. *Zietschrift der Deutchen Geologischen Gesselschaft.* Band 129(2):440.

[423]Hayatsu A. 1979. K-Ar isochron age of the North Mountain Basalt, Nova Scotia, *CE* 16:974.

[424]Armstrong B. L. 1978. K-Ar Dating: Late Cenozoic McMurdo Volcanic Group and dry valley glacial history. Victoria Land, Antarctica. *ZG* 21:692.

[425]Criscione J. J., Davis T. E., and P. Ehlig. 1978. Age of Sedimentation, Diagenesis for Bedford Canyon Formation and Santa Monica Slate in Southern California; Rb-Sr Evaluation. *AG* 62:2352.

[426]Howitt F., Aston E. R., and M. Jacque. 1975. The Occurrence of Jurassic Volcanics in the North Sea (in Woodland A. W. ed. 1975. *Petroleum and the Continental Shelf of North-West Europe.*) John Wiley and Sons, New York, p. 383.

[427]Armstrong. 1978. *op. cit.,* p. 685, 692.

[428]Davidson C. F. 1967. The kimberlites of the USSR (in Wyllie P. J. 1967. *Ultramafic and Related Rocks*) John Wiley and Sons, NY, pp. 259-60.

[429]Speed, R. C. 1979. Collided Paleozoic Microplate in the Western United States. *JG* 87:286-7.

[430]Dobretsov, G. L. and I. A. Zagruzina. 1977. Young Basaltoid Igneous Activity In The Eastern Tien Shan. *DE* 235:68-9.

[431]Sweeney, J. F. 1977, Subsidence of the Sverdup Basin, Canadian Arctic Islands. *GB* 88:41.

[432]Landis, C. A. and D. S. Coombs, 1967. Metamorphic Belts and Orogenesis in Southern New Zealand. *Tectonophysics* 4:505.

[433]Adib, D. 1978. Geology of the Metamorphic Complex at the South Western Margin of the Central-Eastern Iranian Microplate (Neyriz Area). *Neues Jahrbuch For Geologie und Palaeontologie Abhandlungen* 156(3)403-4.

[434]Hynes, A. J., Nisbet E. G., Smith A. G., and M. J. P. Welland. 1972. Spreading and emplacement ages of some ophiolites in the Othris region (eastern Central Greece). *Zietschrift der Deutchen Geologischen Gesselschaft.* Band 123:462-3.

[435]Stocklin, J., 1974. Possible Ancient Continental Margins in Iran (in Burk, C. A. and C. L. Drake. 1974. *The Geology of Continental Margins.*) SV, p. 880.

[436]Richards J. R., White D. A., Webb, A. W., and C. D. Branch. 1966. Isotopic Ages of Acid Igneous Ricks in the Cairns Hinterland, North Queensland. *Australia: Bureau of Mineral Resources Geological and Geophysical Bulletin* 88:10, 13.

[437]Howitt, et. al. 1975. *op. cit.,* p. 386.

[438]Lowry, W. D., 1974. North American Geosynclines—Test of Continental-Drift Theory. *AG* 58:591-2.

[439]Lapointe, P. L., 1979. Paleomagnetismmand orogenic history of the Botwood Group and Mount Peyton Batholith, Central Mobile Belt, Newfoundland. *CE* 16:868.

[440]Shackleton, R. M., Ries, A. C., Coward, M. P., and P. R. Cobbald. 1979. Structure, Metamorphism, and geochronology of the Arequipa Massif of coastal Peru. *GL:* 136:195, 208.

[441]Goldsmith, R., Marvin, R. F., and H. H. Mehnert. 1971. Radiometric Ages in the Santandar Massif, Eastern Cordillera, Colombian Andes. *UG* 750-D46-7.

[442]Vidal, Ph., Auvray B., Charlot R., Fediuk F., Hameurt, J. and J. Waldausiova. 1975. Radiometric age of volcanics of the Cambrian "Krikovlat-Rokycany" Complex (Bohemian Massif) *RU* 64(2)568-9.

[443]Chowns, T. M., 1978. Pre-Cretaceous Geology Beneath Georgia Coastal Plain. *AG* 62:504.

[444]Sainsbury, C. L., Hedge Ce., E., and C. M. Bunker. 1970. Structure, Stratigraphy, and Isotopic Composition of Rocks of Seward Peninsula, Alaska. *AG* 54:2503.

[445]Richards, et. al. 1966. *op. cit.,* p. 8, 21.

The Cephalopods in the Creation and the Universal Deluge

THE CEPHALOPODS IN THE CREATION AND THE UNIVERSAL DELUGE

JOHN WOODMORAPPE*

Received October 24, 1977

"Then God said, 'Let the waters teem with swarms of living creatures' (For He commanded and they were created: Psalm 148:5b . . . calls into being that which does not exist: Romans 4:17d) . . . with which the waters swarmed after their kind: and God saw that it was good . . . And there was evening and there was morning, a fifth day."
—Genesis 1:20-23 (NASB)

The study of claims of Cephalopod evolution reveals many fossil-gaps; but the outstanding result is the discrediting of the Biogenic Law and the discovery of the large degree of similarity in forms considered to be unrelated by evolution.

Much of the stratigraphic order (generic; specific) ascribed to ammonoids is actually due to time-stratigraphic concepts and to taxonomic manipulations. Indeed, "condensed" sequences demonstrate rather mixing with cataclysmic burial.

The known ecological positions of cephalopods independently fit together into a mutually contemporaneous ecologically zones coexistence. The actual stratigraphic order (ordinal; familial) owes its existence to the burial of these ecological zones in the Flood, while physical sorting during burial gave rise to interfamilial stratigraphic order.

Outline

Introduction
I. Creation Versus Evolution of the Cephalopods
   A. Origin of Phylum Mollusca and its class Cephalopoda
   B. The Genesis of Orders and Lower Taxons
   C. Fallacies of Recapitulation as Illustrated by Cephalopods
   D. "Convergence" in Cephalopods as Evidence for Creation

II. Explaining Ammonoid Biochronological Horizons: a Challenge for Diluviology
   A. The Substantial Subjectivity of Fossil "Species" and "Genera"
   B. Procedures Which Eliminate Successional Discrepancies
   C. The Scattered—Not Worldwide—Distribution of Fossil Zones
   D. Biochronologic Ammonoid Zones as Taxonomic Concoctions
   E. "Condensed" Ammonitiferous Deposits Indicate Rapid Burial

*John Woodmorappe, B.A., has studied both Geology and Biology.

III. The Stratigraphically-Ordered Flood Burial of Cephalopods
   A. Ecological Zonation and the Deluge: Preliminary Considerations
   B. The Antediluvian Ecologically-Zones Coexistence of Cephalopods
   C. The Sequential Flood-Burial of Cephalopods

Introduction

The Cephalopods, a Class of the Phylum Mollusca, are a group of predacious marine creatures which have had a long, illustrious fossil-record history but of which only few forms are extant. The Cephalopods deserve Creationists' scholarly attention because (1) They are the most complex, most advanced, and naturally the most studied of all the invertebrates; (2) Most evolutionistic tenets lend themselves to clearer, more revealing examination through Cephalopods than through vertebrates; (3) Some Cephalopods (Ammonoids) play a major role in the intercontinental biostratigraphic "time"-correlation and geologic-"age" claims because of their unrivaled degree of stratigraphically-ordered (wide geographical extent with short vertical range) succession; this group providing the stiffest challenge to Diluviology because of its successional order. Accordingly, this work evaluates claims of their evolution and impeccable successional order, and then provides a carefully-supported ecological explanation for their successional order in the (Flood deposited) Phanerozoic fossil record.

The extant *Nautilus* (Fig. 1) provides a useful reference for a brief synopsis of Cephalopod morphology. The internal organs reveal a high degree of tissue/organ/physiology specialization and efficiency uncommon among invertebrates. The extended tentacles catch food; the radula tears flesh; digestion follows. Reproduction is sexual, from eggs. Some cephalopods have no larval stage (unlike other molluscs). Swimming takes place by means of rocket propulsion when the Mantle Cavity is allowed to fill with water (simultaneously aerating the Gills) and this highly-muscular organ forcibly expels the water out the Funnel (Hyponome), propelling the animal. The conch partly encloses the body.

The conch is the most important part (from the present viewpoint) because only it survives as a fossil. The shell is not molted as in other molluscs, but progressively larger sections are grown (camarae, sealed off by septa). Unlike Gastropods, most cephalopods thus seal off most of their shells by septa during ontogeny. The shell serves as a versatile hydrostatic organ for swimming because (recently discovered) gas amounts in the sealed off camarae are actively regulated by the siphuncle (the only living part of the organism in the conch once that part of it is sealed off).

The Class Cephalopoda is composed of (somewhat controverted) 25 Orders (Fig. 2), 360 Families, 3000 Genera, and 10,000 species. Of these, merely 650 species of 4 Orders survive: the Octopodida (Octopus), Teuthida (Squids), Sepiida (Cuttlefish), and Nautilida (*Nautilus*).

The *Nautilus* (Fig. 1) may serve as a model of this class. The following variations in conch morphology are important: Among subclass Coleoidea (the most radically differing from *Nautilus*), the extinct Belem-

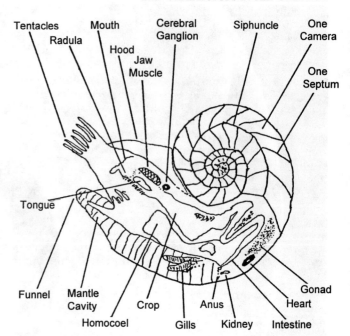

Figure 1. The Nautilus: the sole living representative of the conchiferous Cephalopods. Explained and discussed in the text. Modified after Sweet, Reference 1.

nitida had completely straightened-out (bullet-shaped) conch with a heavy calcareous rostrum ("shield") in front and the conch length devoted to the body, long in relation to camral sections (unlike *Nautilus*). The Order Sepiida went further: the entire ventral (bottom) half of the conch was occupied by the body. Still further goes the extant Squid (Order Teuthida) which has an internal, thin, bullet-shaped shell with only vestigial septa (not supportive of evolution: may be a genetic-code remnant of the more common conchiferous design employed by God elsewhere). It is the largest, fastest, and most advanced of all the invertebrates. The Order Octopodida is a sluggish swimmer with no shell.

Other than the just-discussed Coleoidea, all other Cephalopod subclasses and their orders are quite similar to the *Nautilus* (Fig. 1): differences being primarily of size, shell shape (coiled, uncoiled, tightly or loosely coiled) intraconchoidal deposits, suture shape, etc.

Specifically, the very important (biostratigraphically) subclass Ammonoidea (Orders Anarcestida, Prolecanitida, Clymeniida, Goniatitida, Ceratitida, Phyllocerida, Lytocerida, and Ammonitida) differs in having thinner shells, shell ornamentation (external ribs, keels, spines, nodes, etc.) and—most importantly— corrugated septa. The Endoceratoidea (Endocerida, Intejocerida) are noted for conical sheaths around the siphuncle (for ballast) and uncoiled, crescent-shaped conches. The Actinoceratoidea (Actinocerida) has cameral deposits (for ballast), as do many of Orthocerida, Belemnitida, Discosorida (which is of Subclass Nautiloidea). Most early Paleozoic forms are orthoconic (straightened-out shells) but all other lineages become coiled much like the *Nautilus*, except the aforementioned Coleoids and Bactritida (which remain straight) and a few aberrantly-coiled and torticonic Ammonoids. Most other differences from *Nautilus* are ones of proportion.

I. Creation Versus Evolution of the Cephalopods

A. Origin of Phylum Mollusca and Its Class Cephalopoda

The Cambrian explosion, which immediately eliminates 80% of any supposed (in this case) molecules-to-molluscs evolution, is very striking. In fact, for Fischer[3] the puzzle is "...the simultaneous appearance of exoskeletons in so many different kinds of plants and animals." "To date, the PreCambrian has yielded no molluscs..."[4] and there is only an imagined "...hypothetical ancestral mollusc..."[4]

Dozens of theories have been proposed to explain this explosion; and if Kuhn,[5] a philosopher of science, is correct in stating that the arising of many versions of a theory is a sign of its growing failure to face facts, evolution thus begins to collapse. Just one of many of these subsidiary hypotheses proposes that the Phyla Mollusca, Annelida, and Arthropoda arose from a non-preservable seriated, pseudo-metamorous flatworm[6] which underwent rapid, major (regulatory gene) mutations.[7] Needless to say, such grand-transforming mutations (like gradual mutations/natural selection) have never been demonstrated (much less proved) and "...form a fundamental question in evolutionary biology."[7] The oft-repeated claim that molluscs have arthropod/annelid affinities is based on embryological similarities and the supposedly metameric mollusc *Neopalina*. Many now maintain[8] that this is only a superficial resemblance of true metamerism and therefore not convincing.

Not only is the evidence for an evolutionary origin of mollusca (like other phyla) unconvincing, but so is that of the resulting classes. "The *unrecorded* Precambrian creatures were antecedent to about ten classes of mollusks."[9] Besides, "...survey of the first obscurely crawling molluscs could have afforded *no faintest indication* of possibilities eventually realized in the (among others)...predacious, rapidly darting squids."[10] (emph. added, and so on unless specified.) "The origin of the cephalopods, like that of other mollusks, is shrouded in the darkness of the PreCambrian."[11] "...the roots of the cephalopod tree lie somewhere deep in the PreCambrian."[12] "As with nearly every other group of organisms, *there is no objective record* of the earliest stages of cephalopod history."[13] Many decades ago, the *Volborthella* was proposed[14] as the first cephalopod, but recently it has been pointed out that this fossil is very enigmatic and may be a Tunicate,[15] etc. Recently, the monoplacophoran *Kirenyella*,[16] which is septated (unusual for non-cephalopods) and similar to the cephalopod *Plectronoceras* has been cited as the monoplacophoran-to-cephalopod transition. However, just as there are no (incipient-structure) transitions in vertebrate evolution[17] (no part-fin/part-leg (fish-to-amphibian), no half-scale/ half-feathers (reptile-to-bird), etc., transitions),[17] so likewise the supposedly-transitional *Kirenyella* shows no sign of a siphuncle, or even a partly-evolved one.[18] The origin of siphuncle, like most else, is relegated to speculation.

Although the Cambrian explosion goes contrary to all expectations of evolution, paleontological speculations to explain it away are prevalent, because—as the Soviet paleontologist Sokolov[18] remarks: "I know geologists who regard the whole of Darwin's theory and the present day synthetic theory of evolution (which do in fact have weak spots) as a type of religion; but we may readily imagine the chaos that would face us in geology were the evolution concept to become a myth..."

B. The Genesis of Orders and Lower Taxons

A diagram very similar to Fig. 2 shows dotted lines (gaps) between the Ellesmerocerida rootstock and most subclasses and Nautiloid orders.[19] The long debate as to whether the ammonoids evolved from coiled Nautiloids (Tarphycerida) or the Bactritida continues:[25] reinforced by a conch-coiling gap in the latter alternative[20] as well as the former.[21] Donovan[22] contends for a Belemnitid ancestry for all the Coleoids in contradistinction to Teichert[2] (Fig. 2); Jeletzky[23] simply leaves dotted lines with question marks in portraying how the six Coleoid orders relate to the supposedly-ancestral Bactritids and to each other. Among "advanced" ammonoids: "A phylogenetic classification here breaks down. There are so many ammonites...which cannot reliably be traced back to their parent stock that it is still a practical necessity to retain a polyphyletic suborder Ammonitina for all those ammonites (the vast majority)."[24]

The evolutionist Boucot[26] excellently summarizes the role of transition-lacks: "Since 1859 one of the most vexing properties of the fossil record has been its obvious imperfection. For the evolutionist this imperfection is most frustrating as *it precludes any real possibility for mapping out the path of organic evolution* owing to an infinity of 'missing links'...once above the family level it becomes very difficult in most instances to find any solid paleontological evidence for morphological intergrades between one suprafamilial taxon and another. This lack has been taken advantage of classically by opponents of organic evolution as a major defect in the theory...the inability of the fossil record to produce the 'missing links' has been taken as solid evidence for disbelieving the theory." It is not "taking advantage" to see that, even if time is gratuitously granted to the fossil record, it does not empirically demonstrate bio-transformism: all life—including cephalopods—was directly created by God and some of it buried (why in such stratigraphic order to be discussed in detail) during the Noachian Deluge.

Evolutionists seek to explain the gaps either by claiming that preservation-failures[27] (as from sedimentation breaks, etc.) or "punctuated equilibria"[28] (bursts of evolution too rapid to be fossil-recorded) explain the absent transitions. But the fact remains that admittedly transitions are absent;[31] and no arguments from ignorance or secondary hypotheses (akin to "Epicycles" proposed to "patch-up" the failing Ptolemaic geocentric theory) can remove it. In fact, if anything, evolutionary steps are so deduced to make the smallest gaps: Harper[29] stating, other factors about equal, ancestor-descendant lineages should be constructed that leave "fewer or shorter stratigraphic gaps." All fossil taxonomy is so designed to abet evolutionary speculations: "...any paleontological classification should be...phylogenetical at all levels...the concepts of tie and derivation must, indeed, be brought to define all its categories..."[30]

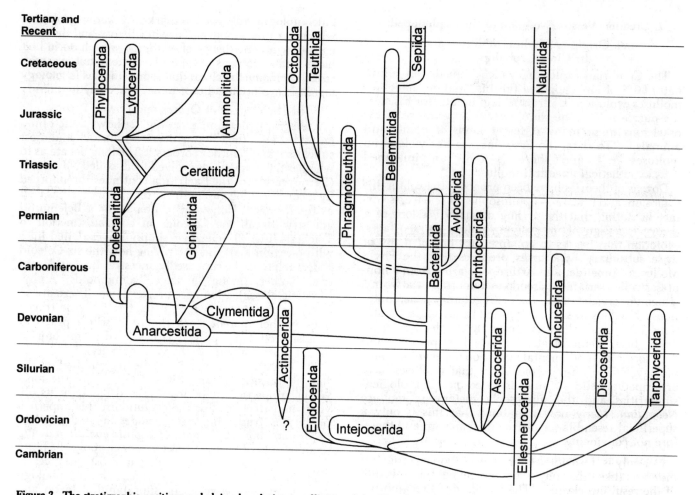

Figure 2. The stratigraphic positions and claimed evolutionary affinities of the 25 orders of Cephalopods. Modified after Teichert, Reference 2.

Not fact, but imaginitive speculation thereby dominates evolution: evolutionary steps can only be "... inferred"[32] and, "Of course, *we can never prove that the sequence we see actually mirrors the evolutionary process.*"[33] The sequence is considered probable if it repeats in different places and "facies," but Flood burial will later be considered an equal or superior explanation. Cephalopods, especially ammonites, specifically yield speculative evolutionary lineages: "The chief obstacle to such studies is that a lineage is an oversimplified concept; *it is impossible to pick out a stratified succession of individuals which can with certainty be said to be genetically connected in the strict ancestor-descendant relationship . . . it is difficult to be sure that our choice of individuals is not guided by preconceptions of what we are looking for.*"[34] Creationists must remember this recent statement by Gould, et. al.:[35] "Paleontologists (and evolutionary biologists in general) are famous for their facility in devising plausible stories; but they often forget that plausible stories need not be true."

Gaps abound at even lower taxon-levels: "... a great number of species of Jurassic ammonites appear and disappear suddenly and have neither known ancestry nor descendants."[36] These are not exceptions: "... explosive evolution of radically new types . . . is so common in the history of the Cephalopoda."[37] Many out-of-nowhere fossils exist.

Many claims, however, are made that a large number of gradual (transition-filled) sequences are now known. Closer examination reveals that abrupt changes of structure actually happen in these "continuous" supposed evolutionary lineages. Reyment,[38] for example, while noting the rarity of transition-filled sequences, claimed that he had some good examples of them: yet he noted that one trend was the disappearance of shell ornament, and this "loss of ventrolateral tubercles" was not gradual; suddenly it was completely gone in the otherwise identical superjacent "descendant." Many apparent lineages, as indeed the stratigraphic distribution of all cephalopods, are simply the result of the ecologically-controlled Floodwater burial of the cephalopod fossils (a whole area which will be discussed and documented in detail): most others seem continuous because they are so defined taxonomically. Thus cephalopod "species" and "genera" are incredibly subjective designations (will be fully documented) readily manipulable to so designate "species" and "genera" that they arbitrarily single out some morphological-attribute trend as an evolving lineage: ignoring others. Fig. 3(i) using general symbols for fossils, illustrates this; whereby a gradual arm-addition trend is taxonomically singled out (enclosed) by completely disregarding arm patterns which do not fit this "gradually evolving" trend, making this semi-random

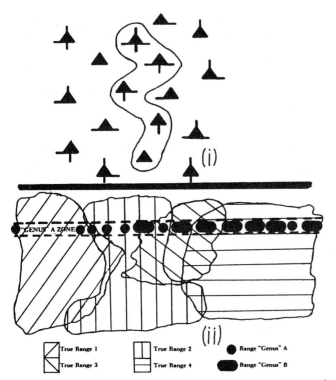

Figure 3. Taxonomic practices contributing to the fallacies of: (i) gradual (transition-filled) evolution within Cephalopod families, and (ii) world-wide zones of ammonoid genera.

distribution of fossils seem evolutionary. Extremely important is this fact: "Paleontologists have observed that rapidly evolving groups appear to have far less phenotypic variation than is true for more slowly evolving groups."[39] This low phenotype variation unmasks the taxonomic splitting which has artifically produced a seemingly transition-filled evolutionary trend. All relevant fossil evidence indicates that cephalopod groups were (and are) always distinct divine creatures (which never evolved).

C. Fallacies of Recapitulation as Illustrated by Cephalopods

The Biogenic Law (ontogeny recapitulates phylogeny) can be readily tested because the cephalopod conch retains juvenile features, since the oldest septa remain throughout lifetime as part of the conch (oldest septa nearest the tip, etc.). Yet it has failed: ". . . 'recapitulation,' embraced uncritically by Hyatt and Buckman, c. 1870-1930, but found to be unworkable, and now generally abandoned by ammonitologists."[40] Yet this matter is most definitely not merely historical, as many still believe it (for example Wiedman,[41] many Russian paleontologists, and sporadically others: a specific use of it reviewed by Haas[42]).

". . . some lineages of Mesozoic ammonites exhibit little or no recapitulation."[43] Not recapitulation, but a whole spectrum of supposed relations between ontogeny and phylogeny is observed: Clark,[44] studying ammonoids, noted paedomorphosis[44] (adults having characteristics of ancestors' juveniles), acceleration[45] (juveniles with characteristics of adult ancestors), and caenogenesis[46] (animals differ when young, but are similar when adult). Cenogenesis also implies dissimilarity to related forms, opposite of palingenesis (recapitulation). Yet supposed ". . . evolution could be either cenogenetic or palingenetic . . ."[47] But they admittedly cannot be distinguished; and: "Clearly one is involved in a *circular argument*; one cannot safely assume that palingenesis has occurred unless the course of evolution is already known."[47]

Acceptance of recapitulation led to admittedly absurd logical conclusions: "The papers of the 2nd and 3rd decades of the 20th century, written in the recapitulationist faith, make strange reading now. If, as was often found, all the expected stages were not present in ontogeny, they must have been skipped; then the 'fact' that a stage had been skipped became itself of great significance, and forms were classed together because of characteristics which none of them possessed! All this was, of course, wrapped up in a scientific jargon . . ."[48]

"The Biogenetic Law . . . had enough adherents among biologists to provoke repeated refutation, for example, by Sedgwick, . . Garstang, . . De Beer . . . Biological critics argued largely from the *obvious fact that developmental stages do not usually resemble adult types . . .*"[49] (Paleontologists) "Pavlov and Spath opposed the universal application by certain paleontologists of recapitulation theory and drew attention to sequences of fossils which did not support it . . . The theory was abandoned by English ammonite workers by the end of the 1920's."[50]

Yet a limited similarity trend does in fact exist between ontogeny and "phylogeny" (not evolutionary; but in reality the Flood-burial order). The to-be-discussed antediluvian marine ecological zonation with near-shore poorly-swimming, simple-sutured (low hydrostatic pressure resistance), nektobenthonic forms and offshore forms of opposite characteristics. During ontogeny, the first septa had simple sutures because of the hydrostatic capabilities of small septa sufficing; growing required acquisition of complexly-sutured septa for larger adult camerae. This ontogenic trend parallels the Flood-burial order ("phylogeny"): hence the "recapitulation." Since a developing complex cephalopod must be simple before becoming a complex adult, it may well resemble the simple, unspecialized, first-buried nektobenthonic forms; hence "recapitulation" of many other characteristics. Some shell ornament on advanced groups is for camouflage; younger forms being more benthonic and not needing it.[51] The unornamented-then-ornamented-camerae ontogenic trend mirroring the nektobenthonic (juvenile)-then-pelagic (adult) lifestyle change in ontogeny parallels the nektobenthonic-then-pelagic Flood-burial trend: thus yet another common "recapitulation."

D. "Convergence" in Cephalopods as Evidence for Creation

Whenever very dissimilar living forms, obviously regarded as being different evolutionary lineages, resemble one another in some morphological attribute, this is termed "convergent evolution." Such cross-similarities are incredibly common among cephalopods: "The most striking feature that emerges from study of the Mesozoic ammonites from the evolutionary point of view is the frequency with which history repeats itself..."[52] "Examples of *striking resemblance*, both in shape and ornamentation between forms or even groups of ammonites *of quite different geological age* have long been known..."[53]

Two examples of "convergence" at the generic level follow: "...the resemblance of *Euomphaloceras cornatum* (Kossmat) to *Plesiacanthoceras wyomingense* (Reagan) is striking indeed."[54] *Trachyphyllites* resembled Lytoceratina to such an extent that the former is now classed within the latter; formerly the similarity was ascribed to convergence,[55] and the former assigned to Phylloceratina.

An example of family-level convergence is provided by *Idiohamites ellipticoides* Spath, whose ornament is identical[56] to those of family Labeceratidae (*I. ellipticoides* is of family Anisoceratidae). "Even in families widely separated stratigraphically and quite unrelated, shells with round or stout whorl section have similar sutures."[57] "The openly coiled heteromorph ammonites such as the hamitids, crioceratids, and ptychoceratids still remain less well understood than most ammonites ... The great degree of parallelism in ornamentation and homeomorphy among otherwise dissimilar heteromorphic species has been recognized only recently."[58] "In the Silurian, there is again a striking and, indeed a confusing, convergence between *Sthenoceras* of the Phragmoceratidae, and *Danoceras* of the Oncoceratidae."[59]

The following are some examples of ordinal-level convergence: "It is interesting to note that systems of radial lamellae developed repeatedly and independently in the siphuncles of several groups of cephalopods: ... Actinocerida...Intejoceratids... Oncocerida..."[60] "A remarkable case of homeomorphy is the development of constricted ('visored') apertures of the *Phragmoceras*-type several times independently in entirely different lines of descent:... ellesmerocerid... Discorsorid... Oncocerid..."[61] "There is so much external resemblance between many oncoceroid and discosorid genera..."[59] Four-lobed first sutures cannot any longer distinguish between orders since it is now known to appear independently in different ammonite lines.[62] Some of heteromorphs (meathook-shaped uncoiled conches) are of order Lytocerida, while others of Ammonitida.[63] Among subclass Coleoidea, "...sheath-like structures arose quite independently in the Aulocerida, Belemnitida, and Tertiary Sepiida."[64]

Class Cephalopoda converges with Class Gastropoda: "It is very much of a puzzle to separate cause from effect in contemplating the fact that certain only distantly related cephalopods acquired at different geologic times shells that were coiled in a helicoid spire like a gatropod instead of being straight or coiled in a plane as cephalopods normally are. Such situations are found in the nautiloids *Lechritrocheras* and *Trochoceras* of the Silurian and Devonian, respectively, the ceratite *Cochloceras* of the Triassic, and the Cretaceous ammonoids *Turrilites* and *Emperoceras*... the convergence is thus heterochronous between different groups of cephalopods as well as between cephalopods and gastropods."[65] Ward[66] adds that this is a "close convergence." Septation on the monoplacophoran *Kirenyella*[16] is another convergence at the class level.

The convergence between Cephalopods and members of other phyla is without peer. Tests occur not only in Phylum Mollusca, but also Phylum Brachiopoda, Fusulina (of Phylum Protozoa), Bryozoa (Phylum Ectoprocta), and others.[3] One of the most outstanding examples of convergence in the animal kingdom has to be that of the cephalopod eye and the vertebrate eye: "...every feature fundamental to its operation for vision in cephalopods appears to be encountered also in fish."[67] The octopod statocyst (otolith) is much like that in vertebrates.[68] The open circulatory system found in most cephalopods nevertheless approaches the closed circulatory system found in vertebrates: the arterial muscle is difficult to distinguish[69] histologically in some cephalopods from that of vertebrates. Overall, Packard[70] contends that the similarity between cephalopods and fish is among the greatest of higher organisms in different phyla.

All of the above-mentioned examples of "convergence" are but a few examples of this cross-similarity which utterly permeates the Class: "Parallelism and convergence within the cephalopods, especially amongst fossil lines, are sufficiently common ... to have made it difficult for an agreed classification to be reached even at the ordinal level."[71] The amount and extent of cross-similarity is thus so great that a great problem arises in separating primary similarities from secondary ones: "Indeed, relationships within the group are such that no tenable classification can be erected dividing the nautiloids into a few clearly defined morphological groups capable of a succinct definition. It is evident that neither the shape of the shell, the form and structure of the siphuncle wall, the presence or absence of actinosiphonate deposits, endocones, concavosiphonate siphuncles, annuli, or cameral deposits are in themselves necessarily a reliable guide to major taxonomic categories."[72] "Septa and sutures, like all other ammonite characters yet recognized—coiling, whorl shape, aperture, ribs, keels, furrows, etc.—are subject ... to repetition of different variants at all levels in the phylogenetic tree and stratigraphic column. No single key to phylogeny has yet been discovered."[73]

Evolutionists, of course, explain convergence as being the result of evolutionary adaptation to a similar environment; or even direct competition (octopus evolving an eye because the fish with which it competed had them, etc.) Any major scientific-Creationist work may be consulted to document the fact that no grand biotransformism (from mutations with natural selection, etc.) has ever been demonstrated (much less proved): the evolutionary hypothesis is therefore incapable of explaining (other than by unsubstantiated speculation) the origin of even one living morphological pattern, let alone the same pattern several times. Yet even if evolution occurred, "Given the indeterminate nature of inherited variability it would be more natural to expect new characters to be unrepeatable even in the case of the adaptation of closely related organisms to similar environmental conditions."[74]

Pertaining specifically to cephalopod "convergence," evolutionists claim that evolution would repeat because there are only a few theoretically-possible designs. This point is controversial. Raup[75] attributes the narrow parameters of conch geometry to optimum survivability, but this is speculative and he was uncertain as to the cause of limitations on some parameters. Even many theoretical phyla could exist.[76]

It isn't the Creationist who must explain why the Creator would create "such a bewildering variety of similarly-living forms" as some evolutionists have charged. God designed His creatures for differences in glory (1 Corinthians 15:38-41). Life is not evolutionarily diverse, but in reality of a very limited, Divine-designed diversity which strays not far from taxon "blueprints." The convergence among cephalopods reflects a high degree of mixture of design features in all forms: "Heterochronous convergence" suggests that geologic-time designations to the fossil record are false: all fossil/extant forms Created simultaneously and mutually contemporaneous.

II. Explaining Ammonoid Biochronological Horizons: a Challenge for Diluviology

Introduction: Demonstrating that the fossil record does not support evolution is only half of the Creation-Flood paradigm: the other half (which occupies the remainder of this work) is to provide scientifically-based explanations for the order of Cephalopod burial by the Flood as is observed in the fossil-rock column.

It is necessary to determine just how much true stratigraphic order the fossil record has: evolutionists-uniformitarians frequently claiming that fossil succession-order is too precise to be explained by anything other than evolution with long ages. Since ammonoids are indisputably the best index fossils, unrivalled in the Mesozoic and useful in the Upper Paleozoic, their successional order must be carefully examined. The Jurassic Period alone is claimed to have been biostratigraphically divided into 52 worldwide successional ammonoid-genera zones.

"The motives and procedures of biostratigraphy are: 1, the collection and description of taxa; 2, the identification of local assemblages in rock sequences (assemblage-zone); 3, the measurement of the total stratigraphic range of significant taxa (range-zones); 4, the definition and recognition of time units based on the stratigraphic range of fossils (Period, Ages, etc.); and 5, the calibration of the biological time scale by isotopic and other numerical dates."[77] Obviously biostratigraphy, and especially that of ammonoids, is the combination foundation/backbone of the uniformitarian geologic column. One zone not mentioned is the acme-zone, or zone of abundant occurrence of a given taxon.[78] Legendary are the mid-19th century stratigraphers Albert Oppel and Alcide D'Orbigny for setting up the above-mentioned zones from ammonoids in Western European rocks.

A. The Substantial Subjectivity of Fossil "Species" and "Genera"

Many paleontologists now agree that ". . . the assignment of groups of organisms to taxonomic categories involves a large element of subjectivity . . .,"[79] and Shaw[80] states that ". . . the species concept is *entirely subjective* . . ." He (quite radically) advocates that designating fossil species be entirely abandoned and replaced by a stratigraphy of morphological attributes because the designation of fossil species depends on what the individual paleontologist considers significant,[81] frequently lumping objective differences as variation.[82] It is not difficult to see that true species have wide degrees of intraspecific morphological variation,[32] so "fossil species" could hardly be recognized. Although cephalopods preserve their juvenile septa, Mapes[83] nevertheless recently warned that mistaking juvenile and adult forms ". . . can and has led to confusion in the literature."

It is not so much "fossil species" as "fossil genera" of ammonoids that are employed in the hair-splitting subdivision of geologic ages. Yet: "When ammonites are considered in the context of the whole invertebrate fauna, which is rarely done, it becomes apparent that ammonite "genera" frequently have the status in terms of morphological variation, of what generally are regarded as species in other groups."[84] The above-mentioned criticisms of fossil species would then apply to "fossil ammonoid genera." It is not uncommon for "genera" to be recognized, named, and allowed to define zones on the presence of but single specimens, as in a case which was condemned.[85] Many so-called genera have been drawn into synonymy by treating their differences as sexual dimorphism;[86] but this may actually be polymorphism,[87] similar to the type found in some hymenopteran insects. This would then be another major element of speculation, internal inconsistency, and arbitrary practice used in designating ammonoid genera, irrespective of whether the "genera" were so when living or if the generic label is only considered operational.

After reviewing the study of Cretaceous ammonites, Haas[88] called attention to ". . . the indistinctness of the hitherto assumed generic characters . . ." The Cephalopod paleontologist Wiedman[89] noted that counting the number of genera is "surely unreliable." Twenty years of study have reduced the number of Lower Lias (Jurassic segment) ammonite genera from 106 to 76.[90] No isolated instance; the amendation of the very plastic generic designations is routine: "The evolutionary diversity of the ammonoids . . . has been exaggerated by . . . dubious theories of descent which have led to unnecessary multiplication of generic names. Close study of a fossil group *almost always reduces the number of valid species or genera*; for example . . . genera in the family Echioceratidae . . . reduced from nineteen to five."[91] An example of ammonoid species subjectivity is provided by the genus *Sonninia*, of which 70 species have been reduced to only 2,[92] and of these 64 reduced to but 1.[93] Even family-level subjectivity is a fact, illustrated by the shifting of *Fanninoceras* from the family Hildoceratidae to the Oxynoticeratidae.[94]

The bearing of the subjective nature of specific and generic designations upon biochronology is best given by Hess,[95] who asks: "Could not . . . shaky or non-precise nature be attributed to the biostratigraphic methods . . . when stages or zones are defined and mapped, for example in the Middle Jurassic, on the basis of the presence of fragmentary ammonite material, which at the time served as holotypes for a dozen new genera and species just to commemorate the name of the investigator?" How all the subjectivities discussed in this section relate to the assignment of "genera" into zones will be discussed in a later section.

B. Procedures Which Eliminate Successional Discrepancies

This section studies some methods by which fossils found where they are not "supposed" to be are effectively eliminated, making the successional order of ammonoid fossils appear much greater than it really is, and reinforcing claims of consistent successions.

It must be realized, first of all, that "genera" which are used to support uniformitarian claims of a very precise worldwide biostratigraphic "onion skin" system of tens of successions per geologic age are selected from many of varying stratigraphic range. Even if the "genera" used in world-biostratigraphic claims were objective entities, there is still a considerable overlap of cephalopod fossils in the geologic record. Many examples of long-ranging forms exist. The genus *Bactrites*, for example, ranges from Silurian to Permian,[96] and many consider it to be nothing more than a straightened-out ammonoid. "The Phylloceratina are . . . almost unchanged through the Jurassic and Cretaceous . . ."[97] Some significant morphological attributes, such as the ten-arm tentacle structure found in modern squids and shown by trace fossils[98] to have persisted since the early Paleozoic forms, span the uniformitarian geologic column.

Many so-called genera of supposed short stratigraphic range are found to have longer ranges, and some of these are then discarded in future claims of precise world successions. In the Spanish Jurassic "The stratigraphic range of some genera was found to be more extensive than previously known."[99] The French Jurassic zones of *Macrocephalites* and *Reineckeia anceps* now admittedly overlap, and these "biostratigraphic anomalies observed repeatedly in the Callovian cannot always be explained by faunal '*remaniement.*' "[100]. "In the lower Triassic, the *Flemingites* zone is overlain by the *Owenites* zone, but now many characteristic Owenitan "genera" are found with *Flemingites*.[101] *Leiostegium*, once thought to be a distinctive Canadian-stage genus, is now known far into the overlying Demingian stage.[102] Stratigraphic-boundary controversies test accept fossil-ranges and often prove them vastly incorrect, as did the Permian-Triassic boundary perplexity: "*Cyclolobus* is regarded as an indicator of latest Permian rocks . . . but . . . now presented evidence . . . that *Cyclolobus* makes it appearance well below the top of the Permian . . ."[103]

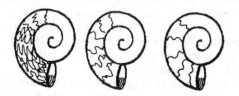

Fossils which are found where they are not "supposed" to be as part of an inviolable biochronological order are likely to be completely ignored. It took nearly 120 years after an original find (1843, Belgian Devonian) before Belemnoids were recognized to appear that early; previously they were believed to be no earlier than the Triassic. Concerning that early find, ". . . For nearly a century this report has been generally ignored or discounted."[104] Pertaining to a century-later Mississippian find: ". . . a report which has met with skepticism, though no good basis for this incredulity has been expressed. One can only conclude that the lack of widespread acceptance . . . stems from a widely-held conviction that there are no belemnoids as old as the Mississippian . . ."[104] Although rare in the Paleozoic, the Belemnites are easily recognized (by their bizarre bullet shape); and since they have no value as index fossils—it is difficult to see why their early appearance was not recognized. This non-recognition is a dramatic example of how fossils which do not fit preconceived notions of their stratigraphic range may be blindly ignored. The successional order of ammonoids (which, in contrast to belemnoids, are not easy specifically to identify) is greatly exaggerated because there are no doubt many "out-of-place" forms ignored.

More commonly they are not ignored, however; the "out-of-place" fossils are given different names. In making identification, there is needed ". . . an evaluation of all characters, *taken in conjunction with stratigraphical evidence*, in making a classification."[105] In fact, classification ". . . depends *absolutely* on stratigraphical information."[106] Clearly then, the identification and classification of "genera" is not independent of their claimed successional status, and the same fossils often are given different names, depending on their stratigraphical position.

An exceedingly common rationalization for "out-of-place" fossils is the concept of "migration"; the claim being that ". . . many different lineages were evolving and migrating simultaneously and so the succession is bound to vary in different places . . ."[107] Pertaining to worldwide correlations, the "migrations" " . . . created a very complex faunal pattern."[108] Another result was contradiction between the successional order of ammonoids and other index fossils: "A number of pelecypods, gastropods, and brachiopods enter the North American succession at a lower horizon than in Europe."[109] The finding of given "genera" among other specific "genera" of different "age" (and explained as "migration") is so common that Hedberg[110] advocates that biostratigraphic and chronostratigraphic designations not be used interchangeably.

C. The Scattered—Not Worldwide— Distribution of Fossil Zones

Arkell[112] categorically states that no "worldwide" ammonoid zone is *de facto* worldwide: Schindewolf[111] adding that it is a time concept which makes it worldwide, not presence of particular genera. ". . . Zones . . . do not apply universally. For instance, no one can recognize the rocks belonging to Kosmoceratan, Quenstedtoceratan, or Cardioceratan 'ages' in the southern hemisphere, where these genera do not exist. It is, however, possible to recognize the Callovian and Oxfordian stages, because those are abstractions, not dependent on occurrence or absence of any particular index species or index genera, but recognized by the general grade of evolution of the ammonite fauna as a whole and by a chain of overlapping correlations carried link by link around the world."[112] Similarly, in the Cretaceous, "The zones of the standard European section cannot be set up in Texas with any great accuracy. Instead, a parallel zonation must be set up in each area, and a correlation estimated, on rarely occurring fossils, stage of evolution, homotaxial superposition of family and generic groups, and intuition."[113] In Montana, *Inoceramus stantoni* Sokolov is an index to the *Scaphites depressus* zone.[114]. Fig. 3(ii) illustrates the fallacy of worldwide "onion skin" zonal claims: the zone of ammonoid "Genus" A seems worldwide only because "Genus" B is considered to be a stratigraphic time equivalent.

"Another interesting example of confusion resulting from lumping two concepts under one set of terms is evident in the common usage of the term 'Fossil zone,' *Fulanus smithi* Zone for example. Thus, one group of paleontologists would interpret *Fulanus smithi* Zone as the body of strata characterized by a certain assemblage of fossils of which *Fulanus smithi* happened to be a prominent member. Another group would understand *Fulanus smithi* Zone to mean the total body of strata in which the species *Fulanus smithi* occurred regardless of its associates. (Moreover, in neither group would there be uniform opinion as to whether actual specimens of either *Fulanus smithi* or the assemblage fossils would have to be present for strata to be included in the zone, or whether simply supposed time equivalence would qualifiy strata for inclusion)."[115] Zonal claims hide under vague terminology.

Not only are there no nearly worldwide "genera" zones and rather a woven quilt of imagined time equivalences; but also fossils tend to be absent in the most unlikely places: ". . . barren segments may intervene between definable zones."[116] Pertaining to Jurassic ammonoids, "It is remarkable that some species abundant in one locality are rare or absent in others of the same region."[117] Speaking of surprising absences in intercontinental ammonoid correlation, von Hillebrandt[118] notes: "These observations illustrate the difficulty of biostratigraphic correction." A criticism of existing Triassic biostratigraphy notes ". . . in large part a hypothetical arrangement of zones defined in widely scattered areas."[119] Just as there are missing geologic "ages" with no unconformity to indicate the supposed tens of millions of years of nondeposition/erosion, so also there are frequent mini-disconformities; for example, between the Toarcian and Sinemurian stages, "Although the section . . . is seemingly continuous, Pliensbachian ammonites . . . appear to be missing."[120]

Zones and even stages are such arbitrary designations that they can easily be dissolved at will. For example, a newly-constructed Cenomian-Toarcian boundary does away with the *Belemnites plenus* zone,[121] whereas the long-held Volgian stage of the Jurassic is recommended to be dropped.[122]

Ammonoids (especially in the Upper Paleozoic) are used with other index fossils, especially conodonts and brachiopods. Conodonts may be found far from where they stratigraphically "belong"; and "reworking" of older rocks' fossils into more "recent" ones is claimed. But in a recent Kansas case of Devonian conodonts in Mississippian rock, "Specimens exhibit *little evidence of reworking*."[123] "Brachiopods are notoriously difficult to use in correlation, and Permian ones especially so because of their provinciality . . ."[124] Not trivial, but major contradictory age-indicators may result from different groups over wide regions, as in this Eastern Siberian example: "The same deposits were long classified as belonging to the Carboniferous on the basis of ammonoids, and to the Permian on the basis of brachiopods and other groups."[125] (The ammonoids proved right.) Biostratigraphic subdivisions are calibrated by radiometric dates to produce the numerical uniformitarian geologic time scale. Yet "Only very few useful Jurassic radiometric dates are available, and the biostratigraphic position of most of these is vague."[126] "Radiometric dates for the Lower Cretaceous are scarce, and nearly all are based on glauconites which become less reliable with increasing age of the section.[127] "The expressed cautions on choosing the radiometric dates and their scarcity only further arouse suspicions that isotopic dates are accepted only if they agree with the biostratigraphic distribution and support and old earth. Biostratigratigraphic/magnetostratigraphic relation is vague.[128]

D. Biochronologic Ammonoid Zones as Taxonomic Concoctions

Not only are the "Worldwide" zones not worldwide and in reality a "patchwork" of supposedly isochronous lateral equivalents, but (as this most-important section will show) the previously considered subjectivities of ammonoid "species" and "genera" are universally manipulated in a way that makes even the lateral-component "patches" appear far more geographically widespread and—more importantly—thinner (shorter stratigraphic range) than actual ammonoid genera were.

The fact that cosmopolitan genera appear over-represented in the fossil record[128] suggests concoction that makes them appear cosmopolitan. Ager,[129] in agreement, states that ubiquity of fossils is often actually paleontologists' imagination, and that geographic distribution of fossils parallels political boundaries. "Taxonomists have long known that geographically widespread species tend to display far more variation than is the case with highly endemic species of the same genus or family."[39] Concerning Texas Cretaceous ammonoids: "The endemic faunas exhibit lower generic diversity than the cosmopolitan ones."[130] "The high variation observed within "cosmopolitan genera" is only the natural result of the lumping of different fossil forms into the so-called genus to make it appear geographically wide-spread.

Similarly, taxonomic splitting makes "generic" successional-order highly exaggerated, both in precision and repetitive consistency. "Thus all 'horizontal' generic, specific, or intraspecific boundaries in paleozoology are *artificial cuts* in uninterrupted 'vertical' evolutionary lines which are *intended to serve the practical ends of biostratigraphy* and geology."[131] "We are... acutely aware... of the arbitrary decisions and disparity of methods among taxonomists. The clade statistics reflect the true history of groups only through these filters. Clades of genera within families for Mesozoic ammonites, for example, are extremely short and fat... *prodigious oversplitting* inspired by *stratigraphic utility* may be the primary cause of these unusual shapes."[132] Hallam,[84] Campbell and Valentine,[133] and others all call attention to this vertical taxonomic splitting of ammonoid genera.

All of the following are examples of ammonoid-"genus" splitting, starting with the Permian: "Definition of *Cyclolobus* involves a progression of evolutionary stages within the family in which *Timorites*... is the immediate predecessor. Separation of these two genera is *arbitrary*."[134] "*Glyptophiceras, Ophiceras,* and *Otoceras* of the lowermost Scythian are extremely plastic stocks. Nearly every researcher who has worked with these genera has testified to this fact."[135] A survey of the Triassic notes: "*Tropites*... the limits of this genus are indefinite."[136] There is in the Cretaceous an (upward) succession of *Scaphites depressus, S. binneyi, S. vermiformis*. In this succession a split occurred because *S. binneyi* was once regarded as just a variety of *S. vermiformis*.."[137] Another important problem that arises in recognizing first occurrences of taxa in an evolutionary sequence and their subsequent use in correlation is the discrepant taxonomic practices of different workers. As an example, one worker may place the first stratigraphic occurrence of a taxon where one-half or more of the sample contains the diagnostic morphologic attribute. A second worker may, with the same data, choose the first occurrence of the diagnostic attribute at the level of its first appearance, no matter how small a part of the sample it is."[138]

All of the considered factors contributing to the illusion of inviolable ammonoid succession-order are summed up in Fig. 3(ii). Selective lateral lumping has fused true genera 1 to 4 into "genera" A and B. A and B, imagined to be time-equivalents, thus comprise the "worldwide" ammonoid zone A. Extreme vertical splitting simultaneously makes both A and B seem very successional (thin, precise, "onion-skins") which in reality are small vertical segments of true genera 1 to 4. Several such zones (all arbitrary, fallacious, taxonomic concoctions) are superposed to define a stage; several stages make up a geologic "period." The fallacies snowball and culminate in the uniformitarian geologic age system. Diluviology need not be burdened with claims of these precise generic successions, as much of this order is imaginary. The order which does exist in the stratigraphic record will be explained in terms of Flood-depositional sequence.

E. "Condensed" Ammonitiferous Deposits Indicate Rapid Burial

Evidences for cataclysmic sedimentation, commonly studied in Diluviology, are excellently manifested among cephalopods. Kranz,[139] following an experimental study of the burial of mollusks, concluded that only rapid burial will preserve a fossil in the first place. "The attractive preservation of many of the ammonites, in particular, is due to pyritization."[140] Pyritization results from reduced-iron conditions caused by bacterial decomposition of pre-fossilized material, the decay of which was halted by rapid burial and culminated in fossilization.

Not only do well-preserved ammonites reflect rapid burial, but poorly preserved, crushed conches do likewise. Crush-fractures often reveal[141] that the conch was flattened before fossilization (before the aragonite recrystallized into calcite). This indicates that conches were not only rapidly buried but also were quickly covered by a heavy overburden of superjacent sediment.[142]

These evidences, however, do not deflate the astronomical-time claims of uniformitarian geology nearly as thoroughly as do thin, ammonite beds containing mixtures of different "age" fossils called condensed beds, a unit of but little thickness in which faunal elements of various ages occur side by side without being any longer separated stratigraphically. Such deposits occur in varicolored, cephalopod-bearing limestones of many Tethyan localities, such as, in the Triassic, at five localities in the Hallstatt limestones of the northern calcareous Alps, one each in Bosnia and Greece, in the Himalayas and on Timor; in the Jurassic, at eight localities of the northern calcareous Alps, four in the southern Alps, seven in Hungary, and one in western Sicily."[143] In Italy, "Paleontological condensation occurs frequently in these deposits..."[144] In the Bajocian (mid-Jurassic) of Midlands, England, are "... thin and condensed sequences... non-sequences... all these are characteristic, even in the basins."[145] These condensed deposits are "... thin layers of clay or limestone crowded with ammonites from more than one horizon."[146]

"Condensed" beds are this thin: one Himalayan bed spans 7 ammonite zones (7-10 million supposed years) in merely 3 feet of sediment.[147] Heim[148] calls attention to a glauconitic sandstone 10-80 cm. thick with ammonites from ten horizons. In Sicily, there are "... 30 ammonite zones represented in one foot of sediment..."[149]

The common uniformitarian explanation is that "Condensed deposits are formed by stagnant sedimentation and reworking causing faunal elements of various ages to occur side by side."[150] Ammonoids of millions of years' duration are imagined to fossilize with little or no sediment and accumulate fossil-by-fossil at the ancient sea-bottom, the sediment accumulating to only a negligible degree (inches or feet in millions of years) and/or being washed away. Ammonoids thus supposedly separated by millions of years' time coexist within inches or are mixed outright.

A number of factors make the uniformitarian explanation incredible. "To maintain an unchanging environment for such long periods suggests conditions that could only be provided in the deep sea."[151] Yet there is evidence for current action,[152] and because of these and other strong evidences, Jenkyns[153] concludes non-tranquil shallow-water deposition. Reworking is claimed on grounds of abrasion, corrosion, etc., of fossils. Many non-condensed, "properly" sequenced ammonoid fossils show these features, and in "condensed" beds actually "... occur in various states of preservation..."[154] "One of the arguments adduced against reworking... for the west Sicilian Jurassic is that the older faunal elements in a 'Condensed' assemblage are often as well preserved as, and sometimes better than, the younger."[155] As a matter of fact, "... the best fossils are in the thinnest sequences...!"[156] While the differences of mineral matrices in "condensed"-bed fossils are used to support reworking, many others are "... indistinguishable by the matrix."[157]

Far more significant are questions of prolonged preservability. Even if ammonites fossilized without sediment, continuously accumulated, and this condition persisted undisturbed for millions of years, they would have miraculously to survive countless episodes of current action, being mixed around many times to incorporate million-years-later ammonites with them. Mass-preservation despite millions of years of turbation associated with reworking would have had to have occurred since "Ammonites may occur in *immense concentrations* in the condensed beds."[158] It is highly reasonable that these highly-fragile fossils would all be ground to powder, incapable of surviving even a few reworkings, let alone countless episodes spanning millions of years.

Many other evidences for short-duration "condensed"-bed deposition exist. If many "condensed" limestones are biogenic in origin, an argument from ignorance is required to explain their scarcity of nanno-organisms.[159] Also, evolution fails: "It is ... remarkable that ... genera, such as *Physodoceras, Amoeboceras, Glochiceras* continue to range through the condensed beds without change."[160] Calcareous concretions in "condensed" limestones may be from decaying organic matter,[161] indicating sudden stoppages in the decay process (as happen during fossilization). "It is worth considering that although the condensed sequences represent vast periods of time, the stromatolitic laminae may be at most an annual or even a *noctidiurnal* phenomenon ... and hence the growth of an algal clump could take place very fast. Hence, some factor must control the growth (or preservation) of the stromatolites since these horizons are of such limited vertical extent."[162] The "factor" is the fact that "Condensed" beds are of short-durational formation.

One of the lesser implications of "condensation" is the wholesale upsetting of elaborate biostratigraphic zonations: "In the Triassic of Europe and Asia, where most Triassic ammonoid genera were first discovered and named, two kinds of problems confront the paleontologist concerned with recognizing natural assemblages ... The Triassic rocks in Tethys ... provide mainly frustration for the ammonoid zonal stratigrapher."[163] Discrepant fossil presences are mitigated by "condensation," as in this Hungarian Jurassic case: "The author is inclined to ... explain the occurrence of 'strange' forms with a mixture of faunas in its lower part and with faunal condensation in the upper."[164] In the Permo-Triassic boundary-controversy, the observed mixing of the *Otoceras* and *Ophiceras* zones is ascribed to "condensation."[165]

The "Condensed" sequences have an infinitely greater significance than that of mixing biostratigraphic horizons. Once "condensed" sequences are seen to be rapidly deposited, the result is nothing less than the complete collapse of all the uniformitarian time-claims ascribed to the fossil record. "Condensed" beds may potentially become the most powerful overall evidence for the cataclysmic, mutually contemporaneous, short-duration burial of the entire fossil record. This is because all the mixed ammonoids must have lived at the same time and have been cataclysmically buried at the same time. Correlation of these "condensed" beds may "condense" most of the Mesozoic, deflating its sedimentation time from hundreds of millions of years to only several weeks (the closing phases of the Noachian Deluge). Without the thin, "condensed" sequences and their mutually-coexisting ammonoids representative of widely-different age-designations, one would need a miraculously long, upright tree trunk extending through miles of sediment of all geologic ages so unequivocally to demonstrate cataclysmic burial with mutual contemporaneity of all fossils.

Weidenmeyer[166] points out that "condensed" beds are often associated with penecontemporaneous tectonism and sedimentation. The high proportion of condensed beds in mountains (especially the Alps) reflects disturbed Flood-burial patterns caused by Floodwater flow-off variability around emerging mountains.

III. The Stratigraphically-Ordered Flood Burial of Cephalopods

A. Ecological Zonation and the Deluge: Preliminary Considerations

The factors causing stratigraphic order in Flood burial most often studied in Diluviology include hydrodynamic selectivity, differential escape possibilities, etc. (emphasized by Whitcomb and Morris[167]) and also ecological zonation (emphasized by Clark[168]). The latter consideration is more likely to have been the major factor in the desposition of the Cephalopods during the Universal Deluge owing primarily to the fact that the cephalopods show a striking stratigraphic eco-pattern.

Extremely significant is this overall fact: "It is worth mentioning that *continuous 'evolutionary' series derived from fossil record can in most cases be simulated by chronoclines*—successions of a geographical cline populations imposed by the changes of some environmental gradients."[169] Thus uniformitarians agree that ecological, not evolutionary factors, can definitely give rise to orderly successions! In speaking of evolutionary vs. ecological fossil-successions, Bell[170] even remarks: "There is, I think, no widely accepted belief in geology that so stultifies paleontologic interpretation as does the belief that successive faunal assemblages in a succession of rocks can be interpreted *only* as comprising species that succeed each other in time." (emphasis his) He also notes that it is very difficult to distinguish between ecologic and evolutionary successions, especially without unconformities, radiometric dates, etc.[171] Once it is accepted that none of these—and other—evidences is valid, and consequently neither geologic time nor biological evolution have any basis, all successions may be viewed as primarily ecological: not merely eco-

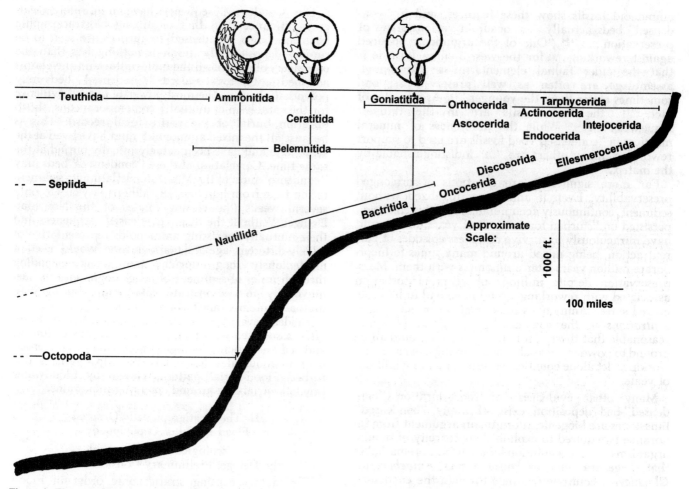

Figure 4. The ecological distribution of Cephalopods in the antediluvian seas: a representative cross-section.

successions, but contemporaneously-living Flood-buried successions. One recent example of non-evolutionary but ecological succession is provided by Thayer,[172] who noted an upper Devonian supposed deltaic progradation causing an ecological succession mimicking evolution.

In mixing evolution-time with ecology (paleoecology), such a practice relies heavily on deductive, *a priori* ecological designations: "... a useful approach is to *assume a given theory* from community and population biology and then to infer what the community would have been like had it obeyed the requirements of the chosen theory."[173] Testing such a designation is not conclusive: "Rather, paleontological tests are often *simply clues* that suggest the likelihood of verification or falsification from evidence that cannot be definitive."[174]

Many evidences suggest that paleoecological designations lumped with supposed time are not real. In noting, for example, hierarchical completeness, "... in general, the completeness decreases at each higher step; at the community level ... many (individuals) are missing, and many populations are missing completely ... at provincial and biospheric levels th³ holes are progressively worse..."[175] These gaps suggest that these are not *in situ* fossil communities but Flood-modified groups. Bambach[176] reviews the widely (but not universally) held view that diversity among all fossils was considerably less than it is today; Teichert and Glenister[177] cite the same for cephalopods. Abnormally low diversity is inevitable as long as actually contemporaneous forms are artificially divided into time-partitioned paleoecological designations.

B. The Antediluvian Ecologically-Zoned Coexistence of Cephalopods

Fig. 4 illustrates the original ecological distribution of the Cephalopods, from the Creation until the Universal Deluge. The Cephalopods, as other marine creatures, were created on the fifth day of the Creation Week (Genesis 1:20-23); entire populations coming into existence out of nothing (Romans 4:17d) at God's command (Psalm 148:5b). Since God designed each part of the individual organism to have a specific function that works as a whole (1 Corinthians 12:14-26), it is easy to envision Him having created different types of Cephalopods, each designed for a specific ecological habitat.

A study of present-day marine ecology notes that by far the main factors involved are the depth to which an organism goes, its distance from the shore, and whether it is a floater (planktonic), active swimmer (nektonic), or bottom-dweller (benthonic).[178] Exactly the same trends are observable in the stratigraphic record. Once the fallacies of evolution and geologic time are rejected and mutual contemporaneity accepted, the stratigraphic-upward ecological trends fit perfectly together as part of one vast ecologically-zoned cephalopod distribution (Fig. 4). Of even greater significance is the fact that the scientifically-determined ecological positions of cephalopods are (going stratigraphically upward) virtually identical with expected stratigraphic-upward trends of Flood burial (nearshore-then-offshore; benthonic-then-nektonic, etc.).

Overall trends are as follows: Fisher[3] calls attention to the stratigraphic-upward trend of heavily-conched to lightly-conched to conchless (squids). "Poorly streamlined shells...do not become common until post-Devonian times. Poorly streamlined and well streamlined shells are both common throughout the late Paleozoic and early Mesozoic, but after the Jurassic well streamlined shells become dominant."[179] (Hence there is benthonic-to-nektonic stratigraphic-upward trend.) Another important trend, noted by Packard,[180] is that the early cephalopods lived in water considerably more shallow than presently-living types; an overall stratigraphic-upward trend towards life in progressively-deeper water is noted. These overall trends are but the general outline of the antediluvian ecological distribution (Fig. 4).

Even greater support for Figure 4 is provided by more specific ecological positions of the cephalopod orders. According to Cowen,[181] Cambrian forms were primarily deposit-feeders; roaming scavengers appear in the Ordovician. The following were the ecologic positions of the Endoceratoids, Actinoceratoids, and Nautiloids: The Ellesmerocerida, Oncocerida, and Discosorida are rated as benthos.[182] The Endocerida, Intejocerida, and Ascocerida are nektobethos.[182] A nektonic rating is given to orders Actinocerida, Orthocerida, and Tarphycerida.[182]. Donovan,[183] agreeing that early Paleozoic forms were benthonic, adds that these forms were capable of neutral buoyancy and therefore capable of some swimming; a conclusion more recently confirmed.[184] Thus these early groups were not totally benthonic and could have been transported onto land during the early Flood stage.

"The earliest nautiloids, with peripheral siphuncle but concave septa, were probably confined to shallow water...as confirmed...from a different line of reasoning."[185] The ammonoids lived further offshore than the above-discussed early forms,[186] justifying the lateral separation portrayed in Fig. 4; the above-discussed ecologic findings already vertically grouping in early forms. The Bactritids were nektobenthonic as were the earliest ammonoids.[187]

A most important stratigraphic-upward trend is that of progressive septal corrugation in the ammonoids: "...the appearance of an intricately curved septum and its corrugation at the point where it articulates with the shell wall (in the places where stresses are the greatest); which are obvious adaptations enabling the septum to withstand high pressures."[188] This is generally regarded as the best explanation for this corrugation-trend.[189] Many independent groups show such a trend.[190] The stratigraphic-progression is as follows: the goniatites, (Figs. 2, 4 conch-suture diagrams going away from the shore) characteristic of the Paleozoic, have a non-corrugated septal morphology (and thus an undenticulated suture-pattern); the ceratites (Triassic) have some denticulation of sutures from slightly-corrugated septa; the ammonites (Jurassic and Cretaceous) have totally denticulated sutures reflecting total corrugation of their septa.[191] This supposedly evolutionary trend may instead by readily veiwed as being mutually-contemporaneous designs for varying depth-capabilities: "...elaboration of sutures would denote adaptation to active swimming habitats in deep water, and simplification of sutures would imply...shallower waters, or sluggishness."[192]

The fact that the complexly-sutured ammonites are heavily ribbed[193] to withstand greater hydrostatic pressure further corroborates the fact of their greater depth capacities as compared with simple-sutured forms.

The calculations of Heptonstall[194] on the weight burden of attached oysters on ammonites strongly indicates that ammonoids must have been capable of adding and removing water from their septa to regulate buoyancy as does the extant *Nautilus*. Overall, "It might...be possible that the primary mode of life of the ammonoids involved the need for continual adjustments to a pressure gradient. There is some indirect evidence for this interpretation."[195] The small teeth and the shape of jaws of ammonoids is one such evidence of plankton-feeding habits that require vertical migration. In addition, "...most ammonoids were fairly efficient at moving themselves vertically but less efficient as swimmers."[196] The heavy vertical lines spanning the water surface and sea-bottom (Fig. 4) illustrate how progressively more complex-sutured goniatites, ceratites, and ammonites were designed for ecologically-zoned lives involving vertical migrations to progressively greater depths.

The aforementioned small, shovel-like jaws of ammonites prompted Lehmann[197] to suggest that they were benthonic. If so, then sutural complication would reflect progressively deeper sea-floor-dwelling habitats. Most, however, regard ammonites as being nektonic: for example, Chamberlain[198] maintains that "...nearly all ammonoids required some swimming proficiency." The near-lack of trace-fossils attributable to ammonoids and the rarity of encrusting animals on them argues strongly against their benthonism, whereas the assymmetry of sutures[199] does not compel belief on ontogeny on the sea floor. The gastropod-like helicoidally-coiled ammonoids had been accepted as being benthonic on the basis of comparison with gastropod habitats; but there is no real evidence for any ammonoid-torticone losing buoyancy.[200] The openly-coiled heteromorphs need not have been benthonic because of their fragile, non-streamlined shells; a deep-water existence would apparently suffice,[201] also corroborating deep-water complex-sutured lives.

The precise depth to which ammonites went has not been settled by modern scholarship. Mutvei,[202] having supported great vertical migrations, contends on the basis of assumed pre-diagenetic conch thickness that ammonoids may have descended to 1000 meters deep, which is many times deeper than the stratigraphically-lowest forms. While most others disagree with the great depth figure, it is nevertheless agreed that the simple-sutured ammonites went only 100 meters deep—scarcely more than the earliest forms—whereas the complexly-sutured ammonites went as deep as the extant *Nautilus*,[203] which is over 600 meters. Thus there is further scientific basis for mutual coexistence (Fig. 4).

Although goniatites (PermoCarboniferous), ceratites (Triassic), and ammonites (entire Mesozoic) with their respective sutural complication trends are so distributed (Figs. 2, 4), there is nevertheless total overlap. Members of Clymeniida have simpler sutures than their supposed ancestors and five-lobed forms precede four-lobed groups.[204] Furthermore: "...highly complex, typically ammonitic sutures are found in some families of the Early Permian; ceratitic sutures appear in some families of the Early Mississippian; goniatitic sutures occur in some Triassic and Cretaceous ammonites ... and more or less ceratitic sutures reappear in both the Jurassic and Cretaceous in numerous families totally unrelated to the Triassic ceratites."[205] This mixing is an indicator of the inevitable overlap of ecological zones and even more inevitable mixture of mutually foreign groups during Flood deposition.

Fig. 4 does not list all cephalopod orders (Fig. 2), but each listed one is also quite representative of adjoining orders. All the ammonoid orders are represented by the above-mentioned three groups. The far-offshore Belemnites are neritic,[206] as are the similar Phragmoteuthids and Aulocerids. The squids (Teuthida) are ten times more efficient as swimmers (far less drag) than the most streamlined coiled-conch forms[207] and 100 times more efficient than non-streamlined conches. The Coleoids, as exemplified by squids, were oceanic forms (Fig. 4) created for rapid, rocket-propelled swimming in contrast to the nearshore groups discussed.

Quantitative pressure-tolerance tests, a major area of cephalopod research in this decade, support Fig. 4. "Assuming that actual habitats ranged to approximately ⅔ of the mechanical limits of the shells, the following maximum depth ranges are indicated by this preliminary survey: Endoceratoidea 100-450 m; Actinoceratoidea 40-150 m; Nautilida, Ellesmerocerida 50-200 m, Orthocerida 150-500 m, Oncocerida <150 m, Discosorida <100 m, Tarphycerida <150 m, Nautilida 200-100 m; Bactritoidea <400 m; Coleoidea, Auloceri-da 200-900 m, Sepiida 200-1000 m, Belemnitida 50-200, exceptionally 350 m.[208] "Particularly noteworthy are the facts of wide range for nektonic forms (in contrast to benthonic forms) and the very great depth capabilities of the Coleoids (except the early-appearing (Fig. 2) Belemnites). The quantitative values, however, are approximate: "The wide range for implosion values and lack of strong correlation between such parameters as septal thickness and implosion indicate that determination of depth ranges for fossil cephalopods may be difficult."[209] Yet, with the major exception of ammonoids (whose range was septally restricted) and Bactritids (nektobenthonic), nektonic (Fig. 2) orders have far longer stratigraphic ranges than do benthonic or even nektobenthonic orders, suggesting greater ecological independence of free-swimming forms as opposed to the narrowly-restricted benthos.

The observed great extent of mutual overlap, expected in marine ecology, nevertheless is restricted by competition; forms flourishing only in marine regions for which they were designed. In fact, when mutual contemporaneity is accepted, patterns of competitive exclusion become evident. Nautiloids are most prevalent (lower Paleozoic) where no ammonoids were; whereas they are very rare in the Mesozoic (when ammonoids flourished). Specifically, "The Silurian ... was perhaps the heyday of the nautiloids..."[210] but "... as compared with other molluscan groups, including ammonoids, 'nautiloid' cephalopods are rare fossils.[211] They are always in minority; outnumbered by other molluscs commonly by 10,000:1 to 1000:1.[211] The gregarious nature of many cephalopods, such as ammonoids,[212] undoubtedly sharpened many ecological boundaries. The lack of larval stages for most cephalopods, notably the ammonoids[213] and belemnoids,[214] further restricted their migrations. The lack of great temperature differences, absence of storms (Genesis 2:6), etc., all characteristic of the antediluvian earth, contributed to a relative lack of sea turmoil (and oceanic currents), further reducing mixture of ecologically-zoned cephalopods—as did the short duration of time between the Creation and the Noachian Deluge (approximately 1,700 years).

C. The Sequential Flood Burial of Cephalopods

Before a description is given as to how the ecologically-zoned cephalopod orders (Fig. 4) were for that reason stratigraphically (Fig. 2) separated during Flood deposition, other factors are noteworthy—which also explain the intra-ordinal stratigraphic order. Examining the original[2] of Fig. 2, it is apparent that most families within an order cover more than half of the stratigraphic range of the entire order; hence ecological zonation was the major factor in family-level as well as ordinal-level stratigraphic ordering. Although fossil cephalopod "genera" and "species" are found to be concocted—not real—entities, somewhat consistent intrafamilial stratigraphic trends are observed. Since these trends are physical (size, shape, mass, etc.), they are evidently caused by Floodwater sorting (see Whitcomb and Morris[167]).

Intrafamilial conch-size increases are very common[215] and are regarded as being an example of Cope's Law of evolutionary size increase, but many exceptions are known.[216] Also "One repeatedly made observation is that the sharp-edged discoidal shell . . . associated with a . . . calcareous and detrital shelly facies."[217] These two trends may be caused by the denser (because smaller and/or mass not spread out in unornamented forms, etc.) forms tending to be buried earlier than ornamented and larger forms; the "facies" separation reflecting sequential increasing-maximum sediment carrying capacity of the Floodwater mass movement. Other separations are caused by differential escape from burial. Since "Properly roughened shells may have conserved as much as 50% of the propulsive power required by smooth shells of the same size and shape,"[218] progressive roughening trends indicate the superior escape-postponement of roughened—as opposed to smooth—forms. The common trend of progressively more ventral siphuncle (greater buoyancy con-

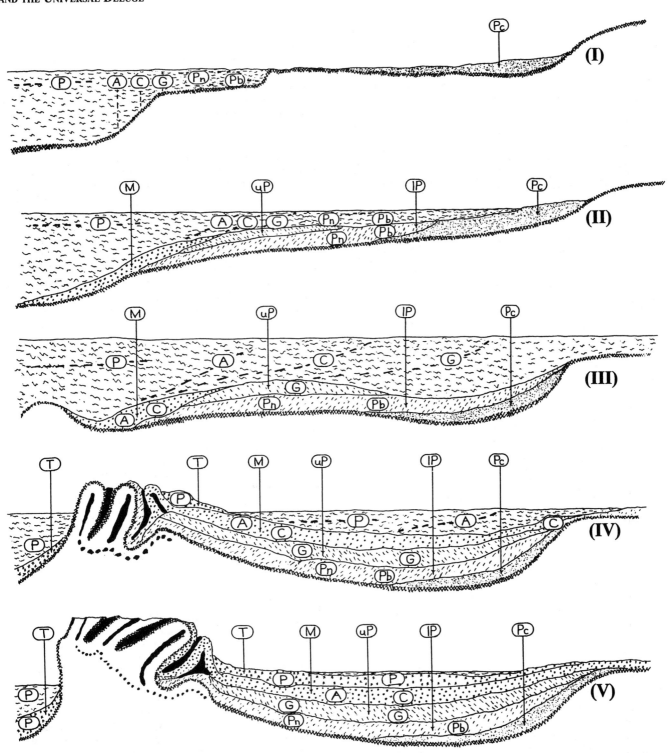

Figure 5. The result of burial in the Flood, according to ecological zonation. The alleged geological ages are indicated as follows: Pc: Precambrian; lP: lower Paleozoic; uP: upper Paleozoic; M: Mesozoic; T: Tertiary. The other symbols, if not obvious, are explained in the text. Note that the left part of (i) is the same situation as that shown in Figure 4.

trol) as exemplified by the *Nicomedites-Gymnotoceras-Frechites* series[219] also mirrors the superior burial-escape of stratigraphically higher forms.

So numerous, however, are major exceptions to these and other trends that Arkell[217] contents that "... there are many obstacles to the acceptance of any generalization." These intrafamilial-order exceptions are to be expected owing to the variability of the physical action of flowing Floodwater. Large-scale sortings are manifest in the Caucasus,[220] whereas in southern Germany "Current orientation of coiled ammonites... indicates fairly strong currents."[221]

Noting that sorting (intrafamilial order) is secondary and ecological zonation (familial and ordinal stratigraphic order) primary in Floodwater sequential burial, and keeping the zonation (Fig. 4) and result (Fig. 2) in mind, the following (Fig. 5) is the indicated scenario of Flood burial:

The earliest stage (i) sees the falling rains and water from the fractured fountains of the deep eroding into the antediluvian mountains and depositing the sediment as the Late Precambrian; the antediluvian seas not immediately overflowing the continental mass(es). (During the entire time span of the deposition, basins subside as they are being smothered by sediment. The antediluvian highland areas (Fig. 5, far right) provide the majority of the clastic sediment in the early (i, ii) and middle (iii) Flood stages; late-Flood uplifts (iv, v, far left) providing sediment during the recessional phases of the Universal Deluge.

The breaking up of the fountains of the deep and all the associated volcano-tectonic results on the ocean floor cause the oceans to rise sharply, overflowing completely the continent(s). The cephalopod habitats are driven unto land, preserving the ecological zonation (Fig. 4) owing to the great lateral extent (several hundred miles) of the zonation in relation to the water depth (a few thousand feet). The ecological zones, separated laterally, become vertically superposed during Flood entombment of their cephalopod constituents, giving rise to the geologic column (Fig. 2).

Thus, as the oceans begin to inundate vast land areas (ii) in the first few weeks of the Deluge, the most frontal waters carry the most-nearshore orders. Accordingly, it is they (Pb-Paleozoic benthos, Pn-Paleozoic nektobenthos and nektos) which are the first deposited in any area as the ocean water moves ever landward, giving rise to the early Paleozoic rock systems. Other systems are laid down contemporaneously not far behind, each area passed-over by ecological zones.

The Flood reaches its greatest depth (covering everything) towards the end of the 150-day prevailing period. At this time (iii), the lower Paleozoic is nearly all laid down, while much of the Upper Paleozoic (with G-Goniatites) is being laid down, followed vertically by the complex-sutured ceratites (C) and then ammonites (A); the last two of which comprise the Mesozoic systems.

The earlier-deposited Paleozoic systems become intensely folded (iv) in the orogenies which mark the Flood—recessional half—year. The recession sees completion of deposition of Upper Paleozoic, most of the Mesozoic, and part of the Tertiary, which contains the pelagos (P), or far off-shore forms which are the sole survivors of the class.

Diluviological research strongly indicates that most (but not all) of the Tertiary is post-Flood. Hence (v) much of the Tertiary is the result of huge inland waterways which persisted perhaps decades after the Flood year. The continental-shelf Tertiary (v, lower left), on the other hand, represents the fixing of the shoreline between ocean and land at the very end of the Flood.

The extinction of cephalopods at various stratigraphic intervals is enigmatic to the evolutionist—uniformitarians, especially the disappearance of the ammonoids at the end of the Cretaceous. The Paleozoic forms and all ammonoids became extinct because they were all driven on the land; the presently-living forms being spared extinction because they are deep-water oceanic forms which were not all forced on land during the Flood. The narrow ecological tolerance of the near-shore forms also contributed to extinction; any sparse ammonoid survivors being incapable of holding their own against the overwhelming populations of deep water forms which had soon begun explosively to proliferate in the Flood-shattered marine ecosystem.

The Flood deposition portrayed by cross-sections in Fig. 5 does not indicate the many exceptions of cephalopod-carrying Floodwaters. Varying local zonations caused observed missing "ages", whereas localized inversions of flow gave rise to "reverse-age" sequences. Interplay with similarly-zoned land fossils and other marine fossils are reflected by the endless variations in stratigraphy. A bed of only ammonites and marine fossils is called "marine Jurassic" whereas the same fossils interspersed with dinosaur bones is called a "continental Jurassic."

In conclusion, this paramount marine-invertebrate class provides an amazing amount of evidences for and ramifications of the Creationist-Diluvialist paradigm. The unifying factor is the way God used combinations of morphological attributes in any given form and placed these forms in different ecological positions.

References

AG—American Association of Petroleum Geologists Bulletin
AJ—American Journal of Science
BP—Bulletins of American Paleontology
BR—Biological Reviews
CE—Canadian Journal of the Earth Sciences
DE—Doklady: Earth Science Sections (translated from Russian)
EV—Evolution
GA—Geological Society of American Abstracts with Programs
GM—Geological Magazine
JG—Journal of Geology
JP—Journal of Paleontology
LE—Lethaia
NA—Nature
NP—Neues Jahrbuch Fur Geologie und Palaontologie
 (M—Monatshefte; A—Abhandlungen)
PA—Palaeontology
PB—Paleobiology
PJ—Paleontological Journal (translation from Russian)
PZ—Palaeontologisch Zeitschrift
TP—Treatise on Invertebrate Paleontology
UP—University of Kansas Paleontological Contributions Paper

[1]Sweet, W. C. 1964. Cephalopoda—General Features. *TP* K7.
[2]Teichert, C. 1967. Major Features of Cephalopod Evolution. *University of Kansas Special Publication* 2, pp. 198-201.
[3]Fischer, A. G. 1965. Fossils, Early Life, and atmospheric history. *Proceedings National Academy of Science USA* 53:1209.
[4]Yochelson, E. L., Flower, R. H., and G. F. Webers. 1973. The bearing of the new Late Cambrian monoplacophoran *Knightoconus* upon the origin of the Cephalopods. *LE* 6(3)276.
[5]Kuhn, T. S. 1970. *The Structure of Scientific Revolutions*. University of Chicago Press, p. 70-1.
[6]Valentine, J. W. and C. A. Campbell. 1975. Genetic Regulation and the Fossil Record. *American Scientist* 63(6):678.
[7]*Ibid.*, p. 674.
[8]Barnes, R. D. 1974. *Invertebrate Zoology*.Saunders & Co., p. 431.
[9]Stasek, C. R. 1972. The Mulluscan Framework. *Chemical Zoology* 3(12):38-40.
[10]Yonge, C. M. 1977. (Review) *NA* 267(5609):379.
[11]Teichert. 1967. *op. cit.*, p. 163.
[12]Vologdin, A. G. 1969. A New Find of Cephalopods in the Middle Cambrian of Czechoslovakia. *DE* 186:258.
[13]Sweet, W. C., Teichert, C., and B. Kummel. 1964. Phylogeny and Evolution. *TP* K106.

14. Ulrich, E. O. and A. F. Foerste. 1933. The Earliest Known Cephalopods. *Science* 78:288.
15. Lipps, J. H. and A. G. Sylaster. 1968. the Enigmatic Cambrian Fossil *Volborthella* and Its Occurrence in California. *JP* 42(2):329.
16. Yochelson, E. H., Flower, R. H., and G. F. Webers. 1972. A Theory of Origin of the Cephalopods. *GA* 4(7):712.
17. Gish, D. T. 1973. *Evolution: The Fossils Say No!* Creation-Life, San Diego. P. 45-70.
18. Sokolov, B. S. 1975. The Current Problems of Paleontology and Some Aspects of Its Future. *PJ* 9(2):137.
19. Teichert, C. and R. C. Moore. 1964. Classification and Stratigraphic Distribution. *TP* K99.
20. Flower, R. H. 1961. Major divisions of the Cephalopoda. *JP* 35(3):571.
21. Teichert. 1967. op. cit., pp. 182-3.
22. Donovan, D. T. 1964. Cephalopod Phylogeny and Classification. *BR* 39(3):266.
23. Jeletzky, J. A. 1966. Comparative Morphology, Phylogeny, and Classification of Fossil Coleoidea. *UP Article 7:Mollusca* p. 20.
24. Arkell, W. J. 1957. Introduction to Mesozoic Ammonoidea. *TP* L103.
25. George, T. N. 1971. Systematics in Paleontology. *Quarterly Journal of the Geological Society of London* 127(3):231-2.
26. Boucot, A. J. 1975. *Evolution and Extinction Rate Controls* Elsevioer, p. 196.
27. Gould, S. J. and N. Eldredge. 1977. Punctuated Equilibria: the tempo and mode of evolution reconsidered. *PB* 3(2):116.
28. Gould, S. J. 1971. Speciation and Punctuated Equilibria: An Alternative to Phyletic Gradualism. *GA* 3(7):585.
29. Harper, C. W. 1976. Phylogenetic Inference in Paleontology. *JP* 50(1):190.
30. Jeletzky, J. A. 1955. Evolution of Santonian and Campanian *Belemnitella* and Paleontological Systematics exemplified by *Belemnitella praecursor* Stolley. *JP* 29(2):489.
31. Boucot. 1975. op. cit., p. 197.
32. Bretzky, S. S. 1977. Recognition of Ancestor-Descendant Relationships in Invertebrate Paleontology. *JP* 51(2) supplement (pt. III of III) p. 4.
33. Murphy, M. A. 1977. On Time-Stratigraphic Units. *JP* 51(2):214.
34. Arkell. 1975. op. cit., p. L113.
35. Gould, S. J., Raup, D. M., Sepkoski, J., T. J. M. Schopf, and D. S. Simberloff. 1977. The shape of evolution: a comparison of real and random clades. *PB* 3(1):34-5.
36. Jeletzky. 1955. op. cit., p. 490.
37. Teichert. 1967. op. cit., p. 194.
38. Reyment, R. A. 1974. Analysis of a Generic-Level Transition in Cretaceous Ammonites. *EV* 28(4):675.
39. Boucot, A. J. 1975b. Standing Diversity of Fossil Groups in Successive Intervals of Geologic Time Viewed in the light of changing levels of Provincialism. *JP* 49(6):1110.
40. Donovan, D. T. 1959. Septa and Sutures in Jurassic Ammonites. *GM* XCIV(2):168.
41. Wiedman, J. 1969. The Heteromorphs and Ammonoid Extinction. *BR* 44(4):588, 590-1.
42. Haas, O. 1971. Recent Literature on Mesozoic Ammonites—Part XII. *JP* 45(3):546.
43. Newell, N. D. 1949. Phyletic Size Increase—An Important Trend Illustrated by Fossil Invertebrates. *EV* 3(2):115.
44. Clark, D. L. 1962. Paedomorphosis, Acceleration, and Caenogenesis in the Evolution of Texas Cretaceous Ammonoids. *EV* 16:300.
45. *Ibid.*, p. 302.
46. *Ibid.*, p. 303.
47. Donovan, D. T. 1973. The Influence of Theoretical Ideas on Ammonite Classification from Hyatt to Trueman. *UP* 62, p. 15.
48. Arkell. 1957. op. cit., p. L110.
49. Donovan. 1973. op. cit., p. 13.
50. *Ibid.*, p. 1.
51. Cowen, R., Gertman, R. and G. Wigget. 1973. Camouflage Patterns in *Nautilus* and Their Implications for Cephalopod Paleobiology. *LE* 6(2):211-2.
52. Arkell. 1957. op. cit., p. L112.
53. Haas, O. 1942. Recurrence of Morphological Types and Evolutionary Cycles in Mesozoic Ammonites. *JP* 16(5):643.
54. Haas, O. 1969. Recent Literature on Mesozoic Ammonites—Part X. *JP* 43(3):787.
55. Tozer, T. E. 1971. One, Two, or Three Connecting Links Between Triassic and Jurassic Ammonoids? *NA* 232(5312):565.
56. Haas, O. 1974. Recent Literature on Mesozoic Ammonites—Part XVI. *JP* 48(5):1005.
57. Arkell, W. J. 1957b. Sutures and Septa in Jurassic Ammonite Systematics. *GM* XCIV(3):235.
58. Ward, P. D. W. and V. S. Mallory. 1977. Taxonomy and Evolution of the Lytoceratid Genus *Pseudoxybeloceras* and Relationship to the Genus *Solenoceras*. *JP* 51(3):606.
59. Flower, R. H. and C. Teichert. 1957. The Cephalopod Order Discosorida. *UP* 21, p. 6.
60. Teichert, C. and R. E. Crick. 1974. Endosiphuncular Structures in Ordovician and Silurian Cephalopods. *UP* 71, p. 11-12.
61. Teichert, C. 1964. Morphology of Hard Parts. *TP* K50.
62. Druschits, V. V. and I. A. Mikhailova. 1974. On the Systematics of Early Cretaceous Ammonites. *PJ* 8(4):477.
63. *Ibid.*, p. 470.
64. Jeletzky. 1966. op. cit., p. 10.
65. Cloud, P. E. 1948. Some problems and Patterns of Evolution Exemplified by Fossil Invertebrates. *EV* 2:329.
66. Ward, P. D. 1976. Upper Cretaceous Ammonites (Santonian-Campanian) From Orcas Island, Washington. *JP* 50(3):459.
67. Packard, A. 1972. Cephalopods and Fish: The Limits of Convergence. *BR* 47:269.
68. *Ibid.*, p. 267.
69. *Ibid.*, p. 263.
70. *Ibid.*, p. 286.
71. *Ibid.*, p. 285.
72. Flower, R. H. and B. Kummel. 1950. A Classification of the Nautiloidea. *JP* 24(5):604-5.
73. Arkell. 1957b. op. cit., p. 246.
74. Tatarinov, L. P. 1976. Current Problems in Evolutionary Paleontology. *PJ* 10(2):123.
75. Raup, D. M. 1967. Geometric Analysis of Shell Coiling: Coiling in Ammonoids. *JP* 41(1):54.
76. Valentine, J. W. and E. M. Morris. 1972. Global Tectonics and the Fossil Record. *JG* 80:178.
77. Runnegar, B. and K. W. S. Campbell. 1976. Late Paleozoic Faunas of Australia. *Earth Science Reviews* 12:247.
78. Harland, W. B. 1977. Essay Review: *International Stratigraphic Guide, 1976*. *GM* 114(3):233.
79. Valentine, J. W. 1969. Patterns of Taxonomic and Ecological Structure of the Shelf Benthos during Phanerozoic Time. *PA* 12(4):688.
80. Shaw, A. B. 1969. Adam and Eve, Palentology and the Non-Objective Arts. *JP* 43(5):1085.
81. *Ibid.*, p. 1094.
82. *Ibid.*, p. 1096.
83. Mapes, R. H. 1976. An Unusually Large Pennsylvanian Ammonoid from Oklahoma. *Oklahoma Geology Notes* 36:47.
84. Hallam, A. 1965. Observations on Marine Lower Jurassic Stratigraphy with Special Reference to the United States. *AG* 49(9):1495.
85. Torrens, H. S. 1969. The Stratigraphical Distribution of Bathonian Ammonites in Central England. *GM* 106(1):67-8.
86. Hall, R. L. 1975. Sexual Dimorphism in Middle Bajocian (Jurassic) Ammonite Faunas of British Columbia. *GA* 7(6):722.
87. Davis, R. A. 1971. Mature Modification and Dimorphism in Selected Late Paleozoic Ammonoids. *BP* 62(272):38.
88. Haas. 1971. op. cit., p. 550.
89. Wiedman, J. 1973. Evolution or Revolution of Ammonoids at Mesozoic System Boundaries. *BR* 48(2):160.
90. Donovan, D. T. and G. F. Forsey. 1973. Systematics of Lower Liassic Ammonoidea. *UP* 64, p. 2.
91. Donovan. 1964. op. cit., p. 270.
92. Donovan. 1973. op. cit., p. 7.
93. Haas. 1969. op. cit., p. 786.
94. Frebold, H. 1967. Position of the Lower Jurassic Genus *Fanninoceras* McLearn and the Age of the Maude Formation on Queen Charlotte Islands. *CE* 4:1145.
95. Hess, P. C. 1972. Essay Review. *AJ* 272(2):193.
96. Teichert. 1967. op. cit., p. 180.
97. Haas, O. 1968. Recent Literature on Mesozoic Ammonites—Part IX. *JP* 42(3):763.
98. Flower, R. H. 1955. Trails and Tentacular Impressions of Orthoconic Cephalopods. *JP* 29(5):857.
99. Haas. 1974. op. cit., p. 1002.
100. Haas, O. 1973. Recent Literature on Mesozoic Ammonites—Part XIV. *JP* 47(3):539.
101. Runnegar, B. 1969. A Lower Triassic Ammonoid Fauna from Southeast Queensland. *JP* 43(3):820.

[102] Flower, R. H. 1968. Cephalopods from the Tinu Formation, Oaxaca State, Mexico. JP 42(3):804.
[103] Tozer, E. T. 1969. Xenodiscacean Ammonoids and their Bearing on the Discrimination of the Permo-Triassic Boundary. GM 106(4):350.
[104] Flower, R. H. and R. Gordon. 1959. More Mississippian Belemnites. JP 33(5):809.
[105] Donovan. 1959. op. cit., p. 169.
[106] Arkell. 1957. op. cit., p. L103.
[107] Arkell, W. C. 1956. Jurassic Geology of the World. Oliver and Boyd, London, p. 6.
[108] Haas. 1974. op. cit., p. 1003.
[109] Hallam. 1965. op. cit., p. 1798.
[110] Hedberg, H. H. 1965. Chronostratigraphy and Biostratigraphy. GM 102(5):451.
[111] Schindewolf, O. H. 1957. Comments on Some Stratigraphic Terms. AJ 255:397-8.
[112] Arkell. 1956. op. cit., p. 12.
[113] Young, K. 1961. Upper Cretaceous Ammonites from the Gulf Coast of the United States. GA-1961, p. 181A.
[114] Corban, et. al. 1958. Scaphites depressus Zone (Cretaceous) in Northwestern Montana. AG 42(3):658.
[115] Hedberg, H. D. 1959. Towards Harmony in Stratigraphic Classification. AJ 257:677.
[116] Wheeler, H. E. 1958. Primary Factors in Biostratigraphy. AG 42(3):651.
[117] Haas, O. 1972. Recent Literature on Mesozoic Ammonites—Part XIII. JP 46(5):755.
[118] von Hillebrandt, A. 1970. Zur Biostratigraphie und Ammoniten-Fauna des sudamerikanischen Jura (isbes Chile). NP A 36(2):167.
[119] Tozer. 1969. op. cit., p. 349.
[120] Hallam. 1965. op. cit., p. 1490.
[121] Haas, O. 1976. Recent Literature on Mesozoic Ammonites—Part XVII. JP 50(5):952.
[122] Ibid., p. 944.
[123] Goebel, E. D. 1969. Late Devonian Age of a Mixed Conodont fauna from a core in Southwestern Kansas. GA 1(6):18.
[124] Grant, R. E. 1970. Brachiopods from Permian-Triassic Boundary Beds and Age of Chhidru Formation, West Pakistan. University of Kansas Special Publication 4, p. 126.
[125] Ruzhentsev, V. Ye. 1975. Carboniferous ammonoids and chronostratigraphy of Eastern Siberia. PJ 9(2):157.
[126] Van Hinte, J. E. 1976. A Jurassic Time Scale. AG 60(4):489.
[127] Van Hinte, J. E. 1976. A Cretaceous Time Scale. AG 60(4):499.
[128] Boucot. 1975b. p. 1108.
[129] Ager, D. 1973. The Nature of the Stratigraphic Record. John Wiley and Sons, New York. p. 17.
[130] Haas, O. 1973. Recent Literature on Mesozoic Ammonites—Part XV. JP 47(5):899.
[131] Jeletzky, J. A. 1950. Some Nomenclatorial and Taxonomic Problems in Paleozoology. JP 24(1):24.
[132] Gould et. al. 1977. op. cit., p. 29.
[133] Campbell, C. A. and J. W. Valentine. 1977. Comparability of modern and ancient marine faunal provinces. PB 3(1):55.
[134] Furnish, W. M. and B. F. Glenister. 1970. Permian Ammonoid Cyclolobus from the Salt Range, West Pakistan. University of Kansas Special Publication 4, p. 185.
[135] Silberling, N. J. 1959. Pre-Tertiary Stratigraphy and Upper Triassic Paleontology of the Union District, Shoshone Mountains, Nevada. United States Geological Survey Professional Paper 322, p. 42.
[136] Corban, et. al. 1958. op. cit., p. 657.
[137] Murphy. 1977. op. cit., p. 218.
[138] Kranz, P. M. 1974. The Anastrophic Burial of Bivalves and Its Paleoecological Significance. JG 82:237.
[139] Haas. 1972. op. cit., p. 752.
[140] Seilacher, A., Andalib, F., Dietl, G., and G. Gocht. 1976. Preservational history of compressed Jurassic ammonites from Southern Germany. NP M 152(3):314.
[141] Ibid., p. 317.
[142] Haas. 1972. op. cit., p. 747.
[143] Ibid., p. 757.
[144] Rayner, D. H. 1967. The Stratigraphy of the British Isles. Cambridge University Press, p. 290.
[145] Ibid., p. 285.
[146] Tozer, E. T. 1971. Triassic Time and Ammonoids: Problems and Proposals. CE 8:995.
[147] Heim, A. 1958. Oceanic Sedimentation and Discontinuities. Eclogae Geologicae Helvetiae 51(3):644.
[148] Ager. 1973. op. cit., p. 40.
[149] Wendt, J. 1970. Stratigraphische Kondensation in tradischen und jurassischen Cephalopodenkalken der Tethys. NP M 1970. Heft 7, p. 433.
[150] Tozer. 1971. op. cit., p. 996.
[151] Jenkyns, H. C. 1971. The Genesis of Condensed Sequences in the Tethyan Jurassic. LE 4(3):333.
[152] Ibid., p. 347.
[153] Ibid., p. 337.
[154] Ibid., p. 330.
[155] Tozer. 1971. op. cit., p. 1008.
[156] Arkell. 1959. op. cit., p. 480-1.
[157] Hallam, A. 1972. Diversity and Density Characteristics of Pliensbachian-Toarchian molluscan and brachiopod faunas of the North Atlantic Margins. LE 5(4):391.
[158] Jenkyns. 1971. op. cit., p. 335.
[159] Ziegler, B. 1959. Evolution in Upper Jurassic Ammonites. EV 13:233.
[160] Fursich, F. T. 1973. Thalassinoides and the origin of nodular limestone in the Corallian Beds (Upper Jurassic) of Southern England. NP M 1973. Heft 3, p. 151.
[161] Jenkyns. 1971. op. cit., p. 343.
[162] Tozer. 1971. op. cit., p. 992.
[163] Haas. 1973. op. cit., p. 535.
[164] Waterhouse, J. B. 1973. An Ophiceratid Ammonoid from the New Zealand Permian and its implications for the Permian-Triassic Boundary. GM 110(4):305.
[165] Wiedenmeyer, F. 1966. Problems in Biostratigraphy and taxonomy of Middle Liassic of Alpine Mediterranean Province. AG 50(3):641.
[166] Whitcomb, J. C. and H. M. Morris. 1961. The Genesis Flood. Baker. p. 273.
[167] Clark, H. W. 1968. Fossils, Flood, and Fire. Outdoor Pictures, California, p. 51-60.
[168] Krassilov, V. 1974. Causal Biostratigraphy. LE 7(3):174.
[169] Bell, W. C. 1950. Stratigraphy: A Factor In Paleontologic Taxonomy. JP 24(4):493.
[170] Ibid., p. 496.
[171] Thayer, C. W. 1973. "Evolution" of Upper Devonian Marine Communities Controlled by Rate of Progradation. GA 5(3):227.
[172] Ibid., p. 441.
[173] Ibid., p. 440.
[174] Bambach, R. K. 1977. Species richness in marine benthic habitats through the Phanerozoic. PB 3:153.
[175] Teichert, C. and B. F. Glenister. 1954. Early Ordovician Cephalopod Fauna from Northwestern Australia. BP 35(150):7.
[176] Cox, C. B. et. al. 1976. Biogeography: An Ecological and Evolutionary Approach. Blackwell Scientific Publications. London, 2nd Edition, p. 48-50.
[177] Chamberlain, J. A. 1973. Phyletic Improvements in Hydromechanical Design and Swimming Ability in Fossil Nautiloids. GA 5(7):571.
[178] Packard. 1972. op. cit., p. 291.
[179] Cowen, R. 1973. Explosive Radiations and Early Cephalopods. GA 5(7):585.
[180] Furnish, W. M. and B. F. Glenister. 1964. Paleocology. TP K120-3.
[181] Donovan. 1964. op. cit., p. 264.
[182] Cowen, R. 1975. Buoyancy Control in Ellesmerocerid Nautiloids. GA 7(7):1041.
[183] Valentine, F. G. and R. M. Finks. 1974. The Functional Significance of the Ammonoid Suture Pattern and the Origin of the Ammonoids. GA 6(7):994.
[184] Hallam. 1972. op. cit., p. 391.
[185] Teichert. 1967. op. cit., p. 185.
[186] Arkhipov, Y. V. and I. S. Barskov. 1970. Nautilids with an Intricately Dissected suture line. DE 195:218.
[187] Raup, D. M. and S. M. Stanley. 1971. Principles of Paleontology. W. H. Freeman and Co., San Francisco. p. 181.
[188] Wiedman. 1973. op. cit., p. 173.
[189] Miller, A. K., Furnish, W. M., and O. H. Schindewolf. 1957. Paleozoic ammonoidea. TP L18.
[190] Arkell. 1975. op. cit., p. L121.
[191] Packard. 1972. op. cit., p. 259.
[192] Heptonstall, W. B. 1970. Buoyancy Control in Ammonoids. LE 3(4):321.
[193] Mutvei, H. and R. A. Reyment. 1973. Buoyancy Control and Siphuncle Function in Ammonoids. PA 16(3):635.
[194] Ibid., p. 623.
[195] Lehmann, U. 1975. Uber Nahrung und Ernahrungsweise von Ammoniten. PZ 49(3):187.

[198]Chamberlain, J. A. 1971. Shell Morphology and the Dynamics of Streamlining in Ectocochliate Cephalopods. *GA* 3(7):524.
[199]Lominadze, T. A. 1966. Assymmetry of the Suture Line in Late Jurassic Ammonites. *DE* 171:241.
[200]Ward. 1976. *op. cit.*, p. 459.
[201]Vermeij, G. J. 1977. The Mesozoic marine revolution: evidence from snails, predators, and grazers. *PB* 3(3):247.
[202]Mutvei, H. 1975. The mode of life in ammonoids. *PZ* 49(3):197.
[203]Packard. 1972. *op. cit.*, p. 292.
[204]Kullmann, S. and J. Wiedmann. 1970. Significance of Sutures in Phylogeny of Ammonoidea. *UP* 47, p. 2.
[205]Arkell. 1957. *op. cit.*, p. L97.
[206]Christensen, W. K. 1976. Paleobiogeography of Late Cretaceous belemnites of Europe. *PZ* 50(3):114.
[207]Chamberlain, J. A. 1976. Flow Patterns and Drag Coefficients of Cephalopod Shells *PA* 19(3):560.
[208]Westermann, G. E. G. 1973. Strength of Concave Septa and depth limits of fossil Cephalopods. *LE* 6(4):383.
[209]Saunders, W. B. and D. A. Wehmann. 1977. Shell Strength of *Nautilus* as a depth limiting factor. *PB* 3(1):83.
[210]Miller, A. K. 1949. The Last Surge of the Nautiloid Cephalopods. *EV* 3:231.
[211]Teicher, C. and R. C. Moore. 1964. Introduction. *TP* K2
[212]Scott, G. 1940. Paleoecological Factors Controlling the Distribution and mode of life of Cretaceous Ammonoids in the Texas area. *JP* 14(4):308.
[213]Haas. 1973. *op. cit.*, p. 543.
[214]Barskov, I. S. 1974. Structure of the Protoconch and Ontogeny of the Belemnites (Coleoidea, Cephalopoda). *DE* 208(1-6):218.
[215]Hallam, A. 1975. Evolutionary Size increase and longevity in Jurassic bivalves and ammonites. *NA* 258(5534):493.
[216]Wiedman. 1969. *op. cit.*, p. 1002.
[217]Arkell. 1957. *op. cit.*, p. L119.
[218]Chamberlain, J. A. and G. E. G. Westermann. 1976. Hydrodynamic Properties of Cephalopod Shell Ornament. *PB* 2:330.
[219]Tozer. 1971. *op. cit.*, p. 1002.
[220]Sakharov, A. S. and T. A. Lominadze. 1971. Ecological Interrelationships of Middle Callovian Ammonites from the Northeastern Caucasus. *DE* 196:227.
[221]Brenner, K. 1976. Ammoniten-Gehause als anzeiger von PaleoStromungen. *NP A* 151(1):101.

A Diluvian Interpretation of Ancient Cyclic Sedimentation

A DILUVIAN INTERPRETATION OF ANCIENT CYCLIC SEDIMENTATION

JOHN WOODMORAPPE*

Received 9 June 1977

"No more fascinating field for research and speculation exists within the entire domain of stratigraphy."
—J. M. Weller

"But God remembered Noah and all the beasts and all the cattle that were with him in the ark: and God caused a wind to pass over the earth, and the water subsided. Also the fountains of the deep and the floodgates of the sky were closed, and the rain from the sky was restrained; and the water receded steadily from the earth, and at the end of the 150 days the water decreased."
—Genesis 8:1-3

The earth's water-laid rocks contain a repetitive type of stratigraphic layering known as cyclothems. Uniformitarian geologists have long debated the processes which formed them; but every theory proposed has failed to explain many basic properties.

In fact, uniformitarianism, which is not a scientific law or principle, has fostered an unnatural interpretation of cyclothemic rocks. Numerous properties of cyclothems are explained far more fully and simply in terms of catastrophic sedimentation. These properties include their world-wide distribution and "age"-transcendence, the shallowness of the deposition, vertical gradation, and evidence in them of contemporaneous tectonism. A detailed account is presented of the genetic relation between cyclothemic sedimentational and stratigraphic properties and the recessional phases of Noah's Flood.

Outline

I. Introduction
II. Uniformitarian Theories Proposed to explain the Causes of Cyclic Deposition
   A. Differential Settling Theory[9]
   B. Intermittent Subsidence Theory[10]
   C. Differential Uplift Theory[11]
   D. Precipitation Control Theory[12]
   E. Diastrophic Control Theory[13,14]
   F. Glacial Control Theory[15]
   G. Plant Control Theory[17]
   H. Compaction Control Theory[18]
III. The Role of Uniformitarianism in Geological Interpretation of Sedimentary Rock Phenomena
   A. The Antitheistic, AntiBiblical Nature of Uniformitarianism
   B. Chronostratigraphic Facets and Cyclothems
     1. The Uniformitarian Geologic Column and Cyclothems
     2. Unconformities Versus Contiuous Deposition
     3. Radiometric Dating and Cyclothems
IV. Ramifications Of The Deluge In Cyclothem Formation
   A. The Worldwide Occurrence and "Age"-Transcendence of Cyclothems
   B. Paleohydrological Parameters and Cyclic Sedimentation
     1. Shallow, Intracratonic, Epicontinental Seas Versus Flooding of Continents
     2. Specific Evidences of Cataclysmic Sedimentation in Cyclothems
     3. A Critique and Re-Interpretation of Uniformitarian "Sedimentary Environment" Concepts as Applied to Cyclothems
V. The Diluvian Tectonosedimentary Generation of Cyclothems
   A. The Tectonic Component of Cyclothem Genesis
   B. Specific Evidences of Cataclysmic Tectonism Contemporary With Cyclothem Formation
   C. The Sedimentational Component of Cyclothem Genesis
   D. The Relationship Between Individual Cyclothem Members and The Flood
     1. The Coals (Member 5)
     2. The Underclay (Member 4)
     3. The Limestones (Members 3, 7, 9)
     4. The Sandstones (Member 1)
     5. The Shales (Members 6, 8A-B, 10A)
VI. Epilogue

Figure 1. This shows an ideal Illinois cyclothem (See Reference 131.) Note that at no locality are all 10 listed members present; the most commonly found cyclothem has 1 (and, or 2), 4, 5, 8, 9, and 10. Nearer to the source area (Pennsylvanian and Ohio) fewer limestones and greater thicknesses of coal prevail, with thicker, more massive clastic material. In Kansas (farthest from the source area) there are thinner coals, and only think fine clastics and massive limestone beds.

The key to the letters and numbers is as follows: 4, underclay, nonbedded; root impressions, slickensides. 3, claystone, with or without limestone nodules and sheets. It grades into 2, siltstone or sandy shale. This grades into 1, sheet or channel sandstone, which is EITHER in sharp erosional contact with, OR grades into 10C, siltstone. This grades into 10B, sandy or silty shale. 10A, shale, gray, usually sandy in the upper parts, with ironstone nodules. 9, limestone, marine, fossiliferous or calcareous shale. 8B, gray shale. 8A, shale, black, fissile, with brackish or restricted marine fauna. 7, limestone, marine, fossiliferous, but rarely present. 6, shale, gray with plant fossils. 5, coal.

I. Introduction

In beginning this article, may I make a suggestion, and then proceed to follow it myself? The suggestion is this: that in the future Flood geologists call themselves Diluvialists, not catastrophists, because of the fact that

*John Woodmorappe, B.A., has studied both Geology and Biology

many orthodox uniformitarian geologists accept rapid sedimentation. Many others (Howorth, Velikovsky, etc.) accept catastrophes in the earth's past but do not accept the universality of the Deluge, or else do not accept the Divine Inspiration and complete factuality of the Biblical accounts, or else adulterate Scripture with claims of many other catastrophic occurrences—thus too often twisting Scripture to fit their presuppositions.

The term Diluvialist therefore refers to that scientist who accepts the historicity and factuality of the Biblical accounts of the Creation-Flood period (Genesis 1-10), without adding or subtracting any tenet which would constitute a twisting of the Inspired Word to fit any other world view or construction of claimed knowledge.

There exist among sedimentary rocks certain types which have very many types of rock, in thin layers, which lie one on top of another and repeat in a regular sequence. Much of the world's coal is found in such repeating layers. Each repetitive sequence (between coals and including one coal) is called a cyclothem. A diagram of an "ideal" or "complete' cyclothem found in Illinois is found in Figure 1. (The reader of this paper should continually refer to Figure 1 whenever the number or the lithological identity of a given member is described in this text.) The numbering and termination of the cyclothem differ, reflecting the disagreement among Pennsylvanina stratigraphers as to where one cyclothem "ends" and a superjacent one "begins." [Thus, some consider a new cyclothem to begin at the basal sandstone, member 1, while others consider a new cyclothem to begin with the coal (member 5)].

It must be hastily added that almost never in the earth does a "complete" cyclothem occur at any location as shown in Figure 1. A real field situation as exists might have this type of layering: members 1, 2, 4, 5, 10, 1, 2, 4, 5, 8, 10, 1, 2, 4, 8, 9, 10, 1, 2, 3, 4, 5, 6, 7, 8, 10, etc. The important fact to realize is that the relative order of the members always exists and that these members do repeat themselves consistently.

The cyclothems are asymmetrical, which means that the coal or shale (or any other member), may be vastly thicker or thinner than the corresponding member of the cyclothems above and below it. Furthermore, even within one cyclothem, the thickness or thinness of one member does not guarantee the thickness, thinness, or even presence of another member in that cyclothem at all. Six or more of the then members are usually found at any given locality and their relative order is always preserved. "The average thickness of a cyclothem is the central states is less than 50 feet[1]..."

Although cyclothems and their valuable coal beds are found in many parts of the world, this paper will concentrate on the cyclothems found in Illinois as far as their morphology and specifics are concerned. "The Pennsylvanian sediments in the basin cover an area of approximately 55,000 square miles, chiefly in Illinois, Indiana, and Kentucky, with minor areas in Missouri and Iowa.

The maximum thickness of Pennsylvanian sediments, more than 2800 feet, occurs in the southern part of the basin in Kentucky. Shale is the predominant lithological unit of the sequence with subordinate amounts of sandstone and much smaller amounts of limestone and coal. The presence of ordered lithological sequences or cycles is the most characteristic lithological feature of the Pennsylvanian sediments. More than 50 such sequences are recongized."[2]

Cyclothems are variable not only in terms of presence of members and thickness of members, but also in terms of lateral extent of each of the members. Some members can be traced for hundreds of miles while others wedge out (& thin out) in only a few miles or else grade into members elsewhere. "Many of the cyclothems are nearly as varied within a single county as within the entire state of Illinois.

For this reason, a detailed study of only a small area may leave the impression that the beds vary greatly, whereas a more general survey of almost the entire Eastern Interior Basin has revealed that the Pennsylvanian system throughout this region is remarkably uniform."[3] Cyclothems, like humans, are-paradoxically-so similar yet so different from each other.

The reason why the area of cyclic sedimentation has been chosen for this study is because "No more fascinating field for research and speculation exists within the entire domain of stratigraphy."[4] This writer has engaged in both "research and speculation" in the field of cyclic sedimentation to try to understand how the Universal Deluge caused cyclic sedimentation.

The uniformitarians have proposed a good number of theories to explain how the cyclothems formed, but "None of these theories has gained much following."[5] This is because "Recognition of the stratigraphic facts of cyclical repetition and distribution, and deduction of the causes responsible for them are unrelated, one to another. Evidence pertinent to the historical interpretation of cyclothems is incomplete, scattered, and not fully understood. Some of it has suggested different conclusions to different persons.

Also, several possible avenues of attack upon this problem have not been adequately explored. It is not surprising, therefore, that there is much disagreement regarding the basic cause or causes responsible for the development of cyclothems."[6] The Diluvialist will add that one of the "Avenues of attack" that has not been explored is that of the Deluge.

The very fact that uniformitarianists are having difficulty coming up with a universally appealing theory to account for the formation of cyclothems is evidence that the case for alternative, non-uniformitarian theories is far from closed. Therefore, a Diluvialist is completely justified intellectually in proposing his own theory.

II. Uniformitarian Theories Proposed to Explain the Cyclic Deposition[7,8]

Of the theories mentioned in the Outline, the first four, which are commonly admitted to encounter grave difficulties, will not be discussed here. The Diastrophic Control Theory contains some truth, and, when understood in the Flood setting, it is workable since logistic problems of timing are overcome, The Glacial Control Theory, since it works on a world-wide scale, deserves some thought; and the other two will be mentioned briefly.

F. GLACIAL CONTROL THEORY: This theory is speculative because of the near impossibility of demonstrating cause-effect in terms of claimed Paleozoic glaciations. A more serious objection has to do with the wedge shape of cyclothems; "Examination of sedimentary cycles at the margins of Carboniferous basins shows that they do not conform to the pattern that would be expected if the sea-level changes were eustatic. This is because eustatic rise in sea level must equally affect the margins and central part of the area of deposition. Deposits of this type are rare in the geologic column outside of the Quaternary... "They are unknown in the Carboniferous basins of deposition."[25] A modification of this theory postulates basinal subsidence, but this way of thinking only "patches up" the problem and makes the theory so non-falsifyable and plastic as to be completely untestable.

Since the Diluvialist contends that the uniformitarian geologic age system is devoid of reality, supposed glaciations "plugged into" those "times" are also rejected. Whitcomb and Morris in their classic Diluvialist work[16] have written of suggested alternative explanations for claimed Paleozoic glaciation. It is significant that the Glacial Control Theory is the main theory among uniformitarians that attempts to come to grips with the worldwide occurrence of very similar cyclothems, because it elevates the riddle of cyclothem formation from local to worldwide proportions. It is up to the Diluvialist now to unify the worldwide aspects together with local morphology into a coherent Diluvian interpretation which is free from the limitations upon thinking created by the geologic dogma of uniformitarianism.

The two remaining minor theories are considered to carry very little weight. They are:

G. PLANT CONTROL THEORY[17] This theory sees cyclothems being formed by differential blockage of zones of sediment accumulation by means of levees. Needless to say, this theory is incapable of explaining the persistent lateral extent of many cyclothems, and also much less the striking similarity of the cyclothem phenomena across the globe.

However, this theory has value in explaining certain local variations in sediment thickness in therms of Diluvian processes. As Floodwaters cyclically retreated and transgresed vast inland areas during the recessional stages of the Deluge, levees must have played a great role in determining over what regions sediment would be carried and stranded by waning fluvial-like water currents.

H. COMPACTION CONTROL THEORY[18] This theory envisions the sedimentary basin being laden with sediments that can be compressed by vastly different degrees; this differential compaction accounting for sequences. For example, when sand is accumulated, the difficulty of compacting it would mean that it would fill up the basin faster than it would subside.

The basin being filled would lead to the area being constantly above water and consequently of peat accumulation. As peat accumulated it would be easily compressed, leading to the rate of subsidence exceeding the rate of accumulation. This would lead to new marine transgression drowning the swamp. The area being again underwater would lead to a repeat of sand accumulation. this entire process would continually repeat, forming cyclothems.

This hypothesis meets its nemesis because materials could not be compacted sufficiently to allow for superjacent cyclothems to be added. As so many other uniformitarian theories, this one also fails to account for cyclothems in their totality as well as to account for worldwide distributions and large extent of cyclothems.

CONCLUSION. It bears repeating that "none of these individual theories has gained much following."[5] It must be stressd once again that the Diluvialist, not restricted by uniformitarian philosophy, can unite the best elements of all these theories into a coherent view of cyclothem formation in terms of the Deluge. But before that venture is begun, an explanation of uniformitarianism and its control of geological theories, including theories of cyclothem origins, will be presented.

III. The Role of Uniformitarianism in Geological Interpretation of Sedimentary Rock Phenomena

PROPOSITION. The uniformitarian world view, which has dominated geologic thought since the 1830's, is not a scientific fact or deduction, nor a scientific law. It is based on the *a priori* decision to reject as invalid any Biblical, historical, or scientific evidence for any sedimentary process significantly different from those encountered today. The Diluvialist position, by admitting the possibility of catastrophes in the past, offers a fresh, new framework for earth history; and it leads to far more satisfactory explanations of sedimentary and stratigraphic phenomena that any theory which limits itself to consideration of present-day sedimentational phenomena.

A. THE ANTITHEISTIC, ANTIBIBLICAL NATURE OF UNIFORMITARIANISM. There is far more to geology in its implications than the study of our planet, "In its widest sense, geology covers the whole spectrum of human experience and understanding... few other disciplines have had and must have in the future a more profound influence on human thought."[19] Thus, there is room for various interpretations of geological thought.

Uniformitarianism is not a scientific law that was deduced from the study of the earth; it was and is an *a priori* viewpoint, "Strict uniformitarianism may often be a guarantee against pseudoscientific fantasies and loose conjectures, but it makes one easily forget that uniformity is *not a law, nor a rule established after the comparison of facts*, but a methodological principle *preceding* the observation of facts. (Italics added)."[20] The uniformitarian must therefore not claim that the study of rocks has disproved the Deluge and established that present-day sedimentary processes have produced the earth's sedimentary rocks precisely because it is his *a priori* presupposition that rejects that historicity of the Universal Deluge and assigns present-day sedimentation to be the guide of the study of the formation of sedimentary rocks.

Uniformitarian geologists may come to wear blinders which hide everything other than what is going on today in sedimentary environments, "Advocates of methodological uniformitarianism seem to view the classic formulation of the uniformity principle as a tautology: our reconstructions of the past are bound to be determined by whatever we discover to be going on at present."[21] It is therefore obvious that even consideration of the Deluge will not get the slightest chance in uniformitarian geology because there are no worldwide floods going on at present. (And God promised Noah that there would never be another worldwide flood. (Genesis 8:21)] It is an exercise of highest futility to hope for geologists as a whole to ever consider the factuality of the Universal Deluge. "... our interpretations of prior events must necessarily consist of inferences based upon present observations,"[22] insist the uniformitarian geologists.

The religion of atheistic humanism dominates geology. Not only is no process not represented in kind today permitted, but God and His Word are specifically banned from geologic thought today, "What are our assumptions in such a procedure (Uniformitarianism)? Fundamentally, there are two: (1) We assume that natural laws are invariant with time. (2) We exclude hypotheses of the violation of natural laws by Divine Providence, or other forms of supernaturalism."[22] "Indeed, if I am correct, the residual issue between uniformitarianism and catastrophism comes down to the issue between *naturalism and supernaturalism* or at least between naturalism and anti-naturalism."[23]

That is where it is! Not a conflict between science (supposedly uniformitarianism) and religion (supposedly only Diluvianism) but a battle between one religion (atheistic, naturalistic, humanism) and another (Biblical Revelation); at one level. At the scientific level, it is a battle between interpreting non-observational, non-experimental, non-repeatable data in terms of the consequences of naturalistic, atheistic humanism vs. interpreting them in terms of the consequences of Divine action in human history and geologic history as revealed and recorded in Scripture. Both positions fit facts to theories. With Diluvialism, it is obvious; with uniformitarianism, it is also true: "We do not merely find facts and make theories; we fit facts to theories as well as theories to facts."[24]

B. CHRONOSTRATIGRAPHIC FACETS AND CYCLOTHEMS

1. The Uniformitarian Geologic Column and Cyclothems

It is well known that the earth's sedimentary strata is pieced together into a vast system of supposed ages of the earth's past; the fossiliferous rock being alleged to go back 600 million years. Yet the basis for this age-column is evolution. Certain fossil forms are supposed to spell a long sequence of evolutionary events. Various forms of these index fossils are found in fossiliferous rocks and a chronology of rock layers is built upon this and similar fossil-recording-evolution progressions.

Thus, if some fossil form evolves from form A to D, then any rock containing A will be older than B which in turn will be older than C, etc. Yet a rock C lying on top of A will be assumed to either have had B on top of A before C was deposited (B being "eroded away" before C's deposition) or else not have been an area of sedimentation during the time B supposedly evolved.

That is how a geologic column some 130 miles net sedimentary thickness spanning some 600 million years is constructed by the evolutionists-uniformitarians in spite of the fact that at no location on earth is the complete sequence found and nowhere on earth does the sedimentary rock cover exceed some 7 miles.

The uniformitarian geologic column is the exact antithesis of the Diluvialist position, which sees practically all of the fossiliferous sedimentary rock laid down during Noah's Flood or shortly afterward. Diluvialists have long noted how artificial and specious this column is. They have exposed the underlying foundation of evolution, and have particularly stressed the innumerable gaps and distinct, abrupt appearances of complex fossils without evidence of any evolution from "lower" forms. This column, a biologic onion-skin theory as stated by the noted late Diluvialist, George McCready Price, ignores, for the most part, other data. Thus, there is no difference between a Cambrian shale or a Cretaceous shale except the claimed age difference. Only fossils and supposed evolutionary stages delineate the rocks and their ages.[26] There are specific examples of fossils of different "ages" mixed together. One example is the finding of Devonian fossils in Pleistocene sediment.[27] Another is the finding of plant spores in coal (or in rock) outside their "ranges" with the result of there being "no consistent pattern" in these assemblages."[28] In both these cases, the protective concept invoked is that of "reworking" whereby an old sedimentary rock is eroded and its fossils included in much younger rock.

However, the prima facie (raw, face value, pre-interpreted) evidence of mixed fossils indicates one fatal flaw in uniformitarian thought. Another example found is that of Eocene and Pennsylvanian fossils coexisting in a rock in Alaska.[29] The rock was labelled Eocene and the other fossil dismissed as a "striking homeomorph (look alike)" of *Annularia*, a Pennsylvanian fossil. The claim of earlier misidentification may have to be granted, but in any case the weak basis for uniformitarian "age" designations is fairly evident.

Divisions of geologic periods, predictably, are quite weak: "I have already indicated something of the difficulty in fixing both upper and lower boundaries of the Upper Carboniferous rocks... The division between Namurian (lower Pennsylvanian) and Westphalian (middle Pennsylvanian) is quite arbitrary... The boundary between Stephanian (upper Pennsylvanian) and Westphalian is even less objective."[30]

Even whole geologic periods in places are invented to keep fostering the illusion of long ages and evolution: "Definition of the boundary between rocks classed as Pennsylvanian and Permian in the Kansas region has led to much debate and disagreement... (It was)... proposed to avoid the difficulty by not recognizing the Permian at all... This procedure might be defended on the basis of the stratigraphic succession in the midcontinent area, but it is evidently unsuited to world-wide application."[31]

In other words, the rock evidence (stratigraphy, gradational nature of rocks) is deliberately forced to conform to evolutionary-uniformitarian thought. The Diluvialist rejects outright these artificial, unnatural divisions of these FLood-deposited rocks, because, "Judgement has been generally expressed that any adopted boundary is measurably arbitrary."³² The Diluvialist can grasp the idea of continuous, nearly contemporaneous deposition of these cyclothemic rocks by the Universal Deluge.

Unnatural age divisions of Flood-laden rock inevitably lead to flagrant contradictions between the forced rock delineation and its stratigraphic properties which must be resolved by illogically segmenting continuous rock bed masses. Into this situation cyclothems enter: "Paleontological and physical evidence seem to conflict, the paleontological evidence favoring Weller's correlation and the physical evidence favoring that of the writer."³³ Also, ". . . laterally continuous lithologic units may be of different ages in different positions in the basin. Matching of the depositional cycles may demonstrate the lateral facies equivalence despite the chronological difference."³⁴

In the final analysis, the entire uniformitarian geologic column is admitted to be arbitrarily divide rocks: ". . . the stratigraphic column does not naturally divide itself into sharply defined systems of world-wide extent."³⁵ "A . . . source of confusion and controversy lies in preconceived notions on stratigraphic classification handed down from the early stages of the development of stratigraphy . . . As an example is the tendency to believe that the classic time-stratigraphic divisions (systems, series, stages) established largely in Europe during the last century constitute "natural divisions" of the earth's stratigraphic column which can be recognized as such around the world . . . Although few would now openly subscribe to this extreme segmentation of the stratigraphic record, nevertheless many almost unconsciously endow the boundaries of these original time-stratigraphic units with a world-wide significance far beyond their real nature of quite *arbitrary* . . .divisions of . . . the earth's sedimentary strata.³⁶" (italics added).

2. Unconformities Versus Continuous Deposition.

A most important ramification of the Diluvian notion of continuous deposition is the gradational nature of sedimentary rock beds. Scientific Diluviology scrutinizes superposed rock beds of widely different "ages" and notes the gradational nature between them with no evidence of any buried erosional surface and consequent hiatus of deposition. Erosional surfaces, or unconformities, definitely exist but are not as widespread as the uniformitarians would have them be.

As a result, uniformitarianism collapses in its claim of that 130-odd mile thick, 600 million year geologic rock-time continuum. On the other hand, the absence of worldwide erosional surfaces, the gradation of every formation somewhere into another, etc., all strongly argue for the Noachian Deluge.

There are various unconformities related to cyclothems. One is the unconformity (buried erosional surface supposed to indicate a considerable span of time) between cyclothem "stacks" and the underlying, noncyclothemic "older" rock. The other is the unconformity which is supposed to divide one cyclothem from the next one above; considerable time is variously ascribed between individual cyclothems.

The Silurian-Devonian in Kansas underlying the cyclothems is ". . . poorly distinguished³⁷. . ."

The unconfirmity between the cyclothemic Pennsylvanian and noncyclothemic lower Paleozoic material is considered to be ". . . one of the most important unconformities in the Paleozoic of the east-central U.S.³⁸. . ." Yet there are very many locations where there is no compelling evidence for any stoppage of deposition, and erosion, between Pennsylvanian and pre-Pennsylvania rocks.

For example. Lower Carboniferous rocks," . . . overlap the Devonian formations, but no noticeable discordance has been observed between the two formation.³⁹" Eocene cyclothems rest on top of Cretaceous: ". . . these were previosuly thought to be of Cretaceous age. These beds rest with no marked discordance on the Cretaceous⁴⁰. . ."

The Mississippi-Pennsylvanian unconformity in the U.S. is very interesting. In attempting to distinguish Mississippian and Pennsylvanian rocks, Siever⁴¹ notes that: "The chief difficulty is that many Chester (Mississippian) beds resemble Pennsylvanian beds in lithological character, texture, and minerology.⁴¹" "Even with this detailed lithological study there will be many drill holes in which it is impossible to ascertain the exact depth of the contact. Under such conditions, and where the stratigraphy is doubtful, the best method is to make an intelligent *guess* substantiated by any information available.⁴²" (italics added)

The Mississippian-Pennsylvanian in Virginia is now seen as "partly contemporaneous.⁴³" In Oklahoma: ". . . the unconformity separating Mississippian and Pennsylvanian strata disappears into the basin. No evidence of truncation or missing stratigraphic intervals either above or below the unconformity can be demonstrated . . . At other places local unconformities may be identified at the base of a particular sandstone unit, but no regional erosional surface can be recognized.⁴⁴"

Quite recently the uniformitarians have come around to recognizing gradation and intertonguing of different "aged" rocks: ". . . the unconformity concept is largely subjective and is based to some degree upon preconceptions concerning modes of origin.⁴⁵" "Finally, it is evident that the regional systemic unconformity between "Pennsylvanian" and "Mississippian" strata . . . (is) no longer valid.⁴⁶" However, the uniformitarians have come up with a new way of explaining away the gradation-intertonguing of different "aged" rocks. They call it a "time-transgressive relation" where some sedimentary basin is filled in a time spanning geologic periods or from sedimentation of continuous sedimentary environments.

However, no amount of assigning new names can cover up the fact that these ages never existed, and that corroboration of the fact of the Flood is offered by gradational-intertonguing relations in massive sedimentary rock beds. Also noteworthy is the fact that Upper Mississipian cyclothems grade directly unto the Pennsylvanian ones.⁴⁷

Much the same non-division noted in the Mississippian-Pennsylvanian also exists in rocks unnaturally divided into the Pennsylvanian and Permian "periods." In the U.S.S.R., for example," Problems remain with the Devonian-Carboniferous and Carboniferous-Permian boundaries.⁴⁸" Also, "The Carboniferous-Permian boundary has long been a puzzle in the Upper Paleozoic in Japan and in the extensive area covering Korea, Manchuria, and North and South China.⁴⁹" In the U.S., "Cyclo-phase deposition crosses the systemic bounday of the Pennsylvanian-Permian.⁵⁰"

Not only do the sedimentary rocks reveal their *prima facie* nature of being continuously Flood-deposited, but rocks layered in "reverse age" sequences create additional embarrassment for the uniformitarian position: "Another problem, more difficult to explain, is that in some localities carbonates and red and green shales with "Mississippian" faunas overlie "Pennsylvanian" orthoquartzites . . . In such cases, the "Pennsylvanian" orthoquartzes are assigned to the "Mississippian" . . . Thus, however awkward, the tabular-erosional explanation has been consistently applied in absence of any viable alternative explanation.⁵¹" Diluvialists have continually exposed the fantastic explanations offered by uniformitarians to escape the fact of glaring contradictions to their geologic time scale.

In summary, the *de facto* gradation, in some (most) locations of Silurian into Devonian, Devonian into Mississippian, in turn into Pennsylvanian, in turn into Permian, all suggests that this span represents a continuously-deposited sequence of rock. The artificial geologic age designations collapse in futility, some 190 million years of nonexistent time evaporates, and the rocks give powerful testimony to the fact of the Noachian Deluge as causing their origin. It will now be shown that the cyclothem layers within them also were continuously deposited and also show no compelling reason to believe that there was any significant time span between their formation.

The two places where cyclothems are divided are the sandstone and underlying siltstone (members 1 and 10C, Fig. 1) and the coal and overlying member (member 5 and whatever is superjacent).

"The prominence of the basal sandstone unit and its channel aspect in the typically limited outcrop areas of the Eastern Interior have led workers to stress the important of this unconformity. However, unconformities appear to be present in less than 20% of the area of any cyclothem in Illinois, and for some cyclothems no unconformities are known⁵². . ." Hence, in the majority of cases, there is a gradation⁵³ between the basal sandstone and the underlying siltstone. Furthermore, the unconformity vanishes almost completely basinward in the Kansas region.⁵⁴ Even where an erosional surface below the sandstone does exist, "No evidence of weathering was observed beneath the contact.⁵⁵"

The artificiality of this sandstone-siltstone transition being used as a depositional-stop is admitted: "The importance of the discordant surfaces present beneath some sandstone bodies . . . has been exaggerated . . . In such a classification system, the products of a single pulse of clastic deposition are unnaturally split between two adjacent units.⁵⁶"

The other position in the cyclothem where a cyclothem is thought to "end" and the superjacent one "begin" is the coal bed. However, this practice of dividing is subjective like the sandstone-siltstone contact: "Nor does it matter whether the cyclothems are separated at the base of the sandstone . . . or at the top of the coal, as is preferred by some others, including most European stratigraphers. The selection of this boundary is a matter of *opinion*⁵⁷ . . ." (italics added). It must be realized that the coal grades into whatever is above it.

Specifically, the coal grades into or intertongues with limestone⁵⁸ and also often grades into the black shale, even to the point of making it ". . . difficult to remove it from the coal during mining operations.⁵⁹" It could similarly be shown that every cyclothemic member grades into superjacent and subjacent members at least somewhere. It can therefore be concluded that there is no substantial time gap between cyclothemic formation, that uniformitarian notions which claim time intervals between cyclothems can be justifiably rejected, and that cyclothems can be understood in terms of uninterrupted Flood action.

The calling attention to the gradational nature of the rock beds is not meant at all to imply that there are no buried erosional surfaces, on the contrary, there are vast areas possessing unmistakeable erosional surfaces. The point was that there is no universal unconformity that points to any significant stoppages of deposition.

The localized nature of the unconformities can be viewed as regions which were covered with Flood water, exposed, then re-inundated. This is eminently reasonable when one considers how shallow several hundred feet of Floodwater are in relation to tectonic changes in topography and their effects. Some erosional surfaces may result simply from very rapid lithification of Flood-laden rock.

"Certain colloids tend to aggregate very quickly; even in minutes . . . For example, polymerization of silica occurs most rapidly and produces an extremely powerful, irreversible cohesion. Such effects are very important in sediments, particularly the clays⁶⁰. . ." Hence some non-gradations, predictably local as is actually the case, can be accounted for by this rapid lithification and later erosion during the overall continuous deposition of the Flood.

An interesting situation exists between some areas of Mississippian rock and overlying Pennsylvanian. The localized erosional surface that has been chosen as the boundary possesses buried valleys whose walls have collapsed (slumped)⁶¹. In some places this slumping is so profound that it ". . . amounts to virtual collapse of the valley wall.⁶²"

How an ordinary river could produce such slumped valley walls is a puzzle to the uniformitarian: "Presumably stream action erodes and destroys most bank materials subsequent to their failure. The fact that structures produced by stream bank failure are preserved is, thus, seemingly an enigma.⁶³" The Diluvian position can amply explain this situation as being caused by poorly consolidated sediment laid down earlier in the Flood being eroded into during the recessional stages of the Flood, with consequent slumping of weakly consolidated sides of gorges.

Thus localized erosional surfaces, when properly understood, do not justify the uniformitarians' extrapolation of them into regions of gradation. Flood action operates as local erosion superposed over continuous deposition.

3. Radiometric Dating and Cyclothems

The familiar use of radioactive decay parent-daughter element ratios has been taken by many to have proved the uniformitarian contentions of extremely long geologic ages.

Diluvialists, on the other hand, have stated repeatedly that the geologic column and its inherent long-age claims were decided long before even the discovery of radioactivity. Diluvialists also noted the independence of the previously-discussed paleontological dating from radiometric dating, and uniformitarians also concur: "The development of an absolute time scale has been of immense importance to geology but it is not accurate and complete enough to replace the standard methods of classical stratigraphy. The geologist who works within a certain system still relies entirely on relative dating. Indeed, many practising stratigraphers regard absolute time as not very essential and absolute ages are mentioned with great care, if not suspicion.[64]"

Diluvialists often cite examples of how uniformitarians accept only those dates which agree with their preconceived notions. A recent example of this is where "The $^{87}Sr/^{86}Sr$ and $^{87}Rb/^{86}SR$ ratios are positively correlated, but this relationship is the result of mixing and does not reflect the age of the provenance nor the time of deposition of the sediment.[65]" If "mixing" indeed accounts for consistent ratios, then the Flood can account for much mixing.

The scrutiny of radiometric dating methods has uncovered the degree of subjective bending of data: "There is no really valid way of determining what the initial amounts of Sr^{87} in rocks were. There is much juggling of numbers and equations to get results in agreement with the U-Th—Pb "clocks." In all these radioactive clocks, all methods are made to give values that fit the evolutionist's belief as to the age of the earth and the ages of the various geological events.

The reason that the various dating methods give similar ages after "analysis" is that they are made to do so. In the case of the initial Sr^{87}/Sr^{86} ratios, these values can be adjusted so that any age desired is obtainable.[66]" A vivid portrayal of this action is offered by recent radiometric dating of inclusions in cyclothemic sandstones: "An initial Sr(87/86) ratio of 0.706 has been CHOSEN for the Rb-Sr calculations. A higher value would make the Rb-Sr age for M1117a lower than the K-Ar age, a phenomenon very unusual in nature and certainly NOT TO BE EXPECTED from a simple detrital muscovite. If muscovites have been weakly metamorphosed, then the K-Ar age may be lower than the Rb-Sr age because of the greater loss of radiogenic Ar^{40} relative to radiogenic Sr^{87}. If an initial ratio of 0.703 (low for a granite; but hypothetically possible) is CHOSEN ages of 378 and 411 Million years (for M1117a and M1125, respectively) result. This brings the K-Ar and Rb-Sr dates for M1117a to within 1 million years for M1125; hence our *preference.* is for the 0.706 value. The K-Ar ages are the mean of duplicate analysis, whereas the Rb-Sr ages are single determinations only[67]. . ." (emphasis added)

Not only do uniformitarians take great liberties in bending data to suit their presuppositions, but also there is a growing body of radiometric evidence for the extreme youthfulness of the earth. Recent findings of radiohalos in Triassic and Jurassic coals, presumed to be 140 to 230 million years old, are an exciting example: "If remobilization is not the explanation, then these ratios raise some crucial questions about the validity of present concepts regarding the antiquity of these geological formations and about the time required for coalification.[68]" Furthermore, "Such extraordinary values admit the possibility that both the initial infiltration and coalification could possibly have occurred within the past several thousand years.[69]" The notion of coalification only "several thousand" years ago is in complete agreement with Scriptural dating of the Universal Deluge.

CONCLUSION: The uniformitarian position, which is intrinsically anti-Biblical and rejects the Deluge, falters gravely in its major tenets of long-ages, extensive stoppages of deposition, various artificial and arbitrary subdivisions of Flood-laden strata, etc. Figuratively speaking, the rocks testify to the reality of the Flood and cry out vociferously against uniformitarian attempts to avoid it.

IV. Ramifications of the Deluge in Cyclothem Formation

PROPOSITION: A variety of factors which are properties of cyclothems are neglected in uniformitarian circles, but these are extremely pertinent as ramifications of The Universal Deluge. Other sedimentological factors are implied to support ideas of present-day type sedimentation, but these can be understood as types of localized Flood action instead.

A. THE WORLDWIDE OCCURRENCE AND "AGE"- TRANSCENDENCE OF CYCLOTHEMS

Many types of cyclothems exist. Among these are carbonate-evaporite cyclothems[70] which may be ascribed to early Flood action. However, this paper concentrates on coal-bearing cyclothems which are ascribed to the recessional stages of the Deluge.

Some coal-bearing cyclothems have been traced along outcrops for 400 miles or more.[71] Cyclothem members have been claimed to have been correlated between basins (which could imply a cyclothem continuity of nearly 1000 miles) but these correlations aren't observable in continuity and therefore cannot be proven. Nevertheless, cyclothems in North America and Europe, from ". . . Texas to the Donetz (Russia) coal basin[72]. . ." are extremely similar. This is understandable in terms of the worldwide Flood producing similar worldwide results.

Although the Pennsylvanian-Permian cyclothems are the most widespread, there are cyclothems of "ages" between Devonian and Miocene, or even more recent. Because Diluvialists are completely free from the uniformitarian Geologic-Column age designations, all coal-bearing cyclothems of the world can be viewed as having formed virtually contemporaneously as the Flood-waters receded everywhere on earth. Thus, the Pennsylvanian-Permian cyclothems of the East and Midwest could be contemporary with the Tertiary cyclothems of the Phillipines and with the short Mesozoic cyclothems of the Rocky Mountain states.

The differences that do exist between "younger" and "older" cyclothems of different regions can be accounted for by differences resulting from ". . . greater slope of the depositional surface.⁷³ "The situation of Jurassic cyclothems overlying Upper Paleozoic cyclothems in the U.S.S.R.⁷⁴ can also be viewed as being continuous recessional Flood deposition.

Cyclothems exists in every niche of earth's surface. Cyclothems are found in North America, Europe, Africa⁷⁵ and Madagascar,⁷⁶, South America,⁷⁷ Australia;⁷⁸ western Asia inlcuding China,⁷⁹ India;⁴⁰ New Zealand,⁸⁰, Japan, ⁸¹, Antarctica.⁸⁴

The uniformitarians sense the unity and universality of cyclothems, but cannot grasp their testimony to the universality of the Flood and its recessional sedimentary-tectonic operation (because of their belief in geologic ages): ". . . it is held that comparison of the different types of cyclothems developed in rocks of many different ages strongly suggests a common cause.⁸² "The occurrence of some coal-bearing sediments in all Systems from the Devonian to the Late Tertiary might be held to suggest that the pulsations causing the cyclic deposition were continuous or nearly so (and not in themselves periodic) giving rise to rhythmic deposits wherever other conditions were favourable. But the time-distribution of important carbonaceous deposits appears to be too irregular to support this suggestion.⁸³

Thus, if geologic ages are accepted, cyclothems appear in the geologic column too irregularly for any regular cause to be discerned: Diluvialists, obviously rejecting all geologic ages, can therefore consider all "age" cyclothems all over the world being formed contemporaneously by the Universal Deluge with its universal, and similar, effects everywhere on earth.

B. PALEOHYDROLOGICAL PARAMETERS AND CYCLIC SEDIMENTATION

1. Shallow, Intracratonic, Eipcontinental Seas Versus Flooding of Continents.

One of the first things which a testbook on historical geology mentions is the claim that most of Paleozoic sedimentation occurred in "shallow, intracratonic, eipcontinental seas." Cyclothems are likewise believed to have been deposited in very shallow seas: "it is generally agreed . . . that the seas in which thin fossiliferous black shales and limestones formed were very shallow and resulted from short-lived inundations of large areas.⁸⁵"

However, uniformitarianism breaks down in its attempts to find *bona fide* examples of such shallow seas today. Although Hudson's Bay has been proposed, the fact remains the the familiar statement, ""The present is the key to the past,' may be a misleading one when considering epeiric sedimentation. There simply is no existing models of epeiric sedimentation to guide our investigations, and although it is true that many similarities do pertain between the past and the present, it is equally true that many differences exist as well.⁸⁶"

This fact can be used to raise questions as to whether the "epeiric seas" were seas at all. In other words, instead of the present time being considered unusual because of its virtual absence of extensive shallow seas, the past sedimentation can be considered the unusual situation consisting of extensive Flooding of all the continents. Hence, when properly understood, the Paleozoic and some Mesozoic "shallow marine" sedimentary blankets covering major portions of all the continents can be seen as the direct result of the Universal Deluge. The absence of such sedimentation today reflects the fact that Noah's flood is ended and no longer will oceanic waters transgress over continents.

The exact depth at which the epeiric waters covered the continents is a matter of controversy in uniformitarian circles. There is some evidence of extremely shallow water. This is ". . . evidence of local dessiccation, brecciation, and inclusion of clastic grains in the very widespread Ames limestone that indicates shallow water or local emergence. The only conclusions possible at present are that these marine members did represent maximum submergence but that this submergence may be as little as 5 feet or as much as 180 feet.⁸⁷" The blanketing of continents with variably deep, but extremely shallow, water is precisely in accord with the Flood.

2. Specific Evidences of Cataclysmic Sedimentation in Cyclothems.

The time ascribed to cyclothem formation by uniformitarians varies from 20,000 to 350,000 years.⁸⁸ However, even uniformitarians now acknowledge that ". . . average rates of sedimentation are meaningless⁸⁹. . ."

One of the family of *prima facie* evidences for rapid burial is the nearly universal presence of clastic inclusions and plant/tree/coal debris in the cyclothem sandstones: " Log casts, carbonized plant debris, and shale fragments are locally abundant in the sandstone⁹⁰. . ." Diluvialists have noted countless evidences of cataclysmic burial in rocks and that only such burial can preserve fossils; and uniformitarians are beginning to agree: ". . . it is generally accepted that rapid burial in a protective medium is necessary for fossil formation . . . anastrophic events may be considerably more important in the formation of fossil assemblages than currently is recognized.⁹¹"

Many sandstones contain the same type of debris in widely separated geographical areas, such as Illinois,⁹² Virginia-West Virginia,⁹³ Tennessee-where the sandstone debris is stated to have been "transported into the area by currents⁹⁴. . .", and Pennsylvania-where coal inclusions ". . . occur(s) as angular, elongate fragments up to several inches long and as discontinuous seams up to half a foot thick and 8-10 feet long.⁹⁵"

The evidence points to rapid deposition of the shale and sandstone members⁹⁶. . ." "Many sandstones must have accumulated with considerable rapidity.³⁵" The poorly-lithified nature of many cyclothemic sandstones in another evidence supportive of rapid sedimentation because poorly consolidated sediments are indicative of having been rapidly accumulated.⁹⁷ One sandstone breccia "represents a catastropic sedimentary event, such as a flood.⁹⁸" The sandstone-shale transition also shows evidence of rapid sedimentation; ". . . the change from dark shale to sandstone occupies no more than a few inches of strata.⁹⁹"

The presence of polystrate fossils is a classic example of rapid sedimentation.¹⁰⁰ "Tree casts are not rare in coal-bearing rocks.¹⁰¹"

Another example of a polystrate fossil is an upright tree trunk in the cyclothemic Francis Creek Shale in Illinois.[102]

Another family of evidences for rapid burial involves transported rocks interbedded with sediments. rocks have been found all over the world (in coal) including the U.S.S.R.[103] Equally important is the finding of Pre-Cambrian rock pebbles in limestone in Illinois,[104] some 250 miles from their source. Copious limestone fragments also reflect the Flood's violent episodes: "... it is unusual for pebbles or angular fragments of limestones to be transported by streams without being pounded to a powder by abrasion in transport, or dissolved.

Nevertheless, there are instances where this has actually happened, and there has been an accumulation of limestone pebbles and boulders of considerable thickness but of local areal distribution[105]..." Rapid flood erosion-deposition could generate so many such fragments that the waters would not be able to abrade-dissolve them all.

The non-sandstone and non-shale members also show evidence of rapid burial: "Pyrite concretions are often preserved in coal, under clay, or grey shale over coals... Pyrite concretions form as a result of a high concentration of reduced organic matter. This may occur as a result of rapid burial[106]..." The presence of bivalve-escape tunnels in shale (member 8B) is further evidence of the rapid deposition of the shale.[107]

The universal presence of evidences of cataclysmic burial at all cyclothem levels coupled with the aforementioned *de facto* gradation of all cyclothem members along with cyclothem-cyclothem gradation and cyclothemic rock-noncyclothemic rock gradation provides the total picture of Flood action.

3. A Critique and Re-interpretation of Uniformitarian "Sedimentary Environment" Concepts as Applied to Cyclothems.

Not only do uniformitarians segment the Flood-laden strata into a system of vast ages, but also inherent in their thought[108] is the concept of "sedimentary environment." In this scheme, modern-type environments (as rivers, deltas, lakes, seas) are supposed to have produced all of the earth's sedimentary rocks. Cyclothems are ascribed to ancient deltas.[52] Deltas have different laterally-contiguous zones of deposition called facies. As the delta progrades seaward, the facies migrate seawards also, causing a net seaward overlap of facies. The end result is that the lateral facies are preserved as vertical rock units; or so it is claimed by uniformitarians. This is Walther's Law, but "it does not state the the vertical sequence always reproduces the horizontal sequence, but that "... only those facies ... can be superimposed ... which can now be seen developing side by side.[109]

Yet "The interpretation of stratigraphic sections is an intricate mixture of *speculation* and observation ... The model constructor *assumes* some situation in the past and tries to develop the sequences of processes which lead to the rocks which are found in the present.[110]" (emphasis added.)

This situation provides the setting for the first of several uniformitarian circular arguments employed when deducing "sedimentary environments." Here the use of present-day facies relationships is justified in its use on ancient strata because the ancient rocks were formed in present-type environments: and the claim of them forming in present-type environnments is justified on the basis of their facies relationships. (Figure 2A) The Diluvialist sees various conditions of Floodwater action producing different suites of sedimentary rocks.

The Diluvialist School of Geologic Thought should be aware of the subjectivity of uniformitarian thought as pertaining to "sedimentary environments." "The human element is an undeniable, highly subjective component of earth science...,[111]" but it plays a large role in claim "sedimentary environment" identifications. "It is a common failing with geologists that at the conclusion of an endeavor based on data assiduously gathered and assessed, they permit themselves to indulge in ill-founded prognostications, half-true generalizations and even virtual fantasy...[112]" Yet "Recognition of ancient environments is not a mathematical problem; it also involves much *hypothetical* thinking.[113]" (emphasis added.

In claiming that ancient sediments were formed by present-type sedimentation, the uniformitarians see only what they want to see in the rock and ignore contrary evidence: "Even though rocks in Tennessee and in other adjacent areas of the Appalachian coal field have been mapped and studied in considerable detail, they have not been adequately examined for fossils. This is probably largely due to the *unwarranted assumption* that most of these beds are non-marine, hence barren of fossils.[114]" (emphasis added) Hence uniformitarians failed to see the fossils that they didn't want to see.

Controversial factors exist in studying strata: "... realization of the awesome complexities involved, the number and variety of variables which may be invoked to explain, for example, so simple an observation as that of a bed of sandstone... The more experienced the stratigrapher becomes, the greater the wealth of observation at his command, the more rapid is the pace at which unifying principles and generalizations appear to recede from his grasp.[115]" The uniformitarians confidently assert that they can identify sedimentary environments in rock, yet they fail to be able to distinguish between widely different alleged environments: "The state of our knowledge of ancient sedimentation is indicated by the Pennsylvanian sands of the Mid-Continent about which much has been written but on the deposition of which little is known.

The same Pennsylvanian shoestring is interpreted by different writers as an offshore bar or a channel, an ancient river, or a marine feature.[116]" The Diluvialists can appreciate the situation of trying to classify a Flood deposit according to present-day sedimentation. The large, widespread cyclothems (allocyclic) are considered to be deposited by processes explained by theories explained earlier. The short, choppy cyclothems (autocyclic) are thought to be caused by delta-lobe switching. However, the best (by far) studied delta-the Mississippi, fails to provide evidence of cyclothem generation by lobe subsidence: "... the stratigraphic evidence requires rapid subsidence between cyclothems... It is difficult to understand the stratigraphy in terms of steady, slow subsidence. The

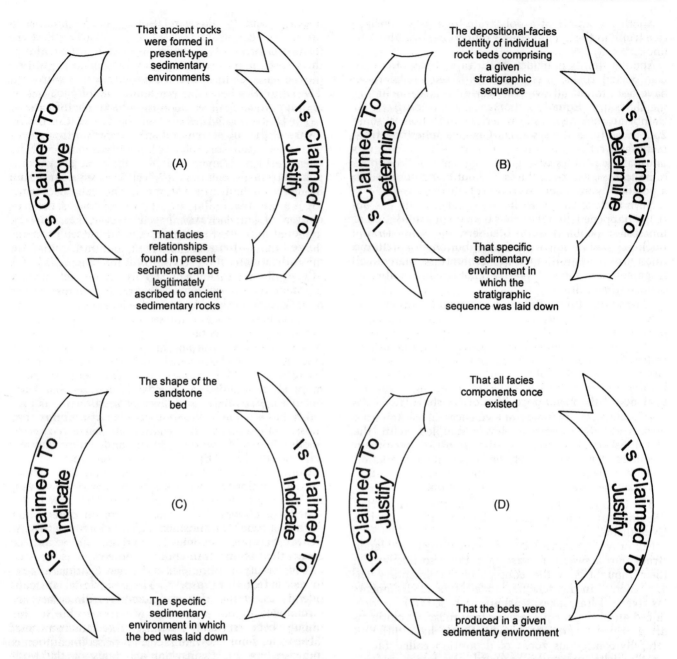

Figure 2. This shows schematically four of the types of circular argument employed by uniformitarians in putting forward their claims to have detected analogies of modern sedimentary environments in the earth's sedimentary rocks.

authors do not consider that there is an analogy to be made with the Mississippi delta.

It is significant that the seven known courses formed during the last 5000 years are all at the surface at the present day and these changes have not caused extensive cycles of sedimentation to be laid down."[117]" The cycles that do exist were glacial-eustatic, and—as previously mentioned—glacial-eustasy can't account for Pennsylvanian cyclothems.

Although the present is supposed to be the key to the past, "... the better known recent sediments may have no equivalents in the geologic record..."[116]" There would be differences between Flood deposits and contemporary sedimentation.

The uniformitarians use circular reasoning in establishing the identity of "sedimentary environments." For example, it is observed that only under "unusual circumstances[118], are entire deltaic sequences prserved. Yet "any stratigraphic correlation based upon the lateral relationship between the various deltaic facies must assume that all facies were originally present and will be preserved, unless removed by erosion.[119]" Circular reasoning (D, Figure 2) results.

Sandstone thicknesses judged to be capable of dilineating components are admitted to be arbitrary[120] and "A misconception of the geometry of the rock unit leads to a misconception or misinterpretation of the environmental conditions of deposition.[121]" Circular reasoning is used (C, Figure 2) when the original shape is unknown[121] due to erosion or poor outcrop-drill hole information, In "sedimentary environment reconstruction," "Individual facies, however, cannot be interpreted by themselves: it is too difficult to draw a unique interpretation from the petrological character of any single facies...[109]" and only can the "deltaic environment" be "reconstructed" by use of "...knowledge of facies relationships drawn from the study of modern environments...[109]" Circular reasoning (B, Figure 2) results.

"Although the term "delta" is not difficult to define in modern sediments, ancient ones are not readily so identified...[122]" and the famous model for an ancient delta, the Catskill Delta,[123] is now questioned as to whether it was a delta at all. That the contention of fossil assemblages proving ancient environments is unprovable is admitted,[124] as is the fact that fossil assemblages are a "...matter of interpretation...[125]" as to what environments they purportedly indicate. A great deal of bioturbation (as traced fossils: "worm burrows, raindrop prints, worm trails,") can be explained as bubble imprints.[126]

That short cyclothems were generated by delta-lobe switching-subsidence can be questioned for various reasons, including the wide area of even short cyclothems.[127] "The wide lateral uniformity of some cyclothem members (particularly the coals) is hardly consonant with the switching of several discreet delta complexes. The existence of all-alluvial cycles also seems difficult to explain by delta switching, as does the sequence of different limestone and shale types in the marine phase.

The development of cyclothems across facies boundaries—e.g., alluvial, deltaic, and lacustrine—appears incompatible with this theory, since the cyclic fluctuations should create the facies, not be imposed upon them.[128]" "Taken individually few of the cyclothem members can be definitely associated with a single environment of deposition: present evidence indicates that each represents some one of a series of environments that produce similar lithologic, structural, and fossil characteristics.[129]" To claim that cyclothems are formed by deltas is therefore to resort to circular reasoning (B, Fig. 2)

The uniformitarians greatly contradict themselves in their claims of sedimentary environment identifications: "Some students have attributed a marine origin to most of the sandstone, underclay, and unfossiliferous shales, but others have considered them to be fluviatile, deltaic, or eolian.[130]" Also: "When Pennsylvanian sandstones are critically examined...their non-marine character is fairly evident...some European stratigraphers continue to regard similar strata as marine.[131]" In fact, "Very few Pennsylvanian strata possess perfectly diagnostic characteristics which are not duplicated at other horizons.[132]" Since few outcrops usually exist in plains regions, drill holes and mechanical log correlations are employed; yet these data are known to contradict other observable data.[133]

The Diluvialist can reject these uniformitarian "sedimentary environment" designations because of illogical reasoning behind them (circular reasoning, also known as *petitio principii*, or begging the question, is one of the classic logical fallacies). The fact that the Diluvialist is free from the assumption that present-type processes are the sole (or even most important) guide to interpreting sedimentary rocks enables him to see the aforementioned circular reasoning, doubtful (if not invalid) extrapolations, contradictory identifications, etc., as clear indicators of the fallacy of the uniformitarian claims of being able to identify modern-type environments (present-day sedimentary processes) in ancient rock.

No erroneous opinion, however, is successful without some major half-truth behind it. The half-truth is the unquestionable similarity that does in fact exist between ancient and modern sediments. However, this similarity does not mean that modern-type sedimentary processes have produced the ancient rock, but rather that the physics of water flow/deposition is the same today as it was during the Flood. The fact that many sedimentary features are present in all modern environments, and that features associated with a given environment may be absent[134] is further indication of the similarity of hydrologic behavior under different conditions.

Two Diluvian conceptual terms are now coined: *Floodwater Mass Movement* (FMM), and *Floodwater Depositional Milieu* (FDM). For illustrative purposes, it may be stated that a long, thin, narrow sandstone (which uniformitarians claim was laid down by an ancient river; hence—by definition—is a fluvial sedimentary environment) was actually laid down by a swift, longitudinal FMM; the sandstone therein deposited in a torrential FDM. The fact that an "ancient river" is described as being "...broad, shallow, and highly sinuous...[135]" accords well with the idea of it not being a "river" at all, but rather a torrential FMM. Also, since it is claimed that "Ancient alluvial plains" may have "...wide lateral extent.,[136]" it can mean that, in reality, it was a wide, swiftly moving FMM which laid down that sedimentary rock.

The previously-discussed "shallow Paleozoic seas" were actually extremely-widespread but stagnant FMM's. Cyclothems were formed in zones of transitional FDM's (that is, the front of torrential FMM's colliding with stagnant FMM's.) The aforementioned absence of some "deltaic facies" reflects the difference between true deltas (operating only before and after the Flood) and transitional FDM's. The fact that cylcothemic sandstones possess certain "marine" and certain "fluvio-deltaic" properties in reality reflects certain "riverlike" and "shallow-sea type" properties coexisting in the FMM. That "...statistical textural studies of ancient sediments have largely proven an unsatisfactory method of environmental diagnosis.[137]" is understandable in terms of variously acting FMM's.

The uniformitarians admit that "It should be apparent...that sedimentary environment analysis is *at best* an inprecise art rather than a deterministic scientific discipline.[138]" Diluvialists must remember that uniformitarian "...sedimentary models remain ill-defined and subjective...Some kind of conceptual model is essential for any IMAGINATIVE kind of interpretation...[139]" (emphasis added). Since "sedimentary environments" in ancient rock are "Imaginative," the

Diluvialist can justifiably reject the entire uniformitarian concept of "Sedimentary Environment," and view ancient rock as being Flood-laden under varying conditions. An example of this, using the two coined Diluvialist concepts, was applied to the cyclothemic Francis Creek Shale/Purington Shale/Oak Grove Limestone[140] facies of North Central Illinois (See Table 1.)

CONCLUSION. The fact of the amazing, world-wide occurrence of cyclothems, together with their transcendence of the specious uniformitarian geologic column, is significant attestation to the universality (in both area and effect) of the Deluge. At the other extreme—the very local-level-uniformitarian ascription of cyclothems to present-type sedimentary environments is based on: graet personal subjectivity, demonstrably illogical reasoning, fantastic extrapolations, and contradictory identifications. The Diluvialist rejects these fallacies and views of rocks as being directly formed by the Deluge.

V. The Diluvian Tectono-Sedimentary Generation of Cyclothems

PROPOSITION. Cyclothems were generated by irregular tectonic activity superposed upon steadily retreating Flood waters.

A. THE TECTONIC OF CYCLOTHEM GENESIS. "The earth strives for gravitational equilibrium, or in other words for a minimum of free potential energy of the rotating globe...[141]" Disturbance of this equilibrium causes spontaneous compensation manifested by crustal tectonics. The Flood would, unquestionably, cause massive disequilibrium. This disequilibrium would be compensated for by mountain building as well as basin formation by downwarp. Only a basin can collect and preserve sediments,[142] protecting them from erosion. The cyclothem-filled Illinois basin subsided irregularly[143] while "pulsatory nature[144]" of mountain building are jointly held responsible for cyclothem formation according to some previously-discussed diastrophic theories. It is obvious that only a minor elevational change will "... inundate vast areas and shift the shoreline by perhaps hundreds of miles.[145]"

A major causal factor is sought for the pulsatory diastrophic events. As mentioned, the chief problem of the diastrophic explanations of cyclothem origin is the incredible nature of "tectonic hiccups;; which would have to repeat regularly, at long intervals of time, and be nearly identical to each other. Continental Drift/Plate tectonics (or Mobilism) has been called upon as an explanation for diastrophic pulses[146] causing cyclothems, but both uniformitarians and Diluvialists are divided on this question, and "... it is too early to choose a single favored theory of global tectonics.[147]"

In any case (whether Mobilism or Fixism), geophysical studies have shown[148] that basinal subsidence can be caused by mantle material flowing into the adjacent newly-uplifting mountains. Pertaining to cyclothem formation, "Recent studies on the strength of the crust suggest that subsidence to form basins commonly takes place in sudden steps along fault lines. Repeated movement along fault lines bordering a basin or shelf region is the primary cause of intermittent subsidence and the formation of sedimentary cycles.[149]" To this may be added that "... the stratigraphic evidence requires rapid subsidence between cyclothems...[117]" and that the Bible (Psalm 29:6) mentions mountains "skipping like calves" during the Flood. See also Psalm 104:6-8.

For every basin-wide cyclothem, there are found many local, short cyclothems or cyclothem parts which wedge into it and/or grade into it. (Fig. 3; Profile, basin-wide coals A, B, C, D, with local cyclothemic coals splintering downward). "Their presence complicates the problem of cyclothemic classification, for what appears to be a typical cyclothem in one area may easily be subdivided into two, three, or four in other areas... They are more common in basins nearer highlands with tectonic activity.[150]"

Cyclothems were most probably formed during the recessional phase of the Flood. Mountains were uplifted while water was receding (Psalm 104:6-9) and their pulsative increase (Psalm 29:6) generated the few basin-wide cyclothems (Fig. 3; Profile, A, B, C, D). The resultant basinal downwarp occurred assymetrically and suddenly, with only local areas subsiding at a time, forming "basinettes" (Fig. 3) which, when filled, became local cyclothems branching out from the basinwide ones. Tectonic megawaves helped cause selective downwarping. Several dozen diastrophic pulses in the last few months of the Flood is reasonable (because both orogenic pulses and selective basinal downwarp operate on threshold values of activation), whereas the notion of a diastrophic pulse every 400,000 years going on for 35 million years is clearly fantastic. If Illinois is taken as the "average" basin tendency with its 50 cyclothems, about 5 of which are extremely widespread; one diastrophic pulse every three weeks would generate the widespread, basinwide cyclothem; with the 9 short ones in between caused by local "Basinette" subsidence. In mountain regions, where there is more tectonism and subsidence, there are 250 cyclothems (in Virginia) but they are short (the same 5 basinwide cyclothems with 30 very short ones in between every basinwide one.)

In sum, diastrophic control of cyclothems as caused by the Deluge in both Scriptural and reasonable (because of time-element, threshold values, etc.) whereas the uniformitarian version stands untenable due to its time encasing. Every cyclothem represents a movement of Floodwater. The short cyclothems form when most basinal areas are not sufficiently downwarped: "... most sediment in transit would by pass the area continuing on until a region of active downwarping is reached...[151]"

B. SPECIFIC EVIDENCES OF CATACLYSMIC TECTONISM CONTEMPORARY WITH CYCLOTHEM FORMATION

The fact of crustal paroxysms associated with the Flood's formation of cyclothems is attested to by the wealth of evidence found amidst cyclothemic rock; much of it clearly occurring contemporaneous with cyclothem deposition. "Deformed rocks interpreted as the result of penecontemporaneous slump and mud flow are common in rocks of Pennsylvanian age in the Appalachian Plateau... Mud flows include strata

ANCIENT CYCLIC SEDIMENTATION

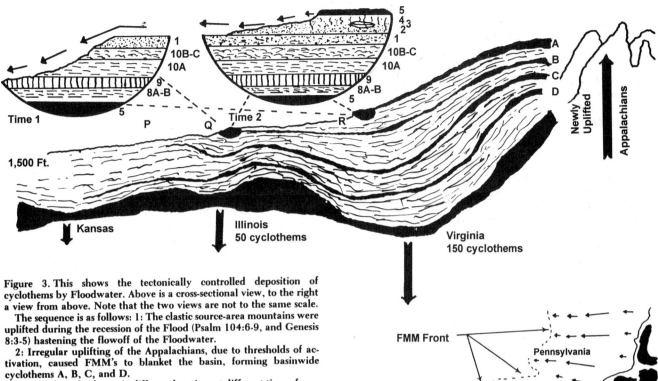

Figure 3. This shows the tectonically controlled deposition of cyclothems by Floodwater. Above is a cross-sectional view, to the right a view from above. Note that the two views are not to the same scale.

The sequence is as follows: 1: The clastic source-area mountains were uplifted during the recession of the Flood (Psalm 104:6-9, and Genesis 8:3-5) hastening the flowoff of the Floodwater.

2: Irregular uplifting of the Appalachians, due to thresholds of activation, caused FMM's to blanket the basin, forming basinwide cyclothems A, B, C, and D.

3: Irregular subsidence, in different locations at different times, formed "basinettes" (the Figure shows P, Q, and R) in which the numerous short-range cyclothems formed between A, B, C, and D.

4: Both basinwide and short-range cyclothems formed as follows: the FMM gradually increased in competence, laying down fining-downward member 10A-10B-10C-1 transition. (Tim 1; vectors show water speed.) After the momentum was spent (Time 2) the fining-upward member 2-3-4 transition was deposited. Then followed tangled masses of vegetation (Member 5), and the flow-off broth (incorporated in member 8A).

which deformed mainly as fluids and partly as plastics.[152]" "Decollement within Pennsylvanian rocks occurred when the rocks were hydroplastic.[153]"

Some volcanic rocks associated with mountain cyclothems were formed with "large amounts of water.[154]" Some widespread, thin clay layers amidst coal may be of volcanic origin.[155] "Coalification patterns . . . reflect (1) Depth of burial . . . and (2) regional thermal disturbances . . .[156]" German cyclothems reveal ". . . an extensive magmatic upwelling . . .[157]" as revealed by coal grade. Most volcanic activity, however, had passed by the time the Floodwaters receded (Genesis 8:2); the sedimentation of cyclothemic rocks occurring during Flood recession supported by the general distribution of volcanic inclusions: ". . . with the exception of the Eocene of the Pacific Northwest, pre-Cretaceous graded sequences tend to have greater volumes of volcanic rocks associated with (and volcanic detritus in) the sandstone and conglomerate . . .[158]"

Even in relatively undeformed basinal cyclothemic strata, there is considerable evidence of tectonism, as growth faults, for example: "growth faults . . . result from tension, caused partly by subsidence of the basin floor, and partly by the *rapid compression* of the *recently deposited* sediments.[159]" Such growth faults, which—by definition—form contemporaneously with sedimentation, are found in southern Illinois cyclothems.[160] In Missouri, ". . . many structures that exist as folds in younger Paleozoic rocks project downward into faults or fault zones in subsurface rocks.[161]" A tectonic graben exists in Illinois.[162] Other clear fault zones occur in Pennsylvania,[163] southern Illinois,[164], Kansas,[165] to name a few locations. Great fissures in cyclothemic rocks occur in Ireland[166]; these are clastic dykes. Such fissures occur also in the Appalachian Plateau,[167] and in Illinois, where a description of these joints states that: ". . . the strata were pulled apart laterally in almost every direction . . . the clay veins exhibit no signs of having been formed at different periods.[168]"

The entire earth bears the scars of God's judgement during the Flood, and violent tectonism during cyclothem sedimentation is attested to by these contemporaneous-with-sedimentation occurrences.

C. THE SEDIMENTATIONAL COMPONENT OF CYCLOTHEM GENESIS. Some Diluvialists have proposed that cyclothems were formed by tidal incursions upon the continents during the beginning stages of the Flood. This paper, on the other hand, proposes a recessional-Flood cyclothem formation and a tectonic mechanism. While the matter is still open, there seems to be no evidence of substantial tidal movement associated with cyclothems,[169] and furthermore, it seems unlikely that tides could form in such shallow water.[169]

The cyclothems reveal the following sedimentational trends as they are traced further from the source areas: Strata thinning of cyclothems,[170] thinning of coals,[170] decrease in grain size of clastics.[170] increase in limestone thickness, repeated presence, etc. The decrease in clastic grain size reflects the net decrease in availability of suspended sediment as Floodwaters flow ever further from the newly-uplifted mountain source areas. Likewise, the progressively basinward thinning and frequency of absence of coals (Fig. 3, Profile A, B, C, D, coal, basinward (Kansas ward) trends reflects the decrease in available (still not stranded) floating plant-tree masses. The limestone increase towards the basin is explicable in terms of the decline of clastics allowing greater freedom of chemical reactions between colliding FMM's possessing different ions in solution.

The *modus operandi* of Diluvian cyclothemic sedimentation is as follows: Receding Floodwaters flow into the regional downwarp of "basinette," (Fig. 3) or else inundate the entire plains area (in the case of the few basinwise cyclothems). The resulting surge of Floodwater passes given geographical points with ever increasing velocity. At first there is no clastic deposition; only the widespread "marine limestone" (Member 9) is formed as the advancing FMM's ions react with those of the stagnant FMM being displaced. When the stagnant FMM is steadily being displaced and the FMM gains progessively greater momentum, this Fore FMM (Table 1) deposits progressively coarser clastics (Member 10A-1]B-10C-1 sequence).

Eventually, the deposition exceeds grade, and erosion begins, forming (in many areas) the characteristic channels (gulleys) Members 10C-1). From then on, the FMM having spent its flow momentum and clastics, the flow becomes progressively weaker and the progressively finer clastics (Members 2, 4) are deposited. The stagnant water grows shallow enough for limestone to precipitate (Mem. e) and for the floating plant-tree debris to settle out and blanket the terrain. The percolating waters emanating from the rotting debris create reducing conditions for the initial shale deposition of the subsequent FMM, and the (Member 8A-8B transition forms.

Varying local conditions cause some missing members of every cyclothem: lack of chemical conditions cause missing limestones (especially members 3 and 7); differential availability of clastics causes varying types of fining-upward, fining-downward clastics to be deposited (with varying shape, thicknesses, and channel erosion from differing grade), and varying amounts of coal (presence, purity, and thickness) are caused by varying amount and presence of floating tree masses. The previously-discussed tectonic mechanism generates varying types of cyclothem: regional tectonic factors being responsible for "extra" members (as coals and limestones (as the uncommon member 7) as well as occasional abrupt thickenings, thinnings, and fade-out of members. Thus is explained the differences superposed upon the profound similarity of all cyclothems.

Every cyclothem is thus the product of increasing FMM velocity competence (Members 9 through 10C) followed by decreasing competence (Members 1 through 4) and ending up with stagnant FDM (Members 6 through 8B). The stratigraphic properties of cyclothem members is explicable in terms of Flood sedimentation. That limestones are "... geographically extensive...[171]" is caused by their independence of clastic supply and reliance upon the chemistry of mixing FMM's.

The fining-upward Member 1-2 transition[172] has probably been overrated at the expense of the much-sharper fining-downward Member 1-10A transition, but now fining-downward sequences are no longer considered uncommon.[173] The author of this article studied 15 drill-hole core logs of Illinois cyclothems and counted 29 (21 sharp and 8 weak) fining-downward sequences and 26 (13 sharp and 13 weak) fining-downward sequences.[174] The change of FMM speed during sedimentation is confirmed by pentrology (detrital interstitial material.)[175]

Clearly, then, the prominent fining-downward trends show "... depositional conditions ranging from low velocity suspension at the base to high velocity traction sedimentation in the upper sandstones.[176]" The fining-upward sequence, which extends to the underclay,[177] on the other hand, "... can be produced ... by deposition in the last phases of a heavy flood ...[178]" The "... extensive mixing of detritus ...[179]" in sandstone reflects certain mixing properties of the FMM, whereas the "... similar sorting ...[179]" of widely-separated sandstone regions attests to the widespread overall similarity of Flood action. "In general lithological respect, the sandstones are homogeneous over several now separated basins from Missouri to Pennsylvania. Thus we have a picture of current activity whose intensity varied greatly in time and space within certain average limits. But the limits and degree of variability were *remarkably uniform* over much of the North.[180];; (emphasis added). That sandstones have various shapes and "... thicken ... basinward ...[181]" reflects their upmost sensitivity to FMM velocities.

At the other extreme (showing the least dependence upon FMM velocities), "The most persistent elements are the coal beds and overlying black shales.[182]" Their widespreadness results from their having been formed from floating tree-plant masses and hence being quite independent of Floodwater velocity. The underclays are somewhat widespread because of the fineness of the particles and their near-universal presence in even the slowest FMM.

Since water flows only downhill, the source areas for the clastics in cyclothems must have been the newly-uplifted mountains: "... the tectonic borderlands of the northern Appalachian mountains ...[183]" Diluvialists must view "source area" claims skeptically, because under "extremely different conditions ...[184]" (from today, as the Flood was) mature and supermature sands can form in one cycle, and because there is some "... intense chemical weathering ...[185]" observed in fact. The Flood must have brought down material from "... a series of point sources, rather than from one uniform source.[186]" Hydrogeochemically, the flood contained silica waters which percolated through Illinois sandstones,[187] cementing them, and caused contemporaneous-with-plant-matter silicification of Antarctic coals.[188]

D. THE RELATIONSHIP BETWEEN INDIVIDUAL CYCLOTHEM MEMBERS AND THE FLOOD.

Every cyclothem member reflects some FDM.

1. The Coals (Member 5)

Member	Outstanding Sedimentary Structure	Hydrologic Characteristic Indicated	Uniformitarian Interpretation (Deltaic facies)			Diluvian Interpretation (FDM facies)	
1	Oscillation ripples Current ripples Ripple drift cross-laminae Scour and fill Small-scale folding Convolute laminations	Orbital wave motion Moving current Continuous traction deposition Turbulent variable current Turbid flow of sediment Fast flow, flood	Delta front sub-aqueous levee Point bars	Front	Middle FMM	FMM at its fastest. Violent, torrential flow capable of carrying sand grains. Localized erosion of gullies, making channel sandstone bodies. Possible wind-generated waves (Genesis 8:1).	
10C	Regular layers Oscillation ripples Small-scale folding Convolute laminations Compactional deformation	Fluctuating velocity Orbital wave motion Turbid flow of sediment Fast flow and/or flood Deposition on irregularities		Delta bottomset beds	Prograding	Subdued velocity of FMM capable of carrying only silt and finer. Possible wind-generated waves (Genesis 8:1).	
10B	Regular layers Irregular layers Mottles	Fluctuations of water velocity Small-scale slump, bioturbation Mostly slumping inclusions	Shallow waters surrounding deltas	prodelta		FMM	Slumping caused by early tectonic activity. Slow but variable FMM velocity
10A	Mottles Homogeneities Concretions (siderite)	As above. Also clastic inclusions No textural variations of sediment Iron-rich water from coal	Tidal channels	Foreset and		Fore	Slowest, most frontal component of FMM. So slow the finest clastics are left; and they can settle out.
9	Irregular layers Mottles Homogeneities Concretions (siderite)	Slumping (tectonic?) bioturbation Mostly slumping clastic inclusions No textural variation of sediment Iron-rich water from coal				Further evidence of early tectonic activity at the start of sedimentation. Most frontal non-clastic component of FMM. Chemical reactions of colliding materials.	

Table 1. Here are proposed the Diluvian interpretations of the cyclothem facies.

Although it is well known that coalification occurs rapidly,[189] the majority of uniformitarians hold on to the autochthonic (in-situ deltaic peat-forming swamp) position of coal formation as opposed to the allochthonic (transported) position. "Some of the arguments for this autochthonous development are: (1) the great lateral extent of many seams; (2) the purity... of many seams; (3) the presence of upright tree stumps in coal measure strata: and (4) the presence of underclay (or "seat earths") beneath many seams. The last two arguments are no longer especially strong ones. Tree stumps are only occasionally found actually penetrating or within a coal seam, but are usually above the seam or in sandstones of shales associated with the seams.

In addition, there is a growing body of evidence... to suggest that "seat earths" are not soils, but are themselves allochthonously derived."[190] The underclays will be discussed separately; the last two arguments are admittedly weak, but the first two in actuality reflect the narrowness of the uniformitarian position. Certainly in any local flood or sedimentary process "... there would be no available source for the vegetation whose detritus was to cover such vast areas.[191]" The global Flood easily denuded the entire earth and blanketed significant portions of continents with layers of floating (on recessional FMM's) vegetation. The purity of coals is not difficult to understand in terms of early-Flood rains washing the floating vegetable detritus free of any (soil) material which would not float.

Furthermore, the coals do show trends of thickness reflecting the thickness of members beneath,[192] suggesting minor FMM's currents nudging floating detritus away from slightly higher areas. More importantly, "The coals thin and become impure over anticlines with maximum structural relief... This phenomena is also observed in other Allegheny cycles.[193]" The Floodrain-cleaned floating vegetal detritus became contaminated in shallower waters over anticlines owing to the greater likelihood of mixing of muddy, underclay-depositing waters with the detritus.

One of the vast lines of evidence for the allochtonic formation of coal is the presence of "... water-worn...[194]" fusain (plant matter) fragments, as is the vast extent of clay layers in coal: "A principal problem to explain in any case is how the forest vegetation of a swamp could be so completely bevelled as to permit accumulation of a continuous layer of clay...[195]" Some coals are magnitudes thicker than any imagined peat swamp condition could accumulate: "Ekibastuz. The thickness of coal seams in this field is unique in the USSR. There is a coal seam which, including thin intercalations of shale is 150M (487.5 ft.), and must have been produced as the result of the accumulation of 450-600m (1462.5-1950 ft.) of peat.

Present-day peats, however, are on average 6-8m (19.5-26 ft.) thick, reaching 20-24m (65-78 ft.) in isolated instances.[196]" Coals follow patterns of sedimentational filling (thickening) in some channel sandstones[197] and possess everwhere stratigraphic properties consistent with local variations of vegetal-floating conditions: "...coals thicken and thin, change in character, and not uncommonly pinch out entirely...[198]" That coals not only have minerological trends similar to underlying underclay[199] but also (in areas without underclay) "...grade down into laminated shales, siltstones, and sandstones.[200]" further confirms their *prima facie* nature of being part of continuous, rapid FMM sedimentation.

2. The Underclays (Member 4)

These were long considered to be fossil soils of deltaic swamps, but recent studies reveal that the most powerful inherent evidence for this position, root impressions, is invalid: "For the last 150 years, *Stigmaria*, the rootlike base of Paleozoic lycopods has been interpreted as occurrence in situ...[201]" "The preferred orientation of specimens of *Stigmaria*... can only be explained by transportation... *Appendices* attached to *Stigmaria* are sometimes found to be cracked, broken, or twisted in a way difficult to explain from a functional point of view... the rapid accumulation of the stigmarian beds in a short time interval, as indicated by the well-preserved upright trunks, rules out the possibility of forest growth in situ.[202]"

The fact of root orientation in current direction observed in superjacent beds[203] reinforces claims to their allochthony. It is most important to note that underclays "lack...a soil profile similar to modern soil...[204]" that "...coals and underclays are not genetically related...[205], that the undisputably-allochthonic "...clay partings...possess most of the characters of underclay...[206], and that underclays contain detritus preservable only by rapid sedimentation: "...well-preserved leaf impressions and tiny coal veinlets are common.[204]"

The homogenity and absence of bedding in underclays is explicable in terms of homogeneity of the finest sediment remaining at the end of the cyclothem-generating FMM flow as an alternative to the root-turbation explanation, and "...the best explanation of slickensides in underclays is the hypothesis of compaction of a sediment deposited in a loose, hydrous condition.[207]" The underclays can therefore be considered to be the finest of the (rapidly-deposited by (waning) FMM) clastics: "It seems reasonable that the gradation of shale or sandstone upward into an underclay may simply represent conditions of transitional sedimentation from coarse to fine particle sizes. It appears that this evidence may be used to explain the detrital origin of underclay.[208]"

3. The Limestones (Members 3, 7, 9)

The cyclothemic limestones formed during clastic-poor FDM periods of stagnation prior to underclay-coal deposition (Mem. 3)) and, more prominently, as a result of chemical reactions between colliding FMM's (Mem, 9, 7): "If carbonate ions are continuously added to a solution containing several metallic ions, the pH of the solution will rise and the metallic ions will precipitate in the order in which the solubility products of their carbonates are exceeded.[209]" The uniformitarian position, in contrast, claims limestone formation by lime-secreting organisms inhabiting the bottoms of shallow seas, but there is admission of the chemical aspects of limestone formation being given "...little or no attention.[210]"

Certain widespread, abrupt, clastic-mixed limestone bed thickenings admittedly mitigate the necessity of alleged marine incursions.[211] "The Pennsylvanian limestones vary greatly in character. Some of them are earthy, shaley, or impure, but others are quite pure and dense.[212]" also is observed the "...massive, blocky nature of some beds and thin, wavy bedding of other limestones.[213]" The Flood-blind uniformitarian position is perplexed by the variations: "The significance of some types of deposits is not understood. For example, the cause of variation in types of limestone...is largely unknown.[213]" The observed limestone variations are much better explained in terms of locally-variable chemically-reacting FMM's rather than monotonous, tranquil shallow seas of old. The gradation of limestones into clastics[212] and "...common...intraformational conglomerates...and...sandstone lenses...[214]" within them further confirm their FMM origin (because of definitely transported clastic lenses in them).

4. The Sandstones (Member 1)

Of the water-sorted clastics, these are the thickest in grain size, Geometrically, the sandstones have several basic shapes; namely, the sheets (relatively wide areal extent but less than 20 feet thick) and elongates (much thicker 20-105 ft.) but narrower (25 ft. to 2-3 miles), filling channels (or gullies)[215] "Virtually all Pennsylvanian sandstones contain both sheet and elongate sand bodies.[92]" The elongates fill channels (gulleys) which "...range from small cut-and-fill structures within the formational boundaries to large channels that were eroded into the underlying formations.[216]" Thus, while Fig. 1 portrays the channel terminating in Member 10A, the channels frequently cut into 2 or more subjacent cyclothems.

As previously discussed, the uniformitarians stumble in attempting to understand the sandstones in terms of presently-operating sedimentary environments: "Opinions have varied as to whether the infilling of the channel was alluvial, part alluvial and part marine, or exclusively marine.[217]" The fact that "...the sandstones have a dendritic pattern, thicken in the direction of the dip...[218]" is claimed to indicate fluvial sedimentaiton whereas "...equidimensional quartz grains...[219]" orientation trends contraindicatively reveal a would-be upstreamward water flow. The channels have been also ascribed to deltaic distributary systems "...but their abrupt entrenchment and the absence of any indication of natural levees suggest that they do not mark distributaries of aggrading streams in a deltaic area.²²⁰"

Just as an individual believing only in apples and cherries would have difficulty telling which of the two a strawberry, is, so analogously the Flood-rejecting, present-process believing uniformitarian has difficulty assigning these Flood deposits to presently-operating sedimentational processes.

FMM dynamics easily explain the different sedimentary/stratigraphic properties of sandstone. For example, the waxing-in-competence cyclothem-generating FMM lays down the fining-downward sequence and then, at maximum competency (contemporaneous with or immediately following sandstone deposition), the FMM torrents sometimes erode gulleys into the just-deposited sandstone (and below), forming the channel sandstones. The observed coexistence of tributary and distributary filled-channel patterns²²¹ and other above-mentioned phenomena are caused by local-slope variations giving rise to differential direction FMM bifurcation or coalescing.

That erosion of the gulleys was admittedly "... rapid ... ²²²" and that neither local increase in slope²²³ nor emergent conditions²²⁴ were necessary for FMM gulley erosion further clarified the nature of the FDM at that part of the FMM flow. Of utmost importance, however, is the fact that local floods readily erode gulleys in (especially unconsolidated) sediment.²²⁵ The sandstone sheets grade into elongates,²²⁶ reflecting the continuous tempo of FMM erosion-deposition, whereas the usual fining-upward trends in filled channels reflect the initiation²²⁷ of the waning phase of the cyclothem-generating FMM flow.

5. The Shales (Members 6, 8A-B, 10A)

"... the shales comprise by far the larger part of the sedimentary lithological column...²²⁸" and also of cyclothems, portraying the awesome magnitude of chemical weathering of the anteDiluvian supracratonic material. That "... shales differ from clays ... only in being bedded or laminated.²²⁹ attests to the fact of rapid-flowing fore-FMM properties in contrast to the stagnancy and particle-sameness of underclay deposition. The shales have a layer which is black and fissile (member 8A), reflecting reducing conditions, but also these "Black shales probably are more a function of *rapid deposition* than of restricted chemical circulation.²³⁰" (emphasis added).

Pertaining to these very-persistent black shales, there are "... several lines of evidence pointing to a widespread mat of vegetation covering the surface of the water.²³¹" and "... these beds extend without appreciable change far to the west and far beyond the coal beds of their cyclothems²³²" These properties suggest that FMM's washed reducing vegetal "broth" (and plants) from the previous FMM-deposited pre-coalified vegetal surficial layer, incorporating it in the shale above the coal and downcurrent into Kansas (forming the black shale stratigraphic equivalents of coal there). The observed grading-upward of black shale into the main gray shale²³³ (members 8B, 10A) marks the point where reducing conditions caused by the subjacent vegetal mass ceased having their chemical effect, whereas the "... fairly well preserved impressions of land plants and somewhat MACERATED land plant re-

mains.²³⁴" found in black shales further corroborates their rapid deposition and origin from precoalified-material flowoff. (Emphasis added.)

CONCLUSION: The basic sedimentary, stratigraphic, and tectonic properties observed in cyclothemic rock provide a picture of the recessional aspects of the Flood.

VI. Epilogue

Diluvialists must always remember that uniformitarianism is not a scientific fact, but an *a priori* atheistic worldview controlling disciplines studying origins. This viewpoint must be balanced by Diluvialists who work from the polar-opposite pro-God worldview. The rise of uniformitarianism (and consequent denial of the Creation and the Flood) is a striking fulfillment of Biblical prophecy (2 Peter 3:3-9). The Diluvian position is just as scientific (if not far more so, because it explains data more fully and simply) as any uniformitarian application. It is hoped that this work will greatly enrich the Diluvian position.

REFERENCES

AAPG—American Association of Petroleum Geologists Bulletin
AJS—American Journal of Science
CONG—18th Interntional Geological Congress (Great Britain) Report, part 4
DIS—Developments in Sedimentology
EPC—Elsevier Scientific Publishing Company: Amsterdam, London, New York
FGM—Fieldiana, Geological Memoir: Natural History Museum, Chicago
GM—Geological Magazine
GSAB—Geological Society of America Bulletin
GSAP—Geological Society of American Abstracts and Programs (Annual Meetings)
GSASP—Geological Society of America Special Paper
GSL—Quarterly Journal of the Geological Society of London
IGS—Illinois State Geological Survey Circular
IGSRI—Illinois State Geological Survey Report of Investigations
JG—Journal of Geology
JSP—Journal of Sedimentary Petrology
JWS—John Wiley and Sons, New York
KGS—Kansas State Geological Survey Bulletin
MM—MacMillan and Co., London
OB—Oliver and Boyd, Edinburgh, London
PGS—Pennsylvania State Geological Survey Bulletin, 4th Series
SEPM—Society of Economic Minerologists and Paleontologists Special Publ.
SV—Springer-Verlag, Berlin, Heidelberg, New York

[1]Weller J. M. 1956. Argument for diastrophic control of late Paleozoic cyclothems. *AAPG* 40(1)38
[2]Potter P. E. 1962. Regional distribution patterns of Pennsylvanian sandstone in Illinois Basin. *AAPG* 46(10)1890
[3]Wanless H. R. and J. M. Weller. 1932. Correlation and extent of Pennsylvanian cyclothems. *GSAB* 43(4)1006.
[4]Weller, 1956. *op. cit.*, p. 48
[5]*Ibid*, p. 17
[6]*Ibid*, p. 18
[7]*Ibid*, pp. 17-50
[8]Beerbower J. R. 1961. Origin of cyclothems of the Dunkard Group (Upper Pennsylvanian, Lower Permian) in Pennsylvania, West Virginia, and Ohio. *GSAB* 72:1029-1050.
[9]Simeons G. 1918. First contribution to the study of the coal measures' classification of the South Wales coal field, based on the theory of sedimentary cycles. *Tonypandy* 83pp..
[10]Stout W. 1923. Theory of The origin of coal formation clays. *Ohio Geological Survey Bulletin* 26, Series 4, pp. 533-568
[11]Hudson, R. G. 1924. On the rhythmic succession of the Yoredale Series in Wensleysdale. *Proceedings Yorkshire Geological Society* 20:125-135.

[12] Brough J. 1928., On rhythmic deposition in the Yoredale Series. *Proceedings of University of Durham Philosophical Society* 8:116-26

[13] Weller J. M. 1930. Cyclic sedimentation of the Pennsylvanian period and Its Significance. *JG* 38:97-135

[14] Weller. 1956. *op. cit.* pp. 44-50.

[15] Wanless H. R. and F. P. Shepard. 1936. Sea level and climatic changes related to late Paleozoic cycles. *GSAB* 47:1177:1206.

[16] Whitcomb J. C. and H. M. Morris. 1961. *The Genesis Flood* Baker, Mich. p. 245-8.

[17] Robertson T. 1948. Rhythm in sedimentation and its interpretation with particular reference to the Carboniferous sequence. *Transactions Edinburgh Geological Society* 14:141-75.

[18] VanDerHeide S. 1950. Compaction as a possible factor in upper Carboniferous rhythmic sedimentation. *CONG* p. 38-45.

[19] Rhodes F. H. T., et.al. 1971. Undergraduate geology: a strategy for design of curricula. *American Geological Institute CEGS Publication* 8, p. 11.

[20] Albritton C. C. J., ed., 1967. Uniformity and simplicity; a symposium on the principle of the uniformity of nature. *GSASP* 89, P. RONDO.

[21] *Ibid.*, p. 1, 2

[22] Hubbert M. K. 1967. Critique of the principle of uniformity *GSASP*, 89, pp. 29-30

[23] Goodman N. 1967. Uniformity and simplicity, *GSASP* 89, p. 95.

[24] *Ibid.*, p. 99

[25] Bott, M.H.P. and G.A.L. Johnson. 1967. The controlling mechanism of Carboniferous cyclic sedimentation. *GSL* 122:421

[26] _____ 1965. American Commission on Stratigraphic Nomenclature; Definition of Geological Systems. *AAPG* 49(10)1966.

[27] Spencer, R.S. and W.S. Rogers. 1970. Reworked Paleozoic Fossils in Pleistocene sediments of southeastern Virginia. *GSAB* 81:263-6.

[28] Peppers, R.A. 1964. Spores in strata of late Paleozoic cyclothems in the Illinois basin. *IGS*. 90, p. 9.

[29] Scholl, D.W., Greene, G.H., and M.J. Marlow. 1970. Eocene age of the Adak "Paleozoic(?)" rocks, Aleutian Islands, Alaska. *GSAB* 81:3598.

[30] Trueman, A.E. 1946. Stratigraphic problems in the coal measures of Europe and North American. *GSL* 102:1i

[31] Moore, R.C. 1949. Division of the Pennsylvanian system in Kansas. *KGS* 83:19

[32] *Ibid.*, p. 21.

[33] Wanless, H.R. 1939. Pennsylvanian correlations in the eastern interior and Appalachian coal fields. *GSASP* 17, p. 71.

[34] Sabins, F.F. 1964. Symmetry, stratigraphy, and petrography of cyclic Cretaceous deposits in San Juan basin. *AAPG* 48(3)316.

[35] Trueman. 1946. *op.cit.*, p. 1i

[36] Hedberg H.D. 1959. Towards harmony in stratigraphic classification. *AJS* 679:257

[37] Moore, R. C., et. al., 1951. The Kansas rock column. *KGS* 89, p. 117.

[38] Wanless H.R. 1955. Pennsylvanian rocks of eastern interior basin. *AAPG* 39:1764-5.

[39] Lee J.S. 1939. *The geology of China*. Thomas Murby & Co., London, p. 276.

[40] Wadia D.N. 1953. *The geology of India*. MM: p. 338.

[41] Siever R. 1951. The Mississippian-Pennsylvanian unconformity in southern Illinois. *AAPG* 35(3)547.

[42] *Ibid.*, p. 548.

[43] Englund K.S. and H.L. Smith. 1960. Intertonguing and lateral gradation between Kentucky, Tennessee, and Virginia. *GSAB* 71:2015.

[44] Vischer G.S., Saitta S.B. and R.S. Phares. 1971. Pennsylvanian delta patterns and petroleum occurrences in eastern Oklahoma. *AAPG* 55(8)1208-9.

[45] Milic, R. C. 1974. Stratigraphy and depositional environments of upper Mississipian and lower Pennsylvanian rocks in the souther Cumberland Plateau of Tennessee. *GSASP* 148, pp. 120-1.

[46] Horne J.C. Ferm J.C. and J.P. Swinchatt. 1974. Depositional model for the Mississippian-Pennsylvanian boundary in northeastern Kentucky. *GSASP* 148, p. 113.

[47] Wanless and Shepard. 1936. *op.cit.*, p. 1180.

[48] Dutro J.T. 1976. Carboniferous geology examined. *Geotimes* 21(3)18.

[49] Takai, et.al., ed. 1963. *Geology of Japan*. Tokyo University Press, Tokyo. p. 45.

[50] Jackson W.E. 1964. Depositional Copography and cyclic deposition in west-central Texas. *AAPG* 48(3)328.

[51] Horne, et.al. 1974. *op.cit.*, p. 98.

[52] Pryor W.A. and E.G. Sable. 1974. Carboniferous of the eastern interior basin. *GSASP* 148, p. 309.

[53] Siever R. 1957. Pennsylvanian sandstone of the eastern interior coal basin. *JSP* 27(3)231.

[54] Moore R.C. 1936. Stratigraphic classification of the Pennsylvanian rocks of Kansas. *KGS* 22, p. 21.

[55] Beutner E.C. et.al. 1967. Bedding geometry in a Pennsylvanian channel sandstone. *GSAB* 78:911.

[56] Swann D.H. 1964. Late Mississippian rhythmic sediments of Mississippi valley. *AAPG* 48(5)652.

[57] Weller. 1956. *op.cit.* p. 30.

[58] Beerbower. 1961. *op.cit.*, p. 1033.

[59] Gentile R.H. 1968. Influence of structural movement on sedimentation during the Pennsylvanian period in eastern Missouri. *University of Missouri Studies* Volume XLV, p. 10.

[60] Dott R.H. 1963. Dynamics of subaqeous gravity depositional processes. *AAPG* 47:107

[61] Bristol H.M. and R.H. Howard. 1974. Sub-Pennsylvanian valleys in the Illinois basin and related Chesterian slump blocks. *GSASP* 148, p. 315.

[62] *Ibid.*, pp. 321-4.

[63] Laury R.L. 1971. Stream-bank failure and rotational slumping: preservation and significance in the Geologic record. *GSAB* 82:1251.

[64] Schwarzacher W. 1975. Sedimentation models and quantitative stratigraphy *DIS* 19, EPC, p. 17.

[65] Schaffer N.R. and G. Faure. 1976. Regional variation of $^{87}Sr/^{86}Sr$ ratios and mineral composition of sediment from the Ross Sea, Antarctica. *GSAB* 87:1491.

[66] Slusher H.S. 1973. Critique of radiometric dating. *ICR Technical Monograph* 2, p. 32.

[67] Brookins D.G. and S.D. Voss. 1970. Age dating of muscovites from Pennsylvanian sandstones near Wamego, Lansas. *AAPG* 54(2)356.

[68] Gentry R.V. 1976. Radiohalos in coalified wood: new evidence relating to the time of uranium introduction and coalification. *Science* 194(4262)317.

[69] *Ibid.*, pp. 316-7.

[70] Winston G.O. 1972. Oil occurrence and lower Cretaceous carbonate-evaporite cyclothems in southern Florida. *AAPG* 56(1)158-60.

[71] Moore R.C. 1950. Late Paleozoic cyclic sedimentation in central U.S. *CONG*, p. 11.

[72] Ager D.V. 1973. The Nature of the stratigraphic record. *JWS* pp. 6-7.

[73] Beerbower. 1961. *op.cit.*, p. 1031.

[74] Nalivkin D.V. 1973. *Geology of the USSR*. OB. p. 271.

[75] Haughton S.H. 1963. *The Stratigraphic history of Africa south of the Sahara*. OB. p. 203, 230, 221.

[76] *Ibid.* p. 235.

[77] Flores-Williams H. 1968. Chilean, Argentine, and Bolivian coals. *GSAP* 1968, p. 209.

[78] Hill D., ed. 1961. Geological results of petroleum exploration in western Papua. *Journal of the Geological Society of Australia* 8(1)10-11.

[79] Lee (1939) *op.cit.*, pp. 139-40.

[80] Kangma J.T. 1974. *The geological structure of New Zealand*. JWS pp. 132-3, 5.

[81] Takai et.al., ed. 1963. *op.cit.* p. 70, 152.

[82] Wells, A. J. 1960. Cyclic sedimentation; a review. *GM* 97:401.

[83] Trueman A.E. 1948. The telation of rhythmic sedimentation to crustal movement. *Science Progress* 36:202-3.

[84] Long W.E. 1965. Stratigraphy of the Ohio Range, Antactica. *American Geophysical Union Publication* 1299, pp. 79, 107-8.

[85] Wanless H.R. Tubb J.B. Jr., Gednetz D.F. and S.L. Weiner. 1963. Mapping sedimentary environments in Pennsylvanian cycles. *GSAB* 74:438.

[86] Irwin M.L. 1965. General theory of epeiric clear-water sedimentation. *AAPG* 49(4)445.

[87] Beerbower. 1961. *op.cit.*, p. 1041.

[88] Duff, M. Mc. L., Hallam, A., and Walton, E. K., 1967 Cyclic sedimentation *DIS* 10, EPC, p. 250.

[89] Carss B.W. and N.S. Neidell. 1966. A Geological cyclicity detected by means of polarity coincidence correlation. *Nature* 212:137.

[90] Beutner, et.al. 1967. *op.cit.* p. 911.

[91] Kranz P.M. 1974. The anastrophic burial of bivalves and its paleoecological significance. *JG* 82:237.
[92] Potter. 1962. *op.cit.*, p. 1894.
[93] Englund K.S. 1974. Sandstone distribution patterns in the Pocahontas formation of southwest Virginia and southern West Virginia and *GSASP* 148, p. 37.
[94] Wilson. C.W. and R.G. Stearns. 1960. Pennsylvanian marine cyclothems in Tennessee. *GSAB* 71:1457.
[95] Schlee J. 1963. Early Pennsylvanian currents in the southern Appalachians. *GSAB* 74:1442.
[96] Wells. 1960. *op.cit.*, 391.
[97] Dott. 1963. *op.cit.*, p. 10.
[98] Milici R. 1969. Nonmarine deposition of basal Pennsylvanian strata in the southeast Cumberland Plateau of Tennessee. *GSAP* 1969, 1(4)53.
[99] Duff, et.al. 1967. *op.cit.*, p. 41.
[100] Wanless, 1939. *op.cit.*, p. 9.
[101] Broadhurst F.M. and D. Magraw. 1959. On a fossil tree found in an opencast coal site near Wegan, Lancashire. *Liverpool Manchester Geological Journal* 2:156.
[102] Shabica C.W. 1970. Depositional environments in the Francis Creek shale. *Illinois State Geological Survey Guidebook Series #8*, pp. 48-9.
[103] Zaritskiy P.V. Erratic boulders in coal of the Donbas. *Doklady: Earth Science Sections* (American Geological Institute Translation) 213:156-7.
[104] Fansa L.F. and A.V. Carozzi. 1970. Exotic Pebbles in LaSalle limestone (Upper Pennsylvanian), LaSalle, Illinois. *JSP* 40:693.
[105] Miller B.L. 1934. Limestones of Pennsylvania. *PGS* M20, p. 57.
[106] Shabica C.W. 1977. Sedimentary structures from the Carbondale formation (lower Pennsylvanian) of northern Illinois. *FGM* (in print), p. 15.
[107] *Ibid.*, p. 27.
[108] Reineck H.E. and I.B. Singh. 1973. *Depositional sedimentary environments* SV p. 154.
[109] Middleton G.V. 1973. Johannes Walther's law of the correlation of facies. *GSAB* 84:983.
[110] Schwarzacher. 1975. *op.cit.*, p. 39.
[111] Frey. R.W., ed. 1975. *The study of trace fossils.* SV., p. 31.
[112] Burton C.K. 1970. Lower Paleozoic rocks of Malay Peninsula: discussion. *AAPG* 54(2)357.
[113] Reineck and Singh. 1973. *op.cit.*, p. 412.
[114] Wilson and Stearns. 1960. *op.cit.*, pp. 1457-8.
[115] Sloss L.L. 1962. Stratigraphic models in exploration. *AAPG* 46(7)1050-1.
[116] Passega R. 1962. Problem of comparing ancient with recent sedimentary deposition. *AAPG* 46(1)117.
[117] Bott and Johnson. 1967. *op.cit.*, pp. 440-1.
[118] Carrigy M.A. 1971. Deltaic sedimentation in Athabasca tar sands. *AAPG* 55:1155.
[119] Glaeser J.D. 1974. Upper Devonian stratigraphy and sedimentary environments in northeastern Pennsylvania. *PGS General Geology Report* 63, pp. 6-7.
[120] Manos C. 1967. Depositional environments of Sparland cyclothem (Pennsylvanian), Illinois and Forest City basins. *AAPG* 51)9)1847.
[121] Martin W.D., Henniger B.R. 1969. Mather and Hockingport sandstone lentils (Pennsylvanian and Permian) of Dunkard basin, Pennsylvania, West Virginia and Ohio. *AAPG* 53:296.
[122] Jansa L. 1972. Depositional history of the coal-bearing upper Jurassic-lower Cretaceous Kootenay Formation, southern Ricky Mtsn, Canada. *GSAB* 83:3200.
[123] Walker R.G. 1971. Nondeltaic depositional environments in the Catskill Clastic wedge, (upper Devonian) of central Pennsylvania. *GSAB* 82:1305.
[124] Johnson R.G. 1962. Interspecific associations in Pennsylvanian fossil assemblages. *JG* 70:49-50.
[125] Hedberg. 1959. *op.cit.*, p. 677.
[126] Cloud P.E. 1960. Gas as a sedimentary and diagenetic agent. *AJS* 258A:43.
[127] Shabica C.W. 1971. Depositional environments in the Francis Creek shale and dissociated strata. *University of Chicago PhD Thesis*, p. 174.
[128] Beerbower, 1961. *op.cit.*, p. 1043.
[129] *Ibid.* p. 1036.
[130] Wanless and Shepard. 1936. *op.cit.*, p. 1187.
[131] Weller. 1956.*op.cit.*, p. 28.
[132] Wanless. 1939. *op.cit.*, p. 6.
[133] Roux W.F. Jr. and S.F. Schindler. 1973. Late Mississippian cyclothems of Heath Formation, western North Dakota. *AAPG* 57(5)961.
[134] Vischer G.S. 1965. Use of vertical profile in environmental reconstruction. *AAPG* 49(1)43.
[135] Padgett G.V. and R. Ehrlich. 1976. Paleohydrologic analysis of a late Carboniferous fluvial system, southern Morocco. *GSAB* 87:1104.
[136] Wheeler H.E. and H.H. Murray. 1957. Base-level control patterns in cyclothemic sedimentation. *AAPG* 41(9)1988.
[137] Selley R.C. 1970. *Ancient sedimentary environments.* Cornell University Press, Ithaca, New York., p. 6.
[138] *Ibid.* p. 212.
[139] *Ibid*, p. 217.
[140] Shabica. 1977. *op.cit.*, p. 3.
[141] Van Bemmelen R.W. 1972. *Geodynamic models.* EPC, p. 21.
[142] Dickinson W.R., ed. 1974. Tectonics and sedimentation. *SEPM* 22:2.
[143] Zangerl R. and E.S. Richardson. 1963. The paleoecological history of two Pennsylvanian black shales. *FGM* 4:22.
[144] Trueman. 1948. *op.cit.*, p. 203.
[145] Wells. 1960. *op.cit.*, p. 391.
[146] Johnson J.G. 1971. Timing and coordination of orogenic, epiorogenic, and eustatic events. *GSAB* 82:3263.
[147] Keith M.L. 1972. Ocean-floor convergence: a contrary view of global tectonics. *JG* 80:273.
[148] Bott M.H.P. 1964. Formation of sedimentary basins by ductile flow of isostatic origin in the upper mantle. *Nature* 201 (4924)1082-4.
[149] Bott and Johnson. 1967. *opcit.*, p. 433.
[150] Wanless H.R. 1964. Local and regional factors in Pennsylvanian cyclic sedimentation. *KGS* 169(2)604.
[151] *Ibid.* p. 603.
[152] Ferm J.C. and J.W. Huddle. 1955. Slump and mud flows of Pennsylvanian age in the Appalachian plateau. *GSAB* 66:1557.
[153] Stearns R.G. and R.C. Milici. 1967. Structure of the southern Appalachian coal field and its origin. *GSAP* 1967, p. 212.
[154] Koenig J.B. 1956. The petrography of certain igneous dikes of Kentucky. *Kentucky Geological Survey Bulletin* 21, Series IX, p. 52.
[155] Spears D.A. 1970. A kaolinite mudstone (tonstein) in the British coal measures. *JSP* 40:386.
[156] Damberger H.H. 1974. Coalification patterns of Pennsylvanian coal basins of the eastern U.S. *GSASP* 153, p. 53.
[157] *Ibid.* p. 56.
[158] Dott. 1963. *op.cit.*, p. 124.
[159] Koinm D.N. and P.A. Dickey. 1967. Growth faulting in McAlester basin of Oklahoma. *AAPG* 51(5)718.
[160] Hunt S.R., Nelson J.W., and G.G. Treworgy. 1976. Soft-sediment deformation in the Pennsylvanian strata of southern Illinois. *GSAP* 1976 18(6)933.
[161] Gentile. 1968. *op.cit.*, p. 46.
[162] McGinnis L.D. 1966. Crustal tectonics and PreCambrian basement in northeast Illinois. *IGSRI* 219:20.
[163] Koppe E.F. 1967. Petrography of coal in the Houtzdale quadrangle, Clearfield County, Pennsylvania. *PGS* M55, p.3.
[164] Weller J.M. and A.H. Sutton. 1940. Mississippian border of easter interior basin. *AAPG* 24(5)852.
[165] Frye J.C. 1950. Origin of Kansas great plains depressions. *KGS* 86(1)3.
[166] Brandon A. 1972. Clastic dykes in the Namurian shales of County Lutrim, Republic of Ireland. *GM* 109:366.
[167] Nichelson R.P. and V.N.D. Hough. 1967. Jointing in the Appalachian plateau. *GSAB* 78:624.
[168] Gresley W.B. 1898. Clay veins vertically intersecting coal measures. *GSAB* 9:50.
[169] Shaw A.B. 1964. *Time in stratigraphy.* McGraw Hill, New York., p. 7.
[170] Wanless and Shepard. 1936. *op.cit.*, p. 1186.
[171] Wanless. 1955. *op.cit.*, p. 1800.
[172] Weller. 1956. *op.cit.*, p. 34.
[173] Bishop D.G. and E.R. Force. 1969. The reliability of graded bedding as an indicator of the order of superposition. *JG* 17:351.
[174] Unpublished Illinois Geological Survey drill-bore stratigraphic logs of cyclothems in north-central Illinois.
[175] Broadhurst F.M. and D.H. Loring. 1970. Rates of sedimentation in the upper Carboniferous of Great Britain. *Lethaia* 3:1-9.
[176] Shabica. 1977. *op.cit.*, p. 32.
[177] Weller. 1956. *op.cit.*, p. 35.
[178] Reineck and Singh. 1973. *op.cit.*, p. 104.

[179]Siever. 1957. op.cit., p. 247.
[180]Ibid. p. 246.
[181]Wheeler and Murray. 1958. op.cit., p. 445.
[182]Moore. 1950. op.cit., p. 8.
[183]Potter P.E. and W.A. Pryor. 1961. Dispersal centers of Paleozoic and later clastics of the upper Mississippi valley and adjacent areas. GSAB 72:1195.
[184]Ibid, p. 1227.
[185]Williams E.G. 1960. Relationship between the stratigraphy and petrography of Pottsville sandstone and the occurrence of high-alumina Mercer clay. Economic Geology 55:1301.
[186]Schlee. 1963. op.cit., p. 1448.
[187]Siever R. 1959. Petrology and geochemistry of silica sedimentation in some Pennsylvanian sandstones. SEPM 7:78.
[188]Schopf J.M. 1970. Petrified peat from a Permian coal bed in Antarctica. Science 169(3942)276.
[189]Davis, et.al. 1960. High-temperature-pressure studies of wood. GSAP 1960, p. 80.
[190]Cohen A.D. 1970. An allochthonous peat deposit from southern Florida. GSAB 81:2477.
[191]Wanless H.R. 1931. Pennsylvanian cycles in western Illinois. IGS 60, p. 173.
[192]Gray H.H. 1962. Sedimentary barriers and the deposition of coal-bearing rocks. GSAP 1962, p. 62A.
[193]Swinehart J.B. and E.G. Williams. 1970. Influence of the continental and marine paleotopography on the continental and marine cycles of the lowest Conemaugh and Allegheny, western Pennsylvania. GSAP 1970 2(1)38.
[194]Kope E.F. 1963. Petrography of the upper Freeport coal. PGS M48, p. 29.
[195]Wanless. 1931. op.cit., p. 172.
[196]Nalivkin. 1973. op.cit., p. 470.
[197]Koppe. 1967. op.cit., p. 15.
[198]Weller. 1930. op.cit., p. 108.
[199]Gluskoter A.J. 1965. Clay minerals in Illinois coals. GSAP 1965. p. 63.
[200]Wilson and Stearns. 1960. op.cit., p. 1455.
[201]Rupke N.A. 1969. Sedimentary evidence for the allochthonous origin of stigmaria, Carboniferous, Nova Scotia. GSAB 80:2109.
[202]Ibid, p. 2112.
[203]Rupke N.A. 1970. (same title): Reply. GSAB 81:2536.
[204]Schultz L.G. 1958. Petrology of underclays. GSAB 69:391.
[205]Weller. 1930. op.cit., pp. 123-4.
[206]Weller. 1956. op.cit., p. 33.
[207]Schultz. 1958. op.cit., p. 382.
[208]O'Brien N.R. 1964. Origin of Pennsylvanian underclays in the Illinois basin. GSAB 75:829.
[209]Zeller E.J. and J.L. Wray. 1956. Factors influencing the precipitation of calcium carbonate. AAPG 40(1)143.
[210]Ibid, p. 150.
[211]Harbaugh J.W. 1964. Significance of marine banks in southeast Kansas in interpreting cyclic Pennsylvanian sedimentation. KGS 169(1)199.
[212]Weller. 1930. op.cit., p. 127.
[213]Moore. 1950. op.cit., p. 14.
[214]Miller, 1934. op.cit., p. 40.
[215]Potter. 1962. op.cit. p. 1910.
[216]Mudge M.R. 1956. Sandstone and channels in upper Pennsylvanian and lower Permian in Kansas. AAPG 40(4)654.
[217]Duff, et.al. 1967. op.cit., p. 108.
[218]Friedman S.A. 1957. Distribution, thickness, and origin of sinuous sandstone lenses of the Allegheny series, Vigo County, Indiana. GSAB 68:1731.
[219]Wanless. 1955. op.cit., p. 1778-1800.
[220]Weller. 1956. op.cit., p. 29.
[221]Potter P.E. 1962B. Shape and distribution pattersn of Pennsylvanian sand bodies in Illinois. IGS 339:14.
[222]Weller. 1930. op.cit., pp. 116-7.
[223]Trueman, 1946. op.cit., p. 1x
[224]Beerbower. 1961. op.cit., p. 1040.
[225]Scheideggar A.E. 1975. Physical aspects of natural catastrophes. EPC, pp. 197-8.
[226]Potter. 1962B., op.cit. p. 8.
[227]Ibid, p. 3.
[228]Krumbein W.C. 1947. Shales and their environmental significance. JSP 17(3)101.
[229]Grim R.E., Bradley W.F. and W.A. White, 1957. Petrology of the Paleozoic shales of Illinois. IGSRI 203, p. 5.
[230]Krumbein. 1947. op.cit., p. 106
[231]Zangerl and Richardson. 1963. op.cit., p. 30.
[232]Weller. 1956. op.cit., p. 32.
[233]Gentile. 1968. op.cit. p. 25.
[234]Moore. 1950. op.cit., p. 8.

Study Questions

The information in parentheses indicates the article in the book, and the page number(s) therein, where the answer can be located. Frequently-asked-questions are listed first.

1. Why do certain marsupials only occur in Australia if all land animals came from Noah's Ark on Ararat? (Causes for the . . . , p. 10).
2. Do "bad" radiometric dates actually amount to a tiny number of malfunctioning watches? (Radiometric Geochronology . . . , p. 158).
3. Based on consistency tests, etc., can it be said that dating methods are self-checking? (Radiometric Geochronology . . . , pp. 160-161; 164-166).
4. Is the use of dinosaur fossils for dating strata an exercise in circular reasoning? (An Anthology . . . Report 2, p. 136).
5. What are the creationist implications of the asteroid-impact theories? (A Diluviological Treatise . . . , p. 50, 56; An Anthology . . . Report 1, p. 137).
6. Can a random set of fossils be "assembled" by evolutionists into a seemingly transition-filled sequence? (The Cephalopods . . . , p. 183).
7. How can ecological zonation explain the order in which ammonoids (an ancient shelled marine animal) appear in rock strata? (The Cephalopods . . . , pp. 190, 192-194).
8. How can the Flood offer a better explanation for cyclic coal-bearing strata than conventional geology does? (A Diluvian Interpretation . . . , pp. 212-217).
9. How much of the standard geologic column actually exists on earth? (The Essential Nonexistence . . . , pp. 105-130).
10. How many different kinds of fossils can actually be found "stacked" at any one spot on earth? (A Diluviological Treatise . . . , table, p. 46-47).
11. Why do we fail to find human remains throughout the bulk of the sedimentary rock record if it had been deposited during the Genesis Flood? (A Diluviological Treatise . . . , pp. 57-61).
12. Why are fossils more and more unlike modern living things the lower one goes into rock strata? (A Diluviological Treatise . . . , p. 56-57).
13. How do "living fossils" fit into a global Flood context? (A Diluviological Treatise . . . , p. 57).
14. Assuming that God did not create birds with nonfunctioning wings, is there evidence that birds can lose their ability to fly in less than several thousand years? (Causes for the . . . , p. 11).
15. Sequence Stratigraphy: Are unconformity-bound units used by geologists objective features? (A Diluvian Interpretation . . . , p. 205-206).
16. Are there far too many vertebrate fossils in rock to have been all alive at the same time? (The Antediluvian Biosphere . . . , p. 18).
17. Aren't fossils which occur in growth position proof for long periods of time and thus evidence against a global Flood? (An Anthology . . . Report 2, p. 85-86).
18. Do clastic dykes, which are evidence that great thicknesses of sediment were still soft before overlying sediment was deposited, common in coal-bearing sequences all over the world? (A Diluvian Interpretation . . . , p. 212-213).
19. Do fossils frequently occur in "wrong" strata relative to that demanded by evolution? (An Anthology . . . Report 2, pp. 87-94).
20. Do *only* creationists identify ancient footprints as once belonging to animals which, according to evolution, did not appear until much later? (An Anthology . . . Report 1, p. 135).
21. Can it be demonstrated that geologists frequently do not publish dating results that do not agree with their preconceptions? (Radiometric Geochronology . . . , p. 158-159; An Anthology . . . Report 1, p. 142).
22. Is it true that there are far more "good" dating results than "bad" ones? (Radiometric Geochronology . . . , p. 158).
23. Are flat Ar-Ar spectra proof for the correctness of the date obtained? (Radiometric Geochronology . . . , p. 163).

24. Do K-Ar isochrons solve the problems of the K-Ar method? (An Anthology . . . Report 1, p. 142; Radiometric Geochronology . . . , p. 164).

25. Is a good isochron relationship proof of closed-system behavior? (Radiometric Geochronology . . . , pp. 162-164).

26. On conventional geochronologists' own terms, is agreement of results from different methods an unassailable proof for the validity multimillion-year date obtained? (Radiometric Geochronology . . . , p. 165-166).

27. Is uniformitarianism, the approach controlling geology in the last 2 centuries, a value-free scientific methodology or an inherently atheistic philosophical concept? (A Diluvian Interpretation . . . , p. 203-204).

28. Are conventional geologic claims that ancient sediments were laid down in analogs of modern sedimentary environments (e. g., deltas, rivers) at least partly based on circular reasoning? (A Diluvian Interpretation . . . , pp. 209-212).

29. Is there evidence of catastrophic earth movements during the deposition of cyclic coal-bearing strata? (A Diluvian Interpretation . . . , p. 212-213).

30. What kinds of evidences are there for Flood-related catastrophic sedimentation during the deposition of coal-bearing strata? (A Diluvian Interpretation . . . , pp. 208-209; An Anthology . . . Report 1, p. 137).

31. Is there any geologic evidence that so-called evaporites did *not* originate from the drying-up of seas over vast periods of time? (An Anthology . . . Report 1, p. 137).

32. Can it be shown that evolutionists are prone to ignore or discount the existence of fossils which occur in strata deemed too young or too old for their occurrence? (A Diluviological Treatise . . . , p. 59; The Cephalopods . . . , p. 186).

33. Are there any conventional geologists who are skeptical towards radiometric dating *in general*? (An Anthology . . . Report 1, p. 142; Radiometric Geochronology . . . , p. 168).

34. Can the rejection of "bad" dates be independently justified by the alteration seen in the rock sample and/or the geologic context the rock comes from? (Radiometric Geochronology . . . , pp. 160-161).

35. How do shell beds containing mixtures of shells from different so-called geologic ages argue against the validity of organic evolution and geologic time? (The Cephalopods . . . , p. 188-189).

36. Can it be demonstrated that evolution denies both God and human dignity? (An Anthology . . . Report 2, pp. 82-83).

37. Are "ancient reefs" seen in rock strata, and the long periods of time they imply, actually there, or are they the products of geologists' interpretation and imagination? (An Anthology . . . Report 2, p. 84-86; see also Anthology . . . Report 1, pp. 138-139).

38. Do ancient shell beds show evidence of large-scale sorting by flood currents? (The Cephalopods . . . , p. 193).

39. Do non-creationist geologists recognize that many so-called overthrusts (where strata are in "wrong" order) display a sharp contact which often resembles a bedding plane? (An Anthology . . . Report 1, p. 141; An Anthology . . . Report 2, pp. 86-87).

40. Has the late Clifford Burdick's controversial report of finding anomalously-old pollen in the Precambrian of the US Southwest (Grand Canyon) been replicated in the Precambrian of other parts of the world? (An Anthology . . . Report 2, pp. 92-94).

41. Do fossils of one "age" tend to be absent wherever there is other strata containing fossils of another "age"? (A Diluviological Treatise . . . , table, p. 42-43).

42. Is the observed global succession of fossils a self-evident basis for their correlation into time-events? (A Diluviological Treatise . . . , p. 24).

43. How can an association of tectonic units and biogeographic provinces (TABs) explain why fossils tend to occur in different layers all over the earth as a result of one global Flood? (A Diluviological Treatise . . . , pp. 44-57)

44. Is there any independent evidence showing a connection of tectonics and type of fossil, thereby justifying the TAB concept ? (A Diluviological Treatise . . . , p. 48).

45. Where does Precambrian strata fit into the Creation and Flood? (A Diluviological Treatise . . . , p. 50).

46. Aren't the earth's sedimentary rocks much too thick to have been deposited in a single global Flood? (A Diluviological Treatise . . . , p. 53).

47. Can circular reasoning, despite evolutionists' denials to the contrary, be shown to be involved in the use of fossils for dating strata? (A Diluviological Treatise . . . , p. 26, 42).

STUDY QUESTIONS

48. Would not such things as ecological zonation, differential escape, and hydrodynamic sorting need to have been unrealistically efficient to have played a significant role in the stratigraphic separation of organisms (later fossils) during the Flood? (A Diluviological Treatise . . . , p. 44).

49. How can the post-Flood Ice Age help explain why animals are so different from each other on various continents even though they all came from the Ark on Ararat? (Causes for the Biogeographic . . . , pp. 8-10).

50. Is there actually too much organic carbon sequestered in the earth's coal and oil deposits to have formed on an earth that is only several thousand years old? (The Antediluvian Biosphere . . . , pp. 15-16).

51. Could all of the carbonate deposits (including chalks) have formed in one Flood? (The Antediluvian Biosphere . . . , pp. 17-19).

52. Is there any basis for the anti-creationist charge that there are far too many fossils in rock to have all been alive simultaneously (that is, before the Biblical Flood)? (The Antediluvian Biosphere . . . , pp. 18-19).

53. Did the Biblical account stifle research into the geography of living things, or did it, to the contrary, encourage such study? (Causes for the Biogeographic . . . , p. 7).

54. Can it be documented that the order of fossils in the rock record is exaggerated by the fact that the same organisms are given different names depending upon their stratigraphic position? (A Diluviological Treatise . . . , p. 26; The Cephalopods . . . , pp. 185-188).

55. Can claims of "reworking," as an explanation for anomalously-occurring fossils, be justified by the wear-and-tear that such fossils show? (An Anthology . . . Report 2, p. 93).

56. All things considered, are "reworked" fossil very rare overall when compared with the fossils which occur in "proper" stratigraphic order? (An Anthology . . . Report 2, p. 93).

57. Do thick igneous bodies show evidence, within their very textures, for having been emplaced rapidly? (An Anthology . . . Report 2, p. 95).

58. Confronting neo-Cuvierist fallacies: Should we accept a straightforward one-on-one relationship between the conventionally-accepted "geologic age" of the sedimentary rock, and whether or not it was deposited during the Biblical Flood, or after it? (The Essential Nonexistence . . . , p. 128).

59. Can the geologic column be justified by virtue of the fact that strata overlap each other, much like shingles on a roof, thereby justifying the concept of stratigraphic succession? (A Diluviological Treatise . . . , p. 24).

60. Was the geologic column established by genuine creationists, or by half-creationists such as Charles Lyell? (A Diluviological Treatise . . . , p. 24-25).

61. Is it correct to assert that the geologic column rest on the Law of Superposition, and nothing else? (A Diluviological Treatise . . . , p. 24).

62. In what ways is the use of fossils for correlation, and the construction of the standard geologic column, as illogical as the old Wernerian system of correlating strata according to primary lithologies? (A Diluviological Treatise . . . , p. 24).

Index

Note: In order to facilitate the study of this work, the listings below include synonyms and closely-related words to those actually used in the text. For example, "baramin" itself does not appear on page 7, but "created kind" does. Both are indexed below.

absolute age, retreat from term, 147
absolute dating, see dating
abstraction—nature of biozones, 187
actualism, see uniformitarianism
Adam, Eve, human descendants of, 59
admixture, fossils of "different ages," see reworking
age dating, see dating
age, appearance of, soil nutrients, 18-19
age, Earth, 4.5 Ga consensus, fallacy of, 167
agreement, contrived, biostratigraphic/isotopic dating, 161
agreement, dating methods, see concordance
allochthony vs. autochthony, coal deposition, 215
ammonoid fossil successions, ecologic, 189-193
ammonoid phylogeny, ambiguities of, 181
Ammonoidea, basic properties of, 180
ammonoids, Carboniferous, global map 34
ammonoids, Cretaceous, global map, 37
ammonoids, Devonian, global map, 33
ammonoids, Jurassic, global map, 37
ammonoids, Permian, global map, 35
ammonoids, Triassic, global map, 36
analytic criteria don't "self-check" dates, 164
ancient civilizations, 60
Andes Mts., "complete" geologic column, 126
animism, 133
Annelida, Phylum, no evolutionary roots, 181
Annularia, 204
anomalous dates, see dating
anomalous footprints, see footprints
anomalous fossils, ignored, 186
anomalous fossils, see reworking
anomalous human fossils, 57
Antarctica, geology of, 105-106
antediluvian human condition, 59
antediluvian topography, 60
antediluvian, see also pre-Flood
anthropogenic remains, see human fossils
anti-Biblical consequences, evolution, 82
anti-Biblical prejudices, uniformitarians, 140
anticlines, Flood waters and, 215
anti-creationists, bogus arguments of, 15-19, 50, 53
ape-men, see human evolution
apologetics, need for, 134
Apostle Peter's prophecy on state of geology, 217
Ar-Ar plateaus, non-diagnostic, 163
Araucaria, 81
archeology, dating of objects, 142-143
Ark, Noah's, animals released from, 57
Ark, Noah's, on Ararat, 7
Arthropoda, Phylum, no evolutionary roots, 181
Asian Arctic, 8
asteroid impact, 50, 56, 137
atheism, basis of evolution, 83
atheism, basis of uniformitarianism, 97, 134, 203-204

aureoles, contact-metamorphic, 95
Australian faunas, 10
Australopithecus, 81
authigenic minerals, dating of, 159-160
Babel, Tower of, 9
baramin, taxonomy of, 7
bedding planes vs. overthrusts, 86
belemnoids, stratigraphic occurrence, 186
bentonites, dating of, 161
bentonites, within coal-bearing strata, 213
biases, intentional, fossil occurrences, 59, 186
Biblical basis: life zones 45
biogenic law, and cephalopod evolution, 183
biogenic law, persistence of, 82
biogeographic vs. ecological zones, 45
biogeography, post-Flood, 7
bioherms, alleged ancient, 84-85
biomass, pre-Flood, 15
biospheres, pre- and post-Flood contrasted 58
biostratigraphic brackets, ambiguities of, 158
biostratigraphic brackets, backpedaling on, 162
biostratigraphic ranges, circularity of, 136
biostratigraphically-progressive extinctions, cause, 56
biostratigraphy, description of, 185
biostratigraphy, taxonomic artifacts, 26
biotite/hornblende concordance, non-proof, 164-165
biozones, fossil, subjectivity of, 185-188
birds, flightless, origins, 11
black shales, Flood deposition of, 217
blooms, planktonic, pre-Flood, 17
bolide impact, 50, 56, 137
bolide impact, volcanic alternatives to, 137
botulism, 82
brachiopods, Devonian, global map 34
brachiopods, Permian, global map, 35
brachiopods, Silurian, global map, 32
British Isles, see Great Britain
buoyancy regulation, ammonoids, 191
Burdick, Clifford, 93-94, 105
Burlington Limestone, 16

Cambrian explosion, evolution and, 80, 133, 181
carbon, organic, non-problem, 15
carbonates, Flood origins of, 18-19, 214
carbonates, purity, Flood non-problem, 216
carbonatitic volcanics, 61
Carboniferous-Permian boundary, vague, 205-206
carnivores' teeth before Fall, 133-134
catastrophic deposition, evidences of, 17, 137, 188-189, 208-209
catastrophism, current attitudes towards, 134-135
catastrophist vs. diluvialist, distinction between, 201-202
Catskill delta, so-called, 211
caviomorph rodents, 10
cementation, inorganic, "ancient reefs," 86

chalks, non-problem of, 17
chance, role of, fossil succession, 41, 49
chance, see also fortuitous
chemical constraints, life zones and, 48
circular reasoning, alleged reworking and, 93
circular reasoning, dating by fossils, 42, 59, 136
circular reasoning, paleoenvironmental reconstruction, 210
circular reasoning, system boundaries, 26
circulatory system, open, 184
civilizations, ancient, 60
clastic dykes, Flood and, 213
Clement, church father, 94
climate, post-Flood, 8
coal, not only Carboniferous, 207-208
coal, organic carbon non-problem, 15
coccolith productivities, 17
collinearity, isochron points, non-proof validity, 164
comparative biospheres, pre-and post-Flood, 58
completeness, fossil record, 81
compromising evangelicals, 24-25
conches, ammonoid, pressure-tolerance tests, 192
concordance, dating methods, rejected at will, 165
concordance, fortuitous, dating methods, 165
concretions (siderite), 215
condensed beds, deflate geologic column, 188-189
conformable contacts, "overthrusts" and, 86
consecutive geologic periods in place, 127
consensus, 4.5 Ga Earth age, fallacy of, 167
consistency, internal, non-proof date valid, 164-165
contact metamorphism, intrusions, unimpressive, 95
contact, strata, knifesharp, "overthrusts," 86-87
continental drift, 8, 60, 96
continental shelf, North American, 127
continental shelves, 105, 126
contradictions, human evolution, 81
contradictory designations, actualistic analogies, 209
contrived metamorphism, dating rationalization, 160
convergent evolution, cephalopods, *ad hoc*, 183-184
convergent evolution, non-explanation, 133
convolute laminations, 215
cooling, rapid, igneous rocks, evidences, 95
Cope's Law, so-called, 192
corals, productivities of, 18
correlation, by well logs, contradict other information, 211
craters, impact, see bolide
Creation Week, geology of, 136-137
creationism, benefits science, 7
creationism, half-developed, 24-25
crinoidal limestones, non-problem 16
crinoids, productivity of, 16
cross laminae, 215
cross-cutting relationships violated, dates, 162, 166
cross-sections, geologic, TABs and, 52
cruelty of nature, evolution and, 82
Cuba, "complete" geologic column in, 126
current ripples, 215
Cuvier, not true creationist, 25
cyclic coal-bearing strata, lateral variations, 202
cyclothem, "ideal," stratigraphic section, 201
cyclothems, global occurrence of, 208

Darwin's *Origin of Species*, 82
data, selectively included on isochrons, 163

dates, anomalous, analytic equipment non-excuse, 147
dates, correctness of, impossibility of proving, 168
dates, discordance, occur as a *rule*, 166
dates, discrepant, *ad hoc* rejection, 142, 158, 161
dates, discrepant, geologic context of, 160-161, 167
dates, discrepant, *majority* of results, 162
dates, discrepant, *not rare*, 158-159
dates, discrepant, usually not published, 158-159
dates, "good" vs. "bad," relative frequency of, 158
dates, isotopic, not self-checking, 164
dates, minimum/maximum age, fallacy of, 162
dates, Precambrian, ambiguities of, 167
dates, protracted-cooling rationalizations, 163
dates, selection, non-statistical, non-experimental, 158
dates, U-Pb, rarity of concordance, 166
dating, paleontologic, circularity of, 42, 59
DeBeer, 183
decollement, when rocks still soft, 212-213
deformation, strata, paucity, "overthrusts," 86
delta-lobe switching, poor analogue for cyclic strata, 211
Deluge: see Flood
design, repudiated by evolution, 83
differential escape, organisms, Flood and, 55
Diluvial synthesis, and TAB Flood model, 53-56
dinosaur bones, and marine fossils, 194
dinosaurs, *assumed* Mesozoic in age, 136
dinosaurs, circularity of dating by, 136
dinosaurs, differential extinction of, 57
dinosaurs, extinction of, theories, 56
dinosaurs, global map, 37
dinosaurs, possible Tertiary "age," 97
diphtheria, 82
discordant dating results, usual, 168
diversity, fossils, favors young earth, 141
diversity, living things, creationist non-problem, 185
dolomite, see carbonates
D'Orbigny, Alcide, 185
downwash, fossils, alleged, 87
drainage patterns, Flood and post-Flood, 141
Dromiciops, oddly-distributed marsupial, 10
dune deposits, mistaken for "ancient reefs," 86
duplicate names for anomalous fossils, 25, 135-136, 204
dysteleological arguments, contra, 11

early Flood conditions, cephalopods and, 191
Earth, 4.5 Ga consensus, fallacy of, 167
Earth, oldest rocks, arbitrary, 167-168
Earth's age, peripheral issue *not*, 95-96
ecological associations, pre-Flood, survival of, 57
ecological zonation, Flood and, 23, 44, 189-191
ecological zonation, Flood stages, cephalopods, 191-194
ecological zonation, not same as TAB, 44-45
effusives, see tuffs
eggs, Flood survival of, 56
Era boundaries, TABs and, 48, 50
erratics, 96
escape, organisms, Flood, differential, 55, 189
eustacy, cyclic strata, limited explanatory power, 203
evaporites, Flood origins of, 137-138
evolution, its religious nature admitted, 181
evolution, "fact" of, 81
evolution, anti-Biblical nature of, 83
evolution, ecological alternatives, trilobites, 24
evolution, geologic column and, 23-25
evolution, inherently atheistic, 83

INDEX

evolution, presumed mechanism of, 80
evolution, problem of living fossils, 81-82
evolutionary rationalizations, Cambrian life, 80
evolutionary sequences, admitted conjecture, 182
evolutionistic biases, taxonomy, 181
evolutionists, intellectual arrogance of, 81
evolution—phenetic contradictions, 136
exotic pebbles, coal, Flood deposition and, 209
experiments, sorting, skeletal remains, 61
extinctions, Lyellian curves, 25
extinctions, post-Flood, 9
extinctions, progressive, TAB explanation, 56

facies analysis, coal-bearing strata, 212-213
facies changes, significance of, 38
facts, geologic, fitted to theories, 204
Fall, carnivores' teeth before, 133-134
fascism, evolutionary roots of, 83
fault lines, rapid subsidence of sedimentary basins, 212
fish, Siluro-Devonian, global map, 32
fissile shales, Flood origins of, 217
flat spectra, Ar-Ar method, non-diagnostic, 163
flightless bird origins, 11
floating mats, vegetation, Flood and, 214
Flood action, unconformities and, 206
Flood, explanation for fossils, existence, 135
Flood model, fossils, author's own, see TABs
Flood origins of carbonates, 18-19
Flood rejection, prophesied, 217
Flood sediments, thickness non-problem, 53
Flood sorting action, obvious examples, 192-193
Flood stages, ecological zonation, cephalopods, 191-194
Flood stages, TAB model and, 53
Flood theories, Occam's Razor and, 53, 56
Flood waters, oceanic source of most, 60
Flood, "ancient reefs" and, 84-86
Flood, emergence of land, 212
Flood, explanation for fossils, 135
Flood, hydrodynamic sorting, 23
Flood, paleohydrological factors, 208-212
Flood, sedimentary rock volumes of, 59
Flood-deposited sediment, erosional gulleys in, 206
flume experiments, 61
footprints, vertebrate, anomalous, 135
fortuitous concordance, dating methods, 166
fortuitous separation, fossils, 41
fossil record, incompleteness, non-excuses, 80-81
fossil separation, early theories, 23
fossil succession, confluence of factors, 54
fossil succession, degree of, 185
fossil succession, local vs. global 38
fossil succession, onion-skin fallacy, 186
fossil succession, role of chance, 41
fossil succession, stratomorphic intermediates, see stratomorphic intermediates
fossil successions, concocted by artificial taxons, 188
fossilization conditions, sediment eH and, 61, 188
fossils, ancient, modern-appearance, 26
fossils, biofacies of each other, 40
fossils, common to *many* geologic periods, 25-26, 135, 186
fossils, duplicate names, 25, see also 59
fossils, Flood model for, see TABs
fossils, global abundance, non-problem, 19
fossils, human, see human fossils

fossils, out-of-place, see reworking
fossils, overall rarity of, 19
fossils, Precambrian, 50
fossils, progressively unlike today's life, reasons, 56
fossils, unexpected discoveries, 26
founder effect, population genetics, 41
fountains of the deep, 194
framework, "ancient reefs," lacking, 84-85
fudged isochrons, 163, 165
fusain fragments, Flood-transported, 215

gastropods, 180
genera, extant, interbreeding among, 7
genera, fossil, see species
Gentry, Robert V., 207, 218
geochron, 4.5 Ga, fallacy of, 167
geochronology, see dating
geochronometry, see dating
geography, post-Flood, 8
geologic column, "complete," discussion, 126
geologic column, "complete," world map, 121
geologic column, circular reasoning and, 136
geologic column, deflated by condensed beds, 188-189
geologic column, evolution and, 23-25
geologic column, inconsistent successions, 126
geologic column, inverted strata and, 86
geologic column, trends in lithology, 52
geologic context, isotopic dating and, 160-161
geologic cross-sections, TABs and, 52
geologic periods, maps of lithologies, 107-121
geologic time, collapse of, 189
geological sciences, interpretative nature of, 134
geologic-process rates, clarification of, 94
geomagnetic polarity scale, selective data, 161
geomagnetic reversals, evidence against, 142
geometry, "reefs," non-evidence, 86
geosynclinal sediments, volumes, 48
Ginkgo, 81
glacial deposits, Quaternary, uniqueness of, 203
glauconite, dating of, 158-159
glauconites, reworked-rationalization, 165
global successions, index fossils, table, 46-47
global unconformities—nonexistent, 206
God, no place for in evolution, 83
Gondwana formations, 50, 105
gouge material, alleged overthrusts and, 86
Gould, Stephen J., admits *ad hoc* evolution, 182
grabens, 50
Grand Canyon, pollen in rocks, 92
granitic rocks, dating of, 162-164
Great Britain, fossil occurrence map, 40
growth faults, Flood and, 213
growth-in-place, Flood alternatives, 85-86
Gulf of Mexico, 141
Gulf Stream, 8

Haeckel, Ernst, 80
hard parts, lack of, organisms, non-excuse, 80
hardgrounds, not long periods of time, 206
heat, released by subterranean magmas, 95
Himalayas, "complete" geologic column in, 126
historical geology, orogenies and, 140-141
homeomorphy, conodonts, 25, 135-136
homeomorphy, paleobotany, 204
hominids, see human evolution

Homo erectus, amorphous taxon, 81
hornblende/biotite concordance, non-proof, 164-165
horse hoofprints, "pre-horse" strata, 135
human communities, pre-Flood, distribution, 61
human evolution, gaps, non-excuses for, 81
human fossils, rarity of ancient, explanation, 57, 59
humans, pre-Flood, condition of, 59
humans-as-animals, evolution and, 82
Hutton, James, 94
hydrodynamic sorting, Flood and, 23, 44, 55
hydrothermal events, Rb-Sr dating vulnerable, 160

Ice Age, 8
ichnofossils, see trace fossils
igneous bodies, rapid cooling of, evidences, 95
imagination—source of uniformitarian paleoenvironments, 211
impact craters, see bolide
importance of young-earth concept, 95-96
in situ fossils, allegations of, 85
incompleteness, fossil record, non-excuses, 80-81
inconsistent use of concordant-dates argument, 165
index fossils, designation of, TAB synthesis, 55
index fossils, juxtapositional tendencies, 42
index fossils, routine misidentification, 136
index fossils, tentativeness of, 97
Indonesia, geology of, 106, 126
infrequency, discrepant isotopic dates, *not*, 158
intellectual arrogance of evolutionists, 81
internal consistency, non-proof date valid, 164-165
intolerance, minority ideas, geology, 96-97
intolerance, uniformitarian, examples, 134
introductions, animals, 9
intrusions, temperatures of parent magma, 95
intrusives, dating of, 162-164
inverted strata, 86
islands, oceanic, life, 11
isochron plots, collinearity of points, non-proof, 164
isochrons, ease of fudging, 163
isochrons, inherited systematics, 159
isochrons, K-Ar, no panacea, 142
isotopic dating, see dating

just-so stories, evolutionistic, admitted, 182
juxtapositions, fossils, nature of, 38

kangaroos, post-Flood, 10
K-Ar isochrons, have own problems, 142
K-Ar method, conventional, main usage of, 160-163
Karoo (see Karroo;—old spelling), 18
Karroo fossil individual numbers, non-problem, 18
kimberlites, dating of, 162
kind, created, taxonomy, 7

laissez-faire economics, 83
late Flood depositional events, 128
Latimeria, 81
lava flows, dating of, 161-162
lava flows, rapid cooling of, 95
life position, fossils, Flood and, 85
life, origins of, evolutionistic, baselessness of, 81
limestone, see carbonates
lithologic similarities, false correlations, 38, 167
lithologies, cross geologic periods, 205
living fossils, evolutionary problem admitted, 81-82

living fossils, mystery solved, TAB explanation, 57
living fossils, see also specific name
logs, used in correlation, contradict other data, 211
Lomonosov, Mihhail, 94
long-ranging fossils, 186
Lower Paleozoic, marine origins of, 50
Lyell, Charles, lingering effects of his prejudices, 134
Lyell, Charles, never true creationist, 25
Lyellian curves, 25

macromutations, 80
Madagascar fauna, 10
magmas, intrusion temperatures, 95
magnetic field, earth, see geomagnetic reversals
malfunctioning watches, fallacious analogy, 158-159
mammals, advantaged, post-Flood world, 57
mammals, fossil, global map, 38
marine ecology, Flood disturbance of, 137
marsupials, Australian, 10
materialism, evolutionistic nature of, 83
mats, floating, of vegetation, give rise to coal, 214
mechanical well logs, contradict other data, 211
mental thrusts, see overthrusts, alleged
metamorphic rocks, possible rapid origins, 94-95, 137
metazoan origins, evolutionary non-explanation, 80, 133
meteors, see bolide
methodological uniformitarianism, mental box, 203-204
micrites, origins, 18
microfossils, Flood transport of, 92
Middle Ages, science in, 134
mineral-pair concordance, non-proof date, 164-165
missing links, human evolution, non-excuses, 81
missing strata, 127, 187. See also paraconformities
Mission Canyon Limestone, 16
Mississippian-Pennsylvanian unconformity, limited, 205
mixing lines (?), Rb-Sr, 159-162, 207
Mollusca, Phylum, no evolutionary roots, 181
monistic philosophies, evolution and, 83
monkeys typing, evolutionary fallacy of, 82
mutations, effects of, 80
mystery of living fossils, solved, TAB model, 57

nature, red in tooth and claw, evolution and, 82
Nautilus, 180
Nazism, evolutionary roots of, 83
Neopalina, 81
Nevada, fossil occurrence map, 40
nihilism, nature of evolution, 83
Noachian Deluge, see Flood
nuclear winter models, applicability, 8

Occam's Razor, Flood theories and, 53, 128
ocean floors, uplift, Flood and, 60
ocean floors, why Meso-Cenozoic 50
ocean-floor sediments, TABs and, 52
oceans, "ages" of, 127
oil, organic carbon non-problem, 15
old Earth, 4.5 Ga consensus, fallacy of, 167
omphalos: see age, appearance of
ontogeny recapitulates phylogeny, myth, 82
ophiolites, 60
Oppel, Albert, 185
ores, U-Pb dating of, 160
orientation, burial, organisms, Flood and, 85
origins of life, evolutionistic, baselessness of, 81

INDEX

orogenies, geologic column and, 140-141
orogenies, part of Flood, 194
out-of-place fossils, see reworking
overlaps, fossils, see stratigraphic ranges
overthrusts, paucity of independent evidence, 86-87
oxidizing conditions, sediment, fossilization and, 61

paleobathymetric estimates, subjectivity of, 208
paleocurrent directions, persistence of, 141
paleoenvironments, reconstruction, 139, 209-215
paleoenvironments, uniformitarian ideation behind, 209-212
paleomagnetism, see geomagnetic reversals
paleosols, misidentified as such, 215-216
Paleozoic glaciation, alternatives to, 203
palynology, Precambrian rocks and, 87-88, 93-94
paraconformities, *ad hoc* nature of, 140
parallel evolution, cephalopods, *ad hoc*, 184
peat, realistic thicknesses, 16
Peleg, days of, 8
pelmatozoan debris, amounts, 16
peridotites, dating of, 162
Permo-Triassic (PT) boundary problems, 186
Permo-Triassic (PT) boundary violated, 26
petrifaction, plants, coal, 214
petroleum, pre-Flood sources, 15
Phanerozoic time scale, major inconsistencies, 148-158
phenetics—contradict evolution, 136
phylogenies, admitted unproveability of, 182
phylogenies, human, contradictions in, 81
pillow lavas, 136-137
placers, mineral, 140
plate tectonics, 8, 60, 96-97, 105
plateaus, Ar-Ar method, ambiguity of, 163
platform sediments, volumes, 48
plutons, composite, rationalization for dates, 162
Poland, "complete" geologic column in, 126
pollen, Grand Canyon, 92
Polonium radiohalos, 207
polystrate fossils, 137, 208-209
position of life, fossils alleged, 85
post-Flood climate, 8
post-Flood depositional events, 128
post-Flood extinctions, 10
post-Flood life, biogeography, 7
post-Flood sediment, 7
potassium-argon, see K-Ar
Precambrian anomalies, listed, 92
Precambrian fossils, global map, 30
Precambrian geology, "layer-cake" fallacies, 167
Precambrian pollen, support for, 87-88, 93-94
Precambrian sediment, Flood and, 50
Precambrian, dating of, difficulties, 142
Precambrian-Cambrian boundary, circular, 42
preconceptions, evolutionistic, taxonomy, 181
preconceptions, stratigraphic, 59, 93, 186
preconceptions, uniformitarian, 96, 209
predation, "ancient reefs" and, 85
predictions, testable, TABs and, 48
pre-Flood animal population densities, 18
pre-Flood carbon inventory, 15-16
pre-Flood conditions, ammonoids, 192
pre-Flood crinoid meadows, 16
pre-Flood ecology, 45, 48
pre-Flood marine productivity, 17

pre-Pleistocene humans, rarity of, explanation, 57, 59
preservation bias, 44
preservation good, despite "reworking," 93, 136
Price, George McCready, 24, 105, 204
probability, evolutionary non-explanation, 82
PT—see Permo-Triassic
publication, selective, dating results, 158-159
punctuated equilibrium, and human evolution, 81
punctuated equilibrium, general, 80, 181
purity-of-chalks myth, 17
purposeless wings, causes for, 11
purposelessness of existence, evolution and, 83
pyrite, and fossilization process, 188, 209

radiohalos in coals, 207
radiohalos, secondary, coalified wood, 207
radiometric dating, not universally accepted, 142, 168
radiometric dating, see dating
rapid cooling, igneous rocks, evidences, 95
rapid cooling, lava flows, 95
rapid metamorphism, clues of, 94-95
rapid sedimentation, see catastrophic deposition
rarity, discrepant isotopic dates, *not*, 158-159
rationalistic preconceptions, uniformitarianism, 94
Rb-Sr dating, vulnerability, hydrothermal events, 160
Rb-Sr method, main usage of, 162-165
recapitulation, apparent, Flood alternative, 183
recapitulation, persistence of belief in, 82
recycled fossils, alleged, see reworking
redeposited fossils, alleged, see reworking
reducing conditions, sediment, fossilization and, 61
reef models, contradictory uniformitarian, 84, 138
reefs, ancient, alleged, 84-86, 137-138
rejuvenation rationalizations, unwanted dates, 163
reliability criteria, dates, inconclusive, 168
religion, theistic, irreconciliability with evolution, 83
religion-bashing fallacies, 134
reptiles, Permo-Triassic, global map, 36
reptiles, pre-Flood populations, 18
reworking rationalizations, fossils, 26, 87-94, 136, 204
rivers, pre-Flood, 61
rock alteration, discrepant-date non-excuse, 160
roots, fossil, transported origins of, 216

salts, evaporite alternatives, 137
sample purity, discrepant dates nonetheless, 160
sampling biases, fossils and, 60
scale differences, modern and "ancient reefs," 85
scientific inquiry, *not* hindered by religion, 134
seas, antediluvian, conditions, 61
seat earths, Flood origins of, 215
sediment, ocean floors, "age," 50
sediment, post-Flood, 7
sedimentary basins, rapid subsidence of, 212
sedimentary environments, see paleoenvironments
sedimentary rocks, Flood, thickness non-problem 53
sedimentary volumes, ratios, 48
sedimentation rates, discrepancies, 140
sedimentation rates, questionable validity of, 208
sediment-distribution maps, global, 107-121
seeds, Flood survival of, 56
self-checking, isotopic dates, myth of, 164
sequence-stratigraphy, cyclic sedimentation and, 205
shales, black, Flood deposition of, 217
sheet sandstones, Flood action and, 216-217

shells, ammonoid, pressure-tolerance tests, 192
silica, role of, 206, 214
size trends, fossil, non-evolutionary, 141
skeletal remains, sorting of, experiments, 61
skeletons, vertebrate, pitfalls of reconstruction, 57
slickensides, alleged overthrusts and, 86
slickensides, in underclays, 216
Smith, William "Strata," 24
Sodom and Gomorrah, 59
soils, ancient, *not*, 215-216
sorting action, Floodwaters, 192-193
sorting by water, skeletal, experiments of, 61
South America, fauna, 10
specially-created niches, 45
species, genera, fossil, subjectivity of, 26, 136, 185-188
spectra, Ar-Ar method, ambiguity of, 163
Sphenodon, 81
splitting, taxonomic, "fills" gaps, 183
splitting, taxonomic, concocts fossil successions, 188
spores, Flood survival of, 56
squid, 180
Stigmaria, not *in-situ* root traces, 216
strata, 3-D limitations of, 38
strata, inverted, little independent evidence of, 86
strata, long-distance traceability of, 207
strata, missing, 127. See also paraconformities
strata, reef flank, Flood alternative, 84
stratigraphic anomalies, fossils, 25, 93
stratigraphic ranges, fossil, special pleading, 186
stratigraphic ranges, quantification, 25-28, 128
stratigraphic-range extentions, 26, 97, 135
stratomorphic intermediates, inconsistent usage, 26, 135, 186
stratomorphic intermediates, repeated appearances of, 204
strontium ratios, assumed, 207
submarine volcanogenic rocks, 128
sudden appearance, organisms, Cambrian, 80, 133
sunlight, not life-limiting factor, 17
superposition, fossils, nature of, 38
superposition, law of, 136
superposition, law of, violated by dates, 161, 166
superstitions, atheistic, 96
survival of living fossils, TAB explanation, 56-57
swamp theory, coal, inadequacy of, 215-216
system boundaries, artificiality of, 141, 204

TAB, introduction to, 44-45
table, anomalous fossils, 88-92
TABs, *modus operandi*, 52
TABs—causality of, 48
TABs—independent evidences for, 48
TABs—schematic representation of, 51
taphonomic studies, 59-60
taxonomic splitting, concocts fossil successions, 188
taxonomy, baramins, 7
taxonomy, equivalence within categories (?), 26
taxons, fossil, subjectivity of, 26
technical typology, archeological fallacies of, 142-143
tectonically-associated biologic province, see TABs
Tectonics and life zones—see TABs
tectonics, plate, see plate tectonics
tectonism, overthrust non-evidence, 86
teeth, carnivores', pre-Fall, 133-134
teleology, evolutionary prejudices against, 83, 133

teleology, life zones, 45
Tertiary, partly post-Flood, 194
thickness, considerable, of coal layers, 215-216
thin-section analysis, discrepant dates still, 160, 164
tides, Flood and, 213
time-transgressive relations—cover-up for disagreement, 205
tonsteins, volcanic origin of, 213
topography, antediluvian, 60
topography, Flood deposition and, 141
trace fossils, facies confusion, 139-140
trace fossils, Precambrian, 80
transitions, cephalopod evolution, lacking, 181
transitions, evolutionary, bogus, 181
tree trunks, upright, catastrophic deposition, 137, 208-209
trends in lithology, geologic column, 52
trends, younging-up, assumed, "reworking," 93
trilobites, alternatives to evolution, 24
trilobites, Cambrian, global map, 30
trilobites, odd discoveries, 42
trilobites, Ordovician, global map, 30
trilobites, Siluro-Devonian, global map 33
tuffs, dating of, 161
turbidite-like structures, igneous rocks, 95
turbidites, crinoidal, 17
Tyrol, geology of, 84

ultraviolet radiation, solar, 17-18
unaltered rock, discrepant dates nonetheless, 160
unconformities, limited extent of, 205-206
unconformities, Precambrian dating, *ad hoc*, 167
unconformities, see paraconformities
unconformity-bound units, unnatural globally, 205
undeformed strata, "overthrusts" and, 86
underclays, not soils, 215-216
unfossiliferous rocks, assumed Precambrian, 92
uniformitarian problem admitted: crinoids, 16-17
uniformitarian problem: cyclic coal-bearing strata, 202-203
uniformitarian problem: sheet sediments, 139
uniformitarian reef concepts, contradictory, 84
uniformitarianism, proliferated hypotheses, 53
uniformitarianism—*a priori* nature of, 203
uniformitarianism—another name for atheism, 134, 203-204
uniformitarianism—Apostle Peter and, 217
uniformitarianism—non-empirical, 96
Universal Deluge, see Flood
universal unconformities—nonexistent, 206
U-Pb dating, zircons, *ad hoc* contamination, 164
U-Pb method, main usage of, 164
upright fossils, Flood and, 85
Ural Mountains, 8
Ussher, Archbishop, 140
Utah, fossil occurrence map, 40

variation, organic, rapid, 11
vegetal mats, floating, give rise to coal, 214
vertebrate footprints, anomalous, 135
vertebrate skeletons, pitfalls of reconstruction, 57
vestigial, wings, origins, 11
vicariance, post-Flood life, 8
volcanic dust effects, 8
volcanics, "time-transgressive," 141-142

volcanics, carbonatitic, 61
volcanics, dating of, 161-162
volcanogenic rocks, submarine, 128
Voltaire, 96

Walther's Law, 209
water, Flood, ocean-bottom uplift, 60
wave-resistance, ancient deposits, unproven, 84-85
worldwide biozones, *not*, 186-187
worldwide unconformities—nonexistent, 206

xenocrystic contamination rationalizations, 163-164

young earth, church failure to appreciate, 94
young earth, fossil diversity favors, 141
young earth, importance of, 95-96
young earth, volcanic action favors, 141-142
younging-up, *an assumption* in "reworking," 93

zircons, U-Pb, *ad hoc* contamination, 164